Foundations of Engineering Mechanics

Anatoly V. Perelmuter, Vladimir I. Slivker

Springer

Berlin
Heidelberg
New York
Hong Kong
London
Milan
Paris
Tokyo

Anatoly V. Perelmuter, Vladimir I. Slivker

Numerical Structural Analysis

Methods, Models and Pitfalls

With 242 Figures and 38 Tables

Springer

Series Editors:

V. I. Babitsky
Department of Mechanical Engineering
Loughborough University
Loughborough, Leicestershire, LE11 3TU
Great Britain

J. Wittenburg
Institut für Technische Mechanik
Universität Karlsruhe (TH)
Kaiserstraße 12
76128 Karlsruhe / Germany

Authors:

Anatoly V. Perelmuter
13, Chokolovsky bld, room 508
SCAD Group
Kiev 03186 GSP
Ukraine

E-mail: aperel@i.com.ua

Vladimir I. Slivker
7, Yablochkova str
JSC Giprostroymost
Saint-Petersburg 197198
Russia

E-mail: slivker@VS3491.spb.edu

ISBN 3-540-00628-1 Springer-Verlag Berlin Heidelberg New York

Cataloging-in-Publication data applied for
Bibliographic information published by Die Deutsche Bibliothek.
Die Deutsche Bibliothek lists this publication in the Deutsche Nationalbibliografie;
detailed bibliographic data is available in the Internet at <http://dnb.ddb.de>.

Springer-Verlag Berlin Heidelberg New York
a member of BertelsmannSpringer Science+Business Media GmbH

http://www.springer.de

Typesetting: Camera-ready copy by authors
Cover-design: de'blik, Berlin
Printed on acid-free paper 62 / 3020 hu - 5 4 3 2 1 0

Contents

3 Basic Relationships for Discrete Systems 65

4 Finite Element Models 139

5 Mistakes and Pitfalls, Special Techniques to Build Finite Element Models 177

6 Estimating and Interpreting Results 249

7 Uncertainty of Parameters 267

Preface

To our sons, Mike, Andrew, Alex, who did not inherit their fathers' level of interest in applied mechanics, but who became sophisticated in software development and in this regard surpassed their parents.

A.P., V.S.

Hard times came, the gods got angry. Children do not behave themselves and everybody wishes to write a book.

Ancient Babylonian inscription

Preface to the English Edition

The book you are reading is a translation from Russian into English. Within a pretty short term this book saw two editions in Russian. The authors received inspiring responses from readers that both stimulated our continuing and improving this work and made sure it would not be in vain of us to try to multiply our readers by covering the English-speaking engineering community.

When we prepared the present edition, we took into account interests of the Western readers, so we had to make some changes to our text published earlier. These changes include the following aspects.

First, we excluded a lot of references and discussions regarding Russian engineering codes. It seems to us those are of no real interest for Western engineers oriented at Eurocode or national construction design regulations.

Second, in cases (unfortunately not too numerous) when English-language publications were available to us which were equivalent by their results to Russian scientific papers, we changed our bibliographic references. Generally, all references in the book are not intended to establish any priorities. Rather, they appeal to a reader who wants to have more knowledge on a particular special question.

The book is well larded with italicized quotations. When we quoted from Western publications, we tried to use original texts written by their authors wherever it was possible. Though, there were cases (just a few) when we did not have the original works available, so we translated back from Russian into English. We were pressed by circumstances to do so. We understand clearly that the language of the original publications could have been distorted by this double translation. So, we beg the quoted authors to accept our humble apologies for possible (unintentional) distortions of the original texts.

There is a popular opinion that the application of software will enable one to have poorly qualified users do the work that used to be done manually by highly qualified specialists. This might seem a correct stimulation for the employment of computer-aided design software. We pay our readers' attention to this circumstance because we believe this attitude is very often encountered in Western publications.

We cannot agree to this standpoint unquestionably. There is a chance this attitude is correct towards problems of certain standard types. But solving a nonstandard problem should engage highly qualified personnel who know well both the essence of the problem itself and the software being used together with theoretical ideas implemented by it. One of stimulations for one to use the computer-aided design software must be the achievable depth of one's design analysis and the possibility to optimize one's structural solution on the basis of this analysis.

Scientific and research publications in Russian deal with this issue closely all the time. There have been many notes that analysts who use computer programs to solve engineering problems will be demanded to have even more expertise than ever before. The example of such is a quotation (an emotionally categorical one) from such Russian gurus of mechanics as I.I. Vorovich and L.P. Lebedev[1]:

"*Recently we've been hearing more and more often that computers are now capable of solving all problems of mechanics, therefore the traditional mechanics is gradually dying. It looks like an opinion of a boy who has just learned how to compose words of letters and therefore believes he knows everything written in books.*"

Years of our experience have shown that even when solving purely practical problems one takes a great risk of failure if one relies solely upon the power of computers and expertise of software developers. One of the objectives of this book is to make more people adhere to this viewpoint. Our hope is that the reader will agree to it after he reads the book. Though, we are not alone with this standpoint among Western authors, too: the example is an epigraph to Chapter 8.

Together with many thanks given in the preface to the original edition, we should express our appreciation of the translator's labor who has worked hard to render adequately (as we do hope!) both the contents of the book and the flair of the authors' native language and style.

We appreciate the effort of the Springer publishing company that has provided the authors with the possibility to familiarize this engineering and research community new for us with our modest work. We do hope the publication of this book will promote the international scientific cooperation of researchers and enables them to get better acquainted with the scientific and engineering traditions of the former USSR countries (at least Russia and Ukraine where the authors live and work).

---

[1] Vorovich II, Lebedev LP (2002) Some Questions of Continuum Mechanics and Mathematical Problems of the Theory of Thin-Wall Structures. International Applied Mechanics, 4:3–19

Preface to the Second Edition

Once we had published our book, we immediately felt the need to re-work it. We wanted to involve additional facts, refine some statements and formulations. Also, a number of responses and notes from our colleagues showed some issues could have been presented in a better way while other statements needed a more detailed argumentation.

This book, a bit extended comparing to its first edition, is oriented at applications in the sense that it's aimed at solving problems advanced by practical needs of structural engineering where contemporary computer-aided design tools are to be employed. At the same time it was and still is a publication of an analytical nature rather than a reference manual. It is not intended for a reader who needs a compendium of facts or exact instructions on what to do. Great Goethe has written that

"*...one knows precisely only when one knows too little. Knowledge nurtures hesitancy*".

The literature on structural analysis inclines noticeably to pure algorithmic recipes and deals too little with analytic issues when describing involved design models and giving recommendations on their purpose and correct application. The result is that there is a gap between theoretical and empirical knowledge.

When the authors were writing this book and preparing its second edition, they made every effort to narrow this gap somehow and avoid the style of a recipe-book as best they could. Of course, it is not possible to get off giving any recipes, at least for the reason that an analyst engineer has to comply with official regulations built, unfortunately, as recipes. Neither it is possible to escape mistakes and confusions in any intellectual area using just a bundle of simple algorithms. In this regard, Andrew Koenig wrote this in his book with a peculiar title "C Traps and Pitfalls"[2]:

"*Errors cannot be avoided by recipe. If they could, we could eliminate automobile accidents by plastering the countryside with Drive Carefully signs! People learn most effectively through experience – their own or someone else's. Merely understanding how a particular kind of mistake is possible is a big step on the way to avoiding it in the future.*"

The first edition of the book being presented has been sold out not being able to satisfy the readers' demand. Responses of our readers to this publication which the authors accepted with great thanks and satisfaction, proved a truest sign for us that the book did not leave experts indifferent to problems discussed and that our effort spent to write the book was not in vain. A great spread of the expertise and experi-

[2] Koenig A (1989) C Traps and Pitfalls. Addison-Wesley Publishing Company

ence of our respondents made us convinced that this book was not a mere space occupant on shelves of both structural design novices and veterans.

It was impossible to get no criticism in response, and we admitted a lot of readers' critical comments. Due to this, the present edition has undergone necessary corrections. Though, as we warned in the preface to the first edition, it is not any and all criticism that we are ready to accept. For example, one of our critical readers wrote the following when discussing a question from the first edition's preface (how to avoid a mistake in creation of a design model and how to understand results of a computation?):

"I'm no judge whether there can be a positive answer to the first part of the question because it's human to err, therefore one should always check oneself. As for the second part of the question (how to understand results of a computation), my opinion is that only one answer is feasible: you've got what you asked for, so understand it as best you can. And don't ask for what you cannot understand. This applies not only to users, but also to developers of software (primarily to the latter)".

We dare to quote from a private letter because this arrogant opinion about a customer has the right to be presented here but we can by no means agree to it. We are convinced a man can reckon on help even when he makes mistakes. As a matter of fact, it was an analysis of mistakes made by software users, included those of result interpretation, that has determined chiefly both the contents of the first edition and additions we have introduced to the second edition.

It is unreasonable to list changes and additions that have appeared in the second edition of the book, for they are too numerous. Sometimes those are small but essential inserts into the text, sometimes whole sections. As for removed parts, the only things we deemed reasonable to exclude from the first edition is an English translation of the preface and table of contents. The reason for this is that English-speaking readers are now able to get themselves familiar with the full text of the book: simultaneously with the second Russian edition being prepared, the Springer Verlag publishing house is preparing a full translation of the book into English in a well-known Foundations of Engineering Mechanics Series.

After some hesitation we have refused to list, even partially, prominent experts that became our correspondents on topics in question. The reason for this was that any reader's response was a strong accelerant for authors to force their work on the second edition, apart from titles and positions of our correspondents. We stand obliged to express our sincere gratitude to all these persons for their interest to our work. Though, we cannot help breaking our intention at least once to mention a special man. We are talking about an immediate and complimentary response to our book from Y.G. Panovko[3].

---

[3] In April 2002, a prominent mechanician and a man of distinguished personality Yakov G. Panovko passed away at the age of eighty nine. A few generations of engineers and researchers were inspired by his books dedicated to most various branches of mechanics, remarkable for their originality and perfect literary style. One of these generations in-

Among notes and recommendations expressed by Professor Panovko, there is one to change the title of the book. That's how Y.G. Panovko thought the book should have been entitled:

Design Models of Structures
(ideas, principles of choice, analysis, dangers and failures).

Isn't it a master touch that you feel at once! Though, in order to maintain the continuance, we have decided to keep the old title because it is not a new book that we present. It is an updated, extended and edited version of the old book with an already recognizable title, though it might seem dry-as-dust a bit.

And the last thing to say. The book contains a lot of references to widely popular software systems most of which are of commercial purpose. We give a list of the software products here for the convenience of readers. The name of each one is accompanied by its respective Web site address:

SCAD	**www.scadgroup.com**
GTSTRUDL	**www.gtstrudl.gatech.edu**
MicroFe	**www.eurosoft.ru**
Lira Windows	**www.lira.com.ua**
ANSYS	**www.ansys.com**
Robot Millennium	**www.robobat.com**
Straus7	**www.strand7.com**
NASTRAN	**www.mscsoftware.com**
ABAQUS	**www.hks.com**
STAAD	**www.reiworld.com**
COSMOS	**www.cosmosm.com**
LS-DYNA	**www.ls-dyna.com**
LUSAS	**www.lusas.com**
DIANA	**www.diana.nl**
RM2000	**www.tdv.at**
ADINA	**www.adina.com**

In most cases one can find out any details of interest by addressing a correspondent Web site.

A. Perelmuter, V. Slivker November 2002

---

cludes the authors of this book, too. A necrology in memoriam of Y.G. Panovko was published in the journal "Advancements in mechanics ", vol.1, No.2, 2002.

Preface to the First Edition

Thirty years ago an authority on structural mechanics claimed [1] that

"*the art of choosing an analytical model is very important for an engineer. It is not specifically taught anywhere. There are no courses, lectures or university curricula where this subject is emphasized and given appropriate attention.*"

While such cources are still lacking, the introduction of computer software has changed the accessible limits of problems solving and, at least on the surface, has created a superficial picture of succesful treatment by the technical solutions based on engineering analysis. We can see now that these models clearly tend to be constructed with greater complexity and more dimensions. But is the use of complicated and detailed design models beneficial? Are there other methods of achieving quality results? How can we avoid mistakes in creating analytical models and how do we interpret and understand the results of the computer output? Answers to these and other related questions comprise the main theme of this book.

After the introduction of computers, structural mechanics in many aspects became an experimental science. For some inexplicable reason this fact was not appropriately acknowledged. Under normal circumstances such scientific subjects as test planning and statistical processing of test results, methods and approaches developed in them and, most important, the ideology of these subjects, remained on the sidelines in engineering education.

The apparent ease with which a numerical experiment can be conducted leads to the fact that practicing engineers and even some researchers conduct these experiments without an organized strategy. Their results are only partially analyzed, which not only reduces the reliability of such an approach, but may lead to a danger of overlooking an error. The opinion has become widespread that "good software" and accurate input data guarantee a precise result[4]. Here important issues may be overlooked: for which design model is the result obtained; what is the level of its adequacy relative to actual construction versus a given design model? We discover it is rather difficult to answer these questions, because in many cases the theory is not adequately developed, to say nothing that the software does not contain the appropriate functions.

The wide availability of software to apply the method of finite elements (MFE) allows us to narrow the problem mentioned in the title of this book to "analyzing finite element models and their evaluation by means of a computer".

A relatively complete introduction to the sphere of analyzed problems can be seen in the list of analytical steps outlined below and questions, which arise from these steps [2]. The main steps are:

- Creation of a model;
- Choice of software;

[4] Everyone knows what "good software" is, but everyone has a different idea in mind.

- Model check
- The computation itself
- Result verification

When creating a model one should ask such questions as:

- What are the actual condition of support – pin or fixed supports?
- Should the joints sense moments or are they to be considered pin (or elastic support)?
- Does the frame interact with any other parts of the construction or is it an isolated flat system?
- What is the actual system load (its own weight, wind, snow, etc.)?
- What are the second-order effects (buckling of columns, torsional instability, etc.)?

When verifying the results it is necessary to distinguish the assessment of the substantiation pursuant to the inner features of the model proper and the verification, which is aimed at the detection of errors in that model. We call them errors when input data are false. If it is not possible to say that a parameter is false, but its value may be slightly different, then we are talking about uncertainties. The evaluation of the substantiation must lead to the conclusion – "This model is adequate for real construction". Pursuant to the verification, it is reasonable to perform the following:

- Check of data (input data in graphic form);
- Check of common balance – sum of reactions equals sum of loads;
- Check of local balances of the subsystems;
- Check of visual deformation picture in accordance with given conditions of support;
- Check of existing symmetry conditions;
- Evaluation of the general picture of the stress and strain state of the construction (SSS), and comparison of the deformation picture with the distribution of stresses.

The usual documentation of the software (if it exists!) contains a description of the library of the finite elements, formal instructions on preparation of input data, and an interpretation and explanation of the obtained results. It does not contain the issues of methods of software usage in actual engineering or research analysis, ways of creating analytical models and methods of interpreting results.

The following statement is devoted to the consideration of these and other similar issues. For software application examples we will usually refer to the software called SCAD [3], by means of which most of the problems in this book are solved (rare exceptions are mentioned in the text), and which is relatively typical software used for analysis of the approach of modern problem solving. The developers of this software have encouraged the writing of this book and have assisted the authors in various ways. Moreover, they suggest it as a useful component of future documentation.

The book is aimed at a reader who is familiar with the usual course in structural mechanics and who has some experience with computer-based construction analysis. This reader probably already realizes that analysis of actual construction is more difficult than the exercises which he/she did in college using analytical models provided by their professors. In other words, the book is aimed at a prepared reader – the reader who understands that his professional training has not terminated with the receipt of the college diploma, and who needs to continue to learn. It is a textbook which can be assigned to a number of university courses. The considered subject is more of an art, and the only method of teaching an art is when the master instructs: "Do as I do". However, a student becomes a master when he no longer does as his master does. So the authors' appeal to do as they do and not otherwise should be accepted with creative criticizm, and in the future do better.

We used the word "aimed" instead of "for" intentionally because we think that the circle of readers is not limited to software users alone. The authors hope that developers will find useful recommendations and advice on the pages of this book for themselves as well. It is artificial to separate the army of professionals into developers and users. One can not divide engineers into, say, generals and soldiers. They are more parts of cooperating army regiments, such as artillery and infantry. The majority of the developers are users as well. We do not use the term "user" in a pejorative sense. We are users ourselves!

As users (and for users) numerous times on the pages of this book we will bring the attention of the readers to various traps that await an engineer who relies solely on software as a guide, who dared to sail in the structural engineering waters on a newly purchased yacht bearing an intriguing name **BWPAC**[5].

As users (and for users) we will share with readers experiences (ours and our colleagues') to help them avoid the traps and pitfalls that would otherwise plague their work. Often it is sufficient to use "small tricks" to overcome these difficulties. Other cases require serious preliminary research.

As users (but for developers) we will suggest desirable features of engineering software packages. Sometimes there is no need to develop new software, since it is sufficient and useful to upgrade and improve the existing one.

We completely agree with Rosin's opinion expressed in his latest book [4]:

" *A person who uses software for analysis of construction and who is interested in the stress-deformation state is not required to know all mathematical and computational problems in detail. However he needs to have an idea about how mathematical problems are formed and what are the numerical methods to solve them. It is hard to rationally choose an analytical model and correctly evaluate the accuracy of the final results.*"

For this reason we conclude that we can not keep silent about mathematical approaches in structural engineering as well as the algorithms and methods of solving numerical problems.

[5] As you could guess this abbreviation stands for the **B**est in the **W**orld **P**rogram to **A**nalyze **C**onstruction

One more issue needs to be brought up in this preface. There are many specific technical terms in the book. We tried to use accepted (at least in the Russian literature) terms, even where we might not like it, or would prefer another word to define the same notion. In the situation where there were two terms for one notion (for instance *slab* and *bendable plate*), we used the one we prefer. Moreover, we had to introduce several new terms. Sometimes we borrowed them from our earlier published works (*null-element*, *semi-infinitive finite element*, *mono-constraint*, *poly-constraint*, etc.). Sometimes we chose or created new terms especially appropriate for this book (*pseudo stiffness*, *pseudo flexibility*, *bi-member model*, etc.). We did not do this merely for the sake of word creation but rather to enrich the vocabulary of the area of expertise to which this book is devoted. If we do not form a glossary of such terms and do not determine and define the notions concisely and precisely in one or two words, then we would risk a state of verbal anarchy in which specialists in this narrow field of expertise would not be able to understand one another[6].

We accept the criticism that we might have omitted one or more important topics in this book – "*one can not limit an unlimited*" (Kosma Prutkov). We will mention only one thing that is not here that we have omitted intentionally. There are dozens (hundreds?) of engineering software packages available. How from an advertisement can one choose the software that would best fit his needs, the one worth buying and studying in detail, and which is the one to be discarded right away? We do not give any recommendation. This doesn't mean that the authors do not have their opinions about this subject. However the market….. The market is a very delicate thing that painfully reacts to rough interference of even economists, while we are engineers. If a reader is disappointed about this then all we can do to comfort him is to advise him to buy "good software". We hope that after reading our book readers will understand the meaning that authors locked in this phrase.

The idea of this book was discussed with several specialists among whom the following people played a significant role: V.N. Gordeev, A.A. Dykhovichny, M.I. Kazakevitch, E. Z. Kriksunov, S. F. Pichugin, and U.B. Shulkin. The authors express their gratitude to all these people.

Many methods and approaches for solving engineering problems were developed while analysing numerous complicated and important constructions. The authors remember with pleasure the creative encounters with engineers and scientists: E.S. Alexandrovskaja, V.B. Barsky, U.S. Borisenko, K.P. Galasova, S.Z. Dinkevitch, L.G. Dmitriev, M.G. Dmitriev, A.Ja. Driving, A.A. Dykhovichny, M.L. Grinberg, A.I. Konakov, N.B. Krasnopolskaja, V.P. Kryzhanovsky, A.O. Kountsevitch, M.A. Mikitarenko, A.G. Pinsker, A.Ja. Pritsker, S.Ju. Fialko,

[6] This brings to mind the story about two aliens who landed somewhere between Saint Petersburg and Moscow. One of them says: "It seems we are at the cemetery. See, there is a tombstone. The Earthling who lays under it lived for 128 years". "Wow! What was his name?". "His name was….His name was KilometersFromMoscow."

E.B. Freifeld, B.A. Shoikhet, and our other colleagues. Their experience and methods have found their way into our book in one way or another.

The authors would like to express a special gratitude to the reviewers of this book, Professors V. N. Gordeev and L.A. Rosin for many recommendations on clarifying and improving its content. We also thank our colleague F. M. Svoisky, who read the book in the handwritten original, and who brought errors to our attention, and thus enabled us to correct most of them prior to publishing. We thank D.A. Maslov and I.N. Laikina for help in illustrating the book. And finally we must emphasize the special support that we received at all times while writing the book from the SCAD Group and Giprostroimost company, without whose help this book would probably not exist.

Lastly we wish to acknowledge three outstanding authors in structural mechanics who influenced our scientific views more than any others. They are Professors Y. Panovko, L. Rozin and V. Feodosjev. We use this opportunity to express our admiration of their works – their depth, their breadth and their leadership.

We are well aware that our expressions are subjective in some cases and our personal experience and preference determines the choice of material. However the authors support the scientific and technical statements described in the book until they are presented with sufficient refutation.

We welcome comments from our readers, containing arguments and criticism and, most important, experience not covered in this book which engineers meet in their practice and suggestions for additions and enhancements. The easiest way to contact the authors is through e-mail. Here are the addresses:

aperel@i.com.ua and **slivker@VS3491.spb.edu**

References

1. Feodosjev VI (1969) Ten lectures-conversations on strength of materials (in Russian). Nauka, Moskow
2. MacLeod IA (1995) A Strategy for the use of computers in structural engineering. The Structural Engineering, Vol. 73, 21:13-21.
3. Karpilovky VS, Kriksunov EZ, Perelmuter AV, Perelmuter MA (1998) Integrated system of analysis of constructions (in Russian). Structure (SCAD) for Windows 95/98/NT, SAPR and graphics, 10:15-18.
4. Rosin LA (1998) Problems of the theory of elasticity and numerical methods (in Russian). SPbGTU Publishing house, St.Petersburg

1 Object of Analysis and Problem of Modelling

Airiness and a bit of irony best facilitate serious development of a serious science. No one should treat one's own person too seriously. Certainly there are people who think everything that is done with an air of seriousness is very reasonable. But they are, as the English say, not clever enough to go crazy. Actually, the deeper the problem, the more probably it's going to be solved in a funny bizarre way lacking any animal graveness.

Daniil Granin

1.1 Multidimensionality and Multifunctionality

Modern buildings and structures are, most often, complex multi-element systems created in order to implement a lot of various functions. During their life cycle a great deal of operation states take place. The specificity of structural engineering is such that its final product (a structure or a building) should meet three, often contradictory, requirements: functionality, aesthetics and designability.

From an analyst's point of view, the most important issue is the object's structural properties, which influence its load-carrying capacity analysis. But its pure structural (load-carrying) function cannot always be distinguished easily. One can easily point out principal structural elements in a framed building (though there may be some problems here too), but it can be not so easy to do with structures of a different type: and perhaps only after a preliminary consideration of a few competing hypotheses. Moreover, the list of load-carrying elements of the building in question can differ in different modes of operation. For instance, neglecting the contribution of "nonstructural" partitions during a limit state analysis or the structure's load-bearing frame design is quite reasonable, but neglecting the same partitions when analysing small-amplitude oscillations of the building can result in a substantial inaccuracy.

Multifunctionality and a multitude of possible modes of operations of a modern advanced structure make it hardly possible to allow for all mentioned factors in a single design model.

The second peculiarity worth paying attention is a multidimensionality of design models used in the modern structural design practice. Tens and hundreds of thousands of degrees of freedom taken into account have long ceased exciting the imagination of an analyst because this amount is no longer the obstacle for computer facilities available to him today. The design model of a structure shown on Fig. 1.1 that numbers 4,416 nodes and 6,632 finite elements is a characteristic example (and a fairly typical one in its size!). And the fact is that the finite element grid of this structure is pretty rough on every wall panel or floor slab.

Among all presently solved problems of engineering there is one worth the Guinness record book by its dimensionality: the modelling of thermal-hydraulic-gas-dynamic-mechanical-strength behaviour of a nuclear reactor performed by the Adapco company in 1998 with the **STAR-CD** software. The dimensionality of the problem was 5.78×10^6 finite elements and hundreds of millions of degrees of freedom. The solution of the problem with a multi-processor system consisting of 64 IBM SP workstations took 55 hours.

The dimensionality of problems to be solved grows further when one has to analyse the structure together with an elastic foundation underneath. Problems of this kind were formulated in the early eighties as an appeal to perform analyses of structures together with their foundations as single spatial systems. And these problems became a common practice in the nineties.

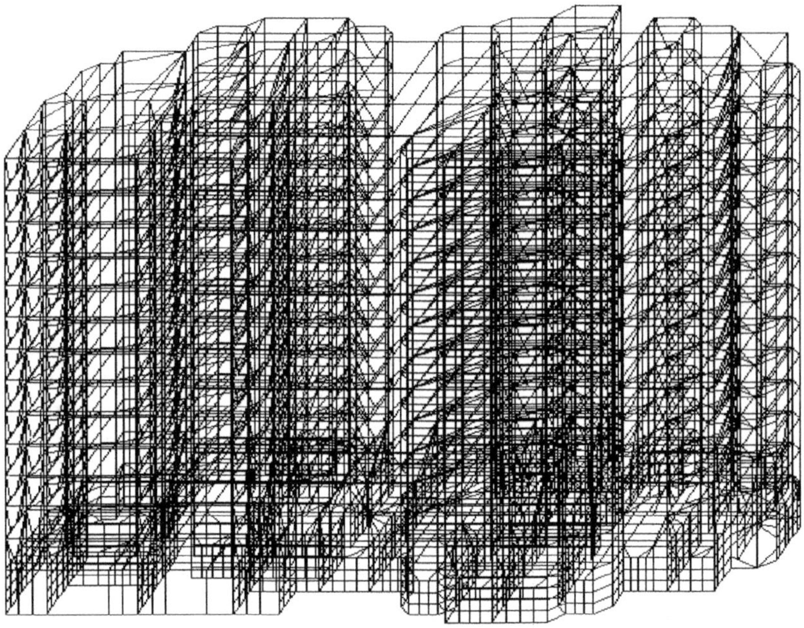

Fig. 1.1. A finite-element model of a structure

1.2 Principal Factors taken into Account during Creation of Design Models

The first step of idealisation is to distinguish a load-carrying part of an object. This step is arbitrary and ambiguous due to several circumstances:

- different roles played by particular structural elements in different regimes of loading: under some loads particular elements can be just enclosures while under other conditions the same elements can affect the stress distribution substantially;
- different transmission of forces under different intensities of the loading: we have already mentioned the changing role of partitions, and other elements can also behave in the like way;
- changes that may occur in different regimes of the object's operation: a good deal of modern structures are transformable, so something that used to be a load-carrying part in one configuration may become a dead weight in another configuration (considering the phases of fabrication, haulage and erection, the transformation of a structure's parts is a rule rather than exception).

After a part of the object to be included in the analysis has been chosen, the process of idealisation of its geometric shape begins – a *geometrical modelling.* This operation can take place either top-downwards when a number of geometric primitives (rectangular or round plates, parallelepipeds, conical, cylindrical or spherical shells etc.) is the basis for the modelling, or down-upwards when the modelling is based on the concepts of a point, line, surface etc. The modelling that involves both types of operation results finally in an idealised geometrical model of the actual structure deprived of some inessential (in the analyst's estimation) details or petty objects such as chamfers or rounding.

In the course of the geometrical modelling one should decide on the issue of a specific idealisation of the object such as making it regular or symmetric, though the object itself may not be exactly regular and its symmetry may be violated to some small extent. But the regularity and symmetry are so powerful methods of reducing the analytical effort that one should not be too picky about minor discrepancies. In the same way one should consider the opportunity of finding some repetitive parts of the object which can be treated as equal subsystems with a certain degree of accuracy. Naturally, the geometrical identity is just one of necessary conditions for making considerations of this kind.

The next stage of the analysis is to *idealise the material* of the structure, or rather the set of its physical and mechanical properties. The most common approach is to treat the material as perfectly elastic or perfectly plastic. Values of the parameters that describe the material (such as elasticity modulus, Poisson ratio, yield point etc.) are to be taken from a reference manual, and they are considered to be constant over sufficiently big parts of the structure (or over the whole structure). In practice, the correspondence between the real values and those accepted for the analysis is a subject of interest of an analyst in very few cases. This tradition results from a classical analytic approach where it was found out and is long known that for some materials such as structural steel the changeability of their physical properties is small and can affect results of an analysis very little. Unfortunately, deductions like this are often used beyond the point they're well-founded. It suffices to say that the foundation soil is often treated as an idealised material homogeneous over the space. But here the assumption of small changeability of the properties is no longer well-founded, and sometimes just inconsistent with data of geological survey. Another typical example is giving to a system some averaged value of the logarithmic decrement of oscillation or another measure of dissipation while the real parameters of this type may change within the same structure tens or hundreds times.

In the course of the material idealisation, often one makes a decision to perform the analysis "taking into account physical nonlinearity", but very seldom this approach makes the analysis more convincing. The matter is that, though "the physical nonlinearity" is supposed to simulate the plastic behaviour, most often (not without exceptions) one reduces this approach to just a nonlinear dependence between stress and strain being the same during loading and unloading, that is, one for an elastic rather than plastic material. For this purpose, one must be confident that the loading is active without any possible unloading, and it is not enough just to assume all external forces to be monotone increasing (let alone the fact this very assumption is far from being true) to have this confidence. Thus it happens so that the said idealisation proves questionably useful in a practical analysis, though its role in a research process may be advantageous.

Finally, the concept of the material idealisation includes sometimes an idealisation of the design. For example, a dense grid of stiffening ribs of a plate or shell is often "smeared" over the object, and the material of the smoothed system is treated as anisotropic (it's so-called "structural anisotropy").

A pretty serious procedure is an *idealisation of loads* applied to the structure in various regimes of its operation. Generally, loads are poorly explored components of structural systems, they are strongly changeable over time and space, and their analytical models operated by practical designers are arbitrary to a great extent. Some of models of loading traditionally used during the creation of design models (uniformly distributed load, concentrated force, impulsive load, harmonic oscillation) are nothing more than physical abstractions, which fact should be well remembered for the purpose of analysis of computational results. A great deal of mistakes in the idealisation of loads are made in their time history specification which may result in an incredible description of the system's dynamical behaviour. The dynamics is the right place to see the feedback between the load and the structure most clearly because the structure's behaviour can change the very nature of dynamically imposed loads (flutter, divergence, mutual influence of different oscillation modes etc.).

The concept of load is a convenient way to describe an interaction of a structure with its environment, but this is not the only form of the interaction. It is often necessary to describe a kinematical interaction rather than that of forces. Some external (with respect to the system of interest) devices may restrain some points against displacement or rotation, or they may enforce their own displacements upon the structure. These conditions called constraints are almost always present in any design model. By the way, it's worth noticing that a given displacement of a structure's point is always implemented as a displacement of the respective constraint, and a conventional support constraint is a particular case of this kinematical influence when the said forced displacement is zero. Of course, an infinitely rigid constraint that forces the system into a certain (possibly zero) displacement is an idealization: actually the interaction with the environment is implemented using some devices of possibly very big but not in any way infinite rigidity.

Generally, one should be speaking about *influences* upon a structure rather than *loads*. If we try to classify influences, then the first approximation can be their division into external and internal ones, on one hand, and force and kinematical, on the other hand. This division is described by Table 1.1 below.

Table 1.1.

	Force influences	**Kinematical influences**
External influences	Loads	Given displacements of support nodes
Internal influences	Controllable pre-stress	Dislocations, temperature displacements

But this classification is arbitrary and incomplete as well because it does not reflect some specific conditions of interaction with the environment such as corrosion, chemical reactions, spalling, change of the structure's interconnections as a result of breakage of constraints (including cases of evil intent) etc.

Idealization of constraints includes, among other things, a description of laws of interaction between particular elements of the system. Most commonly used conditions of complete equality of displacements or mutual rotations in points of junction (a rigid joint) together with their alternatives of absolutely no interaction in these directions of displacement (a hinge or slider) are pretty strong idealizations of the actual interaction behavior.

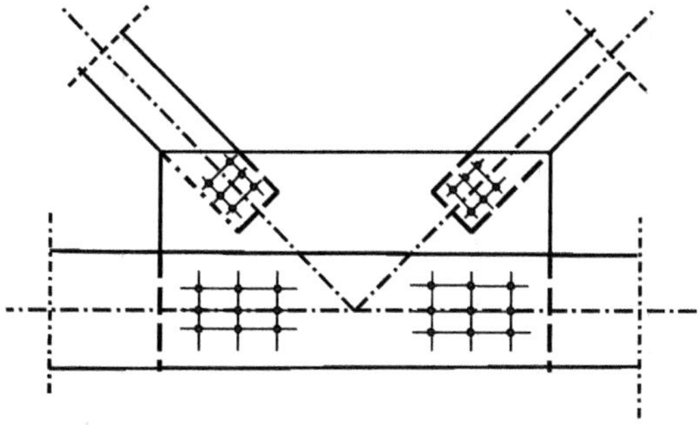

Fig. 1.2. A joint of a truss

Most frequently, one takes into account hypotheses of force interaction rather than kinematical conditions of junction. Looking at the design of a joint of some truss (Fig. 1.2), one can hardly make the decision that absolutely unrestrained mutual rotations of end sections of the bars that meet at this joint should take place. At the same time, there is a hypothesis that leads to this conclusion easily: the hypothesis of negligible bending moments under purely nodal loads sounds quite reasonable for one's intuition.

The analysis of a structure is facilitated very much by the duality of force and kinematical descriptions of constraints described by the following formulation: "*such-and-such mutual displacement is prohibited*" ≡ "*such-and-such force is applied*", or vice versa "*such-and-such mutual displacement is specified*" ≡ "*such-and-such force is not resisted*".

1.3 Incomplete Determinacy of an Object's Knowledge

An idealization of a design model and the impossibility of making it absolutely adequate to an actual structure create a situation of some indeterminacy, in which situation one often has to make design decisions.

Indeterminacy can arise due to both unavailability of all needed information (for example, we do not know in principle all possible future regimes of operation of a structure) and an incompleteness thereof (it is hardly imaginable that we may know exact physical and mechanical properties in any point of a structure). Unavailability of some kinds of data and their incompleteness are principal issues, they cannot be fully overcome, and in whatever detail we studied the problem posed, we can never say we have taken everything into account in our design model.

Though, the situation of indeterminacy may arise not only due to unavailability or incompleteness of data. There is also an ambiguity thereof, that is, a possibility of different interpretations of the same factors, which in turn requires an estimation of available alternatives. There are known classical approaches to the problem of indeterminacy, generally divided into the following options for decision making:

- making use of probability theory when a decision being made is based on an objective previous experience;
- making use of expert estimations, i.e. decision making on the basis of subjective experience of an expert (or a panel of experts);
- a minimax estimation when the decision is the best one among those possibly achievable, with the assumption of the worst expected trend of events, that is, the decision is made by the expected result.

All these techniques can be used together or separately. They are intended to facilitate the estimation of credibility of the design model.

There are other factors that define the approximate character of any design model and cause errors, contradictions and distortions. Those are, firstly, errors of computational modeling (approximation errors) that arise either due to the approximate nature of our knowledge, or due to an intentional rough approximation of it. The "errors" of this kind may include the use of simplified mathematical descriptions such as low-power polynomials to describe a displacement field in the finite element method, truncation of a series of the Galyorkin method etc. The same class of errors includes those caused by inconsistency of scientific theories and hypotheses used to describe different parts of the same design model. A typical example is the inconsistency between concentrated forces as models of loads and the use of plate finite elements that are not capable of balancing a concentrated load by finite values of lateral forces. It is natural that mythic values of shear force in a finite element obtained after such an analysis are the results of this inconsistency of the models.

Secondly, we should notice an approximate nature of almost all specified parameters of a model caused by actual tolerances for sizes, weight and other measured values. From the practical point of view, there is little difference between the two inaccuracy types, though in the first case we deal with a limited accuracy of a model representation while in the second case with that of the "prototype".

The last thing to be mentioned in connection with the discussed problem is possible inaccuracies and indeterminacies due to lack of clear boundaries between involved concepts. Presently this problem is intensively addressed to by a so-called theory of fuzzy sets and fuzzy logic [6]. There are examples of application of this theory to problems of analyzing load-carrying structures in order to estimate their condition when the description of existing damage is imprecise and for the purpose of optimization when the effectiveness function is inexactly specified [9].

1.4 Experiment and Practical Experience

Though this book is not intended to discuss issues of experimental investigation of structural behavior, we cannot leave this point without attention, because the experiment and practical experience are means of validating theoretical solutions, and sometimes tools for creating appropriate design models. An in-place or modeling experiment is usually intended to solve the following problems:

- discover new phenomena or facts;
- validate hypotheses;
- perform physical modeling in order to establish qualitative or quantitative values/dependencies for properties which are impossible to find out theoretically.

Other experiments are performed, too, for example, in order to comply with standardized procedures of acceptance tests of responsible, nontraditional or unique objects, or in order to test key samples of production in volume. Independently from the declared objective of the experiment, one needs to solve at least one of the listed problems with this experiment, to be able to expand the data of a single occurrence to other phenomena (possibly similar ones, or maybe a wider scope as well).

Experimental discovery of new phenomena or new facts can be non-local only if the experimental investigation is accompanied by a careful theoretical analysis. It is the comparison of experimental results with theoretical data that gives us the right to talk about newly discovered facts. If one performs one's experimental investigation without any governing theoretical idea, only to avoid a computational analysis, one cannot expect to produce deductions in any way general. The only thing actually done with experiments of this kind is an addition of new observations to a storage of observable data in order to create a ground for empiric considerations; though in particular cases such experiments may be of use.

A characteristic example is a great deal of fatigue tests of various constructions. For many a year those have been performed in research centers involved in various machine design activities. But, having no fundamental theoretical idea (or, exactly, there were too many assumptions pretending to be such [3]), those activities have not still resulted in a well-founded theory of fatigue failure.

Accumulated experimental data can be often a basis for a further scientific generalization. But it is not every kind of generalization of experimental data or practical experience that can be treated as a scientific generalization. The paper [7] gives an example to confirm this thesis: a formula by Siojurnet, once famous in its time, intended to determine the thickness of stone bridge vaults, reflected not only a collective engineering experience but also a personal experience and psychology of an engineer with the following gradation: cowardly, humbly, normally, bravely, defiantly. It is obvious that this kind of empirical dependence has nothing to do with the development of theory of structures, and Siojurnet himself understood it perfectly well.

Apparently, one can believe that experimental foundations of primary idealizations of particular elements of complex structural systems are completely established to a substantial extent, at least those for common materials and conditions of operation. As for experimental investigation of complex systems, one should, most often, have in mind an idea of validating hypotheses laid in the foundation of their analysis. The primary goal of investigations of this kind is to confirm (or to controvert) the applicability of those analytic assumptions which have been used, of course only if the results being validated have followed from these assumptions unambiguously.

Note that, in order to establish the boundaries of applicability of a design model, any simplifications or approximations, as well as omission of small quantities declared inessential, should be estimated from the standpoint of the final result. For a complex system, this estimation is seldom possible, and there are more than a few hypotheses and simplifications which cannot be experimentally validated in place or on a mock-up. Though, in many cases a complex structure can be decomposed into sub-structures for which the experimental investigation is much simpler. Certainly the very feasibility of the said decomposition should be well-founded, and experimentally if possible. If the decomposition has been done soundly, the value of information obtained via experimental investigation of a sub-structure may prove even higher due to the following reasons:

- subsystems of the same kind may be applicable in other objects, which fact enables engineers to build their design model if connections between subsystems are studied in sufficient detail;
- when the whole system is tested, only integral parameters of the structure's behavior are detectable while data regarding a particular sub-structure may be inaccessible.

Of course, nobody performs any experimental validation for most practical calculations performed by numerous developer organizations because those calculations apply to more or less well studied structural designs. Here a traditional approach enforced by scientific, reference, educational and normative literature prevails. Though this approach works, and following a tradition is a convenient way to go, there is also a negative tendency: an engineer is gradually withdrawing from the habit of validating design models on his own, thus completely relying upon someone else's opinion. Using this approach, it is not too hard to make a mistake of "over-generalization of applicability" by using principally true recommendations far beyond the domain of their theoretical and experimental validity (unfortunately, there is an immensely detrimental tradition not to indicate such domains in reference and educational literature).

A typical example is a standardized approach of designing earthquake resistant structures on the spectral theory basis adopted in most countries of the world. Recommendations of the respective standards apply, strictly speaking, only to the case of a perfectly rigid foundation the points of which move synchronously during an earthquake. But an explicit reference to this hypothesis is hard to find (not every scientific book dedicated to the subject mentions it), which can bring one to questionable results when analyzing structures or buildings the extent of which is comparable to the seismic wave length in soil.

Unfortunately, in standard codes where an immense collective experience is generalized, this is the rule to give no information about design models for which respective recommendations are well-founded. Very often the design models given are primitive and can describe peculiarities of the stress and strain state of an element only roughly when the element is included in a more complex model rather than isolated. From the structural reliability standpoint, this roughness is justifiable because other parameters of the design model (loads, coefficients of operation conditions etc.) are defined on the basis of conservative approach. But, the following contradiction arises: refinement of the design model may lead to less economic solutions because the same safety factors are used both for the rough and for the refined models. It happens so that if the analyst takes on the responsibility for the making of a refined model, he must have the right to correct recommendations of a standard code. Unfortunately, this approach may not be used in countries with a strict conformity of design procedures (in Russia, for example), though it is feasible in other countries where the designer responsibility legislation is better developed and the legal status of standard codes is somewhat different.

So, if we are now discussing standard codes regarding the design and analysis of structures, we may ask the following question: has anyone ever had an idea to estimate, at least by a Delphi technique, the amount of person-years spent annually by employees of developer institutions for discussions and considerations of standard code provisions? Which of those have been established on a purely empirical basis and which are theoretically founded, and what exactly is this theoretical foundation? Of course, some things can be guessed, but a great deal of questions cannot be answered even by professionals.

1.5 General Issues of Modeling

A structure's response to influences of the environment is determined by a number of complex physical phenomena. An adequate description of the phenomena on the whole interval of the load application — from the beginning to the failure of the structure — can be problematic. As a rule, there are not enough data available to the analyst, and often there is no satisfactory and sufficiently detailed theory of the phenomenon in question (such as a description of complex loading in the plastic domain of the structure's behavior; a complete set of parameters to describe the behavior of the material, etc.). This forces the engineer to replace the physical reality with some approximating analytical description, usually called a design model, that possesses idealized properties.

General issues of modeling have often been an object of study, and researchers have formulated a variety of rules for the construction of design models. For example, Ghersevanov [4] in his day advanced three principles of creation of structural design models:

- methods of analysis should be based on forms of failure and deformation confirmed by the experience of constructional practice;
- an analytic hypothesis, which may be arbitrary enough, should subject the model to harder conditions than those the actual structure is subjected to;
- the set of analytic hypotheses should provide economical efficiency of the structure in addition to its strength and stability.

In the opinion of authors of a book specially dedicated to modeling [5], these principles are in no way comprehensive, and some more should be added:

- it is reasonable to have a system of approximating models of a structure's behavior with their respective limitations of applicability rather than a single model;
- the requirement that an approximating model of the structure's behavior should describe the behavior of the actual object correctly and in sufficient detail is not the only one: the model should be also simple enough in order not to make the analysis too cumbersome[1].

[1] A wise man said that "*any theory has the right to exist only if it is a reasonable compromise between mathematical simplicity and physical validity*".

Without arguing these theses in any way, we would like to add a few considerations to be reckoned with when both creating and using design models of structures.

Thus, a special role belongs to general models of structures and elements thereof — a bar, a plate, a shell etc. — using which full design models of structures or parts of models of more complex objects can be constructed. These models are *basic*, they must be investigated with all strictness possible, to make a foundation for further consideration of other design models. After such investigation, having a complete knowledge about these models makes their use quite attractive to an engineer who may then easily guess results of analyses (or at least their qualitative features). The engineer begins to "understand" results of the analysis — in the sense discussed by Panovko who writes that there is a semantic difference between the verbs "to know" and "to understand" and "*... only the understanding makes the knowledge active and creative, while a formal knowledge itself is cheap without true understanding*" [8].

The transition from a structure to its design model composed of basic models occurs, most often, at the level of one's intuition, and the first reason for such transition is a consideration of geometry ("similar" shape). Though, even here there is room enough for a maneuver. The design process often involves actions such as the replacement of a perforated latticed construction with a continuous object the shape of which resembles the original structure only in average, or the "smearing" of ribs and other structural parts mentioned above. There are opposite examples as well, when a continuous body is replaced with its bar analogue (see [10] for examples). Here one uses some knowledge about peculiarities of the mechanical behavior of chosen basic models rather than geometric shapes only. For instance, if a user of a structural analysis software chooses a purely flexural plate model to describe a flat section of his design model, then one should understand that the membrane components of the stress field cannot be found in this way. If the said components might prove essential (for example, they may cause a buckling) then the element of choice will be a more general shell model.

The second way of reasoning that also plays a fundamental part in the transition to a design model is a choice of standardized material idealizations (elastic, perfectly plastic, loose medium etc.). These properties are also represented by previously studied basic models [1] and would not be worth special mentioning but for the following consideration: this choice needs an experimental foundation even more than that of the geometric shape but this phase of research is usually omitted by an engineer. It is a common case that an analyst operates known data about physical models of material behavior obtained from experiments with quite different constructional elements and materials. At the best, the reasoning for this kind of operation is that results of the respective investigations can be used in the design model being built on the basis of some ideas of the possible character of deformations and the expected stress intensity under a load of a particular expected value. But the common reasoning is based on tradition and actual computational capabilities of the analyst. Here's a characteristic excerpt from Russian standard codes for steel design:

"*1.8. Analysis of steel structures should, as a rule, be performed with allowance for non-elastic deformation of steel. For statically indeterminate structures where no standard technique has been developed for such non-elastic deformation analysis, design stresses (bending and twisting moments, longitudinal and lateral forces) should be determined with the assumption of elastic deformation of steel in the original geometry of the structure*".

Similar notes can be found in standard codes of other countries too. For example, Eurocode-3 reads as follows:

"*5.2.6.3 (3) However, as an alternative, rigid-plastic analysis with indirect allowance for second-order effects, as given in (4) below, may be adopted in the following cases:*
(a) Frames one or two stories high in which either no plastic hinge locations occur in the column, or the column satisfies 5.2.7.
(b) Frames with fixed bases, in which the sway failure node involves plastic hinge locations in the column at the fixed bases only ..."

A practical impossibility to carry out a detailed nonlinear analysis of every object or every design model whatsoever does not mean that one should completely abandon the idea of such analysis. Apparently, the most powerful strategy is to perform a careful analysis of some typical models at the level of complexity available for modern computer facilities and to compare results of this analysis with a simplified calculation. This computational experimenting will let us determine (for a particular class of problems) whether a substantial discrepancy between an analysis and the respective experiment is a result of a wrong idealization.

A practical research, which includes the analyzing of design models as well, is not always done on a "mathematical level of strictness" but is often based on credible assumptions [2]. In any case one should remember that a non-rigorous solution and an incorrect solution are two absolutely different things.

1.6 Majorant and Minorant Models

Having in mind all indeterminacies and volitional decisions made to choose a design model, the analyst would like to be confident that those decisions have created a "reserve of reliability", i.e. they provide a certain one-sided estimate of a correct but unknown solution of the problem. It is a very seldom case, to be able to prove that the design model used is a majorant for the exact solution everywhere. Most often, this statement holds locally, that is, the respective majorant estimate takes place for only a single property of the system. It is a dangerous tendency to make a further arbitrary conclusion that the majorant is "global". Some properties of majorancy are possessed by a wide range of systems, but those, too, have a particular domain of application with which one should confine oneself.

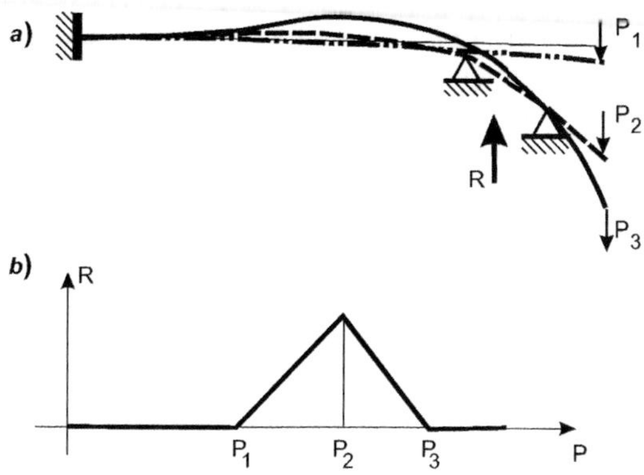

Fig. 1.3. A system with unilateral constraints

A typical example for what has been said is a widely used rule of over-evaluation of loads in order to obtain more reliable solutions. But there are systems (which are not linearly deformable structures) the ultimate internal stresses or forces in which are caused by some intermediate values of loads rather then the latter's extreme values. A confirmation is shown on Fig. 1.3 - *a* that shows a system with unilateral constraints having gaps.

As the load changes in the interval 0...P^+, the reaction of the left support (see Fig. 1.3 - *b*) is 0 at first, then, after the left constraint has come into play, it starts increasing until the right constraint locks, then the reaction begins decreasing and eventually becomes zero again. Thus, a maximum value is achieved inside the interval of the possible loads.

Obviously, it would be perfectly well if the analyst could assuredly operate both majorant and minorant systems. Unfortunately, it can be done in very seldom cases. For example, in the finite element method a finite-dimensional mechanical model is a majorant for the original continuous system by energy at least. Anyway, it's true when only consistent finite elements are used. The same estimation holds for the classical methods by Ritz and Galyorkin. In terms of mechanics, it can be easily explained by the influence of additional constraints upon the respective system. But a minorant system for similar problems can be much harder to build. Sometimes one manages to do so using specific features of a particular problem. An example is a bi-member model of a thin-walled bar discussed in Chapter 8 of this book.

1.7 Posterior Analysis of a Design Model

Though the construction of a design model precedes the solution of the respective problem, some assumptions used during the modeling are based on expected properties of the solution to be found. For example, the assumption of small displacements and, especially, small slope angles is the basis for solving a problem as geometrically linear. Or, the assumption of absence of sharp stress concentration is the basis for the application of a uniform finite element grid. Therefore, one needs to perform a posterior analysis of the solution obtained and, if necessary, make appropriate corrections of the respective design model in accordance with results of the posterior analysis.

Problems of design optimization require a particularly careful attention to the construction of design models. In problems of this type, relationships between particular elements of the design model are not known beforehand, so the solution may belong to the range of values far beyond the domain where design assumptions are true.

For example, if the object to be optimized is a system of bars, and the parameters to be optimized include cross-sections of the bars, it is not impossible that the optimal solution will give the cross-section dimensions such that the possibility of treating the system's elements as bars compliant with the Bernoulli – Euler theory will become questionable.

Adherents of design optimization believe this danger can be avoided by formulating a few additional "structural" limitations. Though, there is no confidence that all dangers of this kind will be averted for any particular problem. Otherwise we would have to assume that the engineer knows the solution of a design optimization problem beforehand (or at least the region in which the extremum takes place).

Analysis of computational results is facilitated by the use of simple and demonstrative models for which an extensive practical experience has been accumulated and expected results are easily predictable. The use of those models is canonized in a perennial tradition, and they seem almost intuitively obvious, and it is this intuitive credibility that may cause a disparagement of a strict posterior analysis.

A typical example is a problem of optimization, say, of the weight of a compressed column. This problem involves the conventional equation of transverse-longitudinal bending in the following form:

$$EIv'' + Pv = 0\,. \tag{1.1}$$

Though, this equation holds only if the longitudinal force P is constant, and only under certain boundary conditions. The solution is obtained nonetheless in the form of a variable-section bar for which the influence of the bar's dead weight on the longitudinal force should have been described by the following equation:

$$(EIv'')'' + (Pv')' = 0\,. \tag{1.2}$$

Of course, inaccuracies of this kind can be comparatively easily detected in the course of the posterior analysis of the solution; unfortunately, this analysis is performed not really often, especially when assumptions laid in the foundation of the design model of the respective optimization problem have not been specified explicitly.

It is worth noticing that a posterior analysis should not lead to a re-calculation of the whole system if any peculiarities inconsistent with the design assumptions have been found in the course thereof.

In cases when properties of such inconsistencies (for example, a localized position thereof in the structure) have been studied in sufficient detail, and there is a sufficient experience of appropriate modifications of the design, one may use the most part of the obtained results easily, only introducing some corrections to the solution. This is the way to do it, for example, if an engineer detects that the edge effect stress in a shell is far beyond the yield point of the material but he understands well that these peaks of stress will be practically smoothed by plastic behavior which is locally allowed by the conditions of the structure's operation.

Generally, one should always remember that a design model is an abstraction created for a particular purpose, therefore the correct attitude to the model must not be its idolization. A bit of sense of humor towards design models is a characteristic feature of an experienced and skilful analyst. Commonly, the same analyst uses a few design models (sometimes implicitly) rather than one, the models being close to one another in some sense, and if results of such analysis of this set of models are close enough, then one feels more confident about their correctness. In this case the posterior validation is intended, in essence, to analyze the stability of the result with respect to variations of the design model. If these variations show only local discrepancies, convincingly explainable by particularities of the used model, then the home free use of the other results is quite justifiable. The said local zones should be subjected to a special analysis, maybe involving other design models.

References

1. Bell JF (1973) Experimental basics of mechanics of deformable solids. In: Trusdell C (ed) Mechanics of Solids I (Encilopedia of Physics, vol. VIa/1). Springer, Berlin
2. Blekhman II, Myshkis AD, Panovko YaG (1983) Mechanics and applied mathematics: Logic and peculiarities of mathematical applications (in Russian). Nauka Publishing House, Moskow
3. Feodosi'ev VI (1969) Ten lectures-conversations on strength of materials (in Russian). Nauka Publishing House, Moskow
4. Ghersevanov NM (1923) Application of mathematical logic to structural analysis (in Russian). ONTI, Moscow
5. Goldenblatt II. et al. (1979) Models of earthquake resistance of structures (in Russian). Nauka Publishing House, Moskow .
6. Kacprzyk J (1986) Zbiory rozmyte w analyze systemowej. Państwowe Wydawnictwo Naukowe, Warszawa
7. Kartvelishvili NA, Galaktionov YI (1976) Idealization of complex dynamical systems with examples from electrical engineering (in Russian). Nauka Publishing House, Moskow
8. Panovko YaG (1965) Mechanics of deformable solids: Modern concepts, mistakes and paradoxes (in Russian). Nauka Publishing House, Moskow
9. Pochtman YM, Fridman MM (1997) Methods of reliability analysis and optimum design of structures operated under extreme conditions. (in Russian). Nauka i obrazovanie, Dnepropetrovsk
10. Trofimov VI, Begoon TE (1972) Structural constructions (investigation, analysis and design) (in Russian). Stroyzdat, Moskow

2 Building a Design Model

Apparently, one has achieved perfection when one can subtract nothing, rather than when one cannot add anything.
Antoine de Saint-Exupery

It is hardly possible to give clear and unambiguous recommendations on the choice and building of design models for structures of various purpose which require different structural solutions to be applied. One can do it, apparently, only by confining oneself to a much narrower class of objects. One should remember that "*any set of rules for building models can have, at the best, only a limited field of application, and at the worst it may even hamper the manifestation of one's intuition*" [13].

In connection with what has been said, firstly, things discussed below should be treated as very mild recommendations only, and secondly, the reader should prepare himself to a fragmentary style of presentment.

Before we go directly to the promised recommendations, we think it's reasonable to draw your attention to the following issue of design modeling. A design model is not itself an object of interest for an engineer; it is a certain simulation of the stress and strain state of a structure that it provides which is of interest. Using the model, the engineer wants to estimate such features of the respective structure as its ability to be erected and reliably operated, be economic etc. In this connection, the engineering analysis can be divided into four stages as regards the process of obtaining the necessary technical information (Fig. 2.1).

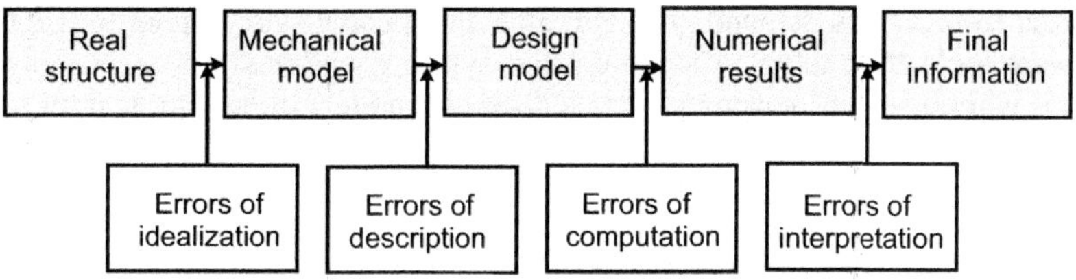

Fig. 2.1. Errors at different stages of analysis

The first stage consists of the transition from a real structure (**RS**) to its mechanical or mathematical model (**MM**). The second stage is to transit from the continuous mechanical/mathematical model to a discrete or design model (**DM**) fitted to capabilities of a particular computational tool. The third stage is to describe this design model with available tools of the chosen software, then to perform the computation itself, and obtain numerical results (**NR**) of the computation. And the final stage is to interpret and analyze the results of the modeling and computation in order to obtain final information (**FI**). So,

$$\mathbf{RS} \Rightarrow \mathbf{MM} \Rightarrow \mathbf{DM} \Rightarrow \mathbf{NR} \Rightarrow \mathbf{FI}\,.$$

At each stage, the role or the extent of participation of the engineer-analyst and the part played by the software are different, as well as different is their responsibility. It is important to notice that every stage of the list includes elements of modeling, thus it introduces its particular contribution to the error accumulated on the way from **RS** to **FI**.

2.1 Determinative Parameters and the Number of Degrees of Freedom

For not a few tens of years there has been a tendency to further complicate design models being used and increase their dimensionality. Developers of modern software products know for sure that all their achievements in quantitative capabilities of software will be quickly exhausted by users. At the same time, the users believe it is a great advantage to be able to analyze the whole building together with its foundation as a single system, so they are anxious to increase the level of detailing and take into account as much as possible.

Computational problems of large dimensionality arise quite naturally when one needs to analyze a three-dimensional problem of continuum mechanics by a finite element technique. If the characteristic number of cells in a finite element grid is m, then the number of unknowns will be of Cm^3 order of magnitude (constant C depends on the character of the problem and on types of the finite elements used), so the dimensionality of the problem will grow very quickly to tens of thousands of unknowns even in physical areas not very much different from canonical shapes. It is these spatial problems that cause the natural demand for high-end software, and the developers are constantly increasing the power of their products in order to meet this demand. Though, after that people start to tend to use these software tools for analyzing structural bar systems, and analysts are tempted to create, without much reasoning, detailed design models of structural frames that include tens of thousands of elastic elements and nodes.

One may think there is no big difference between the two mentioned cases. Though, it is a wrong idea. A problem of continuum mechanics requires only an accurate specification of the respective body's shape and a few material constants, in order to be posed correctly. In the case of a complex bar framework, the array of source parameters that describes the respective problem is very large and heterogeneous. Most parameters of this kind are essentially random values for which the analyst specifies only some of possible realizations. Therefore, as the number of specified parameters increases, the extent of indeterminacy of the whole design model will also grow.

We will postpone the issue of this indeterminacy for a later, more detailed, analysis. Note only, that the increase of the problem's dimensionality will cause a drastic growth of effort needed for preparation of source data and for the analysis of results. Paper [26] presents average data on the distribution of effort during the use of a well-known **ANSYS** software specially intended for solving large-scale problems (Fig. 2.2). This data illustrates the fact that presently the most toilsome activity is the handling of data of design models. In order to save people's effort, one should create a good ground for choosing the reasonable level of complexity of design models.

But the matter is not only with the growth of effort required by data preparation and analysis of results. At times, a formal discretization of a structure's model into a "noodle" of thousands of finite elements will result in only a seeming effectiveness and improvement of the solution's accuracy, while actually the accuracy may fall (even very much!) due to an inevitable accumulation of errors with the growth of the number of unknowns in the discrete model. It is hardly reasonable to, say, divide obvious bars of a caged building frame into three-dimensional finite elements. Returning to our military analogy from the introduction, we may say that shooting flies by cannons may not have been the main idea of the software developers because they know a conventional fly-flap should obviously do much better work to the insects.

The requirement of sufficient simplicity of a model with respect to the chosen description of its properties relates to the extent of its adequacy in a way like that shown on the diagram (Fig. 2.3) taken from the paper [3]. The same paper says that "*there is jocose, but not ridiculous, definition of the value C of a theory as*

$$C = [(k/n) - 1],$$

where k is the number of predictable independent quantities and n is the number of input parameters".

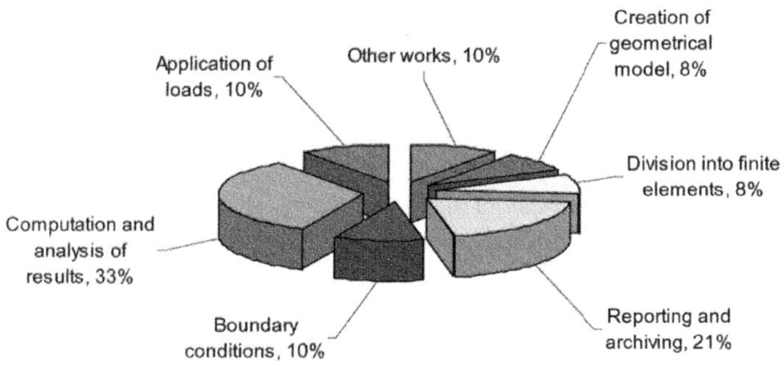

Fig. 2.2. Distribution of effort

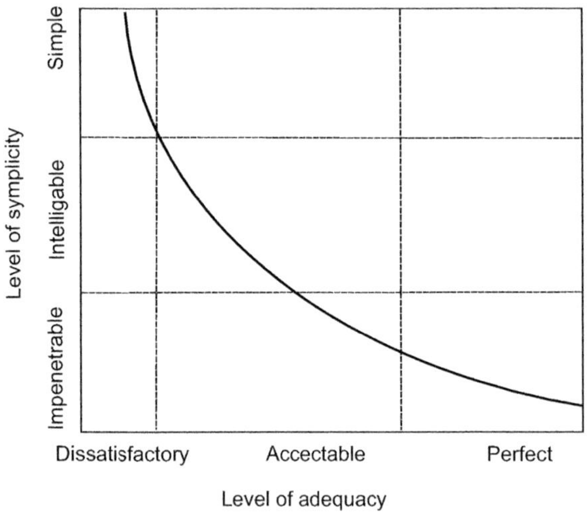

Fig. 2.3. On the estimation of a model's simplicity

The building of a design model is always a kind of compromise. In order to be able to obtain reasonable and practical results with this model, it should be sufficiently detailed and complicated. At the same time, the model should be simple enough to let one obtain the solution under limitations imposed on the result by such factors as deadlines, available software, expertise of analysts, possibility of an analysis and conceptualization of the results obtained. Also, it should be adequate in accuracy to the initial continuous model.

The structural theory owes the majority of its achievements to one's having been able to extract objects with a comparatively small number of determinative parameters and a relatively weak interaction between separate parts. For systems of this kind, one can always pose and solve the problem in such formulation that demands a "full answer", that is, a detailed distribution of stress and strain in all elements of the system. This formulation of the problem is so common for the design practice, that even when the list of relatively simple objects was exhausted, it continued being considered the only possible objective of analysis and was extended to larger systems without any critical reasoning. Optimistic slogans were proclaimed about an unlimited power of modern computer facilities, and thousands of pages of computation results were printed out. The emergence of graphical visualization means has transformed thousands of pages of numbers into hundreds of pages of iso-field images, but the objective of the analysis has remained essentially the same.

At the same time, as one of founders of cybernetics W. Ashby stated [4], there are certain principal limitations of the dimension of analyzable problems, for example the number 10^{100} which is equal to the product of the amount of atoms in the Universe (10^{73}) by the number of microseconds of the Earth crust existence (10^{23}), and "*...the possibility for eliminating this limitation is less likely than, for example, the possibility for eliminating the energy conservation law limitations...*

I am convinced that in the future a system theorist should become an expert on simplification". The authors are gladly joining this appeal.

An expert on simplification of design models can be only a highly qualified specialist. Though, a design software is usually developed for users of different levels of expertise which causes some of the contradictions discussed above. We believe that the possible way out of the situation could be a user's self-identification of his level of expertise (such as "Beginner", "Intermediate", "Expert") quite usual in computer games. Every level will correspond to a certain set of capabilities of the respective software, for example, according to Table 2.1.

Table 2.1.

Level	**Accessibility of the system in the part of:**	
	program setup	**computation**
Beginner	Only standard settings are available	Linear analysis of bar systems without techniques of structural subdivision
Intermediate	Regulation of output of results, specification of forms of onscreen presentation, specification of solution accuracy parameters, setup of accuracy of assemblage and nodal coordinate coincidence	Linear static analysis of arbitrary systems making use of finite element models including plates/shells, dynamical analysis compliant with building regulations (seismic, wind)
Expert	Full access to all settings of the program (styles of toolbars, font, size and color settings of messages, setup of animation modes etc.)	Both linear and nonlinear analysis (physical and geometrical nonlinearity) involving all capabilities of the software, all kinds of dynamic analysis, stability and buckling, technique of fragmentation, all types of document preparation, exporting and importing to/from other systems

The user could be aided to identify himself correctly by a built-in examination subsystem which would be different from conventional "choose-correct-answer" quiz systems in that there would be several correct answers available, with the respective characterization of the level of expertise that corresponds to particular answers, and appropriate comments thereto. Evidently, the development of such quiz would be an interesting and complex educational task.

One of most powerful simplification techniques is the use of some kinematical hypotheses with which the behavior of large groups of "independent" variables can be represented by a comparatively small number of determinative parameters. The typical example (so traditional one that its use is implied rather than declared explicitly) is the use of the plane-sections hypothesis in the Bernoulli – Euler theory of bars.

Similar are hypotheses about the absence of deformations in the floor planes of a multistory building when these are represented by perfectly rigid diaphragms, and the whole building is analyzed under horizontal loads as a system of horizontal disks connected by elastic vertical braces capable of moving only horizontally [14]. All these techniques of aggregation of variables and use of some generalized coordinates are important not so much for reducing the problem's dimensionality (though it is of value as well) as for compressing the information in order to analyze it more conveniently. Also, these techniques are important in cases when data about external influences are not known well enough to be correctly "bound" to a detailed design model.

2.2 Model of Loading as a Part of the Design Model

The fact that idealization of influences/loads is a part of activities of building a design model is not often remembered. There are a lot of reasons for this, and one is that effectual design regulations have been formulated as universal ones and are not tightly connected to the part of the design model that describes the geometry of the respective structure. This approach is reasonable if the load that simulates the environment's influence upon the structure is by no means dependent on the behavior of the structure itself. But sometimes one has to recall that there is not only the *action* of a load *upon* a system but also the *interaction* of the load *with* the system.

Characteristic examples of the interaction between loads and systems are numerous modes of dynamic loading. A structure is a kind of filter which selects a certain part of disturbances that act upon it. This issue has been studied in detail. Distribution of frequencies is necessarily present in descriptions of dynamical loadings, and the modal analysis that gives spectra of natural frequencies and modes of oscillation is a basis for most dynamic analyses.

It is less known that such filtration may take place not only with respect to frequencies (phenomena of resonance) but also with respect to wavelength. We have mentioned that there can be a substantial difference between a seismic influence upon an object with small plane dimensions (a chimney, a tower) and that upon an extensional structure standing on a big foundation plate (a grain elevator's building, a nuclear power plant). In the former case the structure will respond to waves of any length so long as the frequency distribution is such that it fits the area of sensitivity determined by the structure's natural frequency spectrum. In the latter case the structure will respond similarly only to waves approximately twice as long as the foundation's plan size. These waves will act almost in phase along the whole structure while the effect of shorter waves may prove out of phase. Similar effects should be taken into account when analyzing waterworks. Thus, one often manages to reduce the overall level of loading of deep-sea foundations if the latter's general dimensions have been selected so that the wave influence be in anti-phase on the opposite sides of the structures.

Another example of a load's interaction with a structure is an aerodynamic (wind) loading. Its value depends much on the shape of the structure flowed around by wind. One may observe a kind of positive feedback effect here: increasing cross-sectional dimensions would increase the wind load which in turn would increase the cross-section area again. This process is not necessarily convergent, and there have been cases when one failed to design the respective structure appropriately. A similar positive feedback may arise when the dead weight is taken into account, and in cases of a flexible construction loaded by a layer of liquid[1]. In the latter case, the deflection of the structure will increase the thickness of the liquid layer which in turn will increase the deflection once more. In all listed cases a phenomenon of the equilibrium instability may arise somehow, in spite of the absence of compressive loads in the respective system [17].

Often, reactions of eliminated constraints via which the environment interacts with the structure by restraining its displacement and imposing a certain behavior, may be considered as loads upon the system. The structure being analyzed is always a part of a bigger system. When extracting the structure from its environment, we either idealize the latter's influence in the form of an absolute prohibition of certain displacements (the system's attachment to the "earth") or describe this influence as a load upon the system. Using this approach, one should seriously consider the issue of the system's deformation influence upon the loading. In other words, the structure should be in some sense incomparable with its conventionally eliminated environment by stiffness. Only then we can assuredly assume the load to be independent from the system's deformation.

Thus, the standard idealization of the structure's interaction with its environment is a specification of either some (most often, zero) displacements when we impose a constraint or the loading by a force.

Discussing differences between force and deformation influence, V.I. Feodosiev [4] gives an example of a characteristic mistake when a supporting rim of a tank (Fig. 2.4, *a*) is loaded by a force (Fig. 2.4, *b*) equivalent to the dead weight uniformly distributed along the rim.

But the rim cannot be treated as a separate system loaded by forces because it deflects only to the extent allowed by the tank's wall. Practically, the rim is in the condition of a deformation loading, or, better, a kinematical influence.

The difference between a force and a deformation loading is especially evident when analyzing consequences of their swapping. In an elastic system this difference is hardly visible, but the matter is quite different with non-elastic systems. This circumstance is not often addressed. The design practice is deeply bound to the estimation of stresses, and often one forgets about their inequality with deformations when Hooke's law is untrue.

---

[1] In a textbook on mechanics of materials by S.P.Timoshenko and J.M.Gere, the authors call it *a problem of filled tank*.

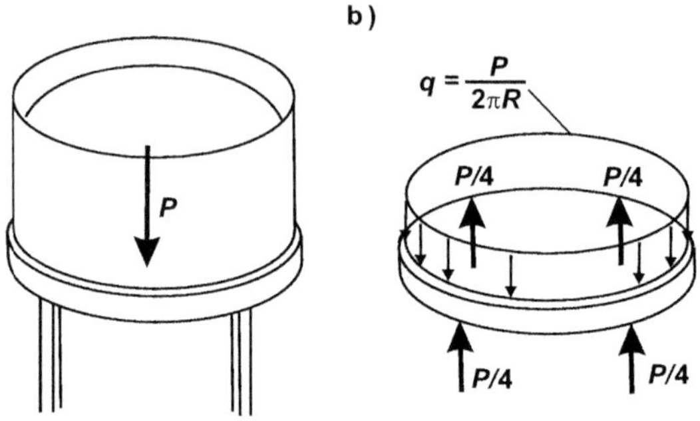

Fig. 2.4. A design model

The mentioned book by V.I. Feodosiev poses the problem very clearly: whether the coefficient of resistance calculated by stress is a sufficient characteristic of reliability or not. The author says it is not, and he discusses two ways of creating a stress in a bar: first, by applying some load P that causes a stress σ in the bar section, second, by forcible deformation that creates a deformation ε exactly corresponding to the same stress σ (Fig. 2.5). In the first case, the increase of stress 1.5 times will cause the rupture of the bar, in the second case (1.5 times increased strain) will cause only the growth of residual deformation.

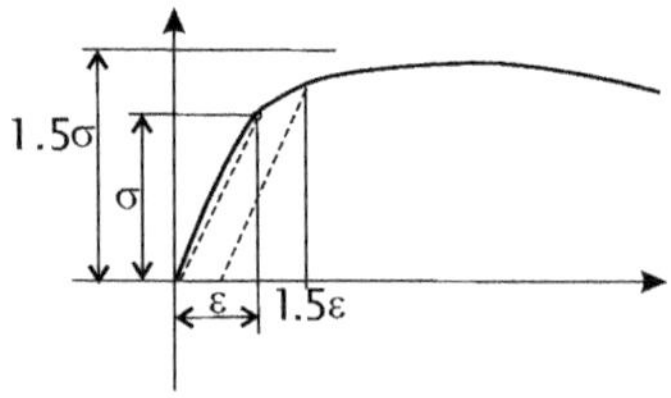

Fig. 2.5. Stress-strain diagram

The difference between a force and a deformation influence is especially dramatic when considering the influence of the load's specification accuracy on results of the analysis. Thus, in a simplest problem of a beam flexure described by the differential equation

$$(EIv'')'' = q(x)$$

one can consider two alternative cases:

- the load is the reason for the flexure, and the deflection is the consequence;
- the deflection is the reason (a template flexure), and the load/reaction is the consequence.

If we assume an absolute accuracy of the source data specification and the method of solution, then there will be no technical difference between these two cases. But in reality this difference is very big because a small change in the load will cause a small change in the deflection, while if the template is specified inaccurately then the dispersion of the resulted load may be really great (we calculate even the first derivative of an inaccurately specified function very poorly). In essence, this is nothing but a nonstandard interpretation of Hooke's law: if in an elastic system big forces cause small deformations, then small deformations may cause big forces. Though, traditions of engineering education do not draw the students' attention to this treatment of the elasticity law.

2.3 Validation and Means of Description of Design Models

An instructive book by A.A.Kosmodemiansky [7] says: "*...when designing and building new structures (such as bridges, dams, aircrafts, rockets, buildings) on the basis of a huge previous experience, specialists are so sure about the correctness of mechanical laws that they treat all deductions from their calculations as an* ***absolute truth****. If in some cases there are discrepancies between the theory and the practice, then during further strict and careful analysis they explain all of those by inaccuracies in input data or arithmetic computational errors*". These words apply mostly to the pre-computer age. Now, when the computer-aided design is a common thing, the principal sources of such discrepancies may be: incorrectly built design models; incorrect use of software; and, most rarely, errors in the software itself.

This situation is well understood by experienced analysts, and they make every effort to use all available methods to validate their design models and check the correlation between computational results and the behavior of the real structure. The validation and checkup techniques involve:

- dimensionality of used quantities;
- character of the result's dependency on changes in some input data, including the check of such properties as an expected symmetry (or anti-symmetry) or insensitivity to some parameters;
- behavior of the system at ultimate values of certain parameters;
- stability of the design model with respect to small changes in its structure or parameter values;
- observance of reciprocal relationships (for displacements, stresses, or works).

Further below, we will discuss some useful techniques of validation and show how they can be practically implemented. Now, we shall start with a trivial checkup of input data.

The probability of errors in input data is much greater in problems of a big dimensionality. Studies on engineering psychology state an approximately power dependence of the human error probability on the amount of information processed by the human. In this regard, the paper [2.10] contains an assumption about a tight connection between the error probability and the size and complexity of the respective structural model.

Any modern computational software operates pretty heterogeneous data which describe properties of finite elements, nodes, loads, influences etc. During solution of problems of big dimension all this information is hard to handle simultaneously, while the use of a standard approach based on fragment extraction is not always convenient. This inconvenience reveals itself particularly when one tries to track relationships between some factor and the system as a whole (for example, localization of elements of certain rigidity in the structure).

Here follows an example of what means of input data analysis are available in the computational software **ANSYS**, one of most powerful products of this kind all over the world:

- color and numeric indication of elements and geometric primitives by their attributes;
- symbolic presentation of loads and boundary conditions;
- construction of any cross-sections and cuts;
- assignment of transparency to surfaces and bodies;
- perspective views;
- positioning of light sources;
- three-dimensional views of the real geometry of sections of beam and shell elements;
- possibility of creating graphical annotations (adding symbols and text to images).

Special tools built in CAD software are commonly used to facilitate the analysis of input information. Below there is a description of such tools included in the **SCAD** software which is pretty typical for modern approaches.

For example, this software suggests the use of so-called *view filters* which help one remove from the screen all information of no current interest, as well as the use of special *visualization tools*. Separately and in any combinations one can filter, say, information regarding elements of particular types or those specifically located in space. When a filter is turned on, only the information of a particular group remains on the screen that is related to this filter (see Table 2.2).

Typical examples of the use of filtering are shown on Figs. 2.6 – 2.9 which show how the amount of information visible on the computer screen is gradually reducing as more and more filters are turned on.

After turning on the filter "Color indication of rigidities", one can specify the part of information (appropriate colors) to be kept for further work. Also, this information can be used to select the respective fragment of the design model, and the rest will just disappear from the screen.

Besides filters, *markers* are used to show additional reference information on the screen when activated. Using markers, one can turn on or off the visualization of various data (nodal loads, distributed loads, constraints, element and node numbering etc.).

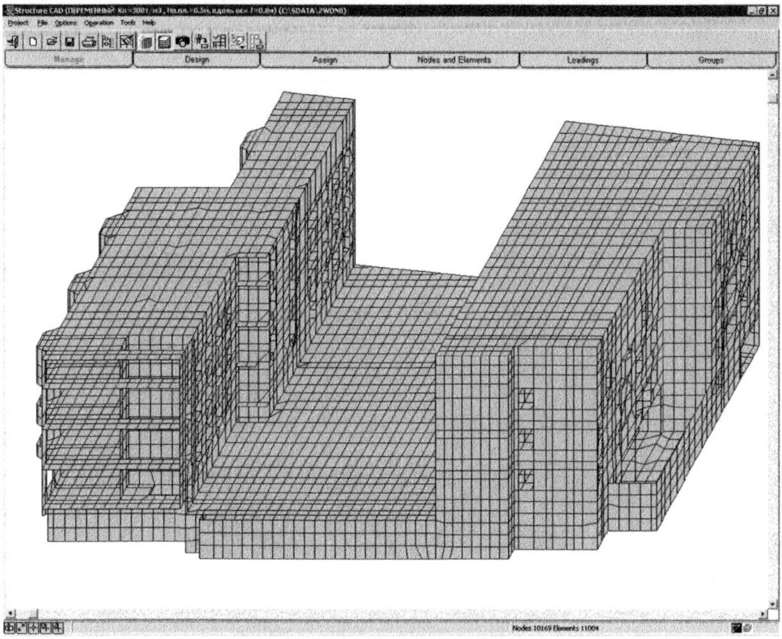

Fig. 2.6. Filtering of information — full view

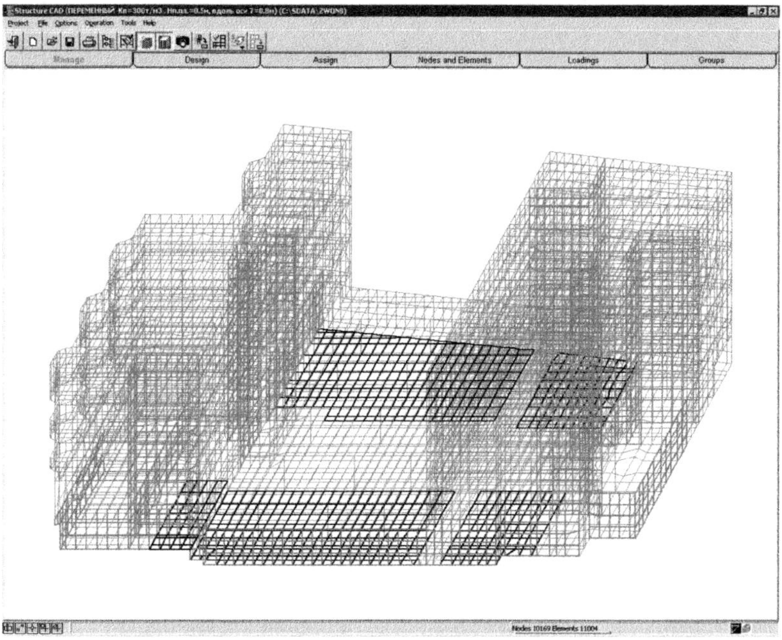

Fig. 2.7. Filtering of information — color indication of rigidity types

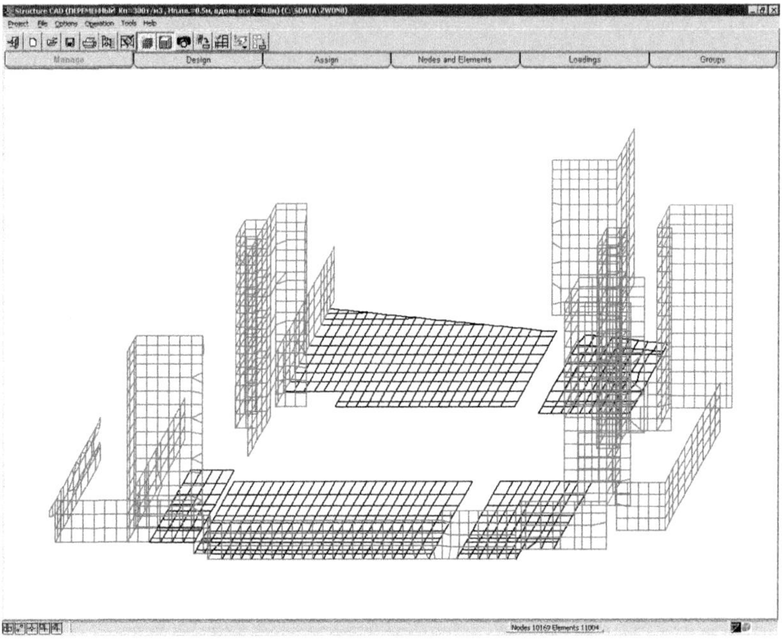

Fig. 2.8. Filtering of information — exclusion of unneeded parts

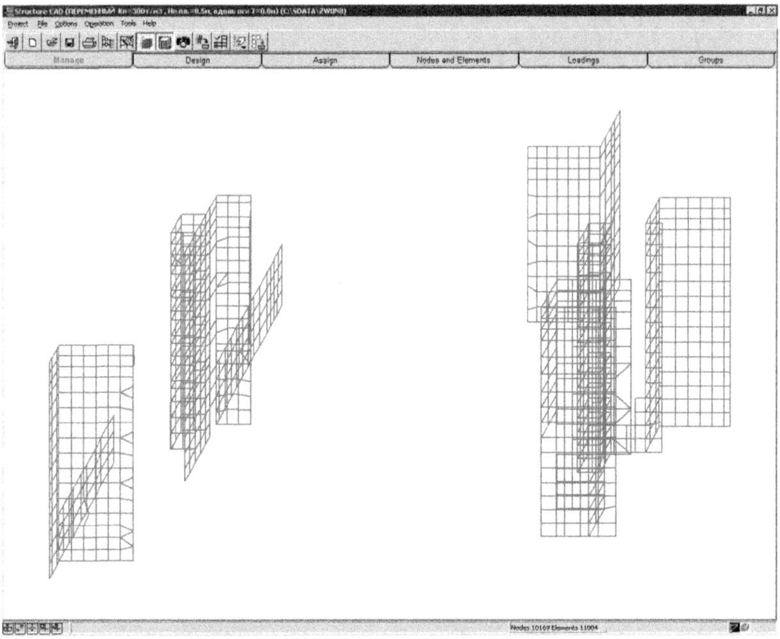

Fig. 2.9. Filtering of information — selection of elements of particular rigidity

Table 2.2.

Group	Filters	Information is skipped related to:
Description of elements	Bars	elements of bar types
	Plates	plate/shell elements
	Three-dimensional elements	three-dimensional elements
	Special elements	elements that simulate constraints
Description of loads	Nodal	nodal loads
	Distributed	loads distributed on lines, areas or volumes
	Concentrated	concentrated off-node loads
Geometric position	Vertical	vertical bars
	Horizontal	horizontal bars
	Oblique	oblique bars
	In the X0Y plane	plates parallel to the plane X0Y
	In the X0Z plane	plates parallel to the plane X0Z
	In the Y0Z plane	plates parallel to the plane Y0Z

Color indication of data of the same type different only in their values is widely used for this purpose. In total, the **SCAD** software offers a lot of various markers some of which are listed in Table 2.3. Note that the markers can be adjusted so that they work on filtered values (for example, mark only horizontal elements).

The **SCAD** software, as well as other similar programs, implements a number of techniques of decomposing a design model in order to facilitate the problem's analysis at the input and output. The decomposition by space (technique of fragmentation), by meaning (technique of grouping), and by build (technique of sub-structuring) is possible.

A fragment can be any part of the design model indicated by a "rubber frame" on the screen; after the indication, elements not completely captured by the frame will be excluded from the fragment. All other elements form the fragment, and only they (while the user needs them) are present on the screen. The part of the model obtained in this way can be then subjected to filtering, and vice versa, the fragmentation can be done over a filtered view. Thus, the decomposition with regard to the fragmentation is a technique for interactive information analysis.

One often needs to fix a fragment of the system in order to address to it for many times with various queries. This operation involves the technique of *grouping* using which any subset of the model's objects (nodes, elements, loads) can be represented as a named group. Finally, any part of the model can be treated as a *substructure* and the whole structure defined as a combination of substructures. The principal difference between a group and a substructure is that the latter itself possesses features of a design model while a group might not be such.

Table 2.3.

Markers	Displayed information
Model description	
Node numbers	Numbering of all nodes visible on the screen
Element numbers	Numbering of all elements visible on the screen
Constraints	Indication of nodes having external constraints
Merged displacements	Visualization of groups of nodes with merged displacements
"Hanging" nodes	Indication of nodes to which no elements are attached
Coincident nodes	Indication of different nodes with the same spatial location
Groups of nodes	Color indication of groups of nodes
Groups of elements	Color indication of groups of elements
Element description	
Element type	Numeric and/or color indication of finite element types
Rigidity type	Numeric and/or color indication of rigidity types
Local axes	Visualization of local coordinates of elements
Hinges	Visualization of end hinges on bars
Stiff inserts	Visualization of stiff inserts in bars
Coincident elements	Indication of different elements having the same spatial location
Load description	
Nodal	Visualization of symbol of nodal load
Distributed	Visualization of symbol of distributed load
Concentrated	Visualization of symbol of concentrated load on the element
Temperature	Visualization of symbol of thermal load
Values of loads	Output of numeric values of loads near the symbols

Another set of checkup tools, besides filters and markers, includes *descriptors*. These can relate to nodes or elements of the system and show, when used, all data referring to a particular object of the design model. Fig. 2.10 shows an example of the descriptor for a node that can display information about a particular node selected on the screen or specified by its number. In the latter case the model is automatically searched for this node, and it is marked to be visible.

This mode of the descriptor lets one browse needed data (coordinates, constraints, loads) and also change them interactively, that is, the descriptor is a means of inputting information as well as a tool for visualization and checkup.

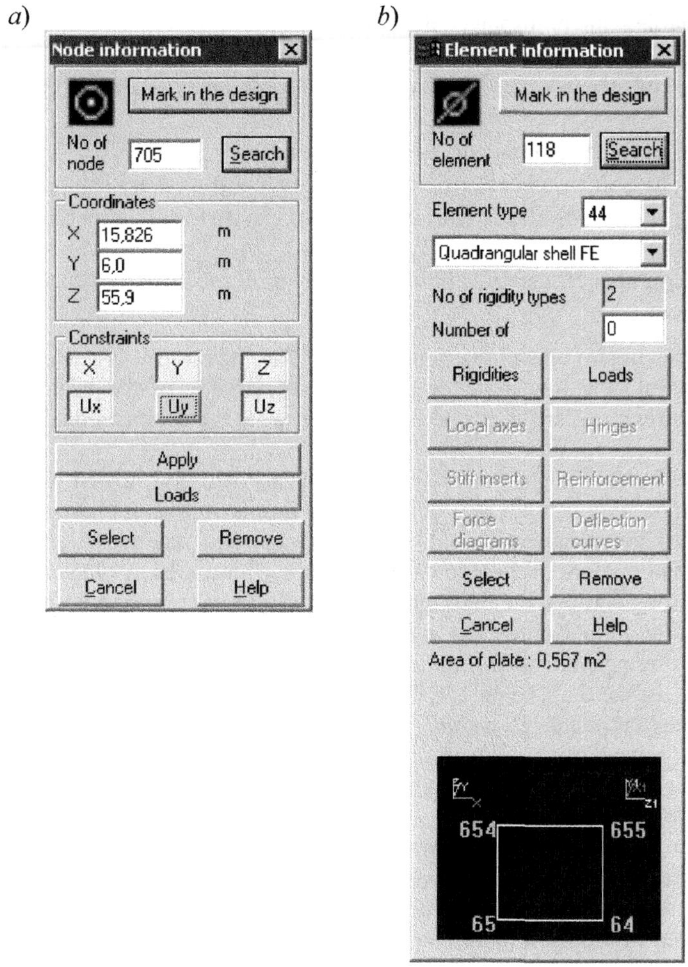

Fig. 2.10. Descriptors: *a)* — of a node; *b)* — of an element

Note that it is of import for a user to be able to set up some elements of the graphical environment in order to create a convenience of work. Among software users there are adherents of both the positive (dark objects against light background) and negative (dark background, light objects) styles of design model representation. The black screen background can be very spectacular for presentations, while the light-gray background is less contrast and tiring, thus more convenient for work. Advantages of these polar approaches can be discussed both long and uselessly. Users who spend long hours before the computer screen should have the right to choose what's best for them.

To conclude this section, we will indicate other tools for improving the visual convenience of results built in modern computational software. Those are:

- controllable scale of displayed deflections;
- animation of displacements;
- building of diagrams of displacements, forces or stresses v. time or load intensity;
- building of force or strain curves for bar elements;
- building of isometric lines or fields of displacements or stress components;
- building of stream-lines for directions of principal stresses in two-dimensional problems;
- calculation of coefficients of limitation utilization specified by structural design regulations.

A skillful use of these functions can facilitate the user's assessment of the obtained result and give him the proper understanding of the behavior of the simulated structure. This is, properly speaking, what the computational analysis is performed for.

Issues of the design model's validation are closely related to tools of description of input data for the computational design software. These tools can be divided into three categories: tables, commands, and interactive graphics. All possible combinations of these tools should, or even must, be available.

For obvious reasons, the description of input information for a design problem in the tabular form (sometimes as sets of *documents*) preceded historically the command-line or interactive tools of data specification. In order to facilitate the data input, reduce their volume in the tabular form, special techniques are widely used such as "*principle of default*" (implicit assignment of predefined values to pieces or groups of data), various *repeaters* of data including nested ones, consolidation of data into named groups with further addressing them by the name of the group etc.

In order to unify the syntax of tabular data, R.A. Reznikov [22] in his day developed a special data description language called **VHOD** (Russian for "*input*"). Presently, another means of source data representation unification is used, one implemented in the **FeMap** software which is a universal preprocessor for such different software products as **NASTRAN**, **ANSYS**, **LS-DYNA**, **ABAQUS**, **STAAD**, **COSMOS** and a number of others.

Command tools of data description (one of typical software representatives widely using the command-based data specification is the **GTSTRUDL** software) are close relatives of the tabular specification as they use the same principles and techniques to reduce data amount, where the data itself is specified as parameters of commands. The principal differences of commands from tabulated data documents are that commands can be entered in any arbitrary order (or almost arbitrary; a few exceptions are strictly specified), and multiple commands with same names can be present in an input data set. Commands themselves are divided into *declarative* and *directive*. Declarations affect the data content while directives manipulate the data. For example, there is a directive for spatial shift of some previously named object (in terms of **SCAD**, an object is a substructure) with data copying and re-definition of the object's name, or with the removal of the copied original. Ordinary Windows users know something like this as a sequence of actions *Copy+Paste+Rename* or *Cut+Paste*. Other directives can initiate a certain type of data processing or have the input data, interim computational results, final results of the analysis arranged and displayed on the screen (and/or printed out and/or sent to another information carrier) at the user's request.

The tabular and command-based means of data input are closely related in the sense that both of them use textual data representations. Though, most advanced capabilities of specifying input information for a computational analysis belong to interactive graphical forms of description and handling of data. The emergence and wide use of the Windows operating system has made the interactive graphical data specification in computational and design software an essential standard, and this approach is implemented in nearly all known structural analysis programs. It is obvious that the possibility of visualizing and handling data in their graphical form facilitates the user's work at the monitor very much, his actions become simpler to learn (more intuitive and easier to remember) and help him check himself. For these reasons, no computational software is potentially competitive on the software market if it does not support interactive graphical tools for data handling.

At the same time, in the authors' opinion, the interactive graphical user interface does not make the tabular and command-based data descriptions unneeded or useless. Firstly, the text data representation is very convenient for the operation of various *converters* — special software utilities to convert data from the structure and syntax intended for one design program to the structure and syntax of data compliant with other software's format(-s). Secondly, the text data representation is relevant for their paper documenting, exchange between people, storage and archiving of input information. Finally, the text form of data enables one to review and check it not at one's computer screen only, but comfortably at one's desk. Some subtle issues of a structural design model appear to be easier and more dependable to analyze by carefully reading its textual data. In this connection, it seems that ideally an interactive graphical data description should be automatically transformable to an equivalent textual form and vice versa. The user should be able to choose his preferable form for a particular purpose, switching when needed from the textual form to the graphical one and backwards.

Along with the data description format, it is very essential to be able to specify different *modes of input* of data. Such modes may be:

- a mode of addition — ADD
- a mode of deletion — DELETE
- a mode of changing — CHANGE

The principal mode activated by default when the respective program is started is the data addition mode which will be the current one until a command is entered to switch to another one. Switching to another mode is performed by entering a command that specifies the name of the new mode that becomes current. It is convenient for a user that the input modes could be switched as many times as needed during the user's interactive session[2]. The data deletion and modification modes can be efficiently used, for example, for the purpose of multiple re-calculation of a structure with its geometry modified or different external influences applied.

The next tool for the user to describe and enter data which is relevant in a computational software, is the possibility of specifying a data status by *activation* or *deactivation* of separate pieces or whole groups of data. Again, the data should be treated as active, that is, present in the model, by default. If some data, say, a group of bars and nodes, is deactivated then it is actually removed from the design model. This capability is especially useful for the purpose of analysis of structures with modified geometry that may occur in the course of erection or assemblage of the structure. By activating or deactivating groups of data in different modes of loading, one can apply combinations of loads to elements which belong to different parts of the model in different modes.

There is another trifle which is hard to do without as soon as one gets used to it. We are talking about identifiers (names) of nodes, bars, finite elements, loads and, generally, any objects of a design model. Numbering (say, numbering of nodes, to be particular) is not the most general, thus not the best way of identification. A simple but useful idea implemented in **GTSTRUDL** is that the name of a node may include an alphabetic or alphanumeric prefix followed by a number, and there is no limitation to the numbering. The numbering may contain omissions, begin with other number than one etc. The numeric suffix of the name can be used to organize repeaters. Finally, the checkup of data and the understanding of the model by an engineer that checks the data can be facilitated very much by choosing convenient and intuitive names. Having some experience, one can always achieve this level of convenience.

2.4 Some Tricks

Computational error caused mainly by data rounding will make coefficients and constant terms of a governing system of equations, formed for the analysis, approximate to some degree. This circumstance and further rounding in the course of solution of the system will force the result to be somewhat different from the accurate one. The solution of an equation system is called stable and its matrix of coefficients well-conditioned if a small disturbance of coefficients of the system will cause only a small change in the components of the solution.

In most cases it is not known beforehand whether the matrix of coefficients of a governing system of equations is going to be well-conditioned or not. Calculation of various measures (numbers) of conditioning, which are intended ideally to determine the rate of change of the solution with respect to coefficient changes, does not always amend the situation because these numbers will, most often, give

[2] It is useful to indicate the current input mode in the so-called status bar usually placed at the bottom of a graphical screen.

a relative rather than absolute characteristic of this rate, i.e. they say matrix **A** is better conditioned than matrix **B**. There are methods for reducing the conditioning number or improving the accuracy of the solution of an equation system on the basis of the *regularization* idea. In connection with problems of structural analysis, these methods are described pretty well in [25]. Here we think it is reasonable to consider the problem at another angle of view, namely: what possible reasons for poor conditioning are and how these reasons can be interpreted by means of mechanics, and when this phenomenon evidences a poorly designed model of a structure.

It is generally known that poorly conditioned stiffness matrices appear often when elements of drastically different rigidities join in a single node of a design model. Let's demonstrate by a simplest example how this situation can be interpreted in terms of mechanics. As it happens, poor conditioning is shown by "*almost imperfect*" structures.

Indeed, an imperfect system corresponds to a degenerate stiffness matrix **K**, and this degeneration means that there is a displacement vector $\mathbf{u} \neq \mathbf{0}$ such that the homogeneous system of governing equations holds:

$$\mathbf{Ku} = \mathbf{0}\,, \tag{2.1}$$

i.e. the displacements are consistent with zero deformations of elements which correspond to zero loading.

Fig. 2.11 shows a formally stable system with the following stiffness matrix:

$$\mathbf{K} = k\left\|\begin{bmatrix} 1+\alpha & -1 \\ -1 & 1+\alpha \end{bmatrix}\right\|. \tag{2.2}$$

Eigenvalues of this matrix $\lambda_1 = \alpha$; $\lambda_2 = 2 + \alpha$, and the conditioning number is

$$H = \lambda_2/\lambda_1 = 1 + 2/\alpha\,.$$

If the stiffness of the middle spring is very big comparing to the stiffness of the end springs, then parameter α is small and number H is great, which evidences the poor conditioning and possible loss of accuracy during the solution of the system having such matrix.

It's easy to notice that the mechanical behavior of this system approaches the behavior of a changeable (imperfect) system. Indeed, the displacement of the middle spring as a rigid body may occur as the end springs' resistance is negligibly small. The reactions of the latter, due to the described relation between the rigidities, cause only a trifle deformation of the middle component. But if we change the ratio of the rigidities to the opposite, then the stiffness matrix will have the conditioning number $H \approx 1$. Again, elements of drastically different rigidities meet in the same node, but the stiffness matrix is well-conditioned and corresponds to an elastic structure (the middle spring) attached to the ground by nearly non-deformable constraints.

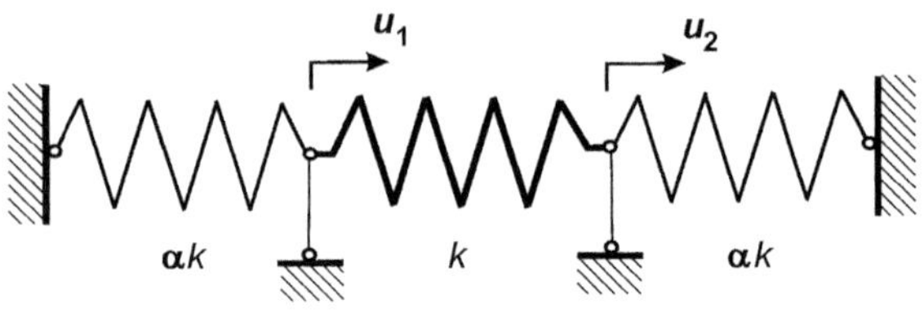

Fig. 2.11. A system with a poorly conditioned stiffness matrix

A classical definition of changeability as the property of a system to have nonzero displacements of nodes that cause no deformation of its elements (a displacement as a stiff frame) was generalized by Y.B. Shulkin [24] and replaced by the following one:

> *a system is considered to be changeable, if it allows small nonzero displacements with which deformations of its elements, even if they are different from zero, are infinitesimals of higher order of smallness.*

Thus, the logic of kinematical changeability has ceased to be black-and-white (yes/no) and has been enriched by the concept of order of smallness of displacements and deformations.

It is worth noticing that detection of "nearly changeability" based on a comparison of orders of smallness between possible deformations may evidence both a lack in the structure and a fault in its modeling. Without considering the first case where the designer should obviously change the structure, we will note that in the second case one often manages to find a simple amendment: declare an element with a very high rigidity to be perfectly rigid, and take this feature into account at the level of the equation system by introducing the appropriate constraint [20].

During analysis of bar systems, one often has to allow for an eccentricity of elements' joining in nodes (Fig. 2.12, *a*).

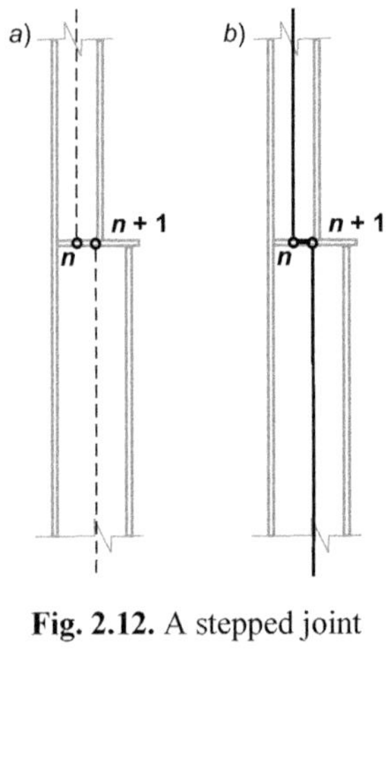

Fig. 2.12. A stepped joint

Inserting a bar of very great but finite rigidity between nodes n and $n+1$, as it seems intuitively right, will cause, as we have demonstrated above, a drastic deterioration of the calculation accuracy due to a poorly conditioned stiffness matrix. In order to circumvent this computational difficulty, CAD programs offer the use of *infinitely stiff inserts* at ends of bar elements. Then the design model has only one node intermediately located anywhere between node n and node $n+1$, and the end sections of the adjoining bars are attached to this node via stiff inserts. No loss of accuracy occurs in this model.[3].

One can go a simplest way by merging this single node N with either node n or $n+1$, then the stiff insert appears at one element only. The payment for this simplification is that internal stresses will be determined only in the elastic part of the bar.

The use of stiff inserts is a common practice when one deals with a plate or shell reinforced by ribs eccentric with respect to its median surface. If the ribs are simulated by bar elements, then the eccentricity can be easily and conveniently allowed for by stiff inserts (Fig. 2.13). Note that the eccentricity of the ribs' position affects the membrane group of stresses, therefore the allowance for eccentricity in a purely flexural structure (constructed of appropriate finite elements) gives nothing special.

One may notice that the description of geometry of stiff inserts placed at ends of bars in the input information of CAD software involves usually projections of the inserts onto the axes of the global coordinate system (X,Y, Z). Meanwhile it is not always convenient for the user. For example, when one analyzes ribbed shell structures with ribs simulated by bars with eccentric attachment of the bars to nodes, the stiff inserts are usually oriented along principal axes of inertia of the bars' cross-sections.

In this case the analyst could take the lengths of projections of the stiff inserts onto axes of a local coordinate system (ξ,η,ζ) associated with the respective bar directly from the drawing and enter them in the input data for the analysis. Though, this user-oriented way of stiff insert specification is seldom available in CAD software.

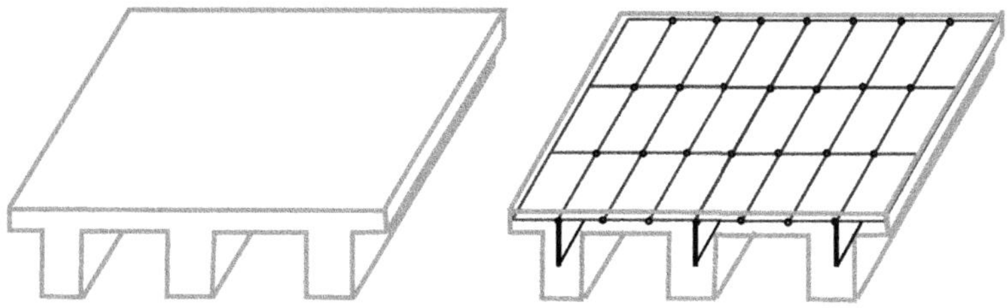

Fig. 2.13. A ribbed plate

[3] It is worth noticing that no loss of conditioning happens as a bar of very small stiffness is inserted between nodes n and $n+1$. It is apparently true in the case when the zero stiffness of the insert keeps the system kinematically unchangeable.

The matter is that, though the purely geometrical problem of mutual orientation of the global and a local coordinate systems seems quite simple, with certain relationships between parameters the solution of this problem may be inexistent (or not unique if existent). A complete analysis of the respective collisions, detailed algorithms and the appropriate procedure have been published in [12]. Note only, that one should be careful about this technique as it can be explained by the following example. Commonly, the analysis of bar systems involves structures with the cross-section depth 1/8…1/10 of the distance between nodes at the most. But there are structures where this ratio comes to 1/5 or even 1/3 (some types of foundations under turbine generators, diaphragms of buildings, hydraulic structures etc.). In these cases a bar model with point nodes located at crossings of element axes is incorrect. To amend the situation, there is a popular approach of taking into account actual sizes of “nodes” by the use of bar elements with stiff inserts. An example of this type of model constructed in accordance with this approach is shown in Fig. 2.14.

This technique has been known and used for such a long time that analysts ask themselves rarely whether the hypothesis of a “non-deformable node” is true in every particular case. Meanwhile, this question is not trivial, as one can see by the consideration of results of a sample problem (Fig. 2.14, *a*). Its bar simulation (Fig. 2.14, *b*) allows no horizontal displacements, so the vertical bar does not bend. A more detailed model, however, indicates the presence of horizontal displacements (Fig. 2.14, *c*) which arise due to a restrained compression along the junction line AB. As there is no such restraint on the opposite side of the “post”, a non-uniform distribution of stress appears which is equivalent to bending.

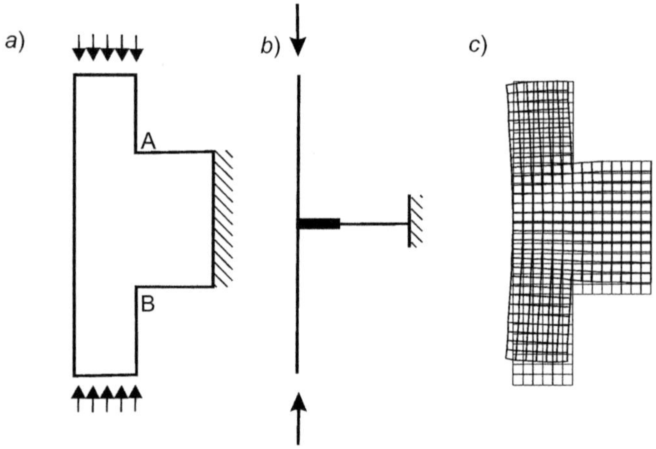

Fig. 2.14. To the analysis of a node of finite size

Another widely popular technique of building design models is the merging of displacements of different nodes of a model, that is, forcing one node’s displacement to be equal to another’s. This technique is used most often to neglect some deformations of the system. For example, if one merges horizontal displacements of nodes in junctions of girders to columns of a single-story frame, this action will be equivalent to the use of the hypothesis of perfect rigidity of the girder with respect to tension/compression.

For a multistory spatial framework, one may often neglect the compliance of floor slabs, treating them as perfectly rigid bodies. If in such model all horizontal displacement components of all nodes of each floor are merged, then the idea of a perfectly rigid disk will be implemented. Though, one should note that no torsion of the floors will be possible, and the floors will be allowed to translate only. This effect may be an unpleasant surprise for the analyst, especially if a structure of an asymmetric build is under consideration. The way out of the situation would be the use of a perfectly rigid body which would simulate the effect of the floors on the behavior of the framework.

It is useful to pay attention to another popular fallacy, best illustrated by a practical example. Thus, they have tried to allow for the role of the structural core in the behavior of a bedplate under a load symmetric with respect to axis X (Fig. 2.15) by merging displacements in nodes of abutment of the structural core to the plate. Linear displacements along axis Z (deflections of the plate) and slope angles θ_x, θ_y were merged. After the analysis, they found out that there were no lateral forces Q_x within the structural core but there were nonzero forces Q_y, which evidenced the bedplate's bending.

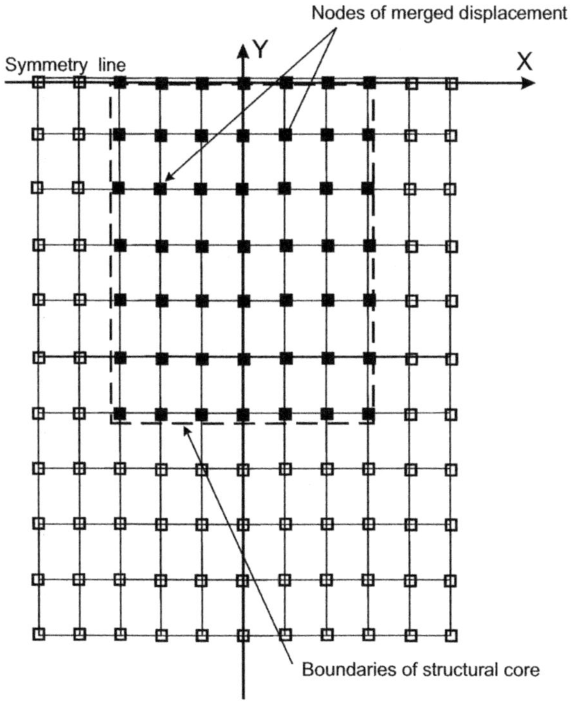

Fig. 2.15. A fragment of a bedplate

The explanation for this result is that no difference was taken into account between restraining a displacement (imposing a constraint) and merging displacements of different nodes. After the linear displacements along axis Z had been merged, the respective part of the plate was able to displace only vertically, and it seemed that the slopes would be zero too. Though, even merging thereof may not assure their equality to zero.

This can be illustrated by a simple model in Fig. 2.16, where nodes 1, 2, 3 ... have the same displacements and *equal but nonzero* slope angles.

In the problem in Fig. 2.15, it would be more correctly to specify a perfectly rigid body in the area of the structural core in order to guarantee the equality of slopes in all nodes of the core with respect to appropriate axes and to assuredly keep the nodes in the same plane in the deformed structure. If in the problem of Fig. 2.15 axis Y is also an axis of symmetry of both the structure and the load, and this axis intersects the structural core, then one can easily do without involving a rigid body. Indeed, in this situation it suffices to restrain the nodes of the structural core against slopes θ_x and θ_y, and merge the linear displacements along axis Z.

A rigid body can be used to simulate such kinematical hypotheses as the Kirchhoff – Love hypothesis in the theory of plates and shells, or the plane sections hypothesis in the theory of bars. If, for some reason, a bar needs to be modeled by a set of finer finite elements which would give a solution in the style of elasticity theory rather than in the style of strength of materials, then one can introduce constraints in the form of rigid bodies to try to restore the latter simpler approach. This idea has been addressed many times by Professor N.N.Shaposhnikov who would declare its didactic value. He implemented the idea in a tutorial software called "Modeling of load-carrying structures behavior" oriented at the training of CAD students at Moscow Institute of Railway Engineers.

The capability of a partial rather than full merging of displacements can be used to form some special boundary conditions in non-framed structures. For example, if one needs to describe the construction of a plate ABCD supported along its contour with its two halves connected by hinges along line EF (Fig. 2.17), one can introduce two systems of nodes (..., 6, 10, 42, ... and ..., 7, 11, 43, ...) attached to the left (ABEF) and right (FECD) halves of the structure, respectively, and merge all displacements but the slope around Y in adjacent node couples.

Another, more direct, way of solving the same problem is to release the places of the finite element attachment to the respective nodes [18]. It does not require an artificial introduction of any listed systems of nodes along the line of the hinged connection between the plates. Unfortunately, somehow most existing CAD programs do not provide the possibility to remove constraints in places of attachment of nodes to finite elements other than bars.

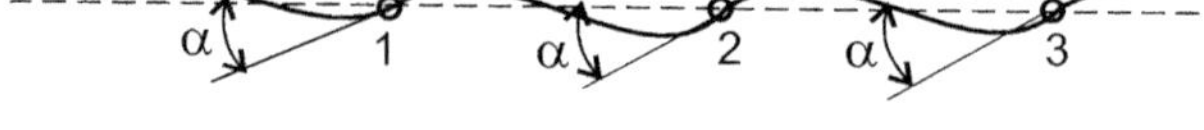

Fig. 2.16. Merging of slope angles

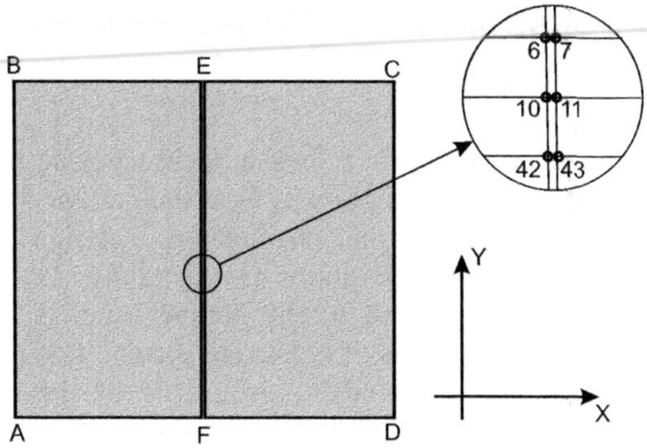

Fig. 2.17. A plate with a cylindrical hinge

Of course, this method of modeling can be used only in cases when the number of involved strip hinges is not too big or they are placed regularly. If the construction is such that the said hinges are arranged parallel to one another and closely spaced, as they may be, for example, in models of sheet pile walls, then it's reasonable to introduce another type of idealization: consider a plate with "smeared" hinges not resisting to bending in one direction (semi-moment-free). It can be easily implemented by using finite element of a non-isotropic material where the non-resistance to bending is specified as zero value of the respective elasticity modulus[4]. It is relevant to notice that the transition from a real compound system to a semi-moment-free one distorts little the distribution of stress and strain, while a continualization of a plate with closely spaced ribs and its representation as an orthotropic system introduces substantial disturbance to the stress field due to a loss of moment peaks in the vicinity of the ribs.

---

[4] One should not think it is an incontrovertible recommendation for all times. If one deals with a shell, one needs to be sure the CAD program used will react correctly to the artificially introduced anisotropy and won't extend it to the physical relationship between membrane (tangential) stress and deformation which will be a mistake. Also, one needs to make sure the specification of zero values for some components of the anisotropy tensor won't cause the problem to degenerate. For example, if a semi-moment-free plate on an elastic foundation (say, a simulation of a sheet pile enclosure) is under consideration, then the system of FEA governing equations will prove degenerate: the semi-moment-free plate won't resist to rotation of nodes around one of coordinate axes. The avoidance of degeneracy by a self-obvious technique of "penalty" (introduction of a nonzero but very small element of the anisotropy tensor) will result in a poorly conditioned system of equations: the effect of a nearly degenerate problem is possible. In order to pass safely between Scylla of a penalty-distorted problem formulation and Charybdis of degeneration of the numerical solution, one should rely upon experienced pilots of computational engineering.

2.5 Mono-Constraints and Poly-Constraints in Design Models

Now it is high time for us to take a look at techniques described in the previous section from a more general standpoint of the structural mechanics.

Among sets of various data that characterize a design model, a special and responsible role is played by descriptions of constraints.

Let vector **u** with components to the number of n is a desirable vector of unknown displacements of nodes of a discrete system. It is apparent not all of the nodal displacements are independent, as a part of them are pre-defined by boundary conditions or, say, conditions of symmetry. Other, more complicated, linear dependencies between components of the vector are also possible. At the mechanical level of consideration all those are constraints imposed upon the displacement of the system.

Mathematically, constraints are represented by linear homogeneous equations (perfectly rigid non-shifted constraints):

$$\sum_{j=1}^{n} c_{ij} u_j = 0 \quad (i = 1,\ldots, p)\,, \tag{2.3}$$

where p is the number of the constraints.

According to [19], let's introduce the following classification of constraints. We will say that a *mono-constraint* is imposed upon a system if the equation of this constraint contains only one nonzero coefficient; otherwise we will refer to this constraint as a *poly-constraint*.

Obviously, this classification reflects a formal mathematical definition of a constraint rather than its mechanical meaning. Moreover, this definition depends on the selection of generalized coordinates in the system (or generalized displacements) and is dictated only by convenience of constraint description and implementation of respective algorithms in structural analysis software.

The fact is that both the description and the implementation of algorithmic operations with mono-constraints are much easier than the same operations with poly-constraints. Therefore, if there is a convenient system of coordinates where a poly-constraint can be transformed to a mono-constraint, be sure to do it. This is one of reasons why the advanced structural analysis software offers their user an option to introduce a local coordinate system (x,y,z) referring to a particular node. Indeed, the restraint of displacements of a node in a certain direction will be a mono-constraint if the generalized coordinates consistent with degrees of freedom of this node are components of its displacements along axes one of which is the direction of the said restraint (Fig. 2.18).

Apparently, in input data the default local coordinate system (x,y,z) for any node of the system should be the same as the global system (X,Y,Z) if other is not stated explicitly. This limited capability is all the user can do about the transformation of poly-constraints to mono-constraints. The reader will undoubtedly notice that in common terms of structural mechanics mono-constraints are associated now with *external* rigid constraints imposed upon displacements of nodes of a system.

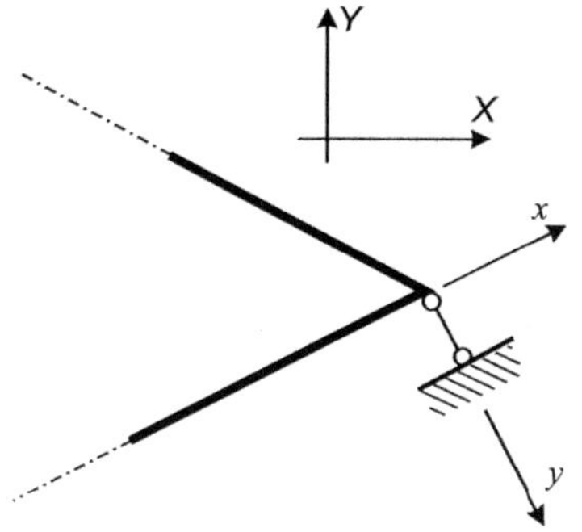

Fig. 2.18. Transforming a poly-constraint to a mono-constraint by introducing a local coordinate system in the node

To finish the discussion of mono-constraints, we will nibble at usable methods of their description in input data for the CAD software. The most popular choice is an indication of nodes attached to the "earth" (this one is used in the **SCAD** software, for example, and in other programs such as **ANSYS** and **COSMOS**). But there is an alternative though less popular approach where firstly a list of all nodes having at least one fixation (these ones become *support* nodes) is to be created, and secondly connections between these nodes and the earth are to be removed either partially or completely where needed. The representative of this approach is the **GTSTRUDL** software. Which alternative would be preferred by a user: "*all is allowed which is not forbidden*" or, vice versa, "*all is forbidden except for what is allowed*"? Though the authors believe the first approach is more convenient, it would be more flexible to allow the user choose this "legislation" principle.

Poly-constraints offer a great variety of mechanical representations, and their presence in a system is often the evidence of perfectly rigid elements introduced into the model, though this is not the only source of poly-constraints. Let's give some examples.

Fig. 2.19 shows a plane frame with an oblique inextensible crossbar.

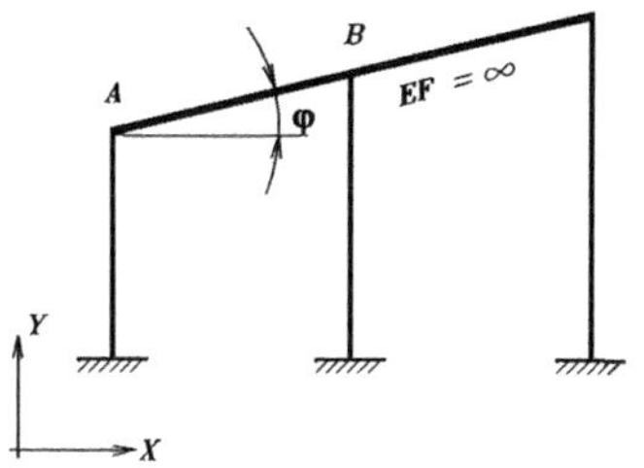

Fig. 2.19. A poly-constraint formed by an inextensible crossbar of the frame

The respective equations of poly-constraints can be written as follows:

$$u_A \cos\varphi + v_A \sin\varphi - u_B \cos\varphi - v_B \sin\varphi = 0,$$

$$u_B \cos\varphi + v_B \sin\varphi - u_C \cos\varphi - v_C \sin\varphi = 0, \tag{2.4}$$

where u and v are linear displacements of the appropriate nodes along axes X and Y respectively.

Fig. 2.20 shows two groups of bars with a rigid connection within each group, while the groups are attached to each other via a hinge.

Assuming that the model includes two separate nodes A and B coincident in their spatial location, we come to two poly-constraints with the following equations

$$u_A - u_B = 0\,, \qquad v_A - v_B = 0. \tag{2.5}$$

By the way, we already know that this and similar cases of poly-constraints are available in CAD software environments such as **SCAD** and **GTSTRUDL** where they can be defined as the merging of displacements (partial) of nodes A and B.

A more detailed discussion on advantages and drawbacks of quadrangular and triangular finite elements (say, in applications to the plane elasticity) will be given further below. Here we would like just to notice that an experienced user of CAD software would prefer a quadrangle to a triangle for the reason of computational accuracy, assuming all other conditions are equal. On the other hand, the triangles are good because they allow one to thicken a finite element grid in places where a big gradient of the structure's stress distribution is expected. Here, poly-constraints can be helpful, allowing one to thicken the grid and to avoid using triangles at the same time.

Fig. 2.21 shows an example of such seemingly unallowable thickened grid. Here we assume for definitiveness that only classical finite elements with a bilinear approximation of displacements are used.

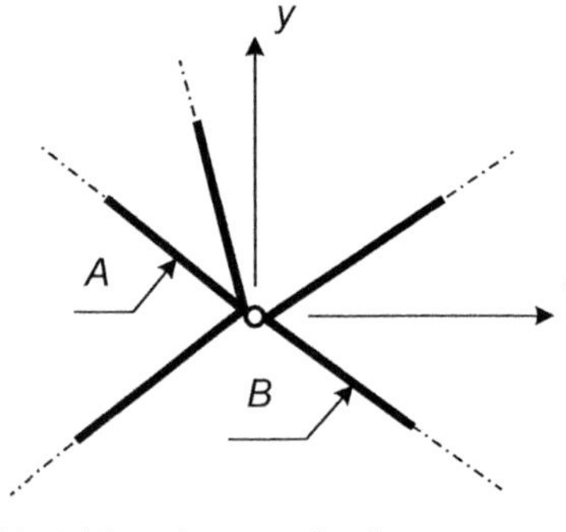

Fig. 2.20. A hinged connection between two nodes

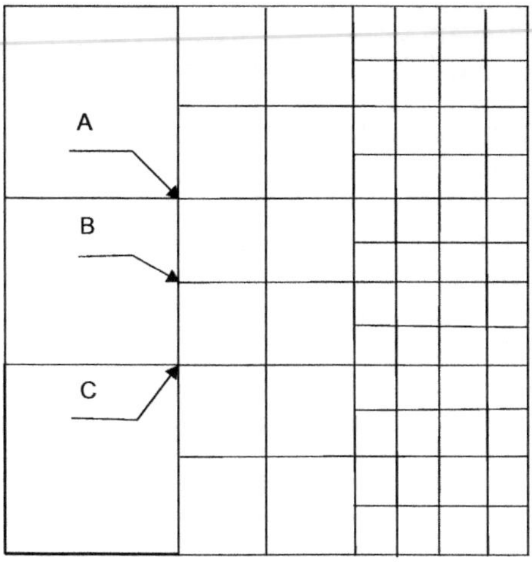

Fig. 2.21. The thickening of a finite element grid

Then, the displacements of node B located, say, in the middle of side AC, should satisfy the following poly-constraint conditions in order to keep the compatibility of displacements along the common boundaries of elements meeting in this node:

$$u_B - 0{,}5(u_A + u_C) = 0\,, \qquad v_B - 0{,}5(v_A + v_C) = 0 \tag{2.6}$$

The reader will easily generalize this case to other similar situations. This last example shows, in particular, that CAD developers may eventually make available the implementation of not only particular, though undoubtedly important, cases of poly-constraining such as displacement merging or perfectly rigid bodies (to be discussed later) but also the most general case of the formation and processing of poly-constraints.

Concerning the form of a general poly-constraint specification, it could be a declarative command with the following consecutive parameters:

- identifier of a node,
- name of a local degree of freedom,
- coefficient in the poly-constraint equation,
- ...
- a command terminating character such as «;»

Thus, for the first poly-constraint of (2.6), this command could look like

```
POLY      B U 1.0,  A U -0.5,  C U -0.5;
```

Certainly this is just one of possible options for a general poly-constraint definition, and we do not profess it to be the best one.

Up to now we have been discussing non-shifted poly-constraints meaning that the respective poly-constraint equations (2.3) are homogeneous. The case of inhomogeneous poly-constraint equations written in the form

$$\sum_{j=1}^{n} c_{ij} u_j = d_i \qquad (i = 1,\ldots, p), \tag{2.7}$$

correspond to dislocation displacements d_i pre-defined in a poly-constraint. It is natural to separate such quantities from poly-constraints in the input data of a problem and treat those as influences, the more so that the right parts of (2.7) can be different in different modes of loading. Technically the source of dislocations d_i could be inaccuracies of the fabrication or erection of the structure, or a self-stressed state of the system — a pre-stress — intentionally created by the designer with poly-constraints.

Design problems arising from the analysis of a structure's states of erection and operation, with allowance for pre-stress, are discussed in a separate section of this book.

2.6 Perfectly Rigid Bodies as Finite Element Types

As we have stated before, a perfectly rigid body is an important component tightly connected with poly-constraints in a design model of a structure. We believe the user should treat the perfectly rigid body from his standpoint as a specific type of finite element. Its relationship to finite elements can be noticed at least from the fact that the rigid body unites nodes attached to it. Roughly, a rigid body can be considered a limiting case of a finite element as the rigidity properties of this element approach the infinity.

Let us first establish the notation and some terms. Regarding rigid bodies, we assume here and further that the number of nodes belonging to one such body will be $r+1$. Also, let us enumerate the nodes within one body sequentially from 0 to r. Next, one of nodes of the rigid body (let it be a node with the local number 0) will be called a master node of the respective body and all the others will be referred to as slave nodes. So, r will be the number of slaves in a rigid body.

Let's introduce a local right-hand Cartesian coordinate system (ξ,η,ζ) associated with the rigid body so that the origin of this system is coincident with the master node and all other parameters of the system are arbitrary and to be set by the user. As usual, there must be a default choice according to which the local coordinate system for a three-dimensional rigid body should be formed from the global coordinate system by translating the latter's origin to the master node.

Components of the vector of linear displacements of the nodes with respect to the local coordinate system (ξ,η,ζ) will be denoted as u, v, w. Components of the vector of slopes of the same nodes (a rotation vector) with respect to the same system will be denoted as θ_ξ, θ_η, θ_ζ. Finally, letters a_i, b_i, c_i will denote coordinates of i-th slave node in the local coordinate of the rigid body.

Now let us try to classify perfectly rigid bodies by their *maximum* geometric dimensionality of the space of nodes belonging to the bodies.

2.6.1 One-Dimensional Perfectly Rigid Bodies

All one-dimensional perfectly rigid bodies shall unite nodes belonging to same straight line and shall merge one-dimensional displacement.

For a one-dimensional rigid body, only one ξ axis of the local coordinate system shall be defined (Fig. 2.22), beginning at the master node 0 (the master's work is to rule) and crossing all slave nodes.

We will distinguish three subtypes of one-dimensional rigid bodies. For the first subtype, the conditions of poly-constraints imposed upon displacements of its nodes will have the form

$$u_i = u_0 \quad (i = 1,\ldots, r)\,, \tag{2.8}$$

and these poly-constraint equations will be formed for all slave nodes, that is, exactly r poly-constraints will appear. Mechanically, a rigid body of this subtype merges the linear displacements of all its nodes along the axis of the body. A rigid body of this type can be imagined as a limiting case of truss bars arrayed in a chain along the ξ axis with their rigidity approaching the infinity.

Representatives of one-dimensional perfectly rigid bodies of the second subtype merge slopes of all their nodes with respect to the ξ axis. Thus, for the second rigid body subtype, the poly-constraint equation shall be:

$$\theta_{\xi i} = \theta_{\xi 0}\ (i = 1,\ldots, r)\,. \tag{2.9}$$

Apparently, this subtype simulates a limiting case of a chain of coaxial shafts of an infinite torsional rigidity.

Finally, the third one-dimensional rigid body subtype is a combination of the first two, forming poly-constraints (2.8) and (2.9) of the total amount $2r$.

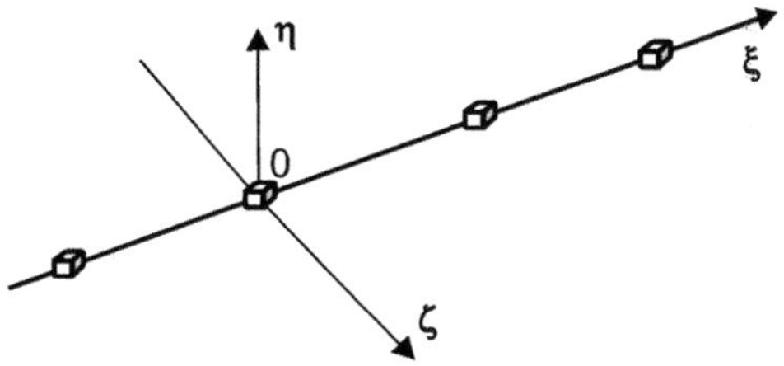

Fig. 2.22. A one-dimensional perfectly rigid body

2.6.2 Two-Dimensional Perfectly Rigid Bodies

All nodes of a two-dimensional rigid body shall be spatially located in the same plane, and the axes (ξ,η) of the local coordinate system shall belong to the same plane with their origin in the master node of the body (Fig. 2.23).

We will distinguish between two subtypes of two-dimensional rigid bodies. The first subtype will bind displacements of nodes of the rigid body in its own plane, that is, in the plane (ξ,η) by the following three equations of poly-constraints for each slave node (slaves shall follow their master according to strict rules just as they must):

$$u_i = u_0 - b_i\theta_{\zeta 0}\,,\quad v_i = v_0 + a_i\theta_{\zeta 0}\,,\quad \theta_{\zeta i} = \theta_{\zeta 0}\qquad (i = 1,\ldots, r)\,. \tag{2.10}$$

A rigid body of this subtype can be interpreted mechanically as a plane frame rigidly bound to all its nodes, in the limiting case of infinitely rigid bars of the frame[5].

The second two-dimensional rigid body subtype will restrain displacements of its nodes as they try to depart from the plane of the body. The poly-constraint equations for this subtype follow:

$$w_i = w_0 + b_i\theta_{\xi 0} - a_i\theta_{\eta 0}\,,\quad \theta_{\xi i} = \theta_{\xi 0}\,,\quad \theta_{\eta i} = \theta_{\eta 0}\quad (i = 1,\ldots, r) \tag{2.11}$$

For convenience, one could imagine a grillage allowed to bend athwart its plane and topologically equivalent to the first subtype plane frame provided with perfectly rigid bars.

These two subtypes won't be merged into a third subtype as it makes no sense: a resulting rigid body would behave exactly in the same way as a three-dimensional body described below.

So, a two-dimensional perfectly rigid body will satisfy $3r$ poly-constraint equations.

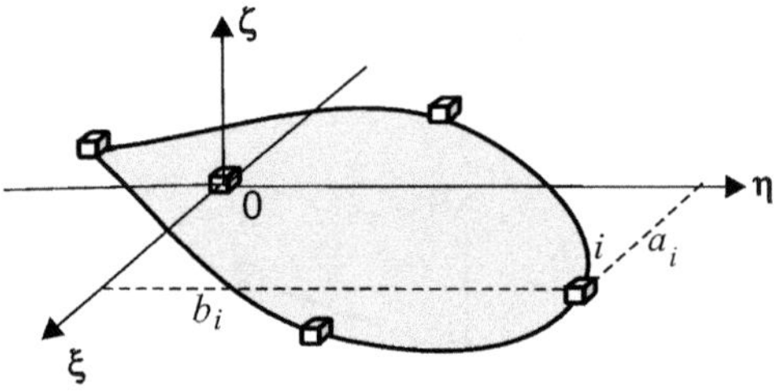

Fig. 2.23. A two-dimensional perfectly rigid body

2.6.3 Three-Dimensional Perfectly Rigid Bodies

A three-dimensional perfectly rigid body (Fig. 2.24) will restrain the slaves of its master to the greatest extent, depriving them of all degrees of freedom by the following set of poly-constraints:

[5] Attention! For the limit to be correct, this frame must be internally statically determinate. The explanation for this requirement will follow later. One may imagine such frame as a "spider" with its body in the master node and r legs going from its body to each slave.

$$
\begin{aligned}
&u_i = u_0 + c_i\theta_{\eta 0} - b_i\theta_{\zeta 0}, &&\theta_{\xi i} = \theta_{\xi 0},\\
&v_i = v_0 - c_i\theta_{\xi 0} + a_i\theta_{\zeta 0}, &&\theta_{\eta i} = \theta_{\eta 0},\\
&w_i = w_0 + b_i\theta_{\xi 0} - a_i\theta_{\eta 0}, &&\theta_{\zeta i} = \theta_{\zeta 0} \qquad (i = 1,\ldots, r).
\end{aligned} \tag{2.12}
$$

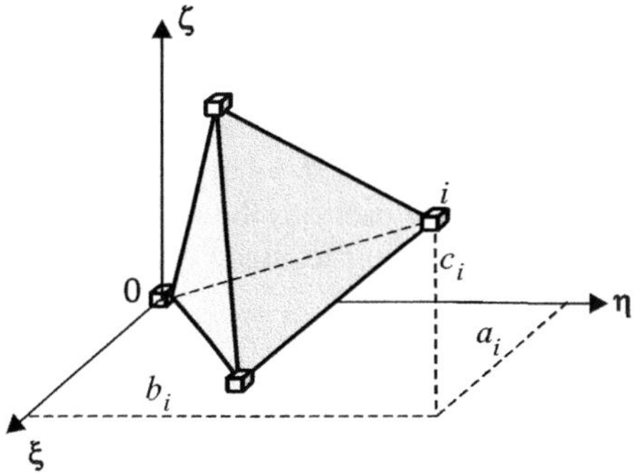

Fig. 2.24. A three-dimensional perfectly rigid body

Formulae (2.10), (2.11) and (2.12) use a convention of signs for rotations according to which positive rotation vectors θ_ξ, θ_η, и θ_ζ are to be formed by the rule of right-hand screw for the respective axes.

Thus, totally 6 types of perfectly rigid bodies will be under consideration, of which 3 are one-dimensional, 2 are two-dimensional and one is three-dimensional.

Let us state a few general provisions of "legislation" regarding relationships between masters and slaves united by rigid bodies:

1. The least number of nodes in any perfectly rigid body shall be two: a master must have at least one slave.
2. Some (or even all) slaves may be given a partial freedom by their master. This (by all means humane) action shall be reflected in the input data for the model by releasing the respective constraints in places where the slave nodes are attached. For example, if a two-dimensional rigid body of first subtype releases constraints that restrain independent rotations of its slaves about the ζ axis, then for each slave node only first two equations of the three poly-constraints (2.10) will hold, that is

$$u_i = u_0 - b_i\theta_{\zeta 0}, \quad v_i = v_0 + a_i\theta_{\zeta 0} \qquad (i = 1,\ldots, r). \tag{2.13}$$

   Obviously, this set of poly-constraints describes a hinged attachment of a flat rigid body to all its slave nodes. It can be used, for example, to simulate a non-compliant floor of a multistory building discussed in Section 2.4.
3. When releasing constraints, keep in mind that a master node may not be detached from the rigid body even partially. In a sense, the master's burden is heavier than that of a slave.

4. The legislation does not forbid a master to have two or more estates with different sets of slaves in each estate. In other words, the same master node can participate in establishing several rigid bodies.
5. A slave shall serve to one master only and in one of the estates solely! (Carlo Goldoni tried to see what would have happened if a servant had tried to serve two masters — and a funny comedy came out).
6. Mono-constraints shall not be imposed on slave nodes: slaves must be fully obedient to their master, and a stubborn slave is no good.
7. We do not know what ancient Romans did about it, but in Russia in times of serfdom there used to be a famous St. George's day when serfs could be given their freedom. In our case, the user has the right to deactivate a perfectly rigid body in desired modes of loading to make every slave just a usual node free from poly-constraints.

This list could be longer, but we will confine ourselves to the given notes. Particular algorithmic implementations could require different limitations for the use of perfectly rigid bodies: just as laws are usually different in different countries. For example, if one legislation allows vassals to have their vassals, then another one may forbid it. Also, it must be apparent that the above stated legislation has been developed by the authors not in order to eternize themselves by making laws for the foretime slave owning. We have tried to formulate a clear set of limitations for the use of perfectly rigid bodies in design models, the violation of which will (or could) cause a CAD software to blow up. Though, we emphasize once more that a lot of things depend on a particular development: the less limitations in the master-and-slave legislation, the more attractive is the program for users.

We have not yet establish any conventions regarding the choice of a master node in a rigid body. Now it's high time to do it. There are two principal opportunities for that:

- a program can select a candidate for the master's position from the list of nodes of a rigid body itself, and then declare all other nodes of the body to be slaves according to the established rules;
- the master can be assigned by the user.

Obviously, the first opportunity must be the default one.

Further, the master node of a rigid body may be either a node of the structure or a separate node which would be isolated but for the rigid body. The design with a master separated from the structure may be needed when the rigid body is attached to partially released nodes of the structure. In this case, a non-isolated master would violate item 3 of the above stated code.

We should note that, on one hand, rigid bodies (or poly-constraints generally) are an extremely powerful tool for building design models that enables a user to be creative and resourceful in most nonstandard situations. On the other hand, this tool is not so safe as it may seem: even experienced analysts may find themselves in a masked and hard to detect trap. Therefore the utilization of complex poly-constrained designs must be the exclusive field of experts (see Table 2.1).

This is one of the mentioned traps: all equations of poly-constraints in a system must be linearly independent. Mechanically, it means that the model must not contain a statically indeterminate subsystem wholly composed of rigid bodies. Otherwise the distribution of stresses in the system would be ambiguous. Mathematically this condition can be formulated via rank of matrix **C** composed of coefficients c_{ij} of the system of equations (2.3) that defines all constraints imposed on the system's nodes:

$$\mathbf{C} = \|[c_{ij}]\| \ (i = 1,..., p; j = 1,..., n), \quad \text{rank}\, \mathbf{C} = p. \tag{2.14}$$

If condition (2.14) does not hold and rank $\mathbf{C} < p$, it means that the constraints are imposed incorrectly and the solution will be problematic. A well-organized software must track this situation and provide the user with warning messages, desirably informing him about locations of possible errors in the input data.

Finally, all that has been said here about poly-constraints is an opinion of the authors of what they would like to see among capabilities of the hypothetic software which the authors would gladly have at their disposal in order to handle poly-constraints correctly. The best of the available programs implement these suggestions only partially.

2.7 On a Nonlinear Analysis

Problems of the analysis of load-carrying structures oriented at an accurate prediction of the peculiar behavior of a system in all phases of its operation including those immediately preceding its failure, are, most often, not solvable by methods of the linear structural theory. The usual "set of nonlinearities" appealed to by most researchers [9, 15] and textbooks [23] includes a non-observance of the Hooke's law (a physical nonlinearity), consideration of the equilibrium in the geometry of a deformed state (a geometrical nonlinearity), an allowance for possible modifications of the design model during its deformation (a structural nonlinearity).

This list is far from being full. It does not include effects of rheological processes in a material (such as creeping) and nonlinear effects of resistance such as a dry friction or others. Also, it does not take into consideration nonlinearities due to an accumulation of stresses and strains in the course of the creation of a structure as the latter is being changed (a *genetic nonlinearity*). The latter nonlinearity type can be treated somehow as a variation of the structural nonlinearity[6], because here systems of a variable build are under consideration. Though, in this case the changes take place intentionally according to the design of a developer rather than spontaneously after the application of a load, so we have a reason to treat and name this case separately.

The authors understand it perfectly well that even a primary discussion of all these problems would require a new and big book to be written (maybe it will

[6] This point of view is expressed by the authors of the study guide [23].

happen once). So, issues of nonlinear analysis discussed below are nothing but fragments.

Without diving too deeply into details, we would like to formulate some general admonitions regarding the nonlinear analysis problems. In particular, it is useful to list questions from [2] which must be asked by a CAD software user to himself before trying to solve a nonlinear problem:

- How accurately are the properties of a material known?
- Is the solution sensitive to the accuracy of the material data specification?
- If so, do we need to perform experimental tests in order to get reliable data of the material properties?
- Does an initial stress and strain distribution, say, a residual stress in the system, affect the final result?
- If so, how can its effect be determined?
- Is there a possibility that an ambiguous solution may exist, for example, different modes in a buckling problem or more than one post-critical state?
- If so, may we need to create any artificial initial imperfections to be able to achieve a single unambiguous solution of interest?
- Does the history of loading affect the behavior of the structure?
- Will the behavior of the material be the same during loading and unloading?
- Assuming big displacements or slopes, will the direction of the load action stay the same or will it follow the distortion of the shape?
- Do assumptions of the nonlinear model correspond to the expected accuracy of the solution?
- Should a greater safety coefficient be used to compensate for a possible inaccuracy or uncertainty about the adequacy of the solution?
- How shall the solution be validated?

These questions arise due to a variety of peculiarities of nonlinear problems, many of which are unusual for specialists grown and bred to linear solutions. Some (by no means all) of the mentioned peculiarities are shown in Table 2.4.

In particular, we should warn a user of CAD systems against the "physical nonlinearity" capabilities of such software. The matter is that many of those deal with problems different from linear ones only in that the functional linear dependence between stress and strain is replaced there by some nonlinear function. The said function is treated as one-valued and is the same for both loading and unloading. But such nonlinearly elastic materials are really rare in the nature, and the non-observance of the Hooke's law by most constructional materials relates to plasticity effects. But is there a plastic material that does not show residual deformations after being unloaded?

The aforesaid note is often parried by the argument that only histories of the structural behavior with monotone increasing loads will be under consideration. But the fact is the monotone increase of a load does not warrant the monotone increase of stresses in all points of a structure, therefore even this self-limitation may not protect one against a mistake.

Table 2.4.

Feature	Linear problems	Nonlinear problems
Displacement vs. load dependence	Displacements depend on the applied load linearly	The dependence between the displacement and the load is nonlinear
Stress vs. strain relationship	A linear relationship between stress and strain is assumed	In problems where physical nonlinearity is essential, the dependence "stress vs. strain" can be a nonlinear function of stress, strain and/or time
Displacement smallness	A change in geometry due to displacements is assumed small and is ignored during validation of the equilibrium	Displacements can be big, a deformed state should be used to validate the equilibrium
Material properties	Linear elastic properties of a material are usually simple to determine	Nonlinear properties of material could be unknown that would require an additional experimental investigation
Reversibility	All deformations are fully reversible and disappear as soon as the system is unloaded	After unloading, the state of the system may differ from the initial one
Boundary conditions	Boundary conditions are not changing during the computation	Boundary conditions may be variable, for example, an area of contact
Sequence of load application	The sequence of the load application does not matter; the final state does not depend on it	The state of the structure may depend on the order in which the loads are applied
Computation time	The computation takes far less time than the nonlinear analysis does	Increments of loads and iterations take much time, especially in cases of a high accuracy required
Stability of computation	Most often, the solution can be obtained without human intervention	In difficult nonlinear problems the solution may converge insufficiently fast, that could require some user's intervention
Handling of results	Results of analysis under different loads allow the summation or multiplication by factors in order to combine modeled states	It is not possible to decompose the problem into components and then combine the results
Initial stress and strain distribution	The initial stress and strain distribution is inessential	The initial stress and strain distribution needs to be specified, especially for a nonlinear material behavior

2.8 Using Several Models at the same Time

Nearly any object of engineering, except for simplest ones, is considered from several standpoints in the course of analysis. So, it would be quite natural to associate the object with multiple, rather than single, design models [4, 6]. Each model would bring its contribution to the understanding of the structure's behavior, and only the complete set of them would give an adequate idea of the actual load-carrying ability of the structure. Also, a coincidence between results obtained from different (independent) models would improve the credibility of the result very much.

Note that the possibility of wielding a variety of design models is a powerful weapon against an excessive complication of the analysis. An over-detailing of a system is a typical response of an *inexperienced* analyst to his need for accurate data regarding an ultimate factor in the case when there is no preliminary information where this ultimate factor may arise. It is exactly when the analyst uses an over-detailed model just in case, with the hope that it's not going to miss the needed result. But the result still can be missed because of a difficulty to understand and analyze a drastically bigger amount of information generated by a too much complicated design model. The fact is that at a certain level of a system's complexity, the ability of a human to formulate accurate and clear statements regarding the system's behavior would fall dramatically. This is a far analogy of the Heisenberg's uncertainty principle concerning the informative content and the level of detailing of an analysis which are complementary. Hence the desire of an *experienced* analyst to use a simple model, but, as the simple model cannot comprise all aspects of a phenomenon, one needs to create and use several design models for one's purpose.

Until recently, even experienced analysts were not at pains to use techniques of working with sets of design models. It was due to a big effort needed for preparation and checkup of a design model for which one would have to spend several man-days and a few weeks of business time. At that time people used to prefer to make a universal detailed model to be validated only once. Now that this work is facilitated very much by advanced diagnosing tools, and interactive design sessions take hours at the most, the said consideration is essentially needless.

One should keep in mind that a series of simple design models can give a more accurate knowledge than one complex (over-complicated) model which yields results hard for a man to comprise. This issue was addressed to by Academician V.V.Novozhilov. He said that the analysis of computational results obtained directly from a complex model resembled a generalization and realization of a large array of experimental data, while an estimation of a phenomenon as a whole (using simple analytic models) would enable one to better understand the process of construction and analysis of a detailed model [16].

Unfortunately, most modern software systems offer very little for this mode of operation. They have neither common strategy nor technical tools for data exchange between several parallel design models of the same object; neither they can exchange computational results between the models. The appropriate tools are yet to be created.

A good CAD software must offer an opportunity to compare results obtained from the analysis of different models. Estimations of this kind must be performed with different types of resulting data both locally (a search for maximum differences achieved in specified elements or nodes of the system) and "in average in the domain" (in the whole structure, in a fragment or a substructure). It seems feasible to construct an averaged estimate of the difference between compared design diagrams by comparing their elastic potentials both in the whole systems and their parts. An exhaustive search through all parts of a design model may prove too complicated (or utterly impossible for a continuous system), so one can involve only a sampled set of such sub-models.

Apparently, elements of the sample set can be defined as finite elements wholly contained in a randomly selected continuous area of space occupied by the structure. This area will contain different parts of the structure in different models, but if the number of attempts is big enough, the "sub-model discretization error" will be small enough, as it seems.

The measure of non-resemblance dK_{rs} of compared sub-models r and s will be the value of a relative difference between their elastic potentials E_r and E_s

$$dK_{rs} = 2(\mathsf{E}_r - \mathsf{E}_s)/(\mathsf{E}_r + \mathsf{E}_s). \tag{2.15}$$

As the sub-models are selected at random, the result of comparison dK_{rs} will be a stochastic quantity. Therefore, the criterion for comparison of the whole models can be a weighted average dK_{rs} taken by all sample sets. The weight factor ω for an arbitrary set will be the relative volume of the respective random area. We assume that after a sufficiently long series of the comparisons the average value will gradually stabilize and from a certain point no substantial change in the result will be expected as the sample set size will continue to grow.

Note that the stochastic sampling does not exclude the likelihood that a very small area of space will be missed. This area may contain a small (maybe zero) part of the design model, and the accumulated characteristics of the comparison will change slightly. Therefore one cannot rely upon the estimate of the current average value increment to establish the stop rule; one should use rules similar to those used in random search algorithms.

If this hypothesis is true, then the algorithm of the comparison between two models r and s may look as follows:

***Phase 0 (preparatory)*:**

$j = 1$;

Calculate elastic potential Eall,r of the whole design model r and elastic potential Eall,s of the whole design model s;

Find initial values of average and standard mean-square deviation:

Krs = 2|(Eall,r – Eall,s)/(Eall,r + Eall,s)|;

Find diameter of hyper-sphere D which the structure belongs to.

/* Loop by samples, j is No of sample */

M: j = j + 1;

Phase 1:

Choose a sub-domain with a random center and random diameter d_j (the location of the center is uniformly distributed within the hyper-sphere circumscribed around the structure, with its diameter evenly distributed at the interval from zero to D/2);

Determine the weight factor $\omega_j = d_j/D$.

Phase 2:

Determine parts of the models contained in the sub-domain.

***Phase 3*:**

For each couple of models being compared, calculate elastic potential of sub-domain (Er(*j*) and Es(*j*));

***Phase 4*:**

Calculate the next term in the series of random comparisons:

dKrs(*j*) = 2ω_j(Er(*j*) – Es(*j*))/(Er(*j*) + Es(*j*));

***Phase 5*:**

Refine values of average and mean-square deviation:

Krs = [(*j* – 1)*Ers(*j*-1) + dKrs(*j*)]|/*j*;

***Phase 6*:**

Check if the loop is over:

If the ending conditions are not satisfied, go to label *M*;

Let us illustrate the described technique by a pretty simple example. Consider two design models shown in Fig. 2.25, *a* and 2.25, *b*.

The first design model is a truss structure for which the elastic potential E_a of any sub-structure at the interval $[X_b, X_e]$ with random ends is equal to the sum of components $N_i^2L_i/EA_i$, where N_i is the internal force, L_i is the length, EA_i is the respective tension/compression rigidity, and the summation is performed over all elements of the sub-structure (Fig. 2.25, *c*). Introducing indicator variables z_i equal to one if the respective bar belongs to the interval $[X_b, X_e]$ fully and zero otherwise, we can write the following expression:

$$E_a = \sum_{i=1}^{n} z_i N_i^2 L_i / EA_i \tag{2.16}$$

where $n = 24$ is the total number of bars.

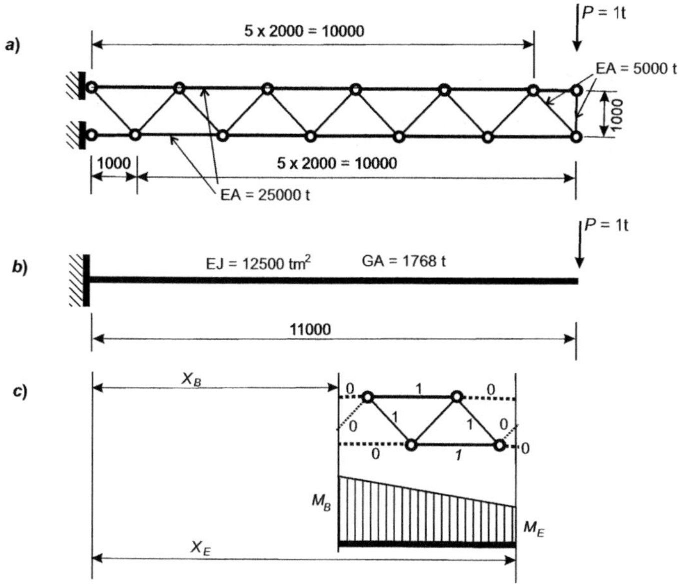

Fig. 2.25. An example of comparison between two design models

For the second design model, shown as a Timoshenko bar with its flexural and shear rigidities corresponding to the rigidity of the truss chords [10], the elastic potential is

$$\mathsf{E} = \int_L [M^2(x)/EI + Q^2(x)/GA]dx\,, \tag{2.17}$$

were integration is performed along the whole length of the bar L.

All calculations are given in Table 2.5 for the whole structure (sample 1) and next four consecutively selected substructures with their boundaries assigned by a random number generator.

Fig. 2.26 shows a diagram of variation of the average K_{rs} estimate for first five hundred samples. The diagram indicates that the estimate stabilizes in the vicinity of $K_{rs} = -0.185$ which confirms the hypothesis stated above (at least for this particular example). The measure of non-resemblance K_{rs} of design models says nothing about which design model is better and which one is worse for the analysis. The measure is not intended for this purpose. This measure can be used only to select more or less close possible models. Though, in order to judge correctly about the relationship between the numerical value and the proximity of design diagrams, one should gain a certain experience. In particular, the value $K_{rs} = -0.185$ found in the above example seems to belong to a range of values that describes a pretty close relation between design models. It follows from the fact that practical engineers have used the transition from models in Fig. 2.25, *a* to those in Fig. 2.25, *b* extensively, and, as far as we know, no trouble has come out of it.

Table 2.5.

	Longitudinal force, [t]	**Elastic potential, [t×m]**	**Indicator variables z_i for the sample:**				
			1	**2**	**3**	**4**	**5**
Upper chord:							
1	10.000	0.00800	0	0	0	0	0
2	8.000	0.00512	0	1	0	0	0
3	6.000	0.00288	0	1	0	1	0
4	4.000	0.00128	0	1	0	1	0
5	2.000	0.00032	0	1	0	0	0
6	0.000	0.00000	0	0	0	0	0
Lower chord:							
1	−11.000	0.00484	0	0	0	0	0
2	−9.000	0.00648	0	0	1	0	0
3	−7.000	0.00392	0	1	0	0	0
4	−5.000	0.00200	0	1	0	1	0
5	−3.000	0.00072	0	1	0	1	0
6	−1.000	0.00008	0	0	0	0	0
Struts:							
1	1.414	0.00057	0	0	0	0	0
2	1.414	0.00057	0	0	1	0	0
3	1.414	0.00057	0	1	1	0	0
4	1.414	0.00057	0	1	0	0	0
5	1.414	0.00057	1	1	0	1	0
6	1.414	0.00057	0	1	0	1	0
7	1.414	0.00057	0	1	0	1	1
8	1.414	0.00057	0	1	0	1	0
9	1.414	0.00057	0	1	0	1	0
10	1.414	0.00057	0	1	0	0	0
11	1.414	0.00057	0	0	0	0	0
Post:	- 1.000	0.00020	0	0	0	0	0
Domain beginning, m		0.00	3.31	1.56	0.26	3.72	5.14
Domain end, m		11.00	5.45	10.57	3.92	9.29	7.74
$E_a \times 100$		4.211	0.057	2.08	0.762	0.973	0.057
$E_b \times 100$		4.171	0.877	2.747	2.562	1.328	0.590
$dK_{rs} \times 100$		0.010	−0.342	−0.227	−0.360	−0.156	−0.389
Accumulated K_{rs}		0.010	−0.166	−0.186	−0.230	−0.215	−0.244

Another measurer of the proximity of design models could be the correlation coefficient of two sequences of elastic potential values $E_{r(j)}$ and $E_{s(j)}$ calculated at randomly generated sub-models. In the example above, this coefficient is 0.928 which evidences sequences highly related in the stochastic sense.

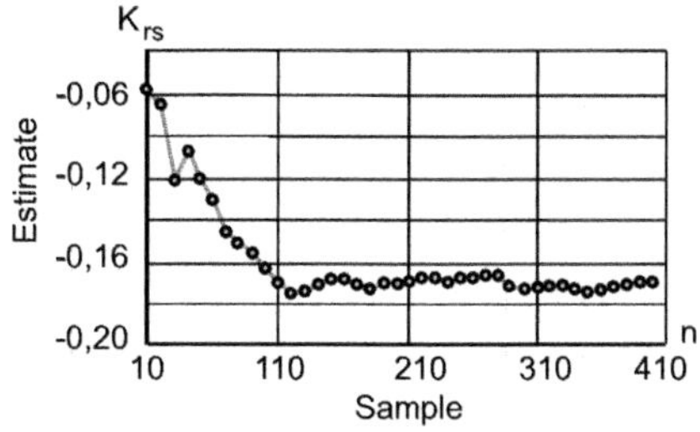

Fig. 2.26. Variation of comparison results

Note that all comparisons are performed in the same particular mode of loading upon the system. The proximity of design models confirmed in one loaded mode may prove incorrect under other loads.

2.9 Comparison between Calculated and Experimental Data

The necessity of an experimental validation of design models has been stated in publications for a great deal of times (see [5, 11, 21] for examples). Nevertheless, works of this type are performed not very often and their methodology leaves much to be desired. An interpretation of experimental results would, most often, include a fragmentary comparison of some measurements with calculated data and a few deductions like "*...a good coincidence can be observed*", "*...data are qualitatively equivalent*" or, at the best, "*...the discrepancy does not exceed so-and-so many percents*".

There are numerous suggestions on the use of the least squares method. This ideology has become dominant and created a tradition to utilize the sum of squares of differences between the experimental data vector components y_i^e and the respective calculated values y_i as the measure of proximity:

$$S_0 = \sum_{k=1}^{n} (y_k^e - y_k)^2 . \tag{2.18}$$

If the experiment measures heterogeneous quantities (such as displacements, rotation angles, relative deformations etc.), then weight factors ω_k should be introduced in the functional (2.18) to determine the measurement credibility level (which is a common reason for the use of weighted sums) and to normalize them to a homogeneous array of data.

On the other hand, absolute differences like $(y_k^e - y_k)$ are seldom used by practical engineers to estimate the accuracy of a measurement. More often they use a relative error

$$\delta_k = (y_k^e - y_k) \ / \ y_k^e, \tag{2.19}$$

which is also called a "constructive correction" [11][7]. It helps build a functional

$$S_\delta = \sum_{k=1}^{n} (y_k^e - y_k)^2 / \ (y_k^e)^2, \tag{2.20}$$

which uses values $1/(y_k^e)^2$ as weight factors and is homogeneous because values δ_k are dimensionless. The difference between experimental and calculated values is showed off even better by maximum relative deviations

$$\Delta_k = \frac{y_k^e - y_k}{\min(|y_k^e|, |y_k|)}. \tag{2.21}$$

If d measurements of each value of interest have been performed, and experimental averages have been found:

$$\bar{y}_k^e = \frac{1}{d} \sum_{j=1}^{d} y_{kj}^e \tag{2.22}$$

then the difference of discrepancy between these averages and the calculated data will be the value

$$S_{ad}^2 = \frac{d}{n-1} \sum_{k=1}^{n} \left[\frac{\bar{y}_k^e - y_k}{\min(|\ \bar{y}_k^e\ |, |\ y_k\ |)} \right]^2, \tag{2.23}$$

called an adequacy dispersion.

To elaborate a well-founded judgment of the adequacy dispersion (2.23) value, one needs to have a comparative scale characteristic for one's structure. This scale can be the measure of dispersion of results of repeated measurement. If this dispersion is big, then the physical model itself is pretty unstable with respect to those random conditions that arise during each repetition of the experiment. In this case the requirement of a good correlation between calculated and experimental data is no longer valid. In other words, the more stochastically the structure behaves, the harder it is to predict its behavior.

A measure for the dispersion of experimental data is a so-called repeatability dispersion

---

[7] Experimentalists believe usually that the difference between one and the average value of constructive corrections is a systematic error while the dispersion of the values is a random error. But this is true only for "well-designed structures" and is not always confirmed by the practice, especially when the validation is performed for two dramatically different modes of loading.

$$S_{cp}^2 = \frac{1}{n(d-1)} \sum_{k=1}^{n} \sum_{j=1}^{d} \left[\frac{y_{kj}^e - \bar{y}_k^e}{\bar{y}_k^e} \right]^2 . \tag{2.24}$$

Under usual assumptions of the error distribution, the adequacy can be validated according to the criterion

$$F_e = \frac{S_{ad}^2}{S_{cp}^2} \le F(\nu_1, \nu_2, \alpha), \tag{2.25}$$

where $F(\nu_1,\nu_2,\alpha)$ is a table value of the Fisher criterion that depends on the number of degrees of freedom of the numerator $\nu_1 = n-1$ and that of the denominator $\nu_2 = n(d-1)$ at the significance level of α.

Note that the use of relative errors δ_k, for example, to refine parameters of the model by the least squares method, will be apparently more effective in cases when components y_k include several values with big absolute deviations $\left(y_k^e - y_k\right)$ that outstand against the average level.

As these quantities contribute most to the functional (2.18), they will be approximated best while the bulk of the experimental data will be approximated worse. The deterioration of the accuracy in the field of small absolute error may even alter the sign of a calculated value in a particular model and thus make it physically meaningless. This drawback is not present in the relative error approach. As for big absolute deviations the approximation of which deteriorates, their presence in a model often indicates the necessity to change the very build of the model rather than its numerical parameters.

For an example let us take results of tests of a framed structure (Fig. 2.27) given in Table 2.6. The structure was made of extended I-beams and tested in Kiev National University of Civil Engineering and Architecture.

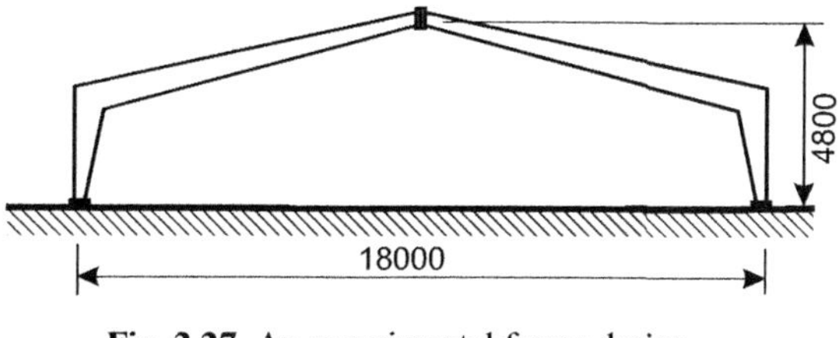

Fig. 2.27. An experimental frame design

Calculations by formulae (2.23) and (2.24) give values of the adequacy dispersion $S_{ad}^2 = 0.60667$ and repeatability dispersion $S_{cp}^2 = 0.48021$. The ratio $F_e = 1.253$ while the Fisher criterion for $\nu_1 = 20$ and $\nu_2 = 21\times 3 = 63$ at the significance level $\alpha = 0.95$ is equal to 1.65.

As $F_e < 1.65$, the correlation between the calculation and the experiment can be considered satisfactory.

Table 2.6.

Values of stress in points of the system, $[kg/cm^2]$				
Calculated	Experimental			
	Exp. 1	Exp. 2	Exp. 3	Exp. 4
–170	–241	–273	–315	–389
–170	–147	–241	–115	–186
–170	–357	–144	42	–170
–170	–116	–205	–220	–99
–822	–777	–756	–633	–435
379	405	158	505	321
–469	–807	–722	–920	–247
2208	2121	2247	1459	1365
–166	168	189	–431	–483
–166	–142	–142	–326	–410
–157	–462	–283	–168	–142
–157	–120	–241	–126	–168
1474	1045	1303	1292	1382
–871	–344	–323	–071	–040
1477	1580	920	1208	929
–871	–323	–981	–428	–386
1477	1525	1378	1219	1819
–939	–984	–784	–462	–941
–833	–462	–662	–568	–357
–155	–225	–101	–251	–272
–155	–168	–126	–378	–347

References

1. Ashby WR (1964) Some notes. In: Mesarović M (ed) Views on general systems theoru. John Wiley and sons, New York London Sydney, pp. 28-34
2. Becker AA (2001) Understanding Non-linear Finite Element Analysis Through Illustrative Benchmarks.— Glasgow: NAFEMS
3. Blekhman II, Myshkis AD, Panovko JaG (1983) Mechanics and applied mathematics: Logic and peculiarities of mathematical applications (in Russian). Nauka Publishing House, Moskow
4. Feodosi'ev VI (1975) Ten lectures-conversations on strength of materials (in Russian). Nauka Publishing House, Moskow
5. Goldenblat II, Bagenov VL (1970) Physical and analytical models (in Russian). Structural mechanics and analysis of structures, 2:23–27
6. Gorodetsky AS, Perelmuter AV, Slivker VI (1989) Intellectual engineering software – prognosis of development (in Russian). In: CAD systems, vol. 6. "Bidivelnyk", Kiev, pp. 43–56

7. Kosmodemjansky AA (1969) Theoretical mechanics and modern technics (in Russian). "Prosvjashtchtnie", Moskow
8. Lind N (1979) Optimization, cost benefit analysis, specification. In: Proc. 3rd Int. Conf. on Applications of Statistics in Soil and Structural Engineering (ICASP 3), Vol.3, Sydney, pp.373 - 384.
9. Lukash PA (1978) Fundamentals of a nonlinear structural mechanics (in Russian). Stroyzdat, Moskow
10. MacLeod IA (1995) A strategy for the use of computers in the structural engineering. The Structural Engineering, Vol.73.— № 21.
11. Mastachenko VN (1971) An adequacy of real and analytical models (in Russian). Structural mechanics and analysis of structures, 4:3–7.
12. Minkovich EI, Slivker VI (1990) One way of geometrical description of 3D-bar (in Russian). In: Rozin LA (ed) Finite elements method and analysis of structures: Proceedings of Leningrad polytehnical institute, vol. 434 (in Russian). Leningrad, pp. 84–94
13. Morris U (1971) Management science (in Russian). Mir, Moskow
14. Nikolaenko NA, Nazarov JuP (1988) Dynamics and seismic stability of constructions (in Russian). Stroyzdat, Moskow
15. Novozhilov VV (1953) Foundations of nonlinear theory of elasticity. Graylock Press, Rochester, N.Y.
16. Novozhilov VV (1979) Mathematical models and accuracy of engineering calculations (in Russian). Shipbuilding, 7:5-12
17. Panovko YaG, Gubanova. II (1965) Stability and oscillations of elastic systems; paradoxes, fallacies, and new concepts. Consultants Bureau, New York
18. Parikov VI, Slivker VI (1983) Stiffness matrix of a finite element non-rigidly connected to its nodes (in Russian). In: Bulletin of B.E. Vedeneyev *VNIIG* , collected scientific papers, vol. 164. Energoatomizdat, Leningrad, pp. 20–28
19. Perelmuter AV, Slivker VI (1976) Peculiarities of algoritmization of the slope deflection method with additional constraints taken into account (in Russian) In: Rozin LA (ed) Finite elements method and structural mechanics: Proceedings of Leningrad Polytehnical Iinstitute, No 349 (in Russian). Leningrad, pp. 28–36
20. Perelmuter AV, Slivker VI (1979) On implementation of complex kinematical conditions in the slope deflection analysis of discrete systems (in Russian). In: Rozin LA (ed) Finite elements method and analysis of structures: Proceedings of Leningrad Polytehnical Iinstitute, No 363 (in Russian). Leningrad, pp. 26–39
21. Podolsky DM (1973) Design models and experimental data (in Russian). Structural mechanics and analysis of structures, 5:71–75
22. Reznikov RA (1971) Solution of problems of structural mechanics using digital computers (in Russian). Stroyzdat, Moskow
23. Rudych OL, Sokolov GP, Pahomov VL (1998) Introduction to the nonlinear structural mechanics (in Russian). Association of structural universities, Moscow
24. Shulkin YuB (1977) The kinematic analysis of beam constructions (in Russian). In: Alekseev SA, Novozhilov VV (eds) Analysis of spaces structures, vol. XVII. Stroyzdat Moskow, pp. 4–31.
25. Shulkin YuB (1984) Theory of elastic bar systems (in Russian). Nauka Publishing House, Moskow

3 Basic Relationships for Discrete Systems

Divide et impera

Latin adage

This simple idea — divide a body and rule its separate fragments — has proved to be immensely fruitful, it has exceeded a lot of deep and delicate ideas in efficiency, it has eliminated a number of previously acknowledged methods and buried the practical application of some beautiful constructions by pure analysts.

M.I.Reitman. Guarantee of strength.

Many facts stated in this chapter are well known and one might think there is no need to discuss them. But we believe it is an interesting idea to reveal certain interconnections between those facts, and this idea is not often expounded as here.

3.1 Governing Equations

For now, we will assume that our system of interest is a structure that includes a certain known number of nodes to which elastic elements are joined. The stress and strain distribution in the elements can be described by a finite set of static and kinematical parameters. So far, in this sense (and only in it) the system is a finite-element model, thus the well-known finite-element analysis technique (the finite-element method) is treated by us here as a method of investigating systems with finite number of degrees of freedom rather than a technique of reducing continual problems to finite-dimensional ones. This approach is advocated, for example, in a known monograph by Rosin [31]. The "internal build" of finite elements will be discussed later.

If we omit unnecessary details as we have already agreed to, then every finite element can be characterized by a set of *internal displacements* (or perhaps *deformations* or *strains* will be more to your liking) $\mathbf{\Delta}_e$ and their respective internal stresses $\mathbf{s}_e$. Here and below the subscript e indicates that the respective quantity refers to a particular finite element. We mean by the internal displacement, for now, a quantity that is characteristic for the element of interest, such as the elongation of a tensile bar. "Their respective" stresses mean that there is an energy conjugation correspondence between the displacements and the stresses commonly used in the structural mechanics: the scalar product of vectors $\mathbf{\Delta}_e$ and $\mathbf{s}_e$ will give the work $\mathsf{A}_e = \mathbf{\Delta}_e^{\mathrm{T}}\mathbf{s}_e$. Vectors $\mathbf{\Delta}_e$ and $\mathbf{s}_e$ in the case we are discussing now (a linearly elastic system) are related to each other through an equation of state (a physical law) expressed by the following equality:

$$\mathbf{s}_e = \mathbf{F}_e \mathbf{\Delta}_e \tag{3.1}$$

having a symmetric positive definite (thus non-degenerate) matrix $\mathbf{F}_e$. The latter means that any nonzero vector of internal displacements $\mathbf{\Delta}_e$ causes a deformation of the finite element (a change in its shape and/or sizes) accompanied by the accumulation of strictly positive internal energy in the element. It means, in particular, that vector $\mathbf{\Delta}_e$ contains no components of movement of the finite element as a perfectly rigid body. This is the reason why we can identify terminologically, as we have done above, internal displacements with strains of finite elements.

The idea of using strains $\mathbf{\Delta}_e$ and internal stresses $\mathbf{s}_e$, related via a non-degenerate matrix $\mathbf{F}_e$, has its roots in an old suggestion by Argyris regarding a *normal* (or *natural*) form of the finite-element method [1]. This definition of strains and stresses within the frames of the finite-element method is of certain methodological advantage, and is also good for computation in nonlinear problems [2], though the respective physical treatment of strains $\mathbf{\Delta}_e$ and stresses $\mathbf{s}_e$ sometimes looks unusual. For example, normal coordinates $\mathbf{\Delta}_e$ for a triangular element used in plane elasticity will be relative elongations of the triangle's sides, and the respective generalized internal stresses will be stresses along the same sides (Fig. 3.1)

The stress field described by these components will be weird enough, therefore if one uses the FEM normal form for computation, one should put the stress field "into alignment" so as to convert it to a conventional form convenient for an engineer.

The element's internal stresses $\mathbf{\Delta}_e$ are coupled with displacements of abutting nodes, which are external with respect to the element, via strain compatibility conditions

$$\mathbf{\Delta}_e = \mathbf{Q}_e^{\mathrm{T}} \mathbf{u}_e + \mathbf{d}_e\,, \tag{3.2}$$

and stresses $\mathbf{s}_e$ must satisfy equilibrium conditions

$$\mathbf{Q}_e \mathbf{s}_e = \mathbf{p}_e\,, \tag{3.3}$$

where $\mathbf{d}_e$ and $\mathbf{p}_e$ are vectors of dislocation (on elements) and force (reduced to nodes) influences respectively[1].

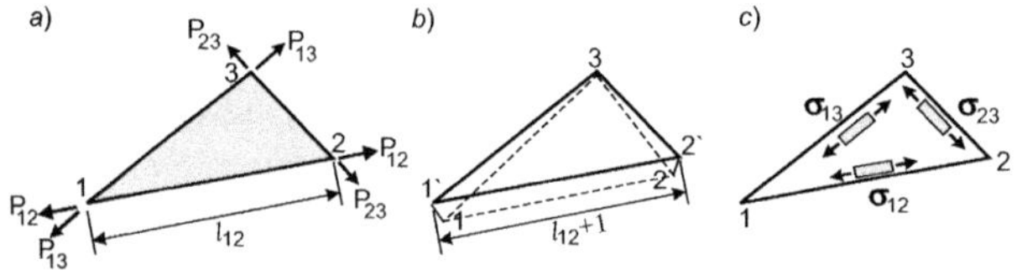

Fig. 3.1. Kinematical unknowns of the natural form of FEM

[1] Recall that *dislocations* are initial strains imposed upon the system as independent influences.

The fact that the transformation matrices in (3.2) and (3.3) are mutually transposed follows from the conjugacy of the strain compatibility equations (3.2) and the equilibrium equations (3.3), which fact in its turn follows from the principle of virtual displacements.

A standard procedure of the assembling of finite elements that form the design model gives a complete system of equations, namely:

$$\begin{aligned}
&\text{equilibrium equations} \ldots\ldots\ldots\ldots\ldots\ldots \mathbf{Qs} = \mathbf{p},\\
&\text{geometrical equations} \ldots\ldots\ldots\ldots\ldots \boldsymbol{\Delta} = \mathbf{Q}^{\mathrm{T}}\mathbf{u} + \mathbf{d},\\
&\text{physical equations} \ldots\ldots\ldots\ldots\ldots\ldots\ldots \mathbf{s} = \mathbf{F}\boldsymbol{\Delta}
\end{aligned} \tag{3.4}$$

with respect to unknowns $\mathbf{s}$, $\boldsymbol{\Delta}$ and $\mathbf{u}$.

The transformations

$$\boldsymbol{\Delta}_e = \boldsymbol{\Gamma}_1\boldsymbol{\Delta}, \qquad \mathbf{d}_e = \boldsymbol{\Gamma}_1\mathbf{d}, \qquad \mathbf{u}_e = \boldsymbol{\Gamma}_2\,\mathbf{u}, \qquad \mathbf{p}_e = \boldsymbol{\Gamma}_2\,\mathbf{p}, \qquad \mathbf{s}_e = \boldsymbol{\Gamma}_3\,\mathbf{s} \tag{3.5}$$

are implemented using matrices that contain only zero and one, and are fully defined by the matrix of the "nodes-to-elements" incidence, except for the order of numbering [10].

The dimensions of vectors $\mathbf{s}$, $\boldsymbol{\Delta}$, $\mathbf{d}$ are the same and equal to m – the total number of *internal* unknown stresses in the whole system. The dimensions of vectors $\mathbf{u}$ and $\mathbf{p}$ are also the same and both equal to the total number of *external* unknown nodal displacements n. The matrix of equilibrium equations $\mathbf{Q}$ has the dimension of $n{\times}m$. The square symmetric matrix $\mathbf{F}$ of the order m has a block-diagonal structure, its number of blocks is equal to the number of elements, the size of each block is the number of internal unknowns of the respective element. In particular, matrix $\mathbf{F}$ will be diagonal for trusses.

According to physical equations in (3.4), $\mathbf{F}$ is a matrix coefficient at internal displacements $\boldsymbol{\Delta}$ in the expression that describes internal stresses $\mathbf{s}$. Therefore Rzhanitsin advanced a reasonable proposal to call this matrix an *internal stiffness matrix*. Naturally, the inverse matrix of $\mathbf{F}$ should be immediately called an *internal compliance matrix*.

3.1.1 Slope Deflection Method

If we accept nodal displacements $\mathbf{u}$ as our principal unknowns, according to a standard procedure of the *slope deflection method*, then we can easily obtain the following, by excluding vectors $\mathbf{s}$ and $\boldsymbol{\Delta}$ from (3.4):

$$\mathbf{QFQ}^{\mathrm{T}}\mathbf{u} = \mathbf{p} - \mathbf{QFd}\,. \tag{3.6}$$

In structural mechanics, the number of unknown components n of the displacement vector $\mathbf{u}$ is called a *kinematical redundancy* of a system.

Suppose that the stiffness matrix of the system

$$\mathbf{K} = \mathbf{QFQ}^{\mathrm{T}} \tag{3.7}$$

is non-degenerate (moreover, positive definite) which can be provided when rank

r of matrix **Q** is equal to the number of its rows n (and this corresponds to the case of a kinematically stable system, as will be shown below)[2]. Thus we obtain

$$\mathbf{u} = \mathbf{K}^{-1}(\mathbf{p} - \mathbf{QFd}) . \tag{3.8}$$

Stresses in elements are calculated as

$$\mathbf{s} = \mathbf{F}(\mathbf{Q}^{\mathrm{T}}\mathbf{u} + \mathbf{d}) = \mathbf{FQ}^{\mathrm{T}}\mathbf{K}^{-1}\mathbf{p} + (\mathbf{I} - \mathbf{FQ}^{\mathrm{T}}\mathbf{K}^{-1}\mathbf{Q})\mathbf{Fd} , \tag{3.9}$$

and this completes the determination of parameters of the stress and strain distribution in the system.

Returning to formula (3.7), note that this one establishes a dependence of the external stiffness matrix **K** on the internal stiffness matrix **F**.

Following Rzhanitsin, let us try to construct the inverse dependence also, by expressing the internal compliance matrix $\mathbf{F}^{-1}$ through the external compliance matrix $\mathbf{K}^{-1}$. With this purpose, let's substitute the expression for **u** from (3.8) to the geometrical equations of (3.4), replacing **p** with **s** using the equilibrium equations. The result is

$$\boldsymbol{\Delta} = \mathbf{Q}^{\mathrm{T}}\mathbf{K}^{-1}\mathbf{Qs} + (\mathbf{I} - \mathbf{Q}^{\mathrm{T}}\mathbf{K}^{-1}\mathbf{QF})\mathbf{d} . \tag{3.10}$$

If now we replace $\boldsymbol{\Delta}$ in the left part of formula (3.10) with $\mathbf{F}^{-1}\mathbf{s}$, then some elementary transformations of the formula will give the following:

$$(\mathbf{F}^{-1} - \mathbf{Q}^{\mathrm{T}}\mathbf{K}^{-1}\mathbf{Q})(\mathbf{s} - \mathbf{Fd}) = \mathbf{0} . \tag{3.11}$$

It might seem that from (3.11) the desirable dependence follows:

$$\mathbf{F}^{-1} = \mathbf{Q}^{\mathrm{T}}\mathbf{K}^{-1}\mathbf{Q} . \tag{??}$$

This is exactly the form in which A.R.Rzhanitsin establishes the expression for an internal compliance matrix through the external compliance matrix. Though, we have numbered this formula by interrogation marks intentionally because it is erroneous in general case! In order to make sure, let us assume that

$$\mathbf{H} = \mathbf{Q}^{\mathrm{T}}\mathbf{K}^{-1}\mathbf{Q} \tag{3.12}$$

and consider the case of a stable statically indeterminate system for which the strict inequality $m > n$ will necessarily take place. Then the square matrix **H** of the order m will be degenerate (the rank of the product of matrices may not exceed the rank of either cofactor), while matrix $\mathbf{F}^{-1}$ is positive definite and thus equality (??) is impossible.

The mistake hides in the transition from (3.11) to (??) and is due to the fact that (3.11) holds not for any arbitrary vectors **s** but only for those *kinematically admis-*

---

[2] A.R.Rzhanitsin [33] proposed to call this matrix an *external stiffness matrix*. The respective inverse matrix $\mathbf{K}^{-1}$ is called by A.R.Rzhanitsin an *external compliance matrix*. The terms regarding stiffness and compliance matrices, supplemented with the definitions "external" and "internal", enable us to avoid the confusion that sometimes arises because of ambiguity of the concepts of "stiffness" and "compliance".

sible, that is, those satisfying the strain compatibility conditions expressed in terms of stresses.

If the system of interest is statically determinate, then any set of internal stresses is kinematically admissible, as one knows, so the equality $\mathbf{F}^{-1} = \mathbf{H}$ will truly follow from (3.11).

To do justice, it is worth noticing that this delicate though vexatious mistake has not caused any erroneous results in the subsequent course of structural mechanics in the textbook quoted here [33], because, as A.Rzhanitsin himself notices, in practical calculations the matrices of internal compliance $\mathbf{F}^{-1}$ and internal stiffness $\mathbf{F}$ "*can be obtained without any special difficulties*", and the external stiffness matrix can be obtained from formula (3.7).

Matrix $\mathbf{H}$ will play an important role in our further constructions, and its properties are inherited to a great extent, though not thoroughly, from the internal compliance matrix $\mathbf{F}^{-1}$. Thus, components of matrix $\mathbf{H}$, just the same as components of matrix $\mathbf{F}^{-1}$, are inversely proportional to the elasticity modulus of the material the structure is made of. Seeing both the similarity and the difference of these two matrices, we will call $\mathbf{H}$ a *pseudo-compliance matrix*. It can be proved that the defect of the pseudo-compliance matrix is equal to the static redundancy of the system.

Having introduced the concept of pseudo-compliance, we can change formula (3.9) for stresses $\mathbf{s}$ into the following:

$$\mathbf{s} = \mathbf{F}\mathbf{Q}^{\mathrm{T}}\mathbf{K}^{-1}\mathbf{p} + (\mathbf{I} - \mathbf{F}\mathbf{H})\mathbf{F}\mathbf{d}\,. \qquad (3.13)$$

From this formula a conclusion follows immediately (a mechanically obvious one) that no internal stresses are caused in statically determinate systems by dislocations, because matrices $\mathbf{F}$ and $\mathbf{H}$ are mutually inverse in systems of this kind.

Let us illustrate our presentment by an example of a simplest system the structure of which is shown in Fig. 3.2 (see [33], p.106).

Matrix $\mathbf{Q}$ of the equilibrium equations for this system follows:

$$\mathbf{Q} = \begin{bmatrix} -1 & 0 & 0 & -0.7071 & 0 \\ 0 & 0 & -1 & -0.7071 & 0 \\ 0 & -1 & 0 & 0 & -0.7071 \\ 0 & 0 & 1 & 0 & 0.7071 \end{bmatrix}.$$

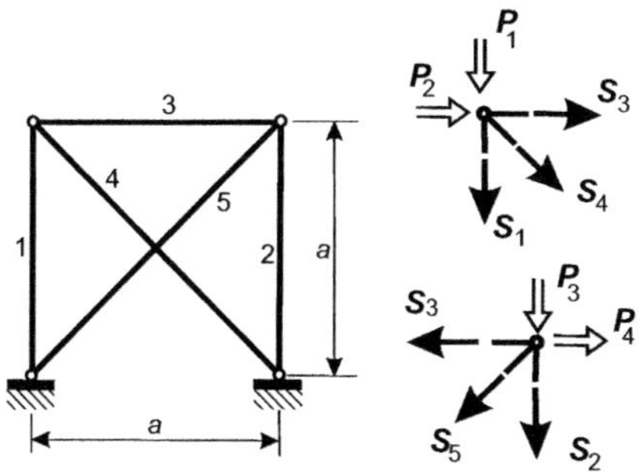

Fig. 3.2. To the example of governing equations

Assuming that the stiffness of all elements EF_i ($i = 1, \dots, 5$) is the same,

$$\mathbf{F} = (EF/a)\times\begin{bmatrix} 1 & 0 & 0 & 0 & 0 \\ 0 & 1 & 0 & 0 & 0 \\ 0 & 0 & 1 & 0 & 0 \\ 0 & 0 & 0 & 0.7071 & 0 \\ 0 & 0 & 0 & 0 & 0.7071 \end{bmatrix}.$$

The external stiffness matrix calculated by (3.7) will be

$$\mathbf{K} = (EF/a)\times\begin{bmatrix} 1.3536 & 0.3536 & 0 & 0 \\ 0.3536 & 1.3536 & 0 & -1 \\ 0 & 0 & 1.3536 & -0.3536 \\ 0 & -1 & -0.3536 & 1.3536 \end{bmatrix}.$$

After making the inverse matrix and using (3.8), we will obtain the vector of nodal displacements in the form

$$\begin{bmatrix} u_1 \\ u_2 \\ u_3 \\ u_4 \end{bmatrix} = (a/EF)\times\begin{bmatrix} 0.8845 & -0.5578 & -0.1155 & -0.4422 \\ -0.5578 & 2.1356 & 0.4422 & 1.6933 \\ -0.1155 & 0.4422 & 0.8845 & 0.5578 \\ -0.4422 & 1.6933 & 0.5578 & 2.1356 \end{bmatrix}\bullet\begin{bmatrix} p_1 \\ p_2 \\ p_3 \\ p_4 \end{bmatrix}.$$

The vector of internal stresses **s** will be obtained from (3.9) in the form

$$\begin{bmatrix} s_1 \\ s_2 \\ s_3 \\ s_4 \\ s_5 \end{bmatrix} = (EF/a)\times\begin{bmatrix} -0.8845 & 0.5578 & 0.1155 & 0.4422 \\ 0.1155 & -0.4422 & -0.8845 & -0.5578 \\ 0.1155 & -0.4422 & 0.1155 & 0.4422 \\ -0.1633 & -0.7887 & -0.1633 & -0.6253 \\ -0.1633 & 0.6253 & -0.1633 & 0.7887 \end{bmatrix}\bullet\begin{bmatrix} p_1 \\ p_2 \\ p_3 \\ p_4 \end{bmatrix}.$$

Using (3.12), we can find the pseudo-compliance matrix H

$$\mathbf{H} = (a/EF)\times\begin{bmatrix} 0.8845 & -0.1156 & -0.1156 & 0.2310 & 0.2310 \\ -0.1156 & 0.8845 & -0.1156 & 0.2310 & 0.2310 \\ -0.1156 & -0.1156 & 0.8845 & 0.2310 & 0.2310 \\ 0.2310 & 0.2310 & 0.2310 & 0.9523 & -0.4622 \\ 0.2310 & 0.2310 & 0.2310 & -0.4622 & 0.9523 \end{bmatrix}$$

thus making it clear that **H** does not coincide with the internal compliance matrix $\mathbf{F}^{-1}$ which is diagonal.

Let us now consider a statically determinate system derived from that shown in Fig. 3.2 by removing the top horizontal bar. The equilibrium matrix **Q** for this system can be derived from the corresponding matrix of the previous redundant system by striking out the third column. The internal stiffness matrix **F** is derived from the previous matrix **F** by striking out the third column and the third row. After necessary matrix operations we will have matrix **K** as follows

$$\mathbf{K} = (EF/a) \times \begin{bmatrix} 1.3536 & 0.3536 & 0 & 0 \\ 0.3536 & 0.3536 & 0 & 0 \\ 0 & 0 & 1.3536 & -0.3536 \\ 0 & 0 & -0.3536 & 0.3536 \end{bmatrix},$$

by inverting which we obtain the external compliance matrix

$$\mathbf{K}^{-1} = (a/EF) \times \begin{bmatrix} 1 & -1 & 0 & 0 \\ -1 & 3.8280 & 0 & 0 \\ 0 & 0 & 1 & 1 \\ 0 & 0 & 1 & 3.8280 \end{bmatrix}.$$

If now we calculate the pseudo-compliance matrix **H** by (3.12), we'll have

$$\mathbf{H} = (a/EF) \times \begin{bmatrix} 1 & 0 & 0 & 0 \\ 0 & 1 & 0 & 0 \\ 0 & 0 & 1.4141 & 0 \\ 0 & 0 & 0 & 1.4141 \end{bmatrix},$$

which is expectedly identical to the internal compliance matrix $\mathbf{F}^{-1}$.

Formulae (3.7), (3.8), (3.12) and (3.13) define the solution of the problem (3.4) by the slope deflection method in application to a finite-dimensional mechanical system.

3.1.2 Force Method

The second way of solution based on the *force method* is also feasible. Here, such integral-valued properties as the above mentioned rank r of the equilibrium matrix **Q** and the redundancy of the system k are of importance. These are related through the following equations

$$r = \text{rank}\,\mathbf{Q}, \qquad k = m - r. \tag{3.14}$$

According to the procedure of the force method, the general solution of equilibrium equations $\mathbf{Qs} = \mathbf{p}$ can be represented as

$$\mathbf{s} = \mathbf{Ax} + \mathbf{Bp}, \tag{3.15}$$

where matrix **A** of the dimensions $m \times k$ is a matrix of stresses in the principal system of the force method caused by unit values of unknowns **x** in the amount of k.

Mathematically, the first term in the right part of (3.15) is a general solution of the respective homogeneous equations of equilibrium, and the second term is a partial solution of the inhomogeneous equations. All columns of matrix **A** are linearly independent and form a so-called *fundamental system* of solutions of the homogeneous equations with the matrix **Q**; thus the rank of **A** is equal to k — the redundancy of the system. Matrix **B** of the dimensions $m \times n$ is a matrix of stresses in the principal system caused by unit values of nodal loads. These two matrices are such that

$$\mathbf{QA} = \mathbf{0}, \qquad \mathbf{QB} = \mathbf{I}. \tag{3.16}$$

where **0** is a rectangular zero matrix of the size $n\times k$, **I** is a unit matrix of the order n.[3]

If again we refer to the example of a statically indeterminate system in Fig. 3.2 of the redundancy one, then we can assume for it

$$\mathbf{A} = \left\|\begin{bmatrix} 1 \\ 1 \\ 1 \\ -1.4142 \\ -1.4142 \end{bmatrix}\right\|, \quad \mathbf{B} = \left\|\begin{bmatrix} -1 & 1 & 0 & 0 \\ 0 & 0 & -1 & -1 \\ 0 & 0 & 0 & 0 \\ 0 & -1.4142 & 0 & 0 \\ 0 & 0 & 0 & 1.4142 \end{bmatrix}\right\|.$$

One can easily make sure that conditions (3.16) hold for these matrices which fact confirms they have been constructed correctly.

By excluding strains $\mathbf{\Delta}$ from the geometrical equations of group (3.4) using the respective physical equations, we will obtain a relationship between the vector of internal stresses **s** and the vector of external displacements **u** in the form

$$\mathbf{F}^{-1}\mathbf{s} - \mathbf{Q}^{\mathrm{T}}\mathbf{u} = \mathbf{d}\,, \tag{3.17}$$

from where we can derive the following by substituting (3.15):

$$\mathbf{F}^{-1}\mathbf{Ax} - \mathbf{Q}^{\mathrm{T}}\mathbf{u} = \mathbf{d} - \mathbf{F}^{-1}\mathbf{Bp}\,. \tag{3.18}$$

After multiplying this equality by $\mathbf{A}^{\mathrm{T}}$ from the left, we have

$$\mathbf{A}^{\mathrm{T}}\mathbf{F}^{-1}\mathbf{Ax} - \mathbf{A}^{\mathrm{T}}\mathbf{Q}^{\mathrm{T}}\mathbf{u} = \mathbf{A}^{\mathrm{T}}\mathbf{d} - \mathbf{A}^{\mathrm{T}}\mathbf{F}^{-1}\mathbf{Bp}\,. \tag{3.19}$$

Due to the first condition of (3.16), the second term in the left part of (3.19) annihilates, and the inversion of the non-degenerate compliance matrix **D** (the non-degeneracy of the compliance matrix **D** directly follows from the equality of the rank of matrix **A** to the number of its columns k)[4]

$$\mathbf{D} = \mathbf{A}^{\mathrm{T}}\mathbf{F}^{-1}\mathbf{A}\,, \tag{3.20}$$

will give us

$$\mathbf{x} = \mathbf{D}^{-1}(\mathbf{A}^{\mathrm{T}}\mathbf{d} - \mathbf{A}^{\mathrm{T}}\mathbf{F}^{-1}\mathbf{Bp})\,. \tag{3.21}$$

---

[3] Here we are interested with properties of matrices **A** and **B** solely, therefore we do not discuss algorithmic aspects of constructing such matrices related to the choice of a principal system of the force method. This issue is addressed to in a lot of publications, of which we'd like to note the mentioned ones [31, 10], and a writing by Busigin and Rosin [5], where they used a technique of pseudo-inverse matrices.

[4] The name appropriate for matrix **D** could be a *compliance matrix of the principal system of the force method*. This entitlement, though very long, emphasizes the dependence of the contents of matrix **D** on the choice of the principal system of the force method, and at the formal mathematical level, on a particular representation of the matrix of fundamental solutions **A**. In cases where no ambiguity may arise, we will use a shorter term: a *compliance matrix* **D**.

The internal stresses are calculated after substituting (3.21) to (3.15), giving

$$\mathbf{s} = \mathbf{A}\mathbf{D}^{-1}\mathbf{A}^{\mathrm{T}}\mathbf{d} + (\mathbf{I} - \mathbf{A}\mathbf{D}^{-1}\mathbf{A}^{\mathrm{T}}\mathbf{F}^{-1})\mathbf{B}\mathbf{p}\,. \tag{3.22}$$

Similarly to the pseudo-compliance matrix **H**, we can introduce a *pseudo-stiffness matrix* **G**

$$\mathbf{G} = \mathbf{A}\mathbf{D}^{-1}\mathbf{A}^{\mathrm{T}}. \tag{3.23}$$

The pseudo-stiffness matrix **G** has the order of m, its rank is equal to the redundancy of the system k, and its defect is equal to r. This matrix is identical with the external stiffness matrix **F** only when $k = m$, that is, in a degenerate case of a system with all nodes fixed, thus $n = 0$. In common terms of the structural theory this case corresponds to a system with the *kinematical redundancy* of zero, that is, to a kinematically determinate system.

Using (3.23), we can re-write the solution (3.22) as

$$\mathbf{s} = \mathbf{G}\mathbf{d} + (\mathbf{I} - \mathbf{G}\mathbf{F}^{-1})\mathbf{B}\mathbf{p}\,. \tag{3.24}$$

3.1.3 Duality of the Slope Deflection Method and Force Method. Projectors

The duality of the force method and the slope deflection method is well-known and has been often discussed in scientific literature. According to this duality (that manifests itself, in particular, in the static-geometric analogy), any concept or formula related to, say, the slope deflection method, shall find its symmetric counterpart in the force method, and vice versa. Here we are going to demonstrate only how this symmetry regards the concepts of pseudo-stiffness matrix **G** and pseudo-compliance matrix **H** introduced earlier.

So, we have two matrix expressions for the internal stress vector **s**: one has been obtained by the slope deflection method and is described by formula (3.13), and the other has been derived by the force method according to formula (3.24). By equating these two expressions, we obtain

$$[\mathbf{F}\mathbf{Q}^{\mathrm{T}}\mathbf{K}^{-1} - (\mathbf{I} - \mathbf{G}\mathbf{F}^{-1})\mathbf{B}]\mathbf{p} = [\mathbf{G} - (\mathbf{I} - \mathbf{F}\mathbf{H})\mathbf{F}]\mathbf{d}\,. \tag{3,25}$$

The brackets in the left part of this equality contain a rectangular matrix of the dimensions $m{\times}n$, and those in the right part contain a square matrix of the order m. As vectors of force and dislocation influences **p** and **d** are absolutely independent and may have any values, then equality (3.25) can hold when and only when each of the two pairs of brackets contains a zero matrix of the respective dimension.

Equating the left part of (3.25) to zero, we will obtain a checkup formula for matrix **B**, namely

$$(\mathbf{I} - \mathbf{G}\mathbf{F}^{-1})\mathbf{B} = \mathbf{F}\mathbf{Q}^{\mathrm{T}}\mathbf{K}^{-1}\,. \tag{3.26}$$

At the first glance one may think that a technique for formal construction of matrix **B** follows immediately from formula (3.26) if only one knows the pseudo-stiffness matrix **G**. Though, it is not the case. Formula (3.26) provides nothing but a check for matrix **B**. Indeed, as we will see a little later, the square matrix $(\mathbf{I} - \mathbf{GF}^{-1})$ of the order *m*, that is a left factor before **B** in equation (3.26), is a degenerate matrix, that's why **B** cannot be found unambiguously from (3.26)[5].

The most interesting result can be found by equating the expression from the right part of equality (3.25) to zero. This will give an important identity

$$\mathbf{FH} + \mathbf{GF}^{-1} = \mathbf{I}, \tag{3.27}$$

that indicates a close connection between the pseudo-stiffness and pseudo-compliance matrices, and confirms the well-known symmetry of the slope deflection method and force method.

Identity (3.27) enables us to reduce formula (3.26) to the expression below, if taking into account the non-degeneracy of **F**,

$$\mathbf{HB} = \mathbf{Q}^{\mathrm{T}}\mathbf{K}^{-1}, \tag{3.28}$$

and, as we know that the pseudo-compliance matrix **H** is degenerate for any redundant system, it follows from this that matrix B cannot be derived from (3.28). Thus, expression (3.26) cannot be used for this purpose either.

Multiplying both parts of equality (3.28) from the right by matrix **Q**, we obtain another interesting identity

$$(\mathbf{FH})(\mathbf{BQ}) = \mathbf{FH}. \tag{3.29}$$

We need also to determine the external displacement vector **u** in order to close the solution of the problem by the force method. The geometrical and physical equations of group (3.4) give us $\mathbf{Q}^{\mathrm{T}}\mathbf{u} = \mathbf{F}^{-1}\mathbf{s} - \mathbf{d}$, and the multiplication of this equality by $\mathbf{B}^{\mathrm{T}}$ from the left with the use of (3.16) will yield

$$\mathbf{u} = \mathbf{B}^{\mathrm{T}}(\mathbf{F}^{-1}\mathbf{s} - \mathbf{d}), \tag{3.30}$$

which will lead us to the following formula for **u** after replacing vector **u** with its expression from (3.24) and making appropriate transformations:

$$\mathbf{u} = \mathbf{B}^{\mathrm{T}}\mathbf{F}^{-1}(\mathbf{I} - \mathbf{GF}^{-1})(\mathbf{Bp} - \mathbf{Fd}). \tag{3.31}$$

Thus, formulae (3.20), (3.23), (3.24) and (3.31) define the stress and strain distribution in a finite-dimensional system analyzed by the force method.

Again, as with vector **s**, one can equate the expressions for the external displacement vector **u** derived by the slope deflection method and the force method, that is, by formulae (3.8) and (3.31). Having done all this and taking into account identity (3.27), we come to the following relation:

$$(\mathbf{K}^{-1} - \mathbf{B}^{\mathrm{T}}\mathbf{HB})\mathbf{p} = (\mathbf{K}^{-1}\mathbf{Q} - \mathbf{B}^{\mathrm{T}}\mathbf{H})\mathbf{Fd}. \tag{3.32}$$

---

[5] As for the possibility of formal construction of partial solutions of the equilibrium equations, i.e. matrix **B**, one may use the formula $\mathbf{B} = \mathbf{Q}^{\mathrm{T}}(\mathbf{QQ}^{\mathrm{T}})^{-1}$, which can be validated by its direct substitution to the second equation of (3.16).

By equating the matrix factor at $\mathbf{p}$ in the left part of (3.32) to zero, we obtain another identity that establishes a relationship between matrix $\mathbf{B}$ and the external compliance matrix $\mathbf{K}^{-1}$ and pseudo-compliance matrix $\mathbf{H}$:

$$\mathbf{K}^{-1} = \mathbf{B}^{\mathrm{T}}\mathbf{H}\mathbf{B}\,. \tag{3.33}$$

Having considered the right part of (3.32), we have

$$\mathbf{K}^{-1}\mathbf{Q} = \mathbf{B}^{\mathrm{T}}\mathbf{H}\,, \tag{3.34}$$

which is the same as formula (3.28) after transposing both parts of the equation.

One should note that the choice of generalized unknown stresses $\mathbf{s}$ and strains $\boldsymbol{\Delta}$ is arbitrary to a certain extent. Indeed, if we choose another set of generalized internal unknowns $\mathbf{s}^*$ and $\boldsymbol{\Delta}^*$

$$\mathbf{s} = \mathbf{L}\mathbf{s}^*\,,\ \boldsymbol{\Delta} = (\mathbf{L}^{-1})^{\mathrm{T}}\boldsymbol{\Delta}^* \tag{3.35}$$

with an arbitrary invertible square matrix $\mathbf{L}$ of the order m, we will keep the expression of work in the form of a scalar product of the internal stress vector $\mathbf{s}^*$ by the strain vector $\boldsymbol{\Delta}^*$ intact, on one hand, but, on the other hand, we will replace the governing system of equations (3.4) by the system

$$\mathbf{Q}^*\mathbf{s}^* = \mathbf{p}\,,\quad \boldsymbol{\Delta}^* = \mathbf{Q}^{*\mathrm{T}}\mathbf{u} + \mathbf{d}^*,\quad \mathbf{s}^* = \mathbf{F}^*\boldsymbol{\Delta}^*, \tag{3.36}$$

where

$$\mathbf{Q}^* = \mathbf{Q}\mathbf{L}\,,\qquad \mathbf{F}^* = \mathbf{L}^{-1}\mathbf{F}(\mathbf{L}^{-1})^{\mathrm{T}},\qquad \mathbf{d}^* = \mathbf{L}^{\mathrm{T}}\mathbf{d}\,, \tag{3.37}$$

and the external stiffness matrix $\mathbf{K}$ will remain the same because

$$\mathbf{K}^* = \mathbf{Q}^*\mathbf{F}^*\mathbf{Q}^{*\mathrm{T}} = \mathbf{Q}\mathbf{L}\mathbf{L}^{-1}\mathbf{F}\,(\mathbf{L}^{-1})^{\mathrm{T}}\mathbf{L}^{\mathrm{T}}\mathbf{Q}^{\mathrm{T}} = \mathbf{K}\,. \tag{3.38}$$

As matrix $\mathbf{F}$ is positive definite, we can, in particular, assume $\mathbf{L} = \mathbf{F}^{1/2}$ which is very close to the Robinson – Haggenmacher transformation [30]. Then, instead of (3.4) we will have

$$\mathbf{Q}^*\mathbf{s}^* = \mathbf{p}\,,\quad \mathbf{D}^* = \mathbf{Q}^{*\mathrm{T}}\mathbf{u} + \mathbf{d}^*,\quad \mathbf{s}^* = \boldsymbol{\Delta}^*\,, \tag{3.39}$$

where

$$\mathbf{s}^* = \mathbf{F}^{-1/2}\mathbf{s}\,,\quad \boldsymbol{\Delta}^* = \mathbf{F}^{1/2}\mathbf{D}\,,\quad \mathbf{d}^* = \mathbf{F}^{1/2}\mathbf{d}\,, \tag{3.40}$$

and also

$$\mathbf{Q}^* = \mathbf{Q}\mathbf{F}^{1/2},\quad \mathbf{F}^* = \mathbf{I}\,,\quad \mathbf{H}^* = \mathbf{F}^{1/2}\mathbf{Q}^{\mathrm{T}}\mathbf{K}^{-1}\mathbf{Q}\mathbf{F}^{1/2},$$

$$\mathbf{A}^* = \mathbf{F}^{-1/2}\mathbf{A}\,,\quad \mathbf{B}^* = \mathbf{F}^{-1/2}\mathbf{B}\,,\quad \mathbf{D}^* = \mathbf{D}\,,\quad \mathbf{G}^* = \mathbf{F}^{-1/2}\mathbf{G}\,\mathbf{F}^{-1/2}.$$

Principal identity (3.27) will have the following form

$$\mathbf{L}^{\mathrm{T}}\mathbf{H}\mathbf{L} + \mathbf{L}^{-1}\mathbf{G}(\mathbf{L}^{-1})^{\mathrm{T}} = \mathbf{I} \tag{3.41}$$

with symmetric terms in the left part. Each one of them is an orthogonal projection matrix [35]

$$\mathbf{M} = \mathbf{L}^{\mathrm{T}}\mathbf{H}\mathbf{L} = \mathbf{L}^{\mathrm{T}}\mathbf{Q}^{\mathrm{T}}\mathbf{K}^{-1}\mathbf{Q}\mathbf{L}, \qquad \mathbf{N} = \mathbf{L}^{-1}\mathbf{G}(\mathbf{L}^{-1})^{\mathrm{T}} = \mathbf{L}^{-1}\mathbf{A}\mathbf{D}^{-1}\mathbf{A}^{\mathrm{T}}(\mathbf{L}^{-1})^{\mathrm{T}}. \tag{3.42}$$

One can prove by direct substitution that the idempotency conditions $\mathbf{M}^2 = \mathbf{M}$, $\mathbf{N}^2 = \mathbf{N}$ hold. These characterize any projector matrix together with the symmetry condition [15].

In terms of projectors **M** and **N**, principal identity (3.27) becomes especially simple:

$$\mathbf{M} + \mathbf{N} = \mathbf{I}\,. \tag{3.43}$$

Referring again to the example system in Fig. 3.2, we have

$$\mathbf{L} = \mathbf{F}^{1/2} = (EF/a)^{1/2} \times \left\| \begin{bmatrix} 1 & 0 & 0 & 0 & 0 \\ 0 & 1 & 0 & 0 & 0 \\ 0 & 0 & 1 & 0 & 0 \\ 0 & 0 & 0 & 0.8409 & 0 \\ 0 & 0 & 0 & 0 & 0.8409 \end{bmatrix} \right\|,$$

$$\mathbf{L}^{-1} = (a/EF)^{1/2} \times \left\| \begin{bmatrix} 1 & 0 & 0 & 0 & 0 \\ 0 & 1 & 0 & 0 & 0 \\ 0 & 0 & 1 & 0 & 0 \\ 0 & 0 & 0 & 1.1892 & 0 \\ 0 & 0 & 0 & 0 & 1.1892 \end{bmatrix} \right\|.$$

The product of the matrices

$$\mathbf{G} = \mathbf{A}\mathbf{D}^{-1}\mathbf{A}^{\mathrm{T}} = (EF/a) \times \left\| \begin{bmatrix} 0.1155 & 0.1155 & 0.1155 & -0.1634 & -0.1634 \\ 0.1155 & 0.1155 & 0.1155 & -0.1634 & -0.1634 \\ 0.1155 & 0.1155 & 0.1155 & --0.1634 & -0.1634 \\ -0.1634 & -0.1634 & -0.1634 & 0.2310 & 0.2310 \\ -0.1634 & -0.1634 & -0.1634 & 0.2310 & 0.2310 \end{bmatrix} \right\|$$

lets us calculate the projector matrix **N** by the second formula from (3.42), which gives the following result:

$$\mathbf{N} = \left\| \begin{bmatrix} 0.1155 & 0.1155 & 0.1155 & -0.1943 & -0.1943 \\ 0.1155 & 0.1155 & 0.1155 & -0.1943 & -0.1943 \\ 0.1155 & 0.1155 & 0.1155 & -0.1943 & -0.1943 \\ -0.1943 & -0.1943 & -0.1943 & 0.3267 & 0.3267 \\ -0.1943 & -0.1943 & -0.1943 & 0.3267 & 0.3267 \end{bmatrix} \right\|.$$

Matrices **M** and **N** are orthogonal (symmetric) projectors, their spherical norm is equal to one, while their cubic and octahedric norms do not exceed $n^{1/2}$ for **M** and $(m-n)^{1/2}$ for **N**. For oblique projectors these norms are not limited and an appropriate change of **L** can make them infinitely big. Also, the following relations hold for components of **M** and **N** which we will denote as μ_{ij} and ν_{ij}:

$$\mu_{ii} + \nu_{ii} = 1\,, \qquad 0 \le \mu_{ii} \le 1 \quad 0 \le \nu_{ii} \le 1 \quad (i = 1,2,\ldots, m\,)\,;$$

$$\mu_{ij} = -\nu_{ij}\,, \qquad |\mu_{ij}|,\, |\nu_{ij}| \le 1 \quad (i, j = 1,2,\ldots, m;\;\; i \ne j)\,; \tag{3.44}$$

and the projectors themselves satisfy the orthogonality conditions

$$\mathbf{MN} = \mathbf{NM} = \mathbf{0}\,, \tag{3.45}$$

which follow from identity (3.43) and the idempotency of matrices **M** and **N**. By the way, another interesting relation between the pseudo-stiffness and pseudo-compliance matrices follows from the orthogonality of the projectors, namely:

$$\mathbf{HG} = \mathbf{GH} = \mathbf{0}\,, \tag{3.46}$$

that is, the pseudo-stiffness and pseudo-compliance matrices are mutually orthogonal.

The rank of **M** is equal to that of **Q**, and the rank of **N** to that of **A**, i.e. to the redundancy of the system, which fact can be expressed as

$$\text{rank}\,\mathbf{M} = r\,, \qquad \text{rank}\,\mathbf{N} = k\,. \tag{3.47}$$

It is known from matrix theory (see, for example, [3]), that the rank of a symmetric idempotent matrix is equal to its trace, that is, the sum of all its diagonal elements. As the order of the discussed projector matrices is m, from this we obtain a relation useful for static and kinematical analysis of the system, taking into account principal identity (3.43)

$$m = \text{trace}\,\mathbf{M} + \text{trace}\,\mathbf{N} = r + k\,. \tag{3.48}$$

The use of projectors **M** and **N** is very effective when analyzing such properties of a system as its response to changes in its stiffness properties [36, 23], or its survivability properties [25] et al.

In problems of static analysis, the explicit construction of projectors **M** and **N** is hardly reasonable from the computational standpoint, though in optimization problems where matrix **Q** is fixed and **F** is an arbitrary block diagonal matrix the effort spent for obtaining **M** and **N** may prove profitable.

3.2 Static and Kinematical Analysis

Static and kinematical analysis yields characteristics of a system invariant with respect to a loading, therefore such analysis is done with homogeneous equations of equilibrium and strain compatibility. When strains $\boldsymbol{\Delta}$ and dislocation influences **d** are absent, the homogeneous system of strain compatibility equations

$$\mathbf{Q}^{\mathrm{T}}\mathbf{u} = \mathbf{0} \tag{3.49}$$

allows only the zero solution $\mathbf{u} = \mathbf{0}$, if and only if rank r of matrix **Q** is equal to the number of external unknowns n, and the structure does not allow nodal displacements without deforming its elements. If $r < n$, then equation (3.49) allows a nonzero solution, and the structure will be imperfect, whereas it will be stable at $r = n$.

In cases when the number of internal stresses and displacements m exceeds the number of external unknowns n, the system of equilibrium equations, which has the following form in absence of external loads $\mathbf{p}$

$$\mathbf{Qs} = \mathbf{0}\,, \tag{3.50}$$

allows a nontrivial solution

$$\mathbf{s}_0 = \mathbf{Ax} \tag{3.51}$$

with an arbitrary $(m - r)$-dimensional column $\mathbf{x}$.

Formula (3.51) defines a space of self-balanced stresses $\mathbf{s}_0$, the dimensionality of which $k = (m - r)$ is equal to the redundancy of the system. Stresses found from (3.51) define the self-balanced stressed state at an arbitrary vector $\mathbf{x}$, including values of $\mathbf{x}$ calculated by external influences, for example, using formula (3.21). Thus, besides stresses caused by an external loading, there can be *self-stresses* in the system caused by no external loading, that do not disappear after all loads are removed. These stresses are initial ones (pre-stresses), and they arise, most often, in the course of the erection of the structure.

It is known [15] that the projector matrix $\mathbf{R}$ related to the full rank matrix $\mathbf{Q}$

$$\mathbf{R} = \mathbf{I} - \mathbf{Q}^{\mathrm{T}}(\mathbf{QQ}^{\mathrm{T}})^{-1}\mathbf{Q} \tag{3.52}$$

transforms any vector $\mathbf{d}_0$ to vector $\mathbf{s}_0 = \mathbf{Rd}_0$ that belongs to the kernel of matrix $\mathbf{Q}$, i.e. one that satisfies the homogeneous equations (3.50)[6]. But it means that the pre-stress vector $\mathbf{s}_0$ is obtained through $\mathbf{d}_0$ which can be treated as a vector of arbitrary dislocation perturbations (such as bar elongations in a truss) that cause the pre-stress $\mathbf{s}_0$.

It is easy to see that projector $\mathbf{R}$ is identical with $\mathbf{N}$ if we assume $\mathbf{F} = \mathbf{I}$, and this is quite possible if we study only immanent static and kinematical properties of the system without seeking for its actual stress distribution. Equality $\mathbf{R} = \mathbf{N}$ with the said condition $\mathbf{F} = \mathbf{I}$ being true can be easily proved by appropriate substitutions; actually, it can be reduced to identity (3.27). The advantage of projector $\mathbf{R}$ over projector $\mathbf{N}$ is that the former is invariant with respect to stiffness properties of the system (provided those are nonzero), whereas it is not the case with the latter[7]. That's why projector $\mathbf{R}$ is a more convenient tool for analyzing static and kinematical properties of the system. By the way, besides formula (3.52), the following relation can be used to construct projector $\mathbf{R}$:

$$\mathbf{R} = \mathbf{A}(\mathbf{AA}^{\mathrm{T}})^{-1}\mathbf{A}^{\mathrm{T}}\,, \tag{3.53}$$

which is a particular case of the principal identity in the form (3.43).

For elements of the structure which are *unconditionally necessary*, i.e. the removal of such from the system will cause the kinematical instability, no pre-stress

[6] Apropos, referring to the formula given in footnote [5], one can obtain the following simple expression for the projector: $\mathbf{R} = \mathbf{I} - \mathbf{BQ}$.

[7] The transition to zero stiffness changes static and kinematical properties of the system because it actually means a removal of the respective element from the system thus changing the integral-valued parameter m.

is possible at any disturbance $\mathbf{d}_0$. Indeed, according to its definition, such element must be necessarily included in the principal system of the force method, thus the stress in it cannot be treated as a component of the vector of unknowns $\mathbf{x}$. But it means that a row and a column with all components being zero should correspond to this element in matrix $\mathbf{R}$ (and in $\mathbf{N}$, respectively). Note that it suffices to check only the diagonal element of matrix $\mathbf{R}$ (or $\mathbf{N}$) for being zero. This sufficiency follows from another well-known property of symmetric idempotent matrices, namely: if a diagonal element of a symmetric idempotent matrix is equal to zero, then all elements of the respective row and column are also zeros [3].

The converse is also true: a *conditionally necessary* element (unlike an unconditionally necessary element, such can be removed from a system without the latter's losing its kinematical stability) corresponds to rows and columns in the projector matrix where nonzero elements are always present, and a nonzero component stands on the main diagonal also.

As an example, let us consider a system shown in Fig. 3.3, and let us assume all its stiffness parameters to be equal to one for the sake of simplicity.

Let's use formula (3.53) to build projector $\mathbf{R}$. Then we will have

$$\mathbf{A}^{\mathrm{T}} = \left\| \begin{bmatrix} 0 & 0 & -1 & -1 & -\sqrt{3}/3 & -\sqrt{3}/3 & 2\sqrt{3}/3 & 1 & 0 & 0 & 0 & 0 & 0 \\ -1 & -1 & 0 & 0 & -\sqrt{3}/3 & 2\sqrt{3}/3 & -\sqrt{3}/3 & 0 & 1 & 0 & 0 & 0 & 0 \\ 0 & 0 & 0 & 0 & 0 & 0 & 0 & 0 & 0 & 0 & 0 & 1 & 1 \end{bmatrix} \right\|,$$

$$\mathbf{A}^{\mathrm{T}}\mathbf{A} = \left\| \begin{bmatrix} 5 & -1 & 0 \\ -1 & 5 & 0 \\ 0 & 0 & 2 \end{bmatrix} \right\|, \qquad (\mathbf{A}^{\mathrm{T}}\mathbf{A})^{-1} = \left\| \begin{bmatrix} 5/24 & 1/24 & 0 \\ 1/24 & 5/24 & 0 \\ 0 & 0 & ½ \end{bmatrix} \right\|,$$

The calculation will give:

$$\mathbf{R} = \left\| \begin{bmatrix} 5/24 & 5/24 & 1/24 & 1/24 & \sqrt{3}/12 & -\sqrt{3}/8 & \sqrt{3}/24 & -1/24 & -5/24 & 0 & 0 & 0 & 0 \\ 5/24 & 5/24 & 1/24 & 1/24 & \sqrt{3}/12 & -\sqrt{3}/8 & \sqrt{3}/24 & -1/24 & -5/24 & 0 & 0 & 0 & 0 \\ 1/24 & 1/24 & 5/24 & 5/24 & \sqrt{3}/12 & \sqrt{3}/24 & -\sqrt{3}/8 & -5/24 & -1/24 & 0 & 0 & 0 & 0 \\ 1/24 & 1/24 & 5/24 & 5/24 & \sqrt{3}/12 & \sqrt{3}/24 & -\sqrt{3}/8 & -5/24 & -1/24 & 0 & 0 & 0 & 0 \\ \sqrt{3}/12 & \sqrt{3}/12 & \sqrt{3}/12 & \sqrt{3}/12 & 1/6 & -1/12 & -1/12 & -\sqrt{3}/12 & -\sqrt{3}/12 & 0 & 0 & 0 & 0 \\ -\sqrt{3}/8 & -\sqrt{3}/8 & \sqrt{3}/24 & \sqrt{3}/24 & -1/12 & 7/24 & -5/24 & -\sqrt{3}/24 & \sqrt{3}/8 & 0 & 0 & 0 & 0 \\ \sqrt{3}/24 & \sqrt{3}/24 & -\sqrt{3}/8 & -\sqrt{3}/8 & -1/12 & -5/24 & 7/24 & \sqrt{3}/8 & -\sqrt{3}/24 & 0 & 0 & 0 & 0 \\ -1/24 & -1/24 & -5/24 & -5/24 & -\sqrt{3}/12 & -\sqrt{3}/24 & \sqrt{3}/8 & 5/24 & 1/24 & 0 & 0 & 0 & 0 \\ -5/24 & -5/24 & -1/24 & -1/24 & -\sqrt{3}/12 & \sqrt{3}/8 & -\sqrt{3}/24 & 1/24 & 5/24 & 0 & 0 & 0 & 0 \\ 0 & 0 & 0 & 0 & 0 & 0 & 0 & 0 & 0 & 0 & 0 & 0 & 0 \\ 0 & 0 & 0 & 0 & 0 & 0 & 0 & 0 & 0 & 0 & 0 & 0 & 0 \\ 0 & 0 & 0 & 0 & 0 & 0 & 0 & 0 & 0 & 0 & 0 & ½ & ½ \\ 0 & 0 & 0 & 0 & 0 & 0 & 0 & 0 & 0 & 0 & 0 & ½ & ½ \end{bmatrix} \right\|.$$

From the consideration of projector $\mathbf{R}$ one can see that 10th and 11th bars are unconditionally necessary because zero columns and rows correspond to them.

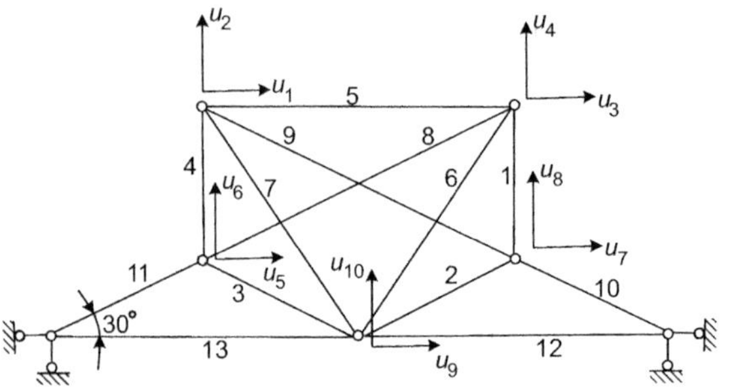

Fig. 3.3. To the example of construction of a projector matrix

Indeed, these bars are involved in no pre-stressed state (see Fig. 3.4), and the removal of at least one of them will make the system infinitesimally changeable.

Calculating the trace of matrix **R** for the purpose of check, we obtain that trace **R** = 3, which is expectedly the same value as the redundancy of the system according to (3.47). The fact that projector **R** can help in analyzing kinematical properties of a mechanical system may be used to resolve the question how the static and kinematical class of a structure changes when an element is removed from it. If k^{th} element is removed from the system, it is required that under any influences (including arbitrary dislocations $\mathbf{d}_0$) k^{th} component of vector $\mathbf{s}_0$ be equal to zero because there can be no stress in the removed element. To achieve this, one needs to exclude variable $s_{0,k}$ from the system $\mathbf{s}_0 = \mathbf{R}\mathbf{d}_0$. To do this, one must make it an independent variable first, because only after that one can change its value, particularly, making it zero. So, one needs to perform a step of Jordan exclusions [39] (see also Appendix A) with the governing element R_{kk}, after which the system $\mathbf{s}_0 = \mathbf{R}\mathbf{d}_0$ becomes

$$s_{0,i} = \sum_{j=1}^{k-1} R^*_{ij} d_{0,j} + R^*_{ik} s_{0,k} + \sum_{j=k+1}^{m} R^*_{ij} d_{0,j} \qquad (i = 1, \ldots, k-1),$$

$$d_{0,k} = \sum_{j=1}^{k-1} R^*_{kj} d_{0,j} + R^*_{kk} s_{0,k} + \sum_{j=k+1}^{m} R^*_{kj} d_{0,j},$$

$$s_{0,i} = \sum_{j=1}^{k-1} R^*_{ij} d_{0,j} + R^*_{ik} s_{0,k} + \sum_{j=k+1}^{m} R^*_{ij} d_{0,j} \quad (i = k+1, \ldots, m), \tag{3.54}$$

where asterisk denotes elements of the transformed matrix. Condition $s_{0,k} = 0$ lets us strike out k^{th} column, and k^{th} row will contain an expression for determining a dislocation (for example, a convergence of nodes previously connected by the removed element) of no interest for us, thus we can strike out k^{th} row as well.

The new matrix **R*** of the order $(m-1)$ is also a projector but for the structure with k^{th} element excluded. If now one can detect new rows and columns in **R*** containing zeros only, this will mean that the respective elements have acquired the status of unconditionally necessary ones after removing k^{th} element (and exactly as a result of this removal)!

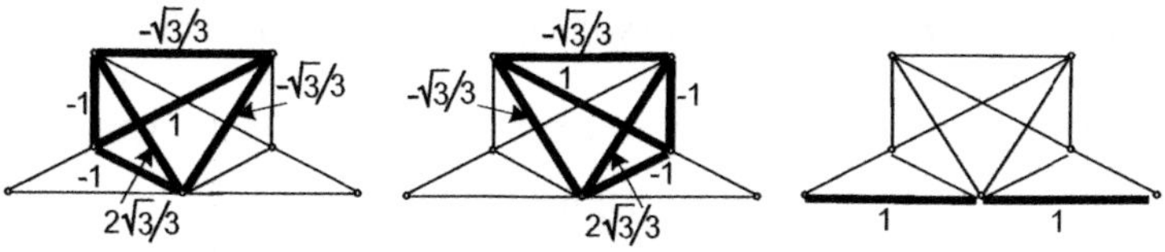

Fig. 3.4. Self-stresses in the system from Fig. 3.3

From (3.13) one can see that components of the projector matrix **N** possess the physical meaning of stresses in the system's elements caused by unit values of dislocation influences. From here a simple procedure for constructing **N** or **R** follows: using any FEM software, analyze the structure under m variants of unit dislocation influences (in the case of trusses, under unit elongations of bars), and the vector of stresses in elements that corresponds to j^{th} variant of loading will form j^{th} column of the respective matrix. This method of calculation may prove much more efficient than the direct implementation of matrix equalities (3.42).

The latter consideration is worth paying attention as it shows how an analysis of mechanical meaning of some formally built matrix can help us actually perform calculations without using formal matrix operations. It has been stated many times that the matrix notation is ideal for computer-aided calculations. It is true only if taking into account the above said note. The matrix notation helps us make mathematical relations clearly evident. As for the calculation itself, one does not necessarily have to operate full matrices reserving additional computer memory for keeping and handling zeros contained in the matrices. There can be a different way, and a more attractive one.

3.2.1 Note on Dislocations

Generally, dislocation influences present an immensely useful tool for an analyst and can be used for a lot of purposes. Besides the possibility of performing a projector-based static and kinematical analysis of a structure discussed above, dislocations can be specified as direct influences caused by deviations of actual sizes of the structure's elements from their designed sizes (either intentionally, in order to create pre-stress, or unintentionally, because of imprecise fabrication and erection). Another pretty important category of useful applications of dislocation influences is the possibility of constructing influence lines and surfaces of internal stresses as displacements caused by these stresses' conjugate unit dislocations, that is, involving a well-known kinematical method of influence line construction. Though, not every CAD software allows arbitrary dislocations to be specified in the source data for an analysis. Hence the possible necessity to simulate dislocation influences by appropriate forces. Formally, the problem can be posed as follows.

Let the only influences upon a mechanical system specified in governing equations (3.4) be dislocations **d**. Let us try to find an equivalent force loading in the form of an external force vector **p** which would create the stress and strain distribution in the system totally identical to that under the dislocation influence. Referring to formulae (3.8) and (3.13), we will equate the respective distributions under the two kinds of influences.

The condition of identity of external displacements **u** will give us

$$\mathbf{p} = -\,\mathbf{QFd}\,. \tag{3.55}$$

The internal stresses in elements of the system caused by these external forces will be

$$\mathbf{s} = -\,\mathbf{FHFd}\,, \tag{3.56}$$

whereas the stresses caused by the given dislocations will be

$$\mathbf{s} = -\,\mathbf{FHFd} + \mathbf{Fd} \tag{3.57}$$

and different from (3.56) by the quantity **Fd**. Therefore, when modeling dislocations by external forces, the displacement distribution is exactly the same while the stresses prove to be shifted by vector **Fd**. Of course, this vector can be added manually to results of the analysis performed by the software, but this would make an additional burden upon the analyst and would hinder the post-processing of the analysis results. Therefore a "good" software provides the opportunity of specification of all possible kinds of dislocations as influences upon the structure.

Talking about arbitrary bar elements (not necessarily truss-type bars), it would be useful to be able to specify concentrated dislocations in bars not only in their junctions at nodes but also in any other (intermediate) sections. It is not very expensive to implement, and the profit can be really big.

To conclude this section, we'd like to notice that the possibility of performing the static and kinematical analysis of a system is an important feature of a commercial software that adds much to its marketability. In this connection there is an issue how to reasonably choose means for displaying static and kinematical properties of a structure. For example, if the structure is imperfect, it is not enough for the user to receive just a respective text message because it is the user who has to seek for reasons that have turned the system of interest into a moveable mechanism. In our opinion, a handy solution implemented in **SCAD, GTSTRUDL** and some other contemporary software products is the use of animation that shows a slow motion of a structure as a mechanism, that is, without any deformation of its elements. Our experience has shown that it may suffice just to take a look at the structure's displacement as a rigid body to quickly detect the reasons for its changeability and subsequently correct the design model.

3.3 Poly-constraints Revisited. Variational Formulation

The governing system of equations of the slope deflection method (3.6) which we will represent in the form

$$\mathbf{Ku} = \mathbf{q} \tag{3.58}$$

with the right part vector $\mathbf{q} = \mathbf{p} - \mathbf{QFd}$, can be interpreted as a condition of the minimum of a functional equal to the difference of potential energy of deformation E and potential of active influences Π

$$\mathsf{L} = \mathsf{E} - \Pi = \tfrac{1}{2}\,\mathbf{u}^{\mathrm{T}}\mathbf{K}\mathbf{u} - \mathbf{q}^{\mathrm{T}}\mathbf{u}\,. \tag{3.59}$$

This functional is called a Lagrange's functional or a full potential energy of a system, and condition (3.58) provides a *necessary* condition of the minimum of full energy L[8]. In the left part of (3.59), strictly speaking, it would be better to add an indication of the functional argument, here as $\mathsf{L}(\mathbf{u})$, to specify the variable with respect to which we are seeking for the minimum. Though, according to a tradition and for the sake of briefness, we will omit the argument if the context makes it clear enough what set of quantities is involved in variation to find the desirable minimum of the functional.

The task of finding the solution of equation (3.58) is equivalent to the task of finding the minimum point of functional L. Though, there is an issue: so far we have said nothing about the range of the functional argument $\mathbf{u}$ at which the minimum is to be found. In this case we have just supposed by default that all components of vector $\mathbf{u}$ can acquire independent values, that is, we are talking about an absolute minimum of functional L.

Recalling the classification of constraints introduced in Section 2.5 that divides those into mono-constraints and poly-constraints, we can easily notice that so far this chapter has dealt with mechanical systems where nodal displacements can be restrained only by mono-constraints. The presence of mono-constraints in the system has been taken into account already in governing equations (3.4) by excluding from vector $\mathbf{u}$ its components that correspond to displacements prohibited by the imposed mono-constraints. It is for this reason that we did not include the displacements of the lower supporting nodes in vector $\mathbf{u}$ in the example shown in Fig. 3.2. Also, it is one of the reasons why it is recommended to introduce local coordinate systems in nodes restrained from linear displacements or/and slopes in any direction: thus one transforms this particular case of a poly-constraint to a mono-constraint.

As for general poly-constraints, so far we have assumed by default that there have been no constraints of this kind in the system of interest, as we have stated above. Now it's high time to deal with this type of constraints including it in the variational formulation of the problem. So, let p poly-constraints be imposed upon the system being considered. These are written in a matrix form as

$$\mathbf{Cu} = \mathbf{0}\,, \tag{3.60}$$

where matrix $\mathbf{C}$ of the dimensions $p{\times}n$ is a poly-constraint matrix with all its rows linearly independent, which requires the following conditions to hold

$$\operatorname{rank}\mathbf{C} = p\,, \qquad p \le n\,. \tag{3.61}$$

[8] The brief entitlement of *Lagrangian* is also usable.

Taking into account the poly-constraints, the kinematical redundancy of the system l is calculated as

$$l = n - p\,, \tag{3.62}$$

and the case of degeneracy $p = n$ returns us to a kinematically determinable system, i.e. the resulting displacement vector **u** will be zero[9].

As regards our variational formulation of the problem, the very expression for the functional L to be minimized according to (3.59) does not change. Only the range of feasible (kinematically admissible) values of vector **u** changes, which we search for the minimum of L.

Geometrically interpreted, this set is a reduction of the original n-dimensional space $\mathfrak{R}^n$ to its subspace $\mathfrak{R}^l$ of dimension l the appropriate term for which is a *space of kinematically admissible displacements*. All elements of it satisfy the poly-constraint conditions. Subspace $\mathfrak{R}^p$ of dimension p orthogonal to the subspace of kinematically admissible displacements will be called a *space of poly-constraints* (Fig. 3.5).

Any vector $\mathbf{v} \in \mathfrak{R}^l$ can be represented in the basis of $\mathfrak{R}^l$, then it will have exactly l components, all independent, or alternatively it may be represented in the basis of the original space $\mathfrak{R}^n$, then it will have n components and satisfy the matrix equality $\mathbf{Cv} = \mathbf{0}$. Further we will deal with third way of representing vector **v**: one in the basis of the original space $\mathfrak{R}^n$ but having only l components with respect to l selected basis vectors of the original space $\mathfrak{R}^n$, all these components being independent. We will explain later how to do it, and now let's discuss what we need it for.

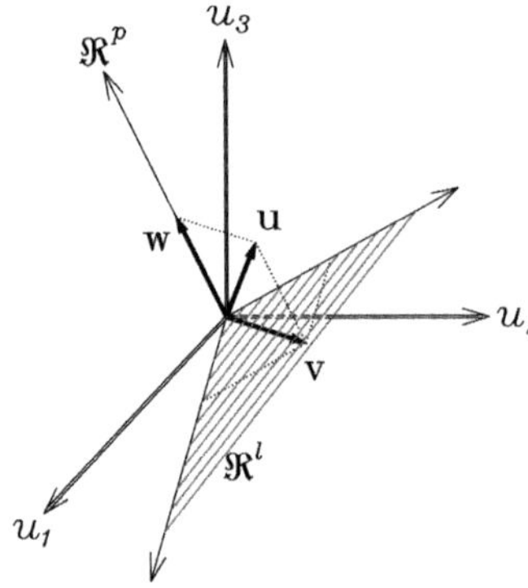

Fig. 3.5. A geometrical interpretation of the decomposition of the original space $\mathfrak{R}^n$ into the sum of the kinematically admissible displacements space $\mathfrak{R}^l$ and the poly-constraint space $\mathfrak{R}^p$

[9] All rows of matrix **C** absolutely must be linearly independent. If there is at least one row in matrix **C** that is a linear combination of the other rows, then mechanically this fact can be interpreted as the presence of a redundant subsystem containing only non-deformable elements in the original system. We exclude this case from our presentment for now intentionally because it causes an ambiguous distribution of internal stresses.

The fact is that components of any vector in the basis of the original space have a clear physical and geometrical sense: they are displacements of nodes of the mechanical system being analyzed with respect to local coordinates (ones referring to particular nodes). As soon as we introduce its own basis in $\mathfrak{R}^l$, this physical clearness will be lost, and the user will treat this fact as a disadvantage of the respective CAD software.

The variational formulation of the problem supposes now the search of subspace $\mathfrak{R}^l$ for the point of minimum of L.

Let us discuss two competing methods of including poly-constraints in the response of a mechanical system. According to the **first method**, the reduced space $\mathfrak{R}^l$ is to be searched for the minimum of the functional by means of changing the functional itself into the form

$$\mathsf{L} = \tfrac{1}{2}\,\mathbf{v}^{\mathrm{T}}\mathbf{K}^{*}\mathbf{v} - \mathbf{q}^{*\mathrm{T}}\mathbf{v}\,, \qquad (3.63)$$

where displacement vector $\mathbf{v}\in\mathfrak{R}^l$, though represented in the basis of the original space $\mathfrak{R}^n$, has only l components all of which are independent. Matrix $\mathbf{K}^*$ is a reconstructed external stiffness matrix of the system of order l. Vector $\mathbf{q}^*$ is a reconstructed vector of active influences upon the system[10].

According to (3.61), poly-constraint matrix $\mathbf{C}$ will necessarily contain p linearly independent columns. Let those be columns with numbers $\alpha_1,\ldots,\alpha_p$ which we will call *basis* columns of the poly-constraint matrix.

Let us form two vectors of displacements $\mathbf{v}$ and $\mathbf{w}$ the first of which has the dimension l and the second p. Components of vector $\mathbf{w}$ present a selection from the full displacement vector $\mathbf{u}$ of those ones that correspond to the basis columns of the poly-constraint matrix, in other words,

$$\mathbf{w}^{\mathrm{T}} = \|[u_{\alpha 1},\ldots,u_{\alpha p}]\|\,, \qquad (3.64)$$

and vector $\mathbf{v}$, on the contrary, is formed from vector $\mathbf{u}$ by striking out exactly the components of vector $\mathbf{w}$.

We will call vector $\mathbf{v}$ a vector of *free displacements*, while vector $\mathbf{w}$ will be called a vector of *dependent displacements*. In the matrix form, the relation between the three vectors: full one $\mathbf{u}$, free one $\mathbf{v}$ and dependent one $\mathbf{w}$, can be represented in the form:

$$\mathbf{v} = \mathbf{P}_{\mathrm{v}}\mathbf{u}\,, \qquad \mathbf{w} = \mathbf{P}_{\mathrm{w}}\mathbf{u}\,, \qquad (3.65)$$

where matrices $\mathbf{P}_{\mathrm{v}}$ of the dimension $l{\times}n$ and $\mathbf{P}_{\mathrm{w}}$ of the dimension $p{\times}n$ are those that perform the said selection. These two matrices, in their turn, present the like selection of rows of a unit matrix of order n. Matrix $\mathbf{P}_{\mathrm{w}}$ includes rows of the unit

---

[10] Actually, in the software implementation one does not have to form the external stiffness matrix $\mathbf{K}$ and vector $\mathbf{q}$ of order n first, and then reconstruct those to $\mathbf{K}^*$ and $\mathbf{q}^*$ of the lower order l. The reduction of the system's order to l is done in the course of assembling the final stiffness matrix and the final loading vector (or vectors if a few loads are present). This is the way most CAD software developers follow: usually, particular cases of poly-constraints are allowed, for example, as merged displacements of different nodes.

matrix with numbers $\alpha_1,\ldots,\alpha_p$, and matrix $\mathbf{P}_v$ includes the other rows.

Let, for example, $n = 5, p = 2, l = 3$, and let the poly-constraint matrix $\mathbf{C}$ be

$$\mathbf{C} = \left\|\begin{bmatrix} c_{11} & c_{12} & c_{13} & c_{14} & c_{15} \\ c_{21} & c_{22} & c_{23} & c_{24} & c_{25} \end{bmatrix}\right\|,$$

the second and fourth columns of this matrix being linearly independent. Then, according to our convention, we will have

$$\mathbf{P}_v = \left\|\begin{bmatrix} 1 & 0 & 0 & 0 & 0 \\ 0 & 0 & 1 & 0 & 0 \\ 0 & 0 & 0 & 0 & 1 \end{bmatrix}\right\|, \quad \mathbf{P}_w = \left\|\begin{bmatrix} 0 & 1 & 0 & 0 & 0 \\ 0 & 0 & 0 & 1 & 0 \end{bmatrix}\right\|,$$

$$\mathbf{u}^T = \|[u_1 u_2 u_3 u_4 u_5]\|, \qquad \mathbf{v}^T = \|[u_1 u_3 u_5]\|, \qquad \mathbf{w}^T = \|[u_2 u_4]\|.$$

It is clear that selection matrices $\mathbf{P}_v$ and $\mathbf{P}_w$ enable us to perform the inverse transition to the full displacement vector $\mathbf{u}$ from the free displacement vector $\mathbf{v}$ and the dependent displacement vector $\mathbf{w}$ as

$$\mathbf{u} = \mathbf{P}_v^T \mathbf{v} + \mathbf{P}_w^T \mathbf{w}. \tag{3.66}$$

Also, these matrices enable us to re-write the poly-constraint conditions (3.60) in another form, more convenient for our purpose, $\mathbf{C}(\mathbf{P}_v^T \mathbf{v} + \mathbf{P}_w^T \mathbf{w}) = \mathbf{0}$ or

$$\mathbf{C}_v\mathbf{v} + \mathbf{C}_w\mathbf{w} = \mathbf{0}, \quad \mathbf{C}_v = \mathbf{C}\,\mathbf{P}_v^T, \quad \mathbf{C}_w = \mathbf{C}\,\mathbf{P}_w^T, \tag{3.67}$$

hence

$$\mathbf{w} = -\,\mathbf{C}_w^{-1}\mathbf{C}_v\mathbf{v}. \tag{3.68}$$

The poly-constraint equations written in the form of linear combinations of free components of the displacement vector present expressions for dependent components in the form (3.68). These will be called *normalized* equations of poly-constraints.

Note that the inversion operation in (3.68) is admissible for the normalization because matrix $\mathbf{C}_w$ consists of basis columns of the poly-constraint matrix (in the example above the basis columns are shaded for better understanding).

Now it should be clear why the full vector of displacements has been divided into a free displacement vector and a dependent displacement one: components of vector $\mathbf{v}$ can be varied independently while components of vector $\mathbf{w}$ cannot due to the poly-constraint conditions (3.68).

After substituting (3.68) to (3.66) we obtain an expression for components of the full displacement vector $\mathbf{u}$ through components of only the free vector $\mathbf{v}$

$$\mathbf{u} = \mathbf{P}\mathbf{v}, \quad \mathbf{P} = \mathbf{P}_v^T - \mathbf{P}_w^T\,\mathbf{C}_w^{-1}\mathbf{C}_v, \tag{3.69}$$

where matrix $\mathbf{P}$ has the dimensions of $n \times l$.

The first formula in (3.69) can be treated as a transition between two representations of the same vector performed by matrix $\mathbf{P}$, $\mathbf{u}$ being the same vector as $\mathbf{v}$

with its components expressed in the full basis of the original space $\Re^n$, hence $\mathbf{Cu} = \mathbf{0}$. One can easily make sure of it using (3.69) directly.

By substituting (3.69) to (3.59) we obtain

$$\mathsf{L} = \tfrac{1}{2}\, \mathbf{v}^{\mathrm{T}}\mathbf{P}^{\mathrm{T}}\mathbf{KPv} - \mathbf{q}^{\mathrm{T}}\mathbf{Pv}\,, \tag{3.70}$$

which yields the following after a comparison with (3.63)

$$\mathbf{K}^* = \mathbf{P}^{\mathrm{T}}\mathbf{KP}\,, \qquad \mathbf{q}^* = \mathbf{P}^{\mathrm{T}}\mathbf{q}\,. \tag{3.71}$$

Here one should notice that the poly-constraint equations given in Section 2.6 have been already represented in the normalized form — equations (2.8) to (2.12) — for a particular but very important case of poly-constraints imposed by a perfectly rigid body. Obviously, the displacements of master nodes will belong to the free displacement vector $\mathbf{v}$, while unreleased degrees of freedom of slaves will belong to the dependent displacement vector $\mathbf{w}$, and no violation of the legislation!

Though, in the general case of any arbitrary poly-constraints one would have to solve an auxiliary problem of extracting linearly independent columns of the poly-constraint matrix $\mathbf{C}$.

The **second method**, as opposite to the first one, is based on expanding the original space to the dimension of $n + p$ rather than on reducing it. According to the rule of Lagrangian multipliers, the problem of finding a conditional extremum of functional (3.59) with constraints (3.60) can be reduced to the problem of finding a stationarity point of the auxiliary functional

$$\mathsf{M}(\mathbf{u}, \boldsymbol{\rho}) = \mathsf{L} + \boldsymbol{\rho}^{\mathrm{T}}\mathbf{Cu}\,, \tag{3.72}$$

where $\boldsymbol{\rho}$ is a vector of Lagrangian multipliers of order p[11]. The stationarity point of functional M is sought in the set of independently varied vectors $\mathbf{u}$ and $\boldsymbol{\rho}$ subject to no limitations. The condition of stationarity of M gives the following system of governing equations:

$$\begin{aligned} &\mathbf{Ku} + \mathbf{C}^{\mathrm{T}}\boldsymbol{\rho} = \mathbf{q}\,, \\ &\mathbf{Cu} = \mathbf{0} \qquad , \end{aligned} \tag{3.73}$$

hence the automatic subjection of vector $\mathbf{u}$ to the poly-constraint equations and thus its belonging to the space of kinematically admissible displacements.

The solution of this system of equations gives us values of components of the full displacement vector $\mathbf{u} \in \Re^l$ (with its components in the basis of $\Re^n$), and the vector of Lagrangian multipliers $\boldsymbol{\rho}$, which is a set of generalized stresses conju-

[11] Often in literature they call functional M a Lagrange's functional which, though historically true, seems to us linguistically infelicitous. As the same great name belongs to the original functional L, from this one may derive something awkward like "Lagrange's functional for Lagrange's functional". Therefore we prefer a more neutral term, an auxiliary functional. By this the historical role played by Lagrange is not forgotten nor depreciated in any way because we mention the Lagrange's multiplier rule when building the auxiliary functional.

gated with the imposed poly-constraints by energy in its physical sense [26]. This solution can be written in the following form:

$$\boldsymbol{\rho} = (\mathbf{C}\mathbf{K}^{-1}\mathbf{C}^{\mathrm{T}})^{-1}\mathbf{C}\mathbf{K}^{-1}\mathbf{q}\,, \quad \mathbf{u} = [\mathbf{I} - \mathbf{K}^{-1}\mathbf{C}^{\mathrm{T}}(\mathbf{C}\mathbf{K}^{-1}\mathbf{C}^{\mathrm{T}})^{-1}\mathbf{C}]\mathbf{K}^{-1}\mathbf{q}\,, \tag{3.74}$$

which is feasible according to the accepted convention of the positive definiteness of the external stiffness matrix **K** and rank limitations (3.61) upon the poly-constraint matrix **C**.

So, there are two approaches to the variational description of the behavior of a mechanical system with poly-constraints imposed. Now let's compare these two approaches from various standpoints.

The first approach is based on a reduction of the original space while the second one on its expansion. The first technique decreases the order of the governing system of equations to $n - p$ (which is good), and the second one increases it to $n + p$ (which is hardly good). But the first method breaks the structure of stiffness matrix **K**, and its reconstructed form $\mathbf{K}^*$ loses (or at least deteriorates) its critically important properties such as its low population and weak connectivity (nonzero elements tend to gather around the main diagonal). Together these properties present most attractive features of finite-element systems, so the lack of them is a great pity.

In the second case the structure of the matrix of the equation system that corresponds to the **K** block remains completely the same (pretty well). Though this useful conservatism is to be paid for by additional blocks in the system, those blocks can be shifted to form a bordering.

Let us agree on a tie 1:1 for now and make a break to discuss the unused opportunities of the first round.

For instance, if a software's computational algorithm uses the band structure of the stiffness matrix essentially, then it may prove reasonable to reduce the bandwidth of the stiffness matrix **K** by increasing its order forcedly, involving fictitious nodes the displacements of which would be defined by poly-constraints thus forcing the system of governing equations to be constructed according to our second approach. Details, examples and quantitative estimations can be found in our paper published as recently as a quarter of a century ago [25].

Returning to our comparison of the two approaches, we'd like to note that a true apologist of the fist method would not be slow to notice the old good conservatism from the other side. Thus, the reconstructed stiffness matrix $\mathbf{K}^*$ not only keeps the positive definiteness of the equation system but even enhances it: it can be proved that the spectrum of eigenvalues of matrix $\mathbf{K}^*$ shifts to the right from that of matrix **K**, and its conditionality improves also. But we can parry this heavy blow dealt to the second approach by as many as three considerations. Firstly, one should not exaggerate the disadvantage of losing the positive definiteness of the full matrix of equation system (3.73). First n steps of the Gaussian direct-pass exclusion procedure or any modification thereof deal with a positive definite block actually, the same matrix **K**. The new matrix $(\mathbf{C}\mathbf{K}^{-1}\mathbf{C}^{\mathrm{T}})^{-1}$ formed in the place of the matrix zero is also positive definite (up to the unimportant factor "minus one"). Secondly, the process of forming the $\mathbf{K}^*$ matrix also requires a bit of computational resources, and the algorithm is more complicated. Thirdly, if the user hits upon the idea to find out the generalized stresses in the poly-constraints (you'd better believe he will, we know for sure): hey presto, here you have the components of the vector of Lagrangian multipliers $\mathbf{\rho}$. If you insist on the first approach, then you'll have to do a hard work of extracting these stresses from the equilibrium equations.

So, the game has ended in a draw, and the final choice is up to the referee, the developer who is to decide which approach to use in a particular software implementation. Being only the users (with rights of touch judges at the most), we'd like to have both options available. As usual, it would be well if one of the options was chosen by the software by default, and the user had the right to enforce his choice. It would be even better if the user could choose what to do with every particular poly-constraint in the same design model. If developers doubt that there exist users with exotic needs like this, our answer will be "just give us the program, and you'll see there will be no lack of users"[12].

Thought, it seems that we have agreed on the tie game too early: the same zealous apologist of the first approach draws an ace from his sleeve and everybody understands the third round is to be played.

Well, so far we have been discussing the situation when the external stiffness matrix $\mathbf{K}$ is positive definite. What if it is not the case? One can easily imagine a structure that becomes moveable as soon as its poly-constraints are removed. The simplest example is a plane perfectly rigid body of the first type connected to the ground by three nonparallel hinged bars that do not cross in one point. Stiffness matrix $\mathbf{K}$ becomes degenerate immediately, though this system that includes a perfectly rigid body is kinematically stable. In other words, poly-constraints as well as conventional elements can be *unconditionally necessary.*

In the first approach, the reconstructed stiffness matrix $\mathbf{K}^*$ is positive definite if and only if the whole system together with its poly-constraints is geometrically stable.

The matter is not so easily settled with the second approach. Indeed, the equation system (3.73) includes $(n + p)$ equations with respect to the same number of unknowns. Based on the fact that the rank of matrix $\mathbf{C}$ is equal to the number of its rows, it can be proved that the determinant of the extended $(n + p)\times(n + p)$ matrix

$$\left\|\begin{bmatrix} \mathbf{K} & \mathbf{C}^{\mathrm{T}} \\ \mathbf{C} & \mathbf{0} \end{bmatrix}\right\| \tag{3.75}$$

can be nonzero even if matrix $\mathbf{K}$ is degenerate but its defect does not exceed the rank of matrix $\mathbf{C}$. It means that the system with no additional constraints is moveable, and the geometrical stability of the structure is provided by these additional constraints. But in this case the solution of system (3.73) with matrix (3.75) by the

[12] It seems that this power should be given only to super-experts for whom there is no place even in Table 2.1. Well, let's refer to the authority of a programming guru Lou Grinzo, who said the following on a similar subject: "*If there is nothing else, remember that if you chose to dance with the devil, be sure you know the steps*" [12].

method of Gauss or Cholesky with the exclusion in the natural order of the unknowns is impossible. In order to overcome this computational difficulty, [27] proposes a technique of establishing so-called *null-compliant constraints*, based on modification of auxiliary functional M into the functional

$$\mathsf{N} = \tfrac{1}{2}\,\mathbf{u}^{\mathrm{T}}(\mathbf{K} + \mathbf{C}^{\mathrm{T}}\mathbf{D}_0\mathbf{C})\mathbf{u} - \mathbf{q}^{\mathrm{T}}\mathbf{u} + \boldsymbol{\rho}^{\mathrm{T}}\mathbf{C}\mathbf{u}\,, \tag{3.76}$$

where $\mathbf{D}_0$ is some (arbitrarily chosen) symmetric positive definite matrix of the order p. Adding the product $\mathbf{C}^{\mathrm{T}}\mathbf{D}_0\mathbf{C}$ to the stiffness matrix $\mathbf{K}$ corresponds to the inclusion of an elastic element (its elasticity properties are defined by matrix $\mathbf{D}_0$) in the system to provide the stability of the shape. But, being "parallel" to a perfectly rigid constraint in the system, this element does not deform and thus does not change the distribution of stresses and strains. As a result, the matrix of the system of equations constructed by the modified functional N acquires the following form instead of (3.75):

$$\begin{Vmatrix} \mathbf{K}+\mathbf{C}^{\mathrm{T}}\mathbf{D}_0\mathbf{C} & \mathbf{C}^{\mathrm{T}} \\ \mathbf{C} & \mathbf{0} \end{Vmatrix} \tag{3.77}$$

and, as is shown in [27], block $\mathbf{K} + \mathbf{C}^{\mathrm{T}}\mathbf{D}_0\mathbf{C}$ becomes positive definite. The status quo of the tie game in the competition that we're in has restored.

Actually, we have performed this general comparison of the two competing approaches not in order to pronounce a final sentence whether to reduce or to expand the original vector space, which is hardly possible. In every particular case, depending upon a problem, the final decision can be to adopt either the first or the second approach. Therefore we advocate a software that could operate both the first and the second technique. Though, we are aware that the implementation of either one is not an easy task for software developers.

Returning to the modified functional N, note that actually we are dealing with a family of functionals rather than a single one. This family depends parametrically on the arbitrarily chosen matrix $\mathbf{D}_0$. Hence the natural wish to dispose of this unexpected arbitrariness so as to have more "fun" with it. This will be discussed in the next section.

Equations (3.73) present certainly one of versions of the *mixed* method of analysis of a finite-dimensional system because both force and kinematical parameters are involved as unknowns. Another version of the mixed method can be derived [34] if you notice that the equality below follows from the physical and geometrical equations (3.4)

$$\mathbf{F}^{-1}\mathbf{s} = \mathbf{Q}^{\mathrm{T}}\mathbf{u} + \mathbf{d}\,. \tag{3.78}$$

and add the equilibrium equations to (3.78). The result is the system of equations of the mixed method

$$\begin{aligned} \mathbf{F}^{-1}\mathbf{s} - \mathbf{Q}\mathbf{u} &= \mathbf{0}\,, \\ \mathbf{Q}^{\mathrm{T}}\mathbf{s} &= \mathbf{0} \quad , \end{aligned} \tag{3.79}$$

the structure of which is the same as that of system (3.73) though the total order of this system is $(n + m)$.

3.4 Null Elements

In the previous section we mentioned null constraints, or, more exactly, null elements. What is this and what is it for?

Now let's break the logic of presentment a little bit and try to answer the second question first. The matter is that the realization of the necessity to handle poly-constraints in structural design models, if even the most popular model of a perfectly rigid body only, has been the social control for software developers. But there is a long time until the control gives a result. And the position of waiting is not very popular among software users, as well as among customers of any kind around the world. Therefore the main incentive for creation of null elements has been the natural wish to invent some rag to temporary cover the indecent nakedness of the contemporary software systems until a better day comes. It had to be something to be able to include perfectly rigid bodies in design models, fix nodes in oblique directions with respect to the global coordinate system, merge displacements of nodes, that is, many of the things that actually must have been done by poly-constraints, and all without a slightest encroachment upon the source code of a software. In other words, from the user's standpoint null elements present just a way to cheat a program to have it get out of its groove.

Definition.

Let us refer to a complex element as a *null-compliant* (or *null-stiff*) if it has a zero compliance (stiffness) matrix with respect to some of its degrees of freedom and at the same time consists of simple elements of a finite stiffness (compliance).

As the further presentment will show, a null-compliant element is at the same time a null-stiff element but with respect to its other degrees of freedom. Therefore we will use a single general shorter term to refer to such elements: null elements. A simplest example of a null element is a package of two consecutively connected springs with the stiffness c and $-c$ equal by magnitude but different by sign (Fig. 3.6).

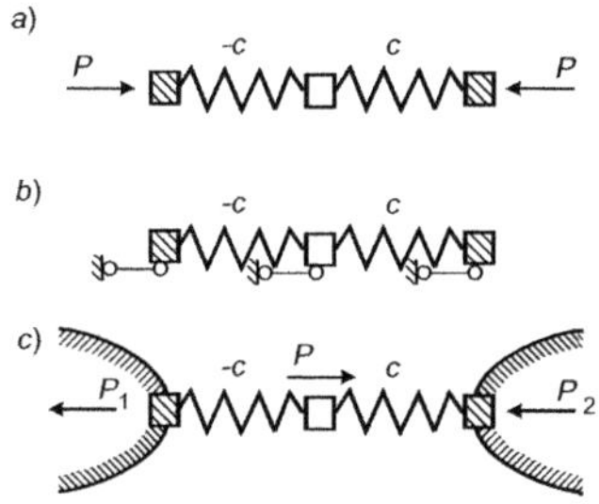

Fig. 3.6. A simplest model of a null-compliant element

It is easy to see that when forces P are applied to the discussed element P (Fig. 3.6 – a) the convergence of the points of application of these forces turns out to be zero; in other words, the element acts as a perfectly rigid one longitudinally. According to the definition given above, the element of interest is a null-compliant element with respect to the convergence of its ends.

If one imposes external constraints on all nodes of this element (Fig. 3.6 – b) and then forces a displacement of the central node, then the reaction of the constraint that corresponds to the displacement of this node, will also be zero. Thus, this element is absolutely compliant with respect to the displacement of the central node, that is, according to the definition it is null-stiff.

Now let us connect two arbitrary nodes of a design model of a structure by the element just discussed, leaving only its central node free from external forces. Then the distance between the nodes connected by this element won't change in the course of the system's deformation and thus the null element will precisely simulate the presence of a perfectly rigid bar.

In practice, users often introduce a bar of a finite but very big rigidity instead of a perfectly rigid bar. Besides the fact that a priori one cannot know well enough what stiffness should be assigned to this bar, this trick is also insecure from the computational standpoint as it deteriorates the conditionality of the system of equations.. In Section 2.4 we already dealt with a similar situation, there with a simple example of a system that consisted of three springs. Then we tried to analyze reasons for possible loss of accuracy of the solution from the mechanical standpoint. Of course, the introduction of a null-element shown in Fig. 3.6 in a design model also has its price because the design model is supplemented with an additional node which is central for the null-element, and thus the order of the governing equation system grows.

Now let us ideally apply an external force P to the central node of the null element, and forces P_1 и P_2 to the nodes of the structure connected by this element, as shown in Fig. 3.6 – c. Let us demand that these forces be mutually balanced, that is, $P = P_1 + P_2$. It is easy to prove that this loading with the forces simulates precisely a dislocation influence upon the system in the form of the convergence of the two nodes of the structure by the distance $d = P/c$. Thus, the null element, as it seems, not only has enabled us to simulate a perfectly rigid bar but also has allowed us to turn a dislocation influence into a force one.

The general idea of constructing null elements out of bars of finite rigidity demonstrated by this simplest example, has enabled a whole series of null elements to be constructed each of which simulates exactly one of versions of poly-constraint equations generated by perfectly rigid bodies. Fig. 3.7 shows 8 types of the bar null elements, and the respective equations of poly-constraints for these null elements are listed in Table 3.1. The same table describes values of external forces that simulate dislocation influences.

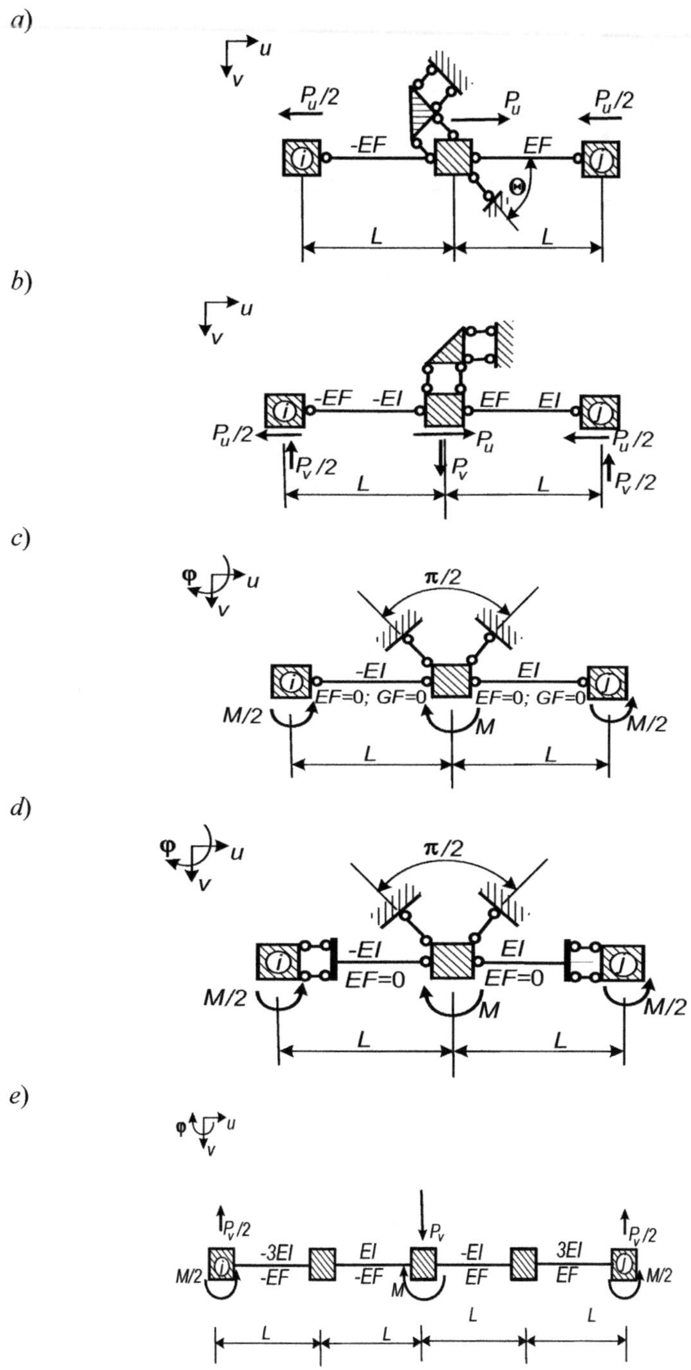
a)
u
v
$P_u/2$
P_u
$P_u/2$
-EF
EF
i
j
Θ
L
L
b)
u
v
-EF -EI
EF EI
i
j
$P_u/2$
P_u
$P_u/2$
$P_v/2$
P_v
$P_v/2$
L
L
c)
φ
u
v
π/2
-EI
EI
EF=0; GF=0
EF=0; GF=0
i
j
M/2
M
M/2
L
L
d)
φ
u
v
π/2
-EI
EI
EF=0
EF=0
i
j
M/2
M
M/2
L
L
e)
φ
u
v
$P_v/2$
P_v
$P_v/2$
-3EI
-EF
EI
-EF
-EI
EF
3EI
EF
i
j
M/2
M
M/2
L
L
L
L

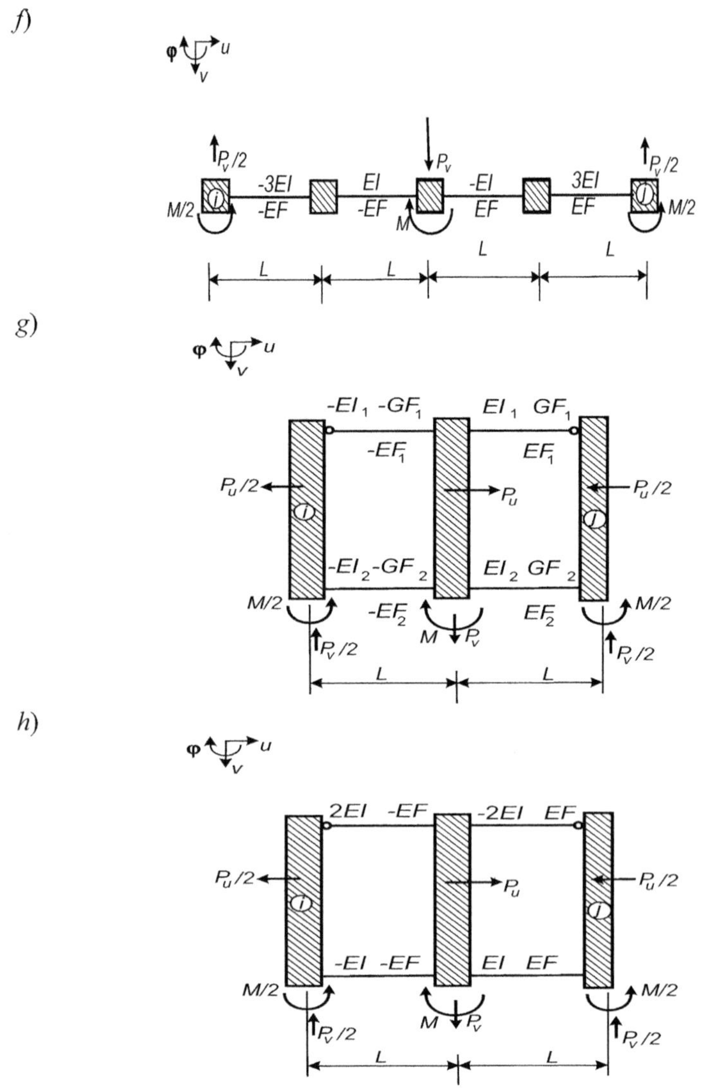

Fig. 3.7. Constructions of null-elements

Together with the notation understood from Fig. 3.7, Table 3.1 uses the following

$$\alpha = 1+12k_1, \qquad k_1 = EI_1/(L^2GF_1), \qquad k_2 = EI_2/(L^2GF_2), \tag{3.80}$$

and the following relation must take place between the rigidities in the null element "*g*":

$$2EI_1/\alpha + EI_2/(1+3k_2) = 0 . \tag{3.81}$$

Table 3.1.

No.	Poly-constraint equations	Dislocations and forces
1	$u_i - u_j = d_u$	$P_u = (EF / L)\, d_u$ $\theta \neq 0$
2	$u_i - u_j = d_u$ $v_i - v_j = d_v$	$P_u = (EF / L)\, d_u$ $P_v = (3EI / L^3)\, d_v$
3	$\theta_i - \theta_j = d_\theta$	$M = (EI / L)\, d_\theta$
4	$\theta_i - \theta_j = d_\theta$	$M = (EI / L)\, d_\theta$
5	$u_i - u_j = d_u$ $v_i - v_j = d_v$ $\theta_i - \theta_j = d_\theta$	$P_u = (EF / L)\, d_u$ $P_v = (3EI / L^3)\, d_v$ $M = (EI / L)\, d_\theta$
6	$u_i - u_j = d_u$ $v_i - v_j + 4L\theta_i = d_v$ $\theta_i - \theta_j = d_\theta$	$P_u = [EF / (2L)]\, d_u$ $P_v = 9EI / (4L^3)(d_v - 2L\, d_\theta)$ $M = -2(EI / L)\, d_\theta$
7	$u_i - u_j = d_u$ $v_i - v_j + 2L\theta_i = d_v$ $\theta_i - \theta_j = d_\theta$	$P_u = (2EF / L)\, d_u$ $P_v = [6EI_1 / (\alpha L^2)](d_v - Ld_\theta)$ $M = [(\alpha - 3)EI_1 / (\alpha L)]\, d_\theta$
8	$u_i - u_j = d_u$ $v_i - v_j + 2L\theta_i = d_v$ $\theta_i - \theta_j = d_\theta$	$P_u = (2EF / L)\, d_u$ $P_v = (6EI / L^2)(d_v - Ld_\theta)$ $M = -(EI / L)\, d_\theta$

Let us give a few illustrative examples to demonstrate some potential applications of various null elements. Fig. 3.8 shows a single-span single-story frame with a perfectly rigid crossbar.

In the condition when the software being used provides no direct specification of rigid bodies, one can use a design model including a null element, "*g*" or "*h*" to one's liking. Null element "*f*" requires three more nodes to be introduced in the design model.

Elements "*g*" and "*h*" have one additional node apiece, but those cannot be involved in the analysis if the software provides an unconditional protection against double connection of the same nodes by different bars. Actually, element "*h*" is a particular case of element "*g*" but uses a classical theory of bars: no shear.

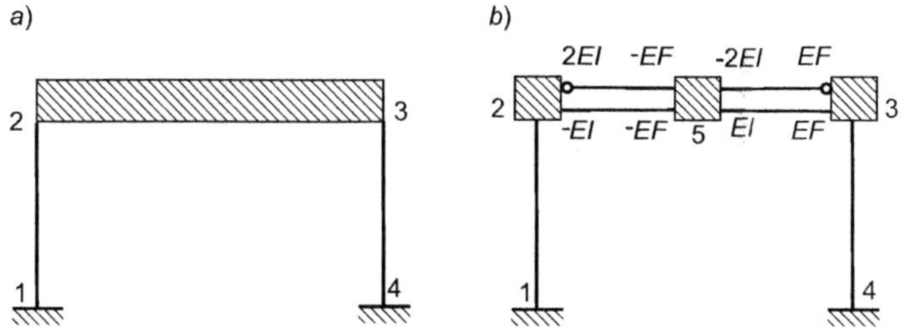

Fig. 3.8. Simulation of a perfectly rigid crossbar by null element "*h*"

The merging of only linear displacements, with the slopes of the two nodes remaining independent, may prove useful for simulating a hinged connection of two groups of bars shown in Fig. 2.16. For this purpose one can use null element "*b*" as it is shown in Fig. 3.9.

Note that the above listed 8 types of null elements do not exhaust all possibilities of constructing new types of such elements in any way. Software users can do it by themselves in order to adjust to capabilities of the software when modeling a particular structure.

Another useful point of the use of null elements is the opportunity of improving the stability of the computation. If originally a system includes elements drastically different in their rigidity properties, this can lead to a quick accumulation of the computation error in the course of solution of the governing equation system of the slope deflection method.

The main idea of fighting this effect by means of null elements will be demonstrated by the example of the same example of the model that consists of three consecutive springs (Fig. 2.8). Fig. 3.10 shows an equivalent model where there are frame posts instead of the edge springs, and a weakly compressive crossbar instead of the central spring.

Introducing the notation

$$\alpha = \frac{3EI_c}{h^3}\frac{l}{EF_p}, \qquad k = \frac{EF_p}{l}, \tag{3.82}$$

we can represent the stiffness matrix of the system shown in Fig. 3.10, *a* by formula (2.2). If the crossbar is very stiff, the dimensionless parameter α will be a small value, and we find ourselves exactly in the unpleasant situation with an almost moveable mechanical system which was considered in Section 2.3.

One of means available to software users to influence the conditionality of the governing system of linear algebraic equations is the replacement of very stiff elements by *almost null-compliant* elements. An almost null-compliant element is different from a full null-compliant element by a small disturbance ε in the rigidities of the positive and/or negative parts of the null element's construction. Parameter ε can be calculated from the condition of equal compliance of the very stiff element to be simulated and the almost null-compliant element.

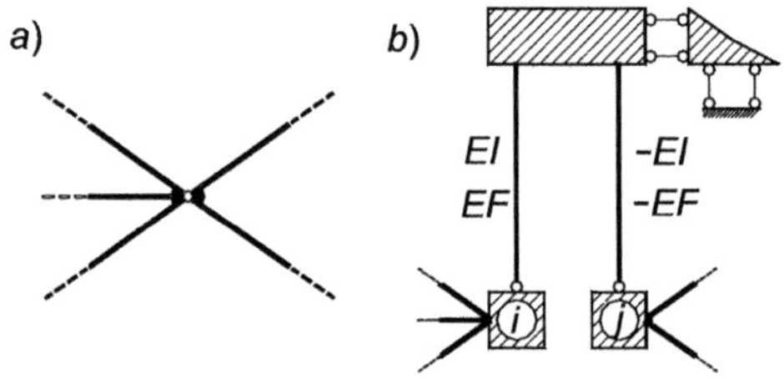

Fig. 3.9. Simulation of a hinged connection of two nodes

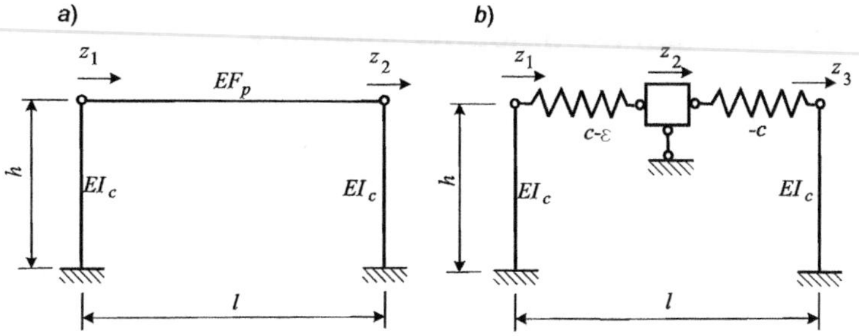

Fig. 3.10. An example of using an almost null-compliant element

In particular, in the case that we are considering the model with an almost null-compliant element is shown in Fig. 3.10, *b*. The small parameter ε will be calculated from the condition of the equal compliance of the frame's crossbar and the almost null-compliant element

$$1/(c-\varepsilon)+1/(-c)=1/k\,, \tag{3.83}$$

hence

$$\varepsilon=c^2/(c+k)\,,\qquad c-\varepsilon=ck/(c+k)\,. \tag{3.84}$$

Assuming $c = k\alpha$ (that's where we happen to use the arbitrariness discussed at the end of the previous section!), we obtain the stiffness matrix for the system shown in Fig. 3.10, *a*:

$$\mathbf{K}=\frac{k\alpha}{1+\alpha}\begin{bmatrix}2+\alpha & -1 & 0\\ -1 & -\alpha & 1+\alpha\\ 0 & 1+\alpha & 0\end{bmatrix}, \tag{3.85}$$

and it may be demonstrated that at any however small values of parameter α the conditionality number of this matrix $|\lambda_{max}| / |\lambda_{min}| < 4$, which is quite satisfactory.

We do not discuss here mathematical aspects that arise in the investigation of null elements. The interested reader can refer to our paper [27]. Let us just note that the appropriate analysis will lead, in particular, to a conclusion of the mathematical equivalency of the analysis of a mechanical system that includes null elements and the variational formulation of the same problem based on a family of modified functionals N. The arbitrarily chosen matrix $\mathbf{D}_0$ for functional N is nothing but a physical characteristics of the respective null element. The general recommendation for software users is that rigidities of null elements should be assigned so that they be values of approximately the same order of magnitude as the rigidities of their surrounding elements of the basic structure.

3.5 Geometrical Nonlinearity. Stability

When we talk about geometrically nonlinear problems, we must define strictly what level of the geometrical nonlinearity we mean. For a CAD software product's documentation to be clear enough (honest, should be added too), it must contain strict explicit unambiguous statements as to what level of the nonlinear analysis is achievable by this software[13]. Without stating it, one may create a dangerous illusion of a wider applicability of the software and declare in the documentation its ability to solve geometrically nonlinear problems without an explicit indication of applicability limitations of a nonlinear theory that is involved.

We need not invent a bicycle here, because a classic by Novozhilov [17] presented a clearly stated, consistent, and now received classification of geometrically nonlinear formulations of mechanical problems about half a century ago. Let us distinguish the following four floors of geometrically nonlinear problem statements in finite-dimensional systems.

3.5.1 Four Floors of Geometrically Nonlinear Problem Statements

The zero, ground floor will naturally shelter all problems where one can totally neglect any geometrically nonlinear effects. In other words, the ground floor is where geometrically linear formulations live.

Before we go up even a single step, it is worth noticing that the geometrical nonlinearity can penetrate to the governing equations of a problem through two channels: the equations that connect displacements with strains, and the equilibrium equations. Therefore in the first floor we will keep problems where only one of the channels will be open to the geometrical nonlinearity, namely one through the equilibrium equations.

The first floor of the geometrical nonlinearity will give a shelter to a weakest version of the geometrically nonlinear theory. In this theory one assumes the equilibrium equations to be composed for the deformed state of a system. As for the strain-displacement relation, this one is assumed to be linear in the first floor of the geometrically nonlinear problems. Sometimes in structural mechanics they call the problems of the first floor of the geometrical nonlinearity a *deformed geometry analysis*. English-speaking authors use the name of a *second-order theory*. Also, the term of *linearized equations* is usable. For constructions used in the structural design practice, the first floor of the geometrical nonlinearity comprises a pretty wide class of problems, primarily, the analysis of bar systems bending moments in which are calculated with corrections introduced by longitudinal forces.

Stepping upstairs, we find ourselves in the second floor of the hierarchy of geometrically nonlinear statements. The difference from the first floor is that the second channel opens for the geometrical nonlinearity: through the geometrical

[13] Again we quote from Lou Grinzo: "*Documentation is the great multiplier for many tools*" [12].

equations that connect displacements and strains. The essential difference from the first floor is that orders of smallness of the strains and slopes are distinguished. Squares of slopes of elements of the model are considered to be quantities of the same order of magnitude as the strain components which, in their turn, are infinitesimal with respect to one. The slopes themselves are also small comparing to one. A characteristic example of problems of this nonlinearity floor is the theory of flexible plates based on von Karman equations [37].

The third floor of geometrical nonlinearity features problems where the strains are small comparing to one while the slopes cannot be assumed to be such. Structures that include flexible ropes/cables are good examples of this type of geometrically nonlinear problems. In their turn, problems of the third nonlinearity floor can be ranked by additional simplifying assumptions. For example, the same theory of flexible cables includes an assumption of flatness of the initial and final configurations of a cable in the course of its loading (a cable of small deflection) that lets one [16] substantially simplify both the mathematical formulation of the problem and, indirectly, its techniques of solution, without going down to the second floor. If you like, problems of this type can be placed in the floor "two and a half", or on an entryway[14].

Finally, at the top of this hierarchy there are most complicated problems: those of the fourth floor of the geometrical nonlinearity where even relative strains cannot be treated as small values comparing to one. Analysts have to storm these hills when they need to analyze constructions made of rubber and the like materials.

For discrete design models discussed in this chapter, the geometrical nonlinearity by no means affects matrices $\mathbf{\Gamma}_1$, $\mathbf{\Gamma}_2$ and $\mathbf{\Gamma}_3$ used in formulae (3.5) because these matrices reflect only topological information about the design model, and the topology is insensitive to relations between orders of smallness of particular strain components. This is critically important because a fortunate possibility follows from it: consider the geometrical nonlinear effects only at the level of a particular finite element and transit to parameters of the whole system using the same formulae (3.5).

3.5.2 Geometrical Nonlinearity in Truss-type Bars

Let's turn to a simplest finite element example to illustrate the four-story geometrical nonlinearity building described above — a truss-type bar (this name will refer to a bar with hinged connections at both ends and no off-nodal loading applied to the bar). The plane case is under consideration for the sake of simplicity.

Let the bar occupy the position MN in its original state, with its beginning in node M and its end in node N. Here the *original* state refers to a state of the bar acquired by it immediately after a pre-tension (dislocation) by the value of d. After

[14] This comparison extorts our recalling the expression "staircase wit" that describes a belated response in a discussion, with the only difference that a wit of this kind is quite constructive in our case.

the respective deformation, the bar, while remaining rectilinear, will occupy the position M*N* as shown in Fig. 3.11. Let's introduce the following notation:

L_x, L_y are projections of the bar onto respective axes in its original state;
L and L^* is the length of the bar in its original and final state, respectively;
u_M, u_N, v_M, v_N are displacements of nodes M and N in the directions of axes x and y, respectively;
e_x, e_y are elongations of the projections of the original bar state onto the coordinate axes;
Δ is the full elongation of the bar;
N is the longitudinal stress in the bar, positive when in tension.

Let us consider a set of vectors **s**, **Δ**, **d**, **u**, **p**, related to a particular bar (element), where subscript "*e*" previously used in formulae (3.1) – (3.3) is omitted for shortness. Then we have

$$\mathbf{s} = \|[N]\|, \quad \mathbf{\Delta} = \|[\Delta]\|, \quad \mathbf{d} = \|[\mathrm{d}]\|, \quad \mathbf{u}^{\mathrm{T}} = \|[\, u_{\mathrm{M}}\, v_{\mathrm{M}}\, u_{\mathrm{N}}\, v_{\mathrm{N}}]\|,$$

$$\mathbf{p}^{\mathrm{T}} = \|[P_{\mathrm{Mx}} P_{\mathrm{My}} P_{\mathrm{Nx}} P_{\mathrm{Ny}}]\|. \tag{3.86}$$

From consideration of the geometry (Fig. 3.11) we obtain

$$e_x = u_{\mathrm{N}} - u_{\mathrm{M}}, \qquad \mathrm{e_y} = v_{\mathrm{N}} - v_{\mathrm{M}}, \qquad \Delta = (L^* - L) + d, \tag{3.87}$$

$$L^2 = L_x^{\,2} + L_y^{\,2}, \qquad L^{*2} = (L_x + e_x)^2 + (L_y + e_y)^2. \tag{3.88}$$

Let λ_x, λ_y, λ_x^*, λ_y^* be the direction cosines of segments MN and M*N* with respect to the fixed system of axes (x, y) (see Fig. 3.11), i.e.

$$\lambda_x = \cos\alpha = L_x / L, \qquad \lambda_y = \sin\alpha = L_y / L, \tag{3.89}$$

$$\lambda_x^* = \cos\alpha^* = (L_x + e_x)/ L^*, \qquad \lambda_y^* = \sin\alpha^* = (L_y + e_y)/ L^*. \tag{3.90}$$

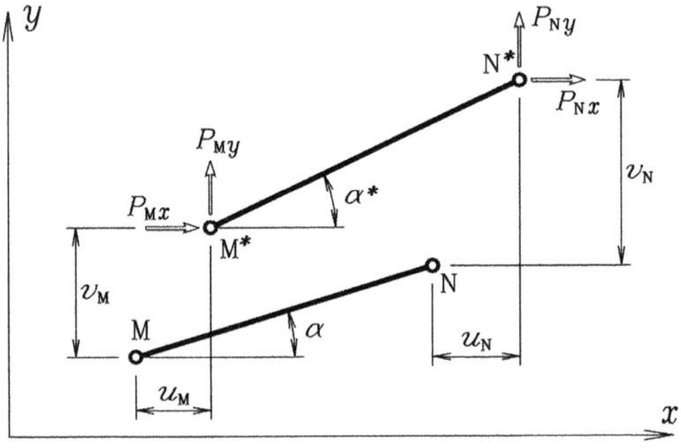

Fig. 3.11. Depiction of displacements of a truss-type bar

Let's introduce the following important geometrical parameters which we will need further below:

$$e = \lambda_x e_x + \lambda_y e_y, \qquad \omega = (\lambda_x e_y - \lambda_y e_x)/L, \qquad \varepsilon = e + ½\,(e^2 + \omega^2 L^2)/L. \tag{3.91}$$

The geometrical meaning of parameter e is evident: it is the projection of the difference of displacements of the bar's ends onto the direction of the bar's axis in its non-deformed state. As for parameter ω, we will see later that it is a slope of the bar provided it is much smaller than one and much smaller than the relative quantity of e/L.

Now let's represent parameter e in the matrix form as

$$e = \mathbf{Q}^{\mathrm{T}}\mathbf{u}, \qquad \mathbf{Q}^{\mathrm{T}} = \|[-\lambda_x\ -\lambda_y\ \lambda_x\ \lambda_y]\|, \tag{3.92}$$

where $\mathbf{Q}$ is the matrix of the equilibrium equations in the geometrically linear formulation.

Note that we can introduce a rotation matrix $\mathbf{\Omega}$,

$$\mathbf{\Omega} = \left\|\begin{bmatrix} 0 & 0 & 0 & 1 \\ 1 & 0 & 0 & 0 \\ 0 & 1 & 0 & 0 \\ 0 & 0 & 1 & 0 \end{bmatrix}\right\|, \tag{3.93}$$

and have $(\mathbf{\Omega Q})^{\mathrm{T}} = \|[\lambda_y\ -\lambda_x\ -\lambda_y\ \lambda_x]\|$, so the slope parameter ω can be represented in the matrix form too:

$$\omega L = \mathbf{Q}^{\mathrm{T}}\mathbf{\Omega}^{\mathrm{T}}\mathbf{u}, \tag{3.94}$$

which will be convenient for us later.

Proceeding from determinative relations (3.87), (3.88) and (3.91), we can derive a formula to connect the bar's strain Δ and displacements of its ends expressed via the geometrical parameters just introduced, namely:

$$(\Delta - d)\left[1 - \frac{\Delta - d}{2L}\right] = \varepsilon. \tag{3.95}$$

The correctness of the formula can be easily proved by direct substitution. Thus, the left part of (3.95) can be turned into the following by a chain of transformations:

$$(L^* - L)\left[1 + \frac{L^* - L}{2L}\right] = \frac{L^{*2} - L^2}{2L} =$$

$$= \frac{(L_x + e_x)^2 + (L_y + e_y)^2 - L_x^2 - L_y^2}{2L} = \lambda_x e_x + \lambda_y e_y + \frac{e_x^2 + e_y^2}{2L},$$

and this is also the form to which the expression for ε from (3.91) can be transformed, seeing that $\lambda_x^2 + \lambda_y^2 = 1$.

From formula (3.95) an expression for the bar's strain follows immediately:

$$\Delta = L(\sqrt{1+2\varepsilon / L} - 1) + d . \tag{3.96}$$

Note that we have used no simplification assumptions regarding the comparison of smallness orders of quantities included in the analysis to derive formula (3.96). Therefore formula (3.96) is the very nonlinear equation that establishes a relation between strain Δ and components of the displacement vector **u** and lives in the highest — fourth — floor of the geometrical nonlinearity.

To go down one floor, we will assume that the relative strain Δ/L as well as the relative dislocation d/L are very small quantities comparing to one, that is

$$(\Delta - d)/L << 1 , \tag{3.97}$$

which enables us to replace the expression that stands in the brackets in the left part (3.95) with one. This leads us to a geometrical equation living in the third floor of the geometrical nonlinearity:

$$\Delta = \varepsilon + d = e + ½\, (e^2 + \omega^2 L^2)/L + d . \tag{3.98}$$

Note that parameters e and ω are linearly dependent on the displacements — see formulae (3.92) and (3.94). Therefore, from the third floor down, strain Δ consists of a part linear with respect to the displacement vector components and a quadratic part. But before we go down to the next floor below, let us discuss the geometrical sense of parameter ω.

From Fig. 3.11 one can see that angle β at which the bar rotates after its ends displace can be determined by the difference of angles α^* and α, i.e.

$$\beta = \alpha^* - \alpha . \tag{3.99}$$

As

$$\mathrm{tg}\alpha^* = (\lambda_y L + e_y)/(\lambda_x L + e_x), \qquad \mathrm{tg}\alpha = \lambda_y / \lambda_x , \tag{3.100}$$

we can use the formula for tangent of difference of two angles and relation $\lambda_x^{\,2} + \lambda_y^{\,2} = 1$, make some simple transformations and thus obtain

$$\mathrm{tg}\beta = (\mathrm{tg}\alpha^* - \mathrm{tg}\alpha)/(1 + \mathrm{tg}\alpha\ \mathrm{tg}\alpha^*) = \omega L/(L + e) . \tag{3.101}$$

At small rotation angles β, their tangents can be equated to the angles themselves, and, seeing that e is small comparing to L, we have

$$\beta = \omega , \tag{3.102}$$

that is, parameter ω presents the angle of rotation of the bar.

Returning to formula (3.98) and going down to the second floor, let's assume that the rotation angle ω, though itself is a small value comparing to one, is not small enough to be neglected in comparison with the relative parameter e/L. More exactly, we will assume these quantities to be of the same order of smallness and both much smaller than one

$$e/L \approx \omega^2 << 1 . \qquad (3.103)$$

As a result, instead of formula (3.98) in the second floor of geometrical nonlinearity, we have the following:

$$\Delta = e + ½\, \omega^2 L + d . \qquad (3.104)$$

Finally, in the first and ground floor we find linear relationships in the equations of the strain-displacement dependence:

$$\Delta = e + d , \qquad (3.105)$$

which is the same as the geometrical equation (3.2) in the linear formulation after taking into account (3.92).

Now let's pay our attention to the equilibrium equations used in all floors of the building that hosts the geometrical nonlinearity. In the most general case, we can obtain the following by writing these equations for the deformed state of the bar:

$$\mathbf{Q}^*\mathbf{s} = \mathbf{p} , \qquad \mathbf{Q}^{*\mathrm{T}} = \|[-\lambda^*_x \; -\lambda^*_y \; \lambda^*_x \; \lambda^*_y]\| , \qquad (3.106)$$

and we have only to find expressions for the direction cosines of the bar's axis in its deformed state.

Let's use formulae (3.90) and (3.99). Then we have:

$$\lambda^*_x = \cos(\alpha+\beta) = \lambda_x \cos\beta - \lambda_y \sin\beta,$$

$$\lambda^*_y = \sin(\alpha+\beta) = \lambda_y \cos\beta + \lambda_x \sin\beta . \qquad (3.107)$$

Sine and cosine of the bar's rotation angle β that figure in formulae (3.107) can be expressed through tangent of the same angle defined by formula (3.101). This leads us to

$$\cos\beta = 1 / \sqrt{1 + \mathrm{tg}^2\beta} = 1 / \sqrt{1 + \omega^2 L^2/(L+e)^2} ,$$

$$\sin\beta = \mathrm{tg}\beta / \sqrt{1 + \mathrm{tg}^2\beta} = [1/\sqrt{1 + \omega^2 L^2/(L+e)^2}\;]\, \omega L/(L+e) , \qquad (3.108)$$

so that

$$\lambda^*_x = [\lambda_x - \lambda_y \omega L/(L+e)] / \sqrt{1 + \omega^2 L^2/(L+e)^2} ,$$

$$\lambda^*_y = [\lambda_y + \lambda_x \omega L/(L+e)] / \sqrt{1 + \omega^2 L^2/(L+e)^2} . \qquad (3.109)$$

Formulae (3.109) that express the direction cosines of the bar's axis in its deformed state are accurate because no assumptions of smallness of the strain parameters have been used to derive them. Therefore these formulae are true in the topmost — fourth — floor of the geometrical nonlinearity building.

It seems that in the third floor of the geometrical nonlinearity we might assume

$$L/(L+e) \approx 1 , \qquad (3.110)$$

but actually we cannot do it because the smallness of relative strains Δ/L comparing to one does not imply the smallness of ratio e/L comparing to one. For example, if the bar rotates by 90° and keeps its length unchanged, then $\Delta/L = 0$, while $e/L = 1$. Therefore the third floor of the geometrical nonlinearity must inherit the full expressions for direction cosines λ^*_x and λ^*_y from the fourth floor.

In the second floor, estimates (3.103) and (3.110) are true, that's why we can derive the following from (3.109) by omitting quantities of a higher order of smallness

$$\lambda^*_x = \lambda_x - \lambda_y\omega\,, \qquad \lambda^*_y = \lambda_y + \lambda_x\omega\,. \tag{3.111}$$

In spite of the fact that slopes ω are small values even in the second floor, we must keep formulae (3.111) intact when going down to the first floor. For example, in the first formula of (3.111), the smallness of the slopes does not imply the smallness of product $\lambda_y\omega$ comparing to λ_x. Only after descending to the zero level we come to linear equilibrium equations

$$\lambda^*_x = \lambda_x\,, \qquad \lambda^*_y = \lambda_y\,, \tag{3.112}$$

that is,

$$\mathbf{Q}^* = \mathbf{Q}\,. \tag{3.113}$$

Now, using the representations of matrices $\mathbf{Q}$ and $\mathbf{\Omega}$ according to (3.92) and (3.93), we are going to obtain an expression for strain Δ in the third floor of the geometrical nonlinearity in the matrix form. Thus we have

$$e^2 = \mathbf{u}^{\mathrm{T}}\mathbf{Q}\mathbf{Q}^{\mathrm{T}}\mathbf{u}\,, \qquad \omega^2L^2 = \mathbf{u}^{\mathrm{T}}\mathbf{\Omega}\mathbf{Q}\mathbf{Q}^{\mathrm{T}}\mathbf{\Omega}^{\mathrm{T}}\mathbf{u}\,. \tag{3.114}$$

So, in the matrix form, in the third floor of the geometrical nonlinearity, strain Δ defined by formula (3.98) will present a sum of a linear and a quadratic forms of the displacements

$$\Delta = \mathbf{Q}^{\mathrm{T}}\mathbf{u} + \frac{1}{2L}\mathbf{u}^{\mathrm{T}}\mathbf{G}\mathbf{u} + d\,, \tag{3.115}$$

where

$$\mathbf{G} = \mathbf{Q}\mathbf{Q}^{\mathrm{T}} + \mathbf{\Omega}\mathbf{Q}\mathbf{Q}^{\mathrm{T}}\mathbf{\Omega}^{\mathrm{T}} = \begin{bmatrix} 1 & 0 & -1 & 0 \\ 0 & 1 & 0 & -1 \\ -1 & 0 & 1 & 0 \\ 0 & -1 & 0 & 1 \end{bmatrix}. \tag{3.116}$$

Respectively, in the second floor we have the following derived from (3.104):

$$\Delta = \mathbf{Q}^{\mathrm{T}}\mathbf{u} + \frac{1}{2L}\mathbf{u}^{\mathrm{T}}\mathbf{\Omega}\mathbf{Q}\mathbf{Q}^{\mathrm{T}}\mathbf{\Omega}^{\mathrm{T}}\mathbf{u} + d\,. \tag{3.117}$$

Also, it is easy to see from (3.111), that equilibrium matrix $\mathbf{Q}^*$ in the first and second floors of the geometrical nonlinearity depends on displacements $\mathbf{u}$ linearly, and

$$\mathbf{Q}^* = \mathbf{Q} + \boldsymbol{\Omega}\mathbf{Q}\omega = \mathbf{Q} + \frac{1}{L}\boldsymbol{\Omega}\mathbf{Q}\mathbf{Q}^{\mathrm{T}}\boldsymbol{\Omega}^{\mathrm{T}}\mathbf{u}\,. \tag{3.118}$$

Now let's gather all obtained results together, placing them in Table 3.2 with the indication of the respective floor of geometrical nonlinearity.

Table 3.2.

Floor	Small quantities	Geometrical equations / Parameters of equilibrium equations
		Geometrical equations
		Parameters of equilibrium equations
IV	—	$\Delta = L(\sqrt{1+2\varepsilon/L}-1)+d$
		$\lambda^*_x = [\lambda_x - \lambda_y\omega L/(L+e)]/\sqrt{1+\omega^2L^2/(L+e)^2}$ $\lambda^*_y = [\lambda_y + \lambda_x\omega L/(L+e)]/\sqrt{1+\omega^2L^2/(L+e)^2}$
III	$\dfrac{\Delta-d}{L} << 1$	$\Delta = e + 0{,}5(e^2+\omega^2L^2)/L + d$ $\Delta = \mathbf{Q}^{\mathrm{T}}\mathbf{u} + \frac{1}{2L}\mathbf{u}^{\mathrm{T}}\mathbf{G}\mathbf{u} + d$
		$\lambda^*_x = [\lambda_x - \lambda_y\omega L/(L+e)]/\sqrt{1+\omega^2L^2/(L+e)^2}$ $\lambda^*_y = [\lambda_y + \lambda_x\omega L/(L+e)]/\sqrt{1+\omega^2L^2/(L+e)^2}$
II	$e/L<<1$ $\omega^2 \approx e/L$	$\Delta = e + 0{,}5\omega^2L + d$ $\Delta = \mathbf{Q}^{\mathrm{T}}\mathbf{u} + \frac{1}{2L}\mathbf{u}^{\mathrm{T}}\boldsymbol{\Omega}\mathbf{Q}\mathbf{Q}^{\mathrm{T}}\boldsymbol{\Omega}^{\mathrm{T}}\mathbf{u} + d$
		$\lambda^*_x = \lambda_x - \lambda_y\omega,\quad \lambda^*_y = \lambda_y + \lambda_x\omega$ $\mathbf{Q}^* = \mathbf{Q} + \frac{1}{L}\boldsymbol{\Omega}\mathbf{Q}\mathbf{Q}^{\mathrm{T}}\boldsymbol{\Omega}^{\mathrm{T}}\mathbf{u}$
I	$\omega^2 << e/L$	$\Delta = e + d$ $\Delta = \mathbf{Q}^{\mathrm{T}}\mathbf{u} + d$
		$\lambda^*_x = \lambda_x - \lambda_y\omega,\quad \lambda^*_y = \lambda_y + \lambda_x\omega$ $\mathbf{Q}^* = \mathbf{Q} + \frac{1}{L}\boldsymbol{\Omega}\mathbf{Q}\mathbf{Q}^{\mathrm{T}}\boldsymbol{\Omega}^{\mathrm{T}}\mathbf{u}$
0	$\omega << 1$	$\Delta = e + d$ $\Delta = \mathbf{Q}^{\mathrm{T}}\mathbf{u} + d$
		$\lambda^*_x = \lambda_x,\quad \lambda^*_y = \lambda_y$ $\mathbf{Q}^* = \mathbf{Q}$

3.5.3 Geometrically Nonlinear Equations in Variations

Generally, after agreeing on the level of achievability, we could build all equations as nonlinear first, then subject these equations to a discretization, and then try to find a method to solve them. Though, we will follow another way. With subsequent stepwise procedures for nonlinear problems in mind, we will construct the needed equations in variations (increments) for finite-dimensional systems at once. Or, more exactly, we will find out the structure of geometrically nonlinear equations in variations that is common for all discrete systems, without turning to particular types of finite elements.

So, let the displacements of nodes **u** that arise in the course of the loading are not small enough to treat the direction cosines and other similar geometrical parameters (meaning slopes or rotations) that figure in the coefficients of the equilibrium equations as the same before and after the deformation. In this case we need to involve equations that take into account geometrical nonlinear effects.

Let some (initial) configuration of the system described by nodal displacements $\mathbf{u}_0$ be known. These displacements can be zero as well, if this reference point is convenient. In this configuration the following equilibrium conditions hold

$$\mathbf{Q}(\mathbf{u}_0)\mathbf{s}_0 = \mathbf{p}_0 \tag{3.119}$$

between initial nodal loads $\mathbf{p}_0$ and initial stresses $\mathbf{s}_0$. Let now the load acquire an increment $\delta\mathbf{p}$, which shall correspond to stress increment $\delta\mathbf{s}$ and displacement increment $\delta\mathbf{u}$. In the disturbed state, the equilibrium equations will look like

$$[\mathbf{Q}(\mathbf{u}_0) + \delta\mathbf{Q}](\mathbf{s}_0 + \delta\mathbf{s}) = \mathbf{p}_0 + \delta\mathbf{p}\,, \tag{3.120}$$

which will change to the following after removal of the parentheses and allowing for (3.119)

$$\delta\mathbf{Q}\mathbf{s}_0 + \mathbf{Q}(\mathbf{u}_0)\delta\mathbf{s} + \delta\mathbf{Q}\delta\mathbf{s} = \delta\mathbf{p}\,. \tag{3.121}$$

If components of the vector of additional displacements $\delta\mathbf{u}$ are relatively small (though require the nonlinear analysis to be performed), then we can compose the equations in variations at the level of the second floor of the geometrically nonlinear problem statement. In this case product $\delta\mathbf{Q}\mathbf{s}_0$ can be represented as a linear function of the displacement increment [18, 19], which leads us to the equations

$$\mathbf{Q}(\mathbf{u}_0)\delta\mathbf{s} + \mathbf{T}(\mathbf{s}_0)\delta\mathbf{u} + \delta\mathbf{Q}\delta\mathbf{s} = \delta\mathbf{p}. \tag{3.122}$$

For example, for a truss-type bar in a plane problem, we can assign the following on the basis of Table 3.2

$$\mathbf{Q}^{\mathrm{T}}(\mathbf{u}_0) = \|[-\lambda_x \; -\lambda_y \; \lambda_x \; \lambda_y]\|\,,$$

$$\mathbf{T}(\mathbf{s}_0)\,\delta\mathbf{u} = \|[\lambda_y \; -\lambda_x \; -\lambda_y \; \lambda_x]\|^{\mathrm{T}}\,\delta\omega\, N_0 = \frac{N_0}{L}\,\boldsymbol{\Omega}\mathbf{Q}\mathbf{Q}^{\mathrm{T}}\boldsymbol{\Omega}^{\mathrm{T}}\delta\mathbf{u}\,, \tag{3.123}$$

where λ_x and λ_y are direction cosines of the bar in the initial configuration of the system, and N_0 is an initial stress in the bar.

The dependence between the element's strain increments $\delta\boldsymbol{\Delta}$ and the increments of its nodal displacements will be defined as follows:

$$\delta\boldsymbol{\Delta} = \mathbf{Q}^{\mathrm{T}}(\mathbf{u}_0)\,\delta\mathbf{u} + \delta\mathbf{Q}^{\mathrm{T}}\delta\mathbf{u}\,. \tag{3.124}$$

Actually, relations (3.124) are varied geometrical equations of the second nonlinearity floor. For example, for a truss-type bar we have

$$\delta\boldsymbol{\Delta} = \delta e + \omega L\,\delta\omega = \mathbf{Q}^{\mathrm{T}}(\mathbf{u}_0)\,\delta\mathbf{u} + \mathbf{u}_0{}^{\mathrm{T}}\boldsymbol{\Omega}\mathbf{Q}\mathbf{Q}^{\mathrm{T}}\boldsymbol{\Omega}^{\mathrm{T}}/L\;\delta\mathbf{u}\,, \tag{3.125}$$

so, in this case

$$\delta \mathbf{Q}^T = \mathbf{u}_0{}^T \boldsymbol{\Omega} \mathbf{Q} \mathbf{Q}^T \boldsymbol{\Omega}^T / L \,. \tag{3.126}$$

Substituting (3.124) to physical equations (3.1), which have the following form in displacements:

$$\delta \mathbf{s} = \mathbf{F} \delta \boldsymbol{\Delta} \,, \tag{3.127}$$

and then introducing the derived expressions for δs into (3.122), we have:

$$[\mathbf{QFQ}^T + \mathbf{T}(\mathbf{s}_0) + \mathbf{QF}\delta\mathbf{Q}^T + \delta\mathbf{QFQ}^T + \delta\mathbf{QF}\delta\mathbf{Q}^T]\delta\mathbf{u} = \delta\mathbf{p} \,. \tag{3.128}$$

In other designations these equations can be written as

$$[\mathbf{K}_0 + \mathbf{K}_1(\mathbf{s}_0) + \mathbf{K}_2(\mathbf{u}_0)]\delta\mathbf{u} = (\mathbf{K}_0 + \mathbf{K}_G)\delta\mathbf{u} = \delta\mathbf{p}. \tag{3.129}$$

Here it is clearly seen that the conventional stiffness matrix $\mathbf{K}_0 = \mathbf{QFQ}^T$ acquires an increment in the form of the matrix of initial stresses $\mathbf{K}_1(\mathbf{s}_0) = \mathbf{T}(\mathbf{s}_0)$, linearly dependent on stresses in the system before the load is incremented, plus the matrix of initial rotations

$$\mathbf{K}_2(\mathbf{u}_0) = \mathbf{QF}\delta\mathbf{Q}^T + \delta\mathbf{QFQ}^T + \delta\mathbf{QF}\delta\mathbf{Q}^T,$$

that includes not more than degree two of the displacements. Equations (3.129) do not include the matrix of initial loading by J.Oden [18] that makes corrections to the load (the typical example that shows why sometimes the load needs to be corrected, is the case of hydrostatic pressure of liquid: when the elementary plane rotates, it causes the load to change). The non-allowance for the initial loading matrix means that we assume the load to be independent from the system's configuration, or we think that appropriate corrections have been already made when forming the load increment vector $\delta\mathbf{p}$. This approach is used because the direct allowance for the initial loading matrix makes the coefficients of the equations asymmetric which is sometimes inconvenient. Matrix $\mathbf{K}_G$ is conventionally called a *geometrical stiffness matrix*.

Let's consider a system equilibrated in the state for which equations (3.119) hold and write equalities to determine the change of the stresses and displacements during trial loading by forces proportional to a single parameters, that is, when the load is incremented as $\lambda\delta\mathbf{p}$. Denoting the respective changes in the stresses and displacements as $\lambda\delta\mathbf{s}$ and $\lambda\delta\mathbf{u}$, we have:

$$[\mathbf{K}_0 + \mathbf{K}_1(\mathbf{s}_0 + \lambda\,\delta\mathbf{s}) + \mathbf{K}_2(\mathbf{u}_0 + \lambda\,\delta\mathbf{u})]\,\delta\mathbf{u} = \delta\mathbf{p} \,. \tag{3.130}$$

Now let's use the new stiffness matrix to find the critical value of the load $\mathbf{p}_{cr} = \mathbf{p}_0 + \delta\mathbf{p}$. Expanding $\mathbf{K}_1$ and $\mathbf{K}_2$ from (3.130) into series and keeping terms linearly dependent on λ, we obtain the equation:

$$\{[\mathbf{K}_0 + \mathbf{K}_1(\mathbf{s}_0) + \mathbf{K}_2(\mathbf{u}_0)] + \lambda\,[\mathbf{K}_1(\delta\mathbf{s}) + \mathbf{K}_2(\mathbf{u}_0,\delta\mathbf{u})]\}\delta\mathbf{U} = \mathbf{0} \,. \tag{3.131}$$

or

$$[\mathbf{K}_0 + \lambda\,\Delta\mathbf{K}_G(\delta\mathbf{s},\mathbf{u}_0,\delta\mathbf{u})]\delta\mathbf{U} = \mathbf{0} \,, \tag{3.132}$$

where $\delta\mathbf{U}$ is an eigenvector that describes a mode of buckling of the system, and

$\Delta\mathbf{K}_G$ denotes the matrix multiplier that stands in the brackets to the right from λ in equation (3.131).

The solution of eigenvalue problem (3.132) is based on a linear approximation in the vicinity of the shifted state of equilibrium. Therefrom we can derive the linear buckling problem assuming all initial stresses and displacements to be zero. If further we omit the quadratic terms in the matrix of initial displacements, then we will come to the result that the left part of (3.132) will linearly depend on the load

$$\{\mathbf{K}_0 + \lambda\,[\mathbf{K}_1(\delta\mathbf{s}) + \mathbf{K}_2(\delta\mathbf{u})]\}\;\delta\mathbf{U} = \mathbf{0}. \tag{3.133}$$

The assumption of independence of the critical load from displacements $\lambda\,\delta\mathbf{u}$ eliminates matrix $\mathbf{K}_2$, and we come to Eulerian equations for the critical load

$$[\mathbf{K}_0 + \lambda\,\mathbf{K}_1(\delta\mathbf{s})]\;\delta\mathbf{U} = \mathbf{0}\,. \tag{3.134}$$

The minimum eigenvalue of this equation λ_1 determines the critical load as $\lambda_1\delta\mathbf{p}$. It is worth noticing that the use of Eulerian equations essentially assumes the unchangeable distribution of the internal stresses in the system as the load's intensity increases. This feature does not always take place, especially in nonlinearly deformable systems. Therefore some software developments determine two critical values of loads: the first one is the least critical value of λ_1 in equations of the (3.134) type, and the second one is determined as the load intensity at which the equations of the (3.129) type lose their positive definiteness in the course of a loading subject to some more or less realistic law (for example, when the live load grows and the dead load remains the same) modeled by a stepwise procedure (see Section 5.6). A nontrivial problem immediately arises how to choose the critical value to be demonstrated as an evidence that the calculated safety coefficient by stability is sufficient.

To conclude this section, it is needless to say that the geometrically nonlinear statement of problems is accompanied, as a rule, by the loss of uniqueness of the solution. Moreover, the presence of multiple solutions can have nothing to do with the buckling of the respective mechanical system. Unfortunately, not every computational software warns its users about such a possibility, and the solution that the user is given is only one of the options chosen either voluntarily by the algorithm or determined by a particular history of loading. We confine ourselves intentionally to this note only, because we believe the analyst *must necessarily* be warned of possible non-unique solutions but the software *does not have to* analyze all branches of the equilibrium state curve. This analysis is far beyond the competence of conventional computational procedures discussed in this book, and it belongs, rather, to the field of scientific research.

3.6 Structural Nonlinearity — Systems with Unilateral Constraints

Systems where limitations imposed upon displacements are defined by inequalities (unilateral constraints) can be encountered more often than it is usually thought. The simplest examples are: a structure simply supported by a surface that restrains the former from penetrating the latter and by no means prohibits its displacement in the opposite direction; flexible cables that allow their end points to converge while forbidding them to diverge by more than the cable's length. Additional typical examples include: a dry masonry, a connection between a structure's foundation and its bedding soil. Loose grounds or perfectly flexible membranes are examples of more complicated systems with unilateral constraints. Though, there is a tradition to treat constraints of this kind as ideal bilateral ones and check a posteriori whether the obtained solution violates the allowed behavior of the constraints. This tradition formed long ago when one could not so easily obtain a solution for a correctly stated problem with nothing but one's "bare hands". Nowadays almost any contemporary software is capable of solving problems like that in the correct formulation[15].

Let's consider an elastic system with unilateral constraints in the number of s, and let's introduce a convention of indexing the constraints of this kind by Greek letters. We choose the rule of signs for reactions R in these constraints so that a stress physically allowed by the respective constraint (compression for an obstacle or tension for a flexible cable) be positive. Let us also treat those displacements U as positive that are not hindered by the respective unilateral constraint (such as the convergence of the ends of a flexible cable). As the unilateral constraint can be only in either of two states: working (the constraint is on) when $R > 0$ and $U = 0$, or idle (the constraint is off) when $R = 0$ and $U > 0$, then for all unilateral constraints in a system the following equalities and inequalities must hold:

$$R_\alpha \geq 0\,; \quad U_\alpha \geq 0\,; \quad R_\alpha U_\alpha = 0 \quad (\alpha = 1, \ldots, s)\,. \tag{3.135}$$

These conditions are written for perfectly rigid unilateral constraints such as a unilateral stop constraint. Considering elastic unilateral constraints (say, for springs of finite stiffness that resist to tension only), we would have to write the following instead of (3.135), seeing our rule of signs:

$$R_\alpha = \tfrac{1}{2}\, k_\alpha(|U_\alpha| - U_\alpha) \qquad (\alpha = 1, \ldots, s)\,, \tag{3.136}$$

where k_α is the stiffness of the elastic unilateral constraint.

To see this, take a positive displacement U_α to which the unilateral constraint does not resist, and have $R_\alpha = 0$ from (3.136), while if the displacement is negative, then the linear dependence $R_\alpha = -\,k_\alpha U_\alpha$ follows from (3.136) with a positive stress (thus resisted by the unilateral constraint). Further we will assume all unilat-

---

[15] Or, exactly, they "are supposed to solve" the problems in the correct formulation, though it would not take you long to find examples of software developments that do not do what they are supposed to.

eral constraints in systems of interest to be rigid. Actually, as will be demonstrated at the end of this section, this circumstance presents no limitation to the class of unilateral constraints because any elastic unilateral constraint can be reduced to a rigid unilateral constraint.

Let us remove all unilateral constraints in our system, replacing their action by stresses R_α ($\alpha = 1,\dots, s$), so far unknown. Let us analyze this system by the slope deflection method making its principal system such that stresses R_α ($\alpha = 1,\dots, s$) could cause no displacements in the directions of the removed unilateral constraints. Also, we will assume that the principal system of the slope deflection method is kinematically stable. In all other aspects the choice of the principal system does not matter.

The set of all kinematical unknowns Z_j ($j = 1,\dots, n$) introduced when composing the principal system of the slope deflection method is sufficient to determine displacements U_α ($\alpha = 1,\dots, s$) in the directions of the removed unilateral constraints that we want to know, or, in other words,

$$U_\alpha = \sum_{j=1}^{n} C_{\alpha j} Z_j + C_\alpha^p \qquad (\alpha = 1,\dots, s). \tag{3.137}$$

where $C_{\alpha j}$ is a displacement in the principal system in the direction of the unilateral constraint with number α caused by unit displacement $\bar{Z}_j = 1$, and C_α^p is the same displacement but caused by external influences.

Equilibrium equations that are conditions of the equality of summed reactions in constraints introduced to build the principal system to zero, will be written as follows:

$$\sum_{j=1}^{n} r_{ij} Z_j + \sum_{\alpha=1}^{s} B_{i\alpha} R_\alpha + r_i^p = 0 \qquad (i = 1,\dots, n), \tag{3.138}$$

where r_{ij} is the reaction in i^{th} additional constraint of the slope deflection method caused by unit displacement $Z_j = 1$, $B_{i\alpha}$ is the reaction in the same constraint caused by unit force $R_\alpha = 1$, and r_i^p is the same reaction caused by the external load. Equations (3.138) are composed to form a conventional linear system, and the coefficients included in these equations are determined by conventional rules of structural mechanics.

On the basis of commonly known displacement and reaction reciprocity relations of structural mechanics we obtain:

$$B_{i\alpha} = -C_{\alpha i} \quad (i = 1,\dots, n;\ \ \alpha = 1,\dots, s). \tag{3.139}$$

The set of expressions (3.135), (3.137) – (3.139) presents necessary conditions of minimization (Kuhn – Tucker conditions [11, 14]) in the following problem of quadratic programming: minimize quadratic functional $f(\mathbf{Z})$

$$f(\mathbf{Z}) = \frac{1}{2}\sum_{i=1}^{n}\sum_{j=1}^{n} r_{ij} Z_i Z_j + \sum_{i=1}^{n} r_i^p Z_i \tag{3.140.a}$$

observing limitations

$$\sum_{j=1}^{n} C_{\alpha j} Z_j + C_{\alpha}^{p} \geq 0 \quad (\alpha = 1, \dots, s). \tag{3.140.b}$$

Properties of solutions of problem (3.140) and methods for its numerical solution have been thoroughly investigated. Interested readers may refer to the literature on this subject (see, for example, [11, 21, 38]), and here we are just going to notice peculiarities of the solutions that must be understood when stating a problem and analyzing its results.

The statement of the problem above regarded systems with ideal constraints and no friction where the only cause of nonlinearity was the presence of unilateral constraints. Systems of this type are often called *structurally nonlinear*. A fully elastic behavior is characteristic for such systems; it is determined by the unique solution of problem (3.128). The property of the solution's uniqueness can be lost by the system when friction forces are involved, even if the history of loading is known [38]. In this regard, one should be very careful about the simulation of elastic-plastic systems by systems with ideal unilateral constraints which is suggested by some researchers from time to time. The identity of the behavior of perfect elastic-plastic systems and systems with friction, as well as the non-uniqueness of the velocity field, is well known [13, 38].

Often enough, developers of software systems avoid using mathematically reliable methods of solving the quadratic programming problem, and use a "physically evident" process of iterations that follows: for a hypothetic state of a system where some unilateral constraints are off and some (working ones) are treated as bilateral, a conventional analysis is performed first. Next, after its results, at the next stage of iteration, all "incorrectly behaving" unilateral constraints are forced to change their state (constraints with unallowable reactions are removed, constraints where an unallowable displacement arises are turned into bilateral ones). This process is then repeated until the system includes no more "incorrectly behaving" unilateral constraints. The slang of professionals has a term for this approach (that seems to be self-evident from the essence of the problem) to finding the actual behavior of the system: a *Rabinovich algorithm*, because it was clearly formulated for the first time about 50 years ago in a paper by I.M.Rabinovich [28], a scripture for the theory of systems with unilateral constraints.

Unfortunately, though this method of solution is widely popular, it does not guarantee the convergence and may lead to an endless loop, as it was shown by a pretty simple example in the paper [24]. The sequence of the modeling states for this example (Fig. 3.12, *a*) is such that the process becomes an endless loop in four iterations (notice the signs of the displacements and stresses in the unilateral constraints)!

In systems with unilateral constraints that are not *redundant* according to the traditional terminology of the force method, there can be other problems too.

It is as likely as not that the system, after successive iterations at some stage of the analysis, will become kinematically changeable, and the software will thus be at its wits' end. The example that evidences this possibility is a perfectly rigid body braced in its plane with five springs of the same stiffness that resist to tension only (Fig. 3.13).

If we start our iteration process with a system where all springs are assumed bilateral, then after the first iteration it appears that the springs 2, 4 and 5 must be turned off. The two other springs No.1 and 3 that are in tension are not enough to provide the kinematical stability of the system. Meanwhile, the working system exists: it is shown in Fig. 3.13 - *b*, and the same figure shows stresses in the springs. We recommend that developers (and users too!) use these two simple problems to test a software that declares in its documentation its ability to handle systems containing unilateral constraints.

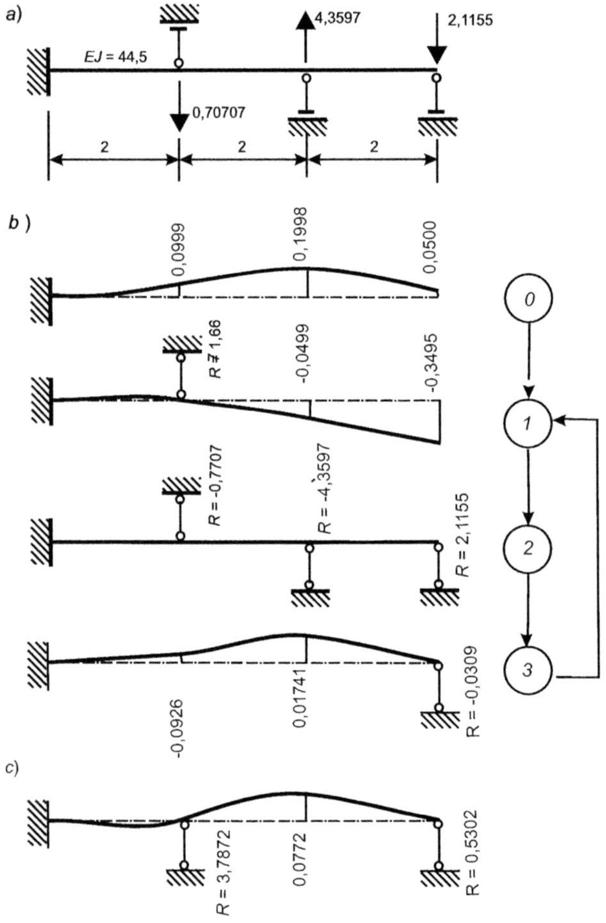

Fig. 3.12. Process of iterating the working system: *a* - the original model; *b* - the sequence of checked states; *c* - the correct solution

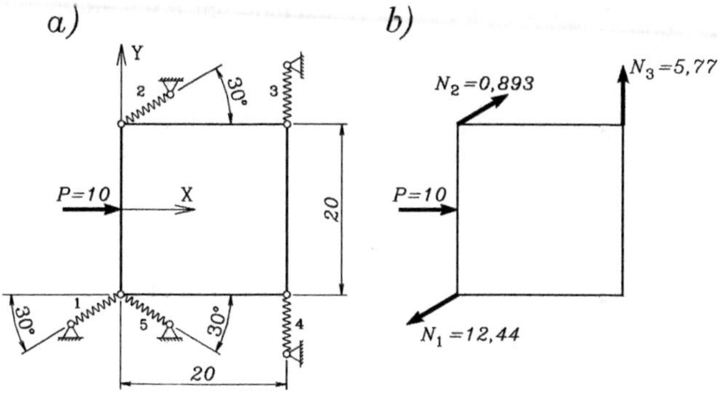

Fig. 3.13. A body braced by springs resisting to tension only

Some software developments capable of working with unilateral constraints provide this capability by specification of special elements: springs that resist to tension only or to compression only. This is the way to specify elastic unilateral constraints, but this does not allow us to handle rigid unilateral constraints. Also, this approach creates difficulties for the analysis of systems where there are unilateral hinges.

Formally, a rigid unilateral constraint can be treated as a particular case of an elastic one (as the stiffness of the respective spring approaches the infinity). But from the computational standpoint this formalism is no good because one has to handle unilateral springs of an infinite stiffness.

Therefore the more fruitful approach is to generalize in the opposite direction: a software capable of operating with rigid unilateral constraints can also handle elastic unilateral ones. The reduction of an elastic unilateral constraint to a rigid one can be easily done by simply shifting the location of the unilateral constraint in the system which is illustrated by Fig. 3.14.

If for the elastic unilateral constraint (Fig. 3.14, *a*) we mean by displacement U the difference of the displacements of points 1 and 2, then in the equivalent system with the rigid unilateral constraint the spring itself is considered to be bilateral, but U should be defined as the mutual displacement of points 2 and 3.

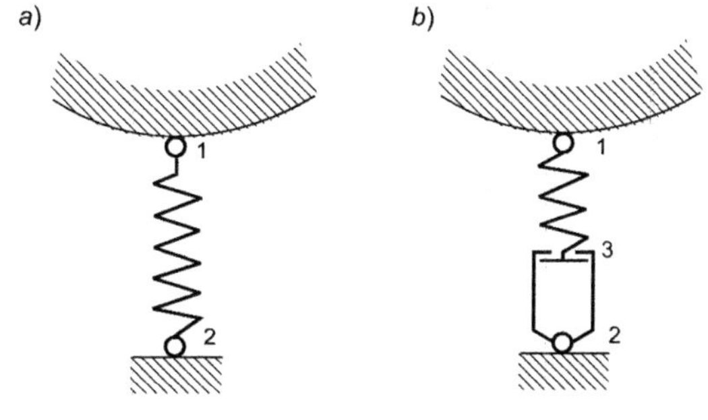

Fig. 3.14. (*a*) a system with an elastic unilateral constraint resisting to tension only, (*b*) an equivalent system with a rigid unilateral constraint

It is worth noticing that from the user's standpoint it is important that the computational software be able to handle unilateral constraints with initial gaps (including negative gaps to simulate an initial stress in unilateral elastic constraints).

Finally, we would like to draw the reader's attention to a trap that waylays even experienced analysts when they try to formally allow for friction forces between contacting surfaces in problems with unilateral constraints. They try to state such problems, for example, for the purpose of the analysis of tunnel lining where the exterior surface can "scale off" the surrounding soil in places to be determined. A similar problem arises with the longitudinal launching of bridge spans where substantial friction forces appear and are carried over to rolling devices.

In all these cases an attempt to use elementary approaches to allow for the Coulomb friction force proportional to stresses in unilateral constraints perpendicular to contacting surfaces will cause the statement of the problem to be incorrect without taking into account the history of loading. Stresses and displacements obtained by this formal analysis have, actually, nothing to do with the real distribution of stresses and strains in the structure. To understand it, note that the presence of friction forces in the design model immediately makes the structure of interest a non-conservative system. According to the definition, in conservative systems the work of all forces is determined solely by the initial and final states of the system, that is, it does not depend on the transition path. And it is easy to see that the friction force violates this condition.

The correct statement of the problem will actually lead to immensely complicated relations, the accurate analysis of which can be found in publications by hydro engineers ([38], pp. 118–122). Alas, this analysis shows that even if the history of loading is known, say, for a simple parametric loading case, the solution of the problem with friction can be non-unique. There is a frank avowal made by the authors of [38], characteristic for the case: "*In order to find a unique solution, we need to introduce additional conditions unknown to us*".

Following the idea of this quotation in connection with the software for structural analysis of systems with unilateral constraints and friction, we will make a similar avowal that we do not know any marketable developments which we could recommend for a wide practical use, though there are programs of a research type that handle some particular problems.

3.7 Cable Elements in Design Models

From the standpoint of structural mechanics, cable systems are structures that necessarily need geometrically nonlinear effects to be allowed for because the cable elements are intensely flexible. When we deal with constructions such as cable-stayed or suspension bridges, flexible mast structures braced by guy-ropes, antennas of various kinds, cable-stayed roofs of buildings etc., we absolutely have to introduce nonlinearly behaving elements in the respective design models. The non-allowance for such effects will drastically distort the real distribution of stresses and strains even from the qualitative standpoint. It is for real, therefore advanced specimens of software enable the user to introduce cable elements in design models directly, with subsequent computational analysis of those including the determination of a self-equilibrium state of the cable structure due to its pre-stressing.

There are two different analytic models of cable elements used in engineering practice. The simplest one is a model of bar that resists to tension and does not resist to compression (Fig. 3.15). This bar can be installed with gap Δ_0 in a unilateral constraint or it can have a pre-stress $H_0 \geq 0$ which can be treated as the presence of a negative gap, that is, $\Delta_0 < 0$.

Fig. 3.16 presents two versions of the diagram of behavior of such element that shows the dependence of longitudinal force H on the divergence of end points Δ. Here, Fig. 3.16, *a* corresponds to the pre-stress case while Fig. 3.16, *b* to the gap case. Tangent of the diagram's slope is equal to EA/L (EA is the tensional stiffness of the bar, L is the bar's length).

It is evident that this (simplified) model of a cable element is actually an elastic unilateral constraint and can be handled by a computational software within the frames of its general algorithm of analyzing unilateral constraints discussed in Section 3.6.

The second (stricter in the statement of the problem) version of the cable element's analytic model is a flexible cable with small slack. In its initial state described by temperature T_0 and tension H_0, it is subjected to a lateral load $q_0(x)$, and its tension becomes H after its chord elongates by Δ, its lateral load changes to $q(x)$ and its temperature to T.

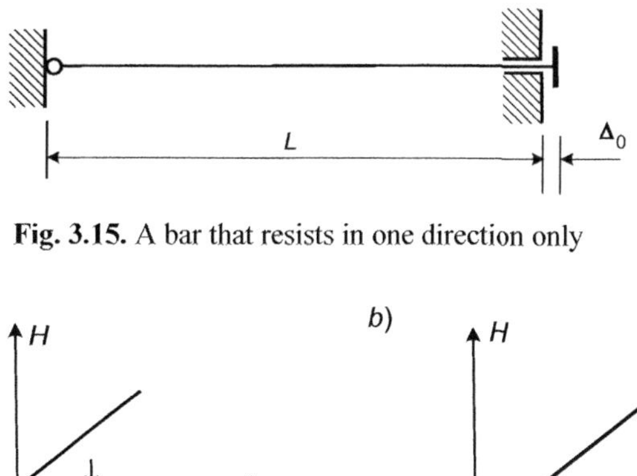

Fig. 3.15. A bar that resists in one direction only

Fig. 3.16. Two versions of the H — Δ diagram

The equation that connects these two quantities has the following form for shallow cables

$$\Delta = \frac{(H-H_0)L}{EA} - \frac{D}{2H^2} + \frac{D_0}{2H_0^2} + \alpha(T-T_0)L \ , \tag{3.141}$$

where

$$D = \int_0^L Q^2(x)dx \, , \, D_0 = \int_0^L Q_0^2(x)dx \, , \tag{3.142}$$

$Q(x)$ is a lateral force that appears in a simply supported beam of span L from the load $q(x)$ directed perpendicularly to the cable element's chord;
$Q_0(x)$ is the same force caused by load $q_0(x)$;
α is the thermal expansion coefficient;
E is the elasticity modulus of the cable's material;
A is the cable's cross-section area.

Equation (3.141) is foundational in the theory of systems with cable elements. It can be referred to as an *equation of state* of a cable element. This equation has been derived in papers dedicated to flexible shallow cables for many times. Here it is given in the form presented in publication [22].

After introducing a dimensionless parameter K that describes the influence of a lateral load upon the cable

$$K = \frac{D}{2L(EA)^2} \tag{3.143}$$

and dimensionless variables

$$\delta = \Delta/L \, , \qquad \delta_0 = \Delta_0/L \, , \qquad t = H/(EA), \tag{3.144}$$

where we denote

$$\Delta_0 = \frac{H_0 L}{EA} - \alpha(T-T_0)L - \frac{D_0}{2H_0^2} \, , \tag{3.145}$$

equation (3.141) becomes

$$\delta = t - \frac{K}{t^2} - \delta_0 \, . \tag{3.146}$$

The diagram of dependence of δ on t at $\delta_0 = 0$ is shown in Fig.3.17, the value of δ_0 can be easily allowed for by shifting the coordinate origin.

This figure shows that, as K decreases, the curve $t = t(\delta)$ tends to cling to its asymptotes which are the abscissa axis at $\delta < 0$ and a straight line with the slope tangent EA/L at positive δ, i.e. lines shown in Fig. 3.16. This is where the connection appears between the two discussed analytic models.

It goes without saying that equation (3.146) should not be relied on unconditionally. An attentive reader will certainly notice that this equation actually takes place only at very small values of δ (about 10^{-3}–10^{-2}) because big strains violate the assumptions at which it has been derived and the conditions of applicability of the cable's state equation (3.141).

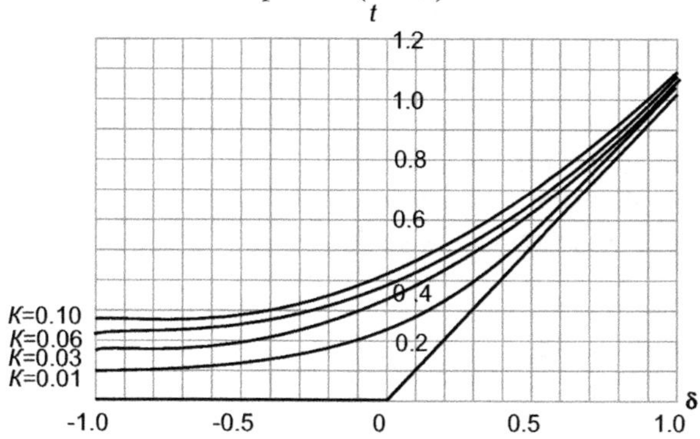

Fig. 3.17. A dimensionless dependence between the tension and the chord elongation

Therefore diagrams shown in Fig. 3.17 describe only formal mathematical properties of the curves defined by equation (3.146). Actually, they reflect the behavior of cable elements only in a small vicinity of point $\delta = 0$. As δ grows, the error of neglecting small terms omitted when deriving the state equation (3.141) will increase. But what we are interested with in practice is exactly small strains of cables, so we do not need to discuss appropriate refinements of the cable state equation.

The instantaneous compliance of a cable element in the vicinity of a stressed state characterized by tension H can be easily determined by differentiating the state equation (3.141)

$$\frac{d\Delta}{dH} = \frac{L}{EA} + \frac{D}{H^3}, \tag{3.147}$$

and thus the instantaneous stiffness matrix of this element is

$$\mathbf{K}(H) = \begin{bmatrix} \dfrac{1}{\dfrac{L}{EA} + \dfrac{D}{H^3}} & -\dfrac{1}{\dfrac{L}{EA} + \dfrac{D}{H^3}} \\ -\dfrac{1}{\dfrac{L}{EA} + \dfrac{D}{H^3}} & \dfrac{1}{\dfrac{L}{EA} + \dfrac{D}{H^3}} \end{bmatrix}. \tag{3.148}$$

Note that when the only distributed lateral load upon the element is the dead weight of the cable, from (3.147) a well-known formula by Ernst follows easily. It is intended to calculate area A_0 of a conventional bar of the same length L with its stiffness in longitudinal deformation to be the same as the instantaneous stiffness of the cable

$$A_0 = A\frac{1}{1+\dfrac{\gamma^2 l^2 E}{12\sigma^3}}, \tag{3.149}$$

where:

γ is the density of the cable's material;
l is the length of the horizontal projection of the cable;
$\sigma = H/A$ is the axial stress in the cable.

3.7.1 Coordinate Axes

A cable element has its local coordinate system X_1, Y_1, Z_1 in which axis X_1 goes from the first node A to the second node B, axis Z_1 is perpendicular to axis X_1 and located in the vertical plane ABB°A°, and axis Y_1 completes the system to a right-hand triple (Fig. 3.18).

A cable element can be subjected to loads similar to those upon bar elements. Local loads taken into account can act only in the directions of Y_1 and Z_1, they can be either distributed — $q_{y1}(X_1)$, $q_{z1}(X_1)$ — or concentrated (P_{y1}, P_{z1}). If the load is specified in components of the global coordinate system, it must be reduced to the components listed above. The component along axis X1 is not calculated, or, exactly, it is reduced to statically equivalent nodal loads applied in the nodes where the cable is attached to the general structure.

If the relation between loads q_{y1} and q_{z1} does not change along the cable (which is the case, as a rule), then these loads can be reduced to resultants that determine the plane of the cable's sag, by conventional formulae

$$q = \sqrt{q_{y1}^2 + q_{z1}^2};\ P = \sqrt{P_{y1}^2 + P_{z1}^2}\ , \tag{3.150}$$

using which the integrals $D = \int_0^L Q^2(x)dx$ can be determined.

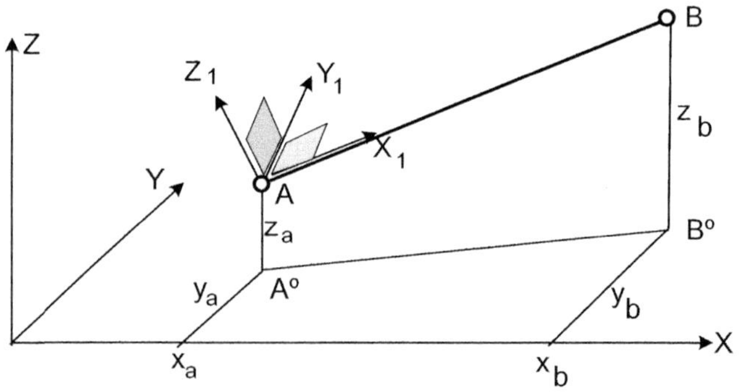

Fig. 3.18. Arrangement of coordinate axes

Table 3.3.

Type of loading	$\int_0^L Q^2(x)dx$
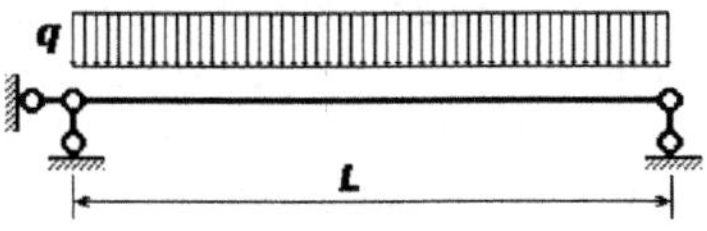	$\dfrac{q^2 L^3}{12}$
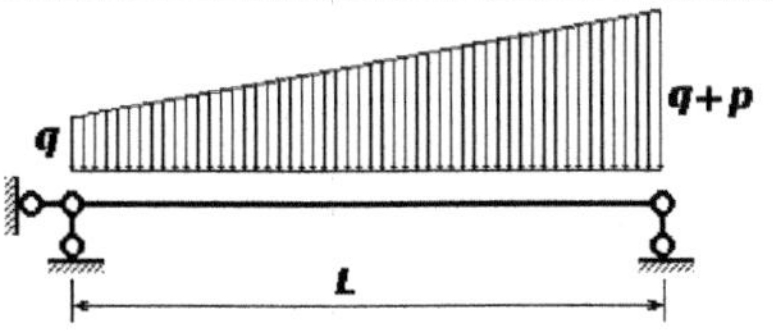	$\dfrac{q^2 L^3}{12} + \dfrac{qpL^3}{12} + \dfrac{p^2 L^3}{45}$
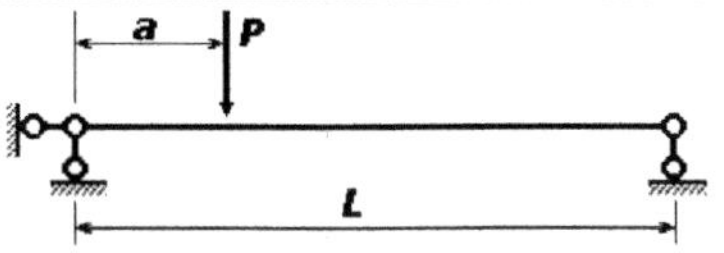	$\dfrac{P^2 (L-a)a}{L}$

Here we give the formulae to calculate values of the integrals

$$D = \int_0^L Q^2(x)dx .$$

These formulae for some popular types of loads are given in Table 3.3.

3.7.2 Specification of Pre-stress

For the purpose of analysis of cable systems, it is very important to specify the reference state correctly. This is characterized by initial values of lateral loads upon cable elements $q_{0,i}$ (as a rule, it is the component of the cable's dead weight perpendicular to the chord of the cable) and tensions $H_{0,i} \geq 0$. Index i is the number of the cable element and varies from 1 to m.

The principal difficulty is that the values of $H_{0,i}$ must correspond to an equilibrium state of the system in which the pre-stress creates a distribution of internal stresses caused by the initial external load $q_{0,i}$, particularly to a zero external load. Usually a lot of feasible equilibrium states of the system exists, and each one corresponds to a particular distribution of pre-stresses in the cables $H_{0,i}$ (i = 1,...,m) where m is the number of the cables. In this connection, an engineer that designs cable structures faces the problem of choosing a reasonable pre-stressed state from a set of feasible ones on the basis of some criteria of optimum and structural limitations. These criteria and limitations may vary depending on a particular design task. For example, a cable-stayed bridge's design usually requires its pylons to be in a bending-free state under the dead load and the pre-stress, and the moments in the stiffening girder caused by these dead loads to be minimized by absolute value. There can be other criteria of optimum and other limiting conditions. Let's confine ourselves to this and not discuss techniques used to resolve particular engineering issues because it is a topic in its own right, far beyond the competence of this book[16]. It suffices to say that a computational software absolutely must be able to handle information about the initial stressed state of a system in order to analyze the distribution of stress and strains in a cable-stayed structure correctly.

Sometimes it is more convenient to use the concept of a *length of initial cable work* $L_{0,i}$ instead of the concept of pre-stress as a set of initial stresses in cables $H_{0,i}$ ($i = 1,\dots,m$). The length of initial cable work L_0 will mean the geometrical length of a non-stressed cable at a fixed temperature T_0. If we use the cable state equation (3.141), then we can easily derive the relationship between the work length and the initial stress H_0. To see this, assume for the cables in their non-stressed state $H = 0$, $T = T_0$, $q(x) = 0$ and thus $D = 0$. It results from (3.141) that

$$\Delta = -\frac{H_0 L}{EA} + \frac{D_0}{2H_0^2}, \tag{3.151}$$

Evidently, the meaning of the quantity Δ as defined by formula (3.151) is the difference between the initial cable work L_0 and the length of the cable's chord L, that is

$$L_0 - L = -\frac{H_0 L}{EA} + \frac{D_0}{2H_0^2}. \tag{3.152}$$

As some programs oriented at the analysis of cable structures provide the user with the capability to specify the relative lengths of initial cable works $L_f = L_0/L$ in source data for the analysis rather than the cable pre-tension values, relation (3.152) can be of use[17].

3.7.3 On Linearized Models of Cable Structures

It should be said that cables are not the only reason for geometrical nonlinearity in design models of structures that we are discussing. The second reason for the geometrical nonlinearity is that cable structures include strongly compressed elements, as a rule (though not necessarily). For example, for cable-stayed bridges those are pylon legs and the span itself compressed by the cable tension, or if the structure is a braced mast, then the mast's shaft is subject to compression. Allowing for longitudinal forces' influence on bending moments in compressed elements of a system, we have to analyze the structure in its deformed geometry.

But, besides the static analysis, structures of this type often require the natural oscillation analysis to be performed because otherwise one cannot determine the pulse component of the wind load, perform a well-founded aerodynamic research or the seismic analysis etc.

[16] Statements of some problems of this kind are discussed in Section 8.2.

[17] For example, the **GTSTRUDL** software is capable of doing this.

At the same time, the spectral analysis of a system (determination of frequencies and modes of its natural oscillation) is sensible only in the linear statement of the problem. Dealing with nonlinear problems, we can only talk about an expansion of the system's motion into a series by modes of free oscillations in the vicinity of its equilibrium state and the linearization of the system's behavior in this vicinity. So, the stress and strain state (SSS) of the system under a static loading must be determined taking into account the said geometrically nonlinear effects. For the purpose of dynamical analysis, knowing the structure's SSS caused by applied static loads, one needs to build a dynamical model of a so-called *linearized system*.

By this term we will mean a structure that behaves geometrically linearly but takes into account the achieved level of stress in its cables in the state that precedes the linearization, and the achieved level of compressive stress in respective compressed elements of the model. Strictly speaking, a linearized system means an original model with all its elements treated as geometrically linear but having tangential (instantaneous) stiffness matrices.

Regarding cable-stayed and suspension bridges, linearized models of these structures are useful (and necessary) for three purposes at least:

- the linearized model is needed to analyze the system for oscillation, including the wind loading, because we want to know the oscillation around the equilibrium state that has been achieved by the system;
- this model is needed to build influence lines which are sensible in linear systems only. Using the obtained influence line, the engineer can determine a pretty close approximation to various patterns of the bridge's loading with a moving load that causes minimal and maximal values of stress and strain factors of interest;
- the linearized model is a pretty useful tool to construct numerical solution procedures for the nonlinear analysis itself. This is especially essential for stepwise processing (see Section 5.7 in this regard).

Below, Fig.3.19 shows a peculiar and a most recent example: a schematic of a cable-stayed bridge across the Neva river in Saint-Petersburg which is being engineered and simultaneously erected as these lines are written. This pretty ambitious object, as for the Russian transport engineering industry, has been designed and analyzed in the *Giprostroymost* Institute (St. Petersburg) and is now being erected.

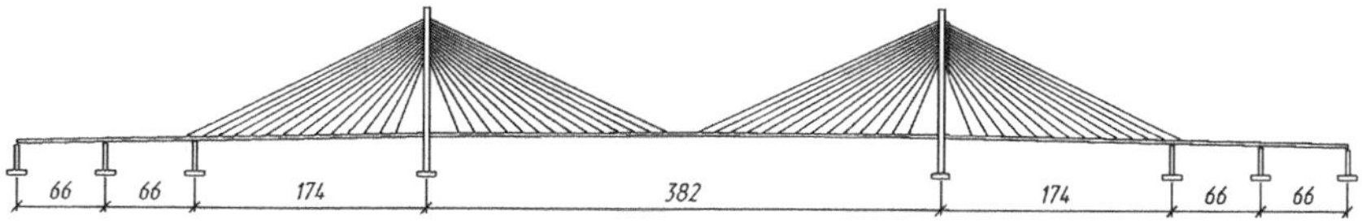

Fig. 3.19. A schematic of a cable-stayed bridge in Saint-Petersburg

This object is worth being a characteristic example not just because its analysis involves the practical application and validation of models discussed below, but also due to the reason that the necessity of a careful analysis of this very structure was one of the incentives for the authors to deal with the said models.

Let's discuss the two said reasons for geometrical nonlinearity separately, from the standpoint of the linearized models.

3.7.4 Linearization of Cable Elements in a Design Model

So, we need to construct a model that allows for the resistance of the cable part of the structure to increments of stresses in the same cable elements. In structural mechanics this procedure is sometimes called an analysis with a tangential stiffness matrix.

To build a linearized model, it is sometimes convenient to treat the system's cable part as a chain of hinged links. Let's consider a separate link of this chain and assume that there is an internal tension N in this link in the current state of the system (Fig.3.20). Here and below, we use the designation of N rather than H in order to unify the stress notation for compressed and tensioned elements.

Let's present some elementary considerations of structural mechanics to help understand the meaning of the linearized model suggested below.

It is easy to see how lateral displacements of nodes of this element by distances of v_i and v_j, as shown in Fig.3.20, cause the resistance forces Q to appear in the nodal points of the element. Equilibrium equations constructed for the deformed geometry of the element give us the following expressions to determine forces Q:

$$Q = \frac{N}{L}\Delta , \qquad \Delta = v_j - v_i \ . \tag{3.153}$$

Thus, in the design model of the linearized system each cable link should be replaced by some (geometrically linear) bar element in which lateral forces Q appear, connected by relations (3.153) with the displacements of the nodes of this element. Fig. 3.21 shows three feasible types of such bar elements.

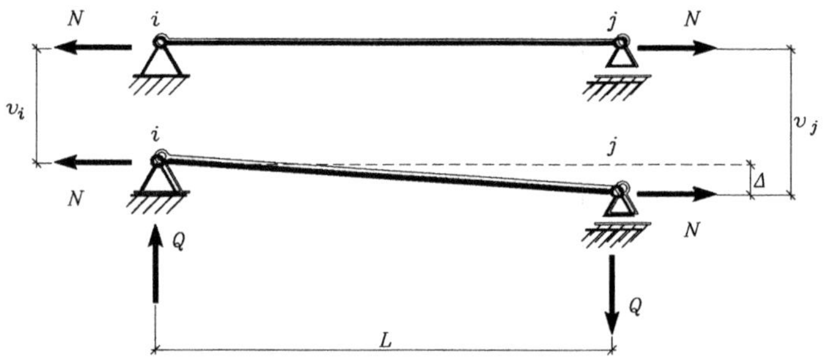

Fig. 3.20. A link of a hinged chain

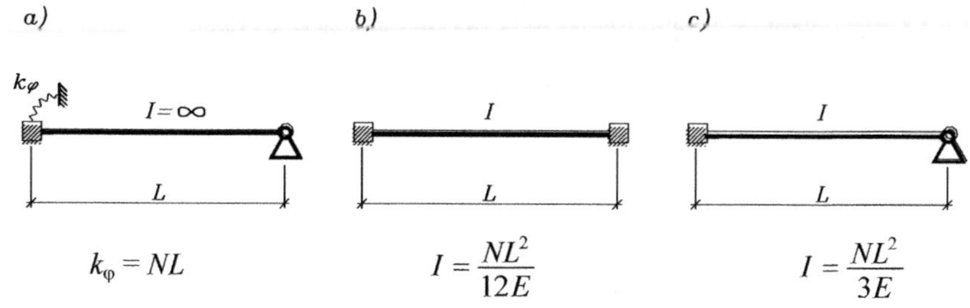

Fig. 3.21. Feasible types of bar elements

The "*a*" type element is a bar perfectly rigid against bending, reinforced in one of its nodes by a clock spring of stiffness k_φ, selected to satisfy relation (3.153). The "*b*" and "*c*" types are bendable bars: "b" features a stiff fixation to both nodes, and "*c*" supposes a hinged connection to one of the two nodes. The "b" and "c" types restrain the end nodes of elements from rotation. The inertia moment of the cross-section of bar *I* is also selected to satisfy relation (3.153). Further we can use every one of the three element types to build a linearized design model. It may be easily made sure that relation (3.153) holds for all three element types shown in Fig.3.21, that connects the lateral force Q with the relative displacement Δ (lateral) of the ends of the bar.

In order to validate the linearized model that has been constructed, we can check it with a simple problem for which an accurate solution is known. This test problem will be a von Mises truss (Fig.3.22). The solution of this problem in a strict geometrically nonlinear formulation is found in a book by Panovko and Gubanova [20].

Let us denote as L and L_0 the length of the von Mises truss's bar in its deformed and non-deformed states, respectively. Then, as can be seen from Fig.3.22,

$$L_0 = \frac{a}{\sin\alpha_0}, \qquad L = \frac{a}{\sin\alpha}, \tag{3.154}$$

and the value of the bar's elongation ΔL will be

$$\Delta L = \frac{a}{\sin\alpha} - \frac{a}{\sin\alpha_0}. \tag{3.155}$$

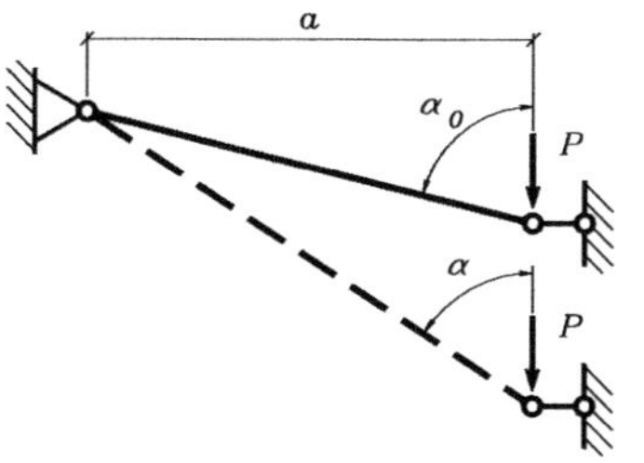

Fig. 3.22. A von Mises truss

The condition of equilibrium of the central node of the truss in the deformed state gives us the longitudinal force N in the von Mises truss bar

$$N = P/\cos\alpha \ ,$$

wherefrom we obtain by involving physical relations (Hooke's law)

$$N = \frac{EA\,\Delta L}{L_0} = EA\left(\frac{\sin\alpha_0}{\sin\alpha} - 1\right) , \qquad (3.156)$$

where A is the area of the bar's cross-section. Thus

$$P = N\cos\alpha = EA(\sin\alpha_0 \operatorname{ctg}\alpha - \cos\alpha) . \qquad (3.157)$$

Geometrical considerations help us easily find the vertical displacement v of the central node of the truss

$$v = a\,(\operatorname{ctg}\alpha - \operatorname{ctg}\alpha_0) . \qquad (3.158)$$

The derived relations give us accurate expressions for the resistance of the system in the central node of the truss, being the ratio of the load increment and the displacement increment, that is, derivative dP/dv. From (3.157) and (3.158) we obtain

$$\frac{dP}{d\alpha} = EA\ \left(-\frac{\sin\alpha_0}{\sin^2\alpha} + \sin\alpha\right), \qquad \frac{dv}{d\alpha} = -\frac{a}{\sin^2\alpha} , \qquad \frac{d\alpha}{dv} = -\frac{\sin^2\alpha}{a} ,$$

and thus

$$\frac{dP}{dv} = \frac{dP}{d\alpha}\frac{d\alpha}{dv} = \frac{EA}{a}(\sin\alpha_0 - \sin^3\alpha) . \qquad (3.159)$$

For the linearized model, we confine ourselves to the use of the bar element shown in Fig. 3.21, *c*, and the result is a design model shown in Fig. 3.23.

Assuming the bar cross-section area in the model to be A_1

$$A_1 = A\frac{\sin\alpha_0}{\sin\alpha} , \qquad (3.160)$$

let's determine the reaction of the system r_{11} to a unit displacement of the central node in the vertical direction. The desirable reaction will be expressed as follows, after projecting the lateral Q and longitudinal T forces in the model bar shown in Fig. 3.23 onto the vertical axis,

$$r_{11} = \frac{3EI}{L^3}\sin^2\alpha + \frac{EA_1}{L}\cos^2\alpha = \frac{N}{L}\sin^2\alpha + \frac{EA_1}{L}\cos^2\alpha . \qquad (3.161)$$

Let's now replace the longitudinal force N in this equation by its expression according to (3.156).

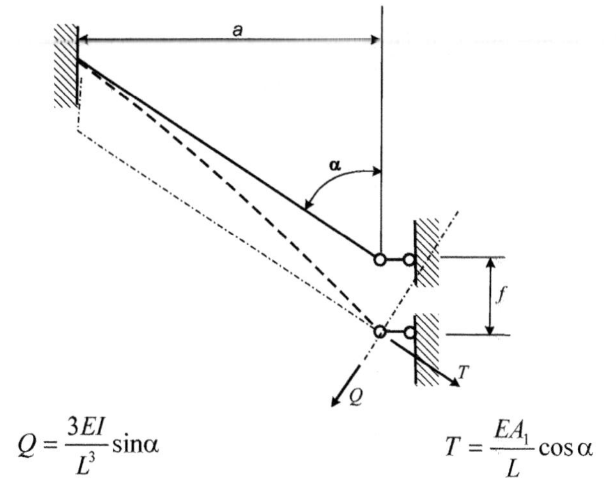

Fig. 3.23. A deformed state

The result is

$$r_{11} = \frac{EA}{L}(\sin\alpha_0 \sin\alpha - \sin^2\alpha + \frac{A_1}{A}\cos^2\alpha), \qquad (3.162)$$

which can be finally transformed to the expression below, taking into account (3.154) and (3.160)

$$r_{11} = \frac{EA}{a}(\sin\alpha_0 - \sin^3\alpha). \qquad (3.163)$$

Comparing (3.159) and (3.163), we can see that the constructed linearized model represents the instantaneous stiffness of the original nonlinear system accurately, in the vicinity of its equilibrium state. Note that in practical calculations one can assume $A_1 \cong A$, because one can easily see that the ratio $\sin\alpha_o/\sin\alpha = 1 + \varepsilon$, where ε is the relative longitudinal strain in the bar in its equilibrium state, and this small quantity can be neglected comparing to one.

So, the linearized model of a cable system can be constructed by replacing the cable elements with any of the bar flexural elements shown in Fig. 3.21.

Regarding the design model of the above mentioned cable-stayed bridge across the Neva river (Fig.3.19), each cable of $14\times8 = 112$ present in its spatial design model was represented by two spatial bar elements as shown in Fig. 3.24. In this figure the following designations are used:

W is the top (at the pylon) node of the cable attachment;
V is the bottom (at the span) node of the cable attachment;
M is an intermediate artificial cable node that geometrically divides segment VW in two halves.

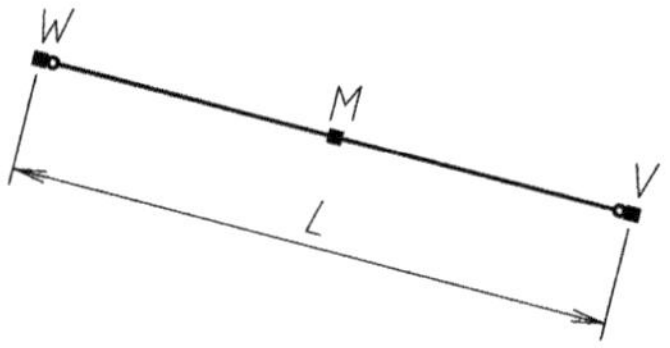

Fig. 3.24. An element of the cable system of the bridge

Two bars are introduced in the linearized design model:

- upper bar KW with its beginning node W and end node M;
- lower bar KV with its beginning node V and end node M.

These bars are connected via hinges to nodes W and V, while they are welded rigidly to node M. The intermediate node M is restrained externally from all slopes, thus it possesses only three degrees of freedom: linear displacements in the directions of three axes of the global coordinate system (X,Y,Z).

Let N be the tension of the cable calculated with this design model. We assume that by the moment of the dynamical analysis the stresses in all cables are known. Then, according to what has been said, the flexural rigidity of these bars will be described by the formula

$$EI = \frac{NL^2}{12}, \tag{3.164}$$

where L is the full length of the cable (or, exactly, the geometrical length of the cable's chord equal to the distance between the points of attachment of the cable W and V).

Note that the full hinge in the attachment nodes causes the model bars not to resist to the torsion of nodes V and W. The tensional rigidity of these model bars EA_0 (a so-called design *effective stiffness* of the cross-section) is determined using the tangential modulus by (3.149). That is, the increment of the longitudinal force in the cable is connected with the divergence of nodes W and V.

3.7.5 Linearization of Compressed and Bent Elements of a Design Model

Let us consider a compressed and bent bar, introducing designations

EJ – the flexural rigidity of the bar;
N – the longitudinal compression force in the bar;
L – the length (span) of the bar;
v – deflection (lateral displacement) of the bar's cross-section;
$q(x)$ – the lateral load applied to the bar.

The differential equation for the deflection of a compressed and bent bar that takes into account the longitudinal forces' effect on the bending moments in the bar, has the following well-known form

$$(EJv'')''+ Nv''= q(x)\,. \tag{3.165}$$

If the longitudinal force N is known beforehand (and does not depend on deflections v), then equation (3.165) is a linear differential equation for the desirable function v. Thus, the differential equation (3.165) with the fixed longitudinal force N is exactly the implementation of our linearized model.

Though, the problem is that not every available computational software can allow for the additional term Nv'' in the left part of equation (3.165) when building the stiffness matrix of a bar. Nevertheless, even in this situation one can allow for this term by building a special discrete model within the competence of the respective standard structural CAD software.

The easiest way to build such a model is to involve energy considerations, requiring that the bar described by equation (3.165) and the discrete model being built be equivalent by energy.

Let us write the expression for the potential energy of deformation E, accumulated by a compressed and bent bar in the course of its bending. We have

$$\mathsf{E} = \int_0^L \frac{EJv''^2}{2}dx - N\int_0^L \frac{v'^2}{2}dx\,. \tag{3.166}$$

The first integral in (3.166) is an energy of deformation accumulated in the bar when bent, while the second integral is the work of longitudinal force N. But this integral can be interpreted in other, mechanically formal sense: namely, as an energy of deformation accumulated in an elastic foundation that resists to the rotation of the bar's cross-sections with a linear negative stiffness coefficient $k_\varphi = -N$ (Fig. 3.25).

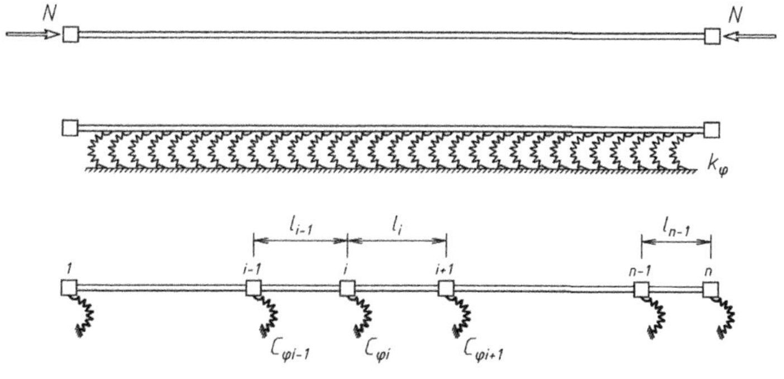

Fig. 3.25. A model of a compressed and bent element

If the finite element library of a software includes a bar element supported by a foundation that resists to rotation elastically, and the stiffness coefficient of this elastic foundation can be assigned a negative value, then all is well, and there is no need to build any additional model[18]. If the violation of at least one of these conditions takes us beyond the applicability of the software, then the elastic foundation can be allowed for approximately, by dividing the original bar into several segments and adding a concentrated torsion (clock) spring of stiffness C_φ to each node that appears after this division. The stiffness is assumed to be

$$C_\varphi = k_\varphi(l_{i-1} + l_i)/2 = -N(l_{i-1} + l_i)/2 \ , \tag{3.167}$$

where l_{i-1} and l_i are respective distances from the node to its two adjacent nodes (Fig. 3.25). This discretization can be interpreted mathematically as the use of numerical integration to calculate the second energy integral in (3.166), or

$$N \int_0^L \frac{v'^2}{2} dx \approx \frac{1}{2} \sum_{i=1}^{n} C_{\varphi i} v_i'^2 \ , \qquad L = \sum_{i=1}^{n-1} l_i \ . \tag{3.168}$$

The discrete model according to Fig. 3.25 is directly available in almost every finite element analysis software.

To conclude the section, it should be noted that a model similar to that shown in Fig. 3.25, was suggested before by Reznikov [29], though he did not involve energy considerations in the derivation of the model.

3.8 Dynamical Analysis

If the load applied to the system changes with time t, i.e. $\mathbf{p} = \mathbf{p}(t)$, then we have to assume the stresses and displacements also to be functions of time, which may force us to introduce velocities $d\mathbf{u}/dt$ and accelerations $d^2\mathbf{u}/dt^2$ in the consideration. If inertia forces that arise

$$\mathbf{J}(t) = \mathbf{M}\frac{d^2\mathbf{u}}{dt^2} \tag{3.169}$$

cannot be treated as negligibly small comparing to the loads upon the system and to the elasticity forces, then they should taken into account in the equilibrium equations which will necessarily become differential equations of motion (or, as sometimes they say, equations of dynamical equilibrium):

$$\mathbf{M}\frac{d^2\mathbf{u}}{dt^2} + \mathbf{K}\mathbf{u}(t) = \mathbf{p}(t) \ . \tag{3.170}$$

---

[18] It's another example that confirms the thesis supported many times in this book, how the prohibition of assigning negative values to stiffness properties limits useful capabilities of structural modeling and design.

If all masses are concentrated in nodes of the system, then the inertia matrix **M** will be diagonal. In other cases, the reduction of the matrix to a diagonal form will be an approximate procedure.

The problem of determining characteristics of natural oscillations of the system (modal analysis) is to find conditions in which the unloaded system performs harmonic oscillations according to the law

$$\mathbf{u}(t) = \mathbf{\Psi}\sin(\omega t + \mu)\,. \tag{3.171}$$

In expression (3.171) vector $\mathbf{\Psi}$ describes the mode of the natural oscillations (relations between the displacements of nodes), ω is the frequency of the oscillations, μ is the initial phase. Substitution of (3.171) to (3.170), seeing that $\mathbf{p}(t) = \mathbf{0}$, will give an equation for *natural* (another term is *free*) oscillations

$$(\mathbf{K} - \omega^2\mathbf{M})\mathbf{\Psi} = \mathbf{0}\,, \tag{3.172}$$

the nontrivial solution of which exists only when values ω_i $(i = 1,...,n)$, called natural frequencies or eigenvalues, turn the determinant of matrix $(\mathbf{K} - \omega^2\mathbf{M})$ into zero. The respective modes of natural oscillations $\mathbf{\Psi}_i$ (eigenvectors) can be calculated only up to an arbitrary multiplier the value of which is to be determined by a chosen technique of mode normalization.

Two methods of normalization are used most often. The first method is to normalize the modes so that the maximum component of vector $\mathbf{\Psi}_i$ be equal to one. The second method is to determine the components of vector $\mathbf{\Psi}_i$ from the condition of reduction of the mode to unit mass, that is $\mathbf{\Psi}_i^{\mathrm{T}}\mathbf{M}\mathbf{\Psi}_i = 1$. Also, the property of orthogonality of eigenvectors is worth noticing, with respect to both the mass matrix and the stiffness matrix. Further we will assume that the natural oscillation modes are normalized by mass which gives us the following relations of orthonormality:

$$\begin{aligned} &\mathbf{\Psi}_i^{\mathrm{T}}\mathbf{M}\mathbf{\Psi}_j = 0 \quad \text{and} \quad \mathbf{\Psi}_i^{\mathrm{T}}\mathbf{K}\mathbf{\Psi}_j = 0 \quad \text{at } i \neq j\,, \\ &\mathbf{\Psi}_i^{\mathrm{T}}\mathbf{M}\mathbf{\Psi}_i = 1 \quad \text{and} \quad \mathbf{\Psi}_i^{\mathrm{T}}\mathbf{K}\mathbf{\Psi}_i = \omega_i^2\,. \end{aligned} \tag{3.173}$$

In the dynamical analysis n, the number of components of vector $\mathbf{u}$ where inertia forces are involved (the number of *dynamical degrees of freedom*) can be often much less than static components. The typical example is presented by nodal slopes which, usually, are much less influential dynamically than the respective linear displacements. If some inertial components of the load are not taken into consideration, then we can divide vector $\mathbf{\Psi}$ into sub-vector $\mathbf{\Psi}_0$ where the inertia forces are zero and sub-vector $\mathbf{\Psi}_1$ where the inertia forces are involved, and write system (3.172) in the form

$$\left\|\begin{matrix} \mathbf{K}_{00} & \mathbf{K}_{01} \\ \mathbf{K}_{10} & \mathbf{K}_{11} \end{matrix}\right\| \left\|\begin{matrix} \mathbf{\Psi}_0 \\ \mathbf{\Psi}_1 \end{matrix}\right\| = \omega^2 \left\|\begin{matrix} \mathbf{0} & \mathbf{0} \\ \mathbf{0} & \mathbf{M}_{11} \end{matrix}\right\| \left\|\begin{matrix} \mathbf{\Psi}_0 \\ \mathbf{\Psi}_1 \end{matrix}\right\|. \tag{3.174}$$

Sub-vector $\boldsymbol{\Psi}_0$ can be excluded from this system. After this procedure of *static condensation*, the dimension of the modal analysis problem decreases dramatically, and the problem becomes stated as follows:

$$(\mathbf{K}_{11} - \mathbf{K}_{10}\mathbf{K}_{00}^{-1}\mathbf{K}_{01} - \omega^2 \mathbf{M}_{11})\boldsymbol{\Psi}_1 = \mathbf{0}\,. \tag{3.175}$$

As the natural oscillation modes are orthogonal, the solution of any dynamical problem represented as an expansion

$$\mathbf{u}(t) = \sum_{i=1}^{n} y_i(t)\,\boldsymbol{\Psi}_i \tag{3.176}$$

will lead to the splitting of the system of differential equations (3.170) into independent equations with respect to generalized coordinates $y_i(t)$. These equations usually contain one more term proportional to velocity, to take into account the viscous resistance to the motion. Thus the equations become

$$M_i \frac{d^2 y_i}{dt^2} + C_i \frac{dy_i}{dt} + \omega_i^2 M_i\, y_i = P_i(t) \qquad (i = 1, \ldots, n)\,. \tag{3.177}$$

Generalized forces

$$P_i(\mathrm{t}) = \boldsymbol{\Psi}_i^{\mathrm{T}}\mathbf{p}(t), \tag{3.178}$$

masses

$$M_i = \boldsymbol{\Psi}_i^{\mathrm{T}}\mathbf{M}\boldsymbol{\Psi}_i \tag{3.179}$$

and viscosity coefficients C_i, together with initial conditions for generalized displacements $y_{0i} = y_i(0)$ and generalized velocities $y'_{0i} = dy_i(0)/dt$, derived from $\mathbf{u}(0)$ and $d\mathbf{u}(0)/dt$ by formulae

$$y_{0i} = \boldsymbol{\Psi}_i^{\mathrm{T}}\mathbf{M}\mathbf{u}(0)/\,M_i\,, \qquad y'_{0i} = [\boldsymbol{\Psi}_i^{\mathrm{T}}\mathbf{M}\; d\mathbf{u}(0)/dt]/\,M_i\,, \tag{3.180}$$

determine the complete solution of the problem. This solution can be written as the expression [7]

$$y_i(t) = e^{-\xi_i\omega_i t}\left\{\left[\left(y_{0i}\xi_{0i}\omega_i + y'_{0i}\right)\frac{\sin\bar{\omega}_i}{\bar{\omega}_i} + y_{0i}\cos\bar{\omega}_i t\right] + \right.$$

$$\left. + \frac{1}{\bar{\omega}_i M_i}\int_0^t P_i(\tau)\; e^{-\xi_i\omega_i(t-\tau)}\sin\bar{\omega}_i\left(t-\tau\right)d\tau\right\},$$

where the first term takes into account the initial conditions and the second one is called a Duhamel integral.

Dimensionless damping parameter ξ_i [7] is related to viscosity coefficient C_i as

$$\xi_i = \frac{C_i}{2\omega_i M_i}\,. \tag{3.181}$$

The frequency of damped oscillations $\overline{\omega}_i$ included in the expression for $y_i(t)$

$$\overline{\omega}_i = \omega_i \sqrt{1-\xi_i^2} \tag{3.182}$$

is little different from ω_i at usual values of the *logarithmic decrement*, therefore

$$\delta_i = 2\pi\xi_i \omega_i / \overline{\omega}_i \approx 2\pi\xi_i \,. \tag{3.183}$$

Most methods of dynamical structural analysis recommended by reference manuals and standards operate the expansion (3.176) by natural oscillation modes. The analysis is performed with the split system of equations each of which determines a particular response of the system that is of interest for the analyst (a displacement, a stress, a force, an inertia force etc.) when the system moves in accordance with its particular natural mode

$$X(t) = \sum_{i=1}^{n} a_i y_i(t) \,, \tag{3.184}$$

where n is the number of dynamical degrees of freedom of the system.

The said responses are functions of time, and their summation should be performed for the same moments of the motion. But this requires an unsteady-state dynamical problem to be solved and all transient processes in the system to be taken into account. To do it, one needs much more detailed information than that usually available to the analyst. Most often, it is only possible to find so-called modal response $|a_i y_i|_{\max}$ for each particular natural mode, which is the maximum value of the system's response at i^{th} mode. The problem of summation of the modal responses arises immediately. Generally, the modal responses are achieved in different moments of time, and the extreme value that is of interest for us

$$X_{\max} = \max_i \left| \sum_{i=1}^{n} a_i y_i(t) \right| \tag{3.185}$$

cannot be easily evaluated. This makes us to try to search for more or less suitable estimates of the value of $X_{\max}$.

There are several methods of the modal response summation. The upper bound is easiest to obtain

$$X_U = \sum_{i=1}^{n} |a_i| |y_i|_{\max} \geq X_{\max} \,. \tag{3.186}$$

The difference between X_U and $X_{\max}$ can be very big, so other estimates have been suggested. In the seismic resistance theory an estimate by Rosenblum is very popular:

$$X_{RMS} = \sqrt{\sum_{i=1}^{n} (a_i |y_i|_{\max})^2} \,, \tag{3.187}$$

which is usually called "the most probable response value"[19]. Assuming a pretty strong hypothesis of all modal responses being normally distributed stochastic quantities with the same correlation coefficients to be true, expression (3.187) can be "proved" [4].

Generally, taking into account variable correlation coefficients ρ_{ij} betweeb i^{th} and j^{th} modes of oscillation, we can derive the following[20]

$$X_{CQC} = \sqrt{\sum_{i=1}^{n}\sum_{j=1}^{n} \left|a_i y_i\right|_{\max} \rho_{ij} \left|a_j y_j\right|_{\max}} \;. \tag{3.188}$$

For example, the **GTSTRUDL** software, with referring to [9], accepts the following values as the correlation coefficients:

$$\rho_{ij} = \frac{8\sqrt{\xi_i \xi_j}\,(\xi_i + h\xi_j)h^{3/2}}{(1-h^2)^2 + 4\xi_i\xi_j h(1+h^2) + 4(\xi_i^2 + \xi_j^2)h^2}\,, \tag{3.189}$$

where ξ_i and ξ_j are dimensionless damping parameters that correspond to respective oscillation modes, and $h = \omega_i/\omega_j$ is the frequency ratio. The same software suggests a wide spectrum of specification of other correlations and other ways of determining the modal responses defined by national regulations. Particularly, in compliance with regulations by the U.S. Commission for nuclear power plant design in seismically dangerous regions [8], they accept

$$\rho_{ij} = \left[1 + \left(\frac{\overline{\omega}_i - \overline{\omega}_j}{\omega_i\zeta_i + \omega_j\zeta_j}\right)^2\right]^{-1}, \tag{3.190}$$

and $\zeta_i = \xi_i + 2/t_d\,\omega_i$, while t_d is the interval of time during which the motion takes place. The interval of seismic influence t_d recommended by these regulations is 10 seconds, though the user can change this default interval.

Formula (3.187) can be treated as a particular case of (3.188) in which the normalizing correlation matrix $\|[\rho_{ij}]\|$ is assumed to be a unit matrix. In this case the modal response correlation coefficients will be zero. This assumption has been confirmed by practice quite satisfactorily when the natural frequencies are far enough. If the frequencies are near, hypotheses are introduced to take into account their correlation. Particularly, a "10% rule" is used: the frequencies are divided into two groups, near to and far from one another. For the first groups where frequencies ω_i and ω_j are different by no more than 10%, that is, $|\omega_i/\omega_j - 1| \le 0{,}1$, the correlation coefficient is assumed to be one, and all other couples are treated as non-correlated.

Another way of the estimation has been also suggested:

---

[19] The subscript of the Rosenblum estimate is the abbreviation of *Root Mean Square.*

[20] The subscript is the abbreviation of *Complete Quadratic Combination.*

$$X_C = X_{DOM} + \sqrt{X_{RMS}^2 - X_{DOM}^2}\,, \tag{3.191}$$

where the dominating modal response is determined by the formula

$$X_{DOM} = \max\left(|a_1|\,|y_1|_{\max}, |a_2|\,|y_2|_{\max}, \ldots, |a_n|\,|y_n|_{\max}\right). \tag{3.192}$$

Estimate X_C is always between estimates X_U and X_{RMS}.

Paper [6] suggests the following estimate from below

$$X_L = 2X_{DOM} - X_U\,, \tag{3.193}$$

that holds at $X_L/X_U > 1/3$. Thus, the available estimates can be arranged in the following chains:

$$X_L \le X_{RMS} \le X_C \le X_U, \qquad X_L \le X_{\max} \le X_U\,. \tag{3.194}$$

3.9 Continual Systems in Finite-Dimensional Representation

At the beginning of this chapter we chose not to consider the "internal build" of finite elements intentionally, in order to concentrate on other analytic aspects. Now let's fill in the blanks and turn to some important details of finite-element description of continual systems. To be particular, let's take a three-dimensional elasticity problem for an isotropic body.

Let the displacement field within an element be described by vector $\mathbf{v}(X)_e = \|[u, v, w]\|^T$ with its components being functions of point X and presenting displacements with respect to a Cartesian coordinate system. For the purposes of the finite element analysis, this field is to be bound to the nodal displacement vector $\mathbf{u}_e$ using interpolation functions (a matrix of shape functions) $\mathbf{U}(X)$, so that we could write

$$\mathbf{v}(X)_e = \mathbf{U}(X)_e\mathbf{u}_e\,. \tag{3.195}$$

Introducing a symbolic matrix   of the differentiation operations [32], we can define the strain vector

$$\varepsilon_e = {}^{T}\mathbf{v}(X)_e = {}^{T}\mathbf{U}(X)_e\mathbf{u}_e\,, \tag{3.196}$$

$$= \begin{Vmatrix} \frac{\partial}{\partial x} & 0 & 0 & \frac{\partial}{\partial y} & 0 & \frac{\partial}{\partial z} \\ 0 & \frac{\partial}{\partial y} & 0 & \frac{\partial}{\partial x} & \frac{\partial}{\partial z} & 0 \\ 0 & 0 & \frac{\partial}{\partial z} & 0 & \frac{\partial}{\partial y} & \frac{\partial}{\partial x} \end{Vmatrix}, \quad \varepsilon_e = \|[\varepsilon_x, \varepsilon_y, \varepsilon_z, \gamma_{xy}, \gamma_{yz}, \gamma_{zx}]\|^T, \tag{3.197}$$

and the stress vector also, using the elasticity matrix **F**

$$\boldsymbol{\sigma}_e = \mathbf{F}\boldsymbol{\varepsilon}_e\,, \tag{3.198}$$

$$\mathbf{F} = \left\|\begin{matrix} \lambda+2\mu & \lambda & \lambda & 0 & 0 & 0 \\ \lambda & \lambda+2\mu & \lambda & 0 & 0 & 0 \\ \lambda & \lambda & \lambda+2\mu & 0 & 0 & 0 \\ 0 & 0 & 0 & \mu & 0 & 0 \\ 0 & 0 & 0 & 0 & \mu & 0 \\ 0 & 0 & 0 & 0 & 0 & \mu \end{matrix}\right\|, \quad \boldsymbol{\sigma}_e = \|[\sigma_x, \sigma_y, \sigma_z, \tau_{xy}, \tau_{yz}, \tau_{zx}]\|^{\mathrm{T}}, \tag{3.199}$$

where elastic Lame characteristics λ and μ are so related to the technical properties of the material, elasticity modulus E and Poisson ratio ν, that

$$\lambda = \frac{E\nu}{(1-2\nu)(1+\nu)}, \qquad \mu = \frac{E}{2(1+\nu)}\,. \tag{3.200}$$

It is easy to notice a resemblance between (3.2) and (3.196), from which it follows that for a continual system the role of matrix $\mathbf{Q}_e^{\mathrm{T}}$ is played by matrix ${}^{\mathrm{T}}\mathbf{U}(\mathrm{X}) = \mathbf{Q}^{\mathrm{T}}(\mathrm{X})_e$.

The differential equations of equilibrium can be written as follows using the introduced notation:

$$\boldsymbol{\sigma}_e + \mathbf{q}(\mathrm{X}) = \mathbf{0}\,, \tag{3.201}$$

where $\mathbf{q}(\mathrm{X})$ is the vector of body forces.

As the generalized displacements chosen for the analysis are the nodal displacements, all body and surface forces must be transformed to nodal ones. Proceeding from the principle of virtual displacements, body force vector $\mathbf{q}(\mathrm{X})$ can be reduced to nodal forces $\mathbf{p}_e$ using the same shape functions $\mathbf{U}(\mathrm{X})$, which gives us

$$\mathbf{p}_e = -\int_{\Omega} \mathbf{U}^{\mathrm{T}}(\mathrm{X})\, \mathbf{q}(\mathrm{X})d\Omega\,, \tag{3.202}$$

where the integration is to be performed over the whole volume of element Ω_e. The reduction of the surface forces is performed exactly in the same way.

Finally, internal elastic stresses arereduced to the nodes similarly [32], and the result is an expression for the element's stiffness matrix:

$$\mathbf{K}_e = \int_{\Omega} ({}^{\mathrm{T}}\mathbf{U})^{\mathrm{T}}\mathbf{F}\; {}^{\mathrm{T}}\mathbf{U}\, d\Omega = \int_{\Omega} \mathbf{Q}(\mathrm{X})_e\mathbf{F}\mathbf{Q}^{\mathrm{T}}(\mathrm{X})_e\, d\Omega\,, \tag{3.203}$$

that corresponds to (3.6).

The procedure of discretization of a continual problem described above is known as the finite element method in the form of slope deflection method. It is by no means the only way of discretization from the standpoint of the finite element method.

3.9.1 A Note on Terminology

It seems curious (also insecure, in our opinion) that the same entitlements such as the slope deflection method, force method, mixed method should be used in two totally different senses. On one hand, a method (such as the slope deflection method) means a procedure of solving a finite-dimensional problem, and on the other hand, this is a discretization procedure. But these are utterly different! Moreover, these two concepts are quite independent. Indeed, one can easily imagine the discretization to be performed as an approximation of the displacement field while the solution of the respective discrete problem may be obtained using the force method. So which one has been used to solve the original continuous problem: the force method or the slope deflection method?

This mess of terminology confuses even skillful analysts not very much educated in the theory. That's why we would prefer not to use the names of the force method, slope deflection method etc. when we talk about the procedure of reducing a continuous problem to a problem having a finite number of degrees of freedom. These entitlements have been reserved earlier in history as indications of a method of solving a finite-dimensional problem. They originated from the classical structural mechanics of bar systems which was created and filled with its specific terminology in days when nobody could even imagine the finite element method would appear.

Unfortunately, these names have succeeded in fixating themselves in the literature on the finite element method.

3.9.2 A Note on Methods of Visualizing Design Models

In the slope deflection method, *elements of a system are considered to connect only to nodes of the design model.* They are not connected to one another directly. This peculiar feature of design models is not often mentioned in educational and reference books. It is not always shown in traditional methods of visualization of design models.

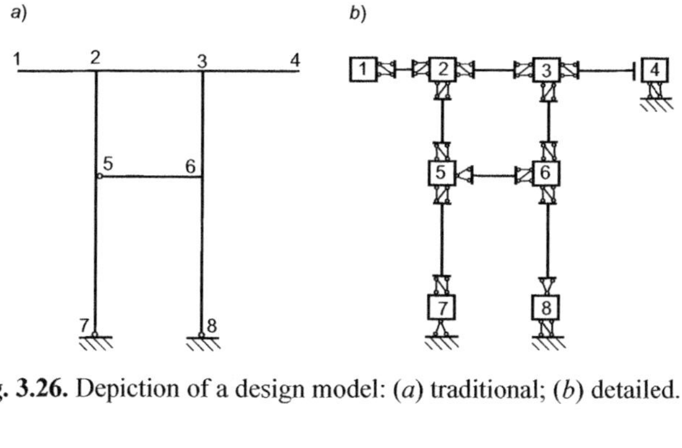

Fig. 3.26. Depiction of a design model: (*a*) traditional; (*b*) detailed.

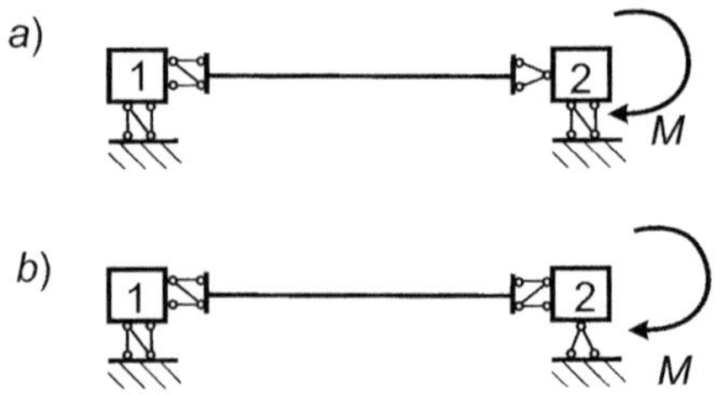

Fig. 3.27. Two ways of representing the same design model

For example, the design model shown in Fig. 3.26, *a* in a traditional manner can give one an idea that its elements are directly connected to one another, while a more detailed picture of Fig. 3.26, *b* helps one avoid this deduction.

Note also that the detailed picture shows other peculiarities of this model. In particular, one can see the possibility for specifying the same kinematical conditions with different sets of constraints.

The non-consideration of the said difference is not always harmless. For example, from the standpoint of kinematical properties of the problem, two versions of a design model shown in Fig. 3.27, are equivalent (a beam clamped at its left end and simply supported at the right end).

But from the standpoint of the load specification these ways are different: the model in Fig. 3.27, *b* transfers the moment to the bar, thus node 2 of this model will have a slope, while the model in Fig. 3.27, *a* transfers no moment and thus node 2 will have a zero slope. In order for the model in Fig. 3.27, *a* to have the bending moment in its bar, one should treat this moment as applied to the bar in a cross-section near the node rather than to the node itself.

References

1. Argyris JH (1964) Recent advanced in matrix methods of structural analysis. Pergamon Press, Oxford London New York Paris
2. Argyris JH (1977) Computer and mechanics. In: Koiter WT (ed) Theoretical and applied mechanics. Proceeding of the 14th UITAM congress. North-Holland publishing company, Amsterdam, pp. 15-99.
3. Bellman R (1960) Introduction to matrix analysis. MacGraw-Hill Inc., New York Toronto London
4. Birbraer AN (1998) Seismic analysis of structures (in Russian). Nauka Publishing House, Sankt-Petersburg
5. Busygin VG, Rozin LA (1976) Force method based on pseudo-inversion In: Strength and stability of structures and their foundations: Proc. of Leningrad Polytechnic Institute, No 354.— Leningrad (In Russian)
6. Butzel LM., Merchant HC (1973) The Use and Evaluation of Shock Spectra in the Dynamic Analysis of Structures Proceedings of ASME, Applied Mechanics 2: 115-121
7. Clough RW, Penzien J 1975) Dynamics of structures. McGraw-Hill, New York

8. Combining Modal Responses and Spatial Components in Seismic Response Analysis // Regulatory Guide, U.S. Nuclear Regulatory Commission Office of Standard Development, Section 1.92, February 1976.
9. Der Kiureghian A (1981) A Response Spectrum Method for Random Vibrations Analysis of MDF systems. Earthquake Endineering and Structural Dynamics, 9:419-435
10. Filin AP et al. (1983) Algorithms of derivation of governing equations of bar system mechanics (in Russian). Stroyzdat, Leningrad
11. Gordeev VN, Perelmuter AV (1967) The analysis of elastic systems with unilateral constraints treated as a problem of quadratic programming In: Rabinovich IM (ed) Research in structural theory, Issue 15 (in Russian). Stroyzdat Moskow, pp. 208–212
12. Grinzo L (1996) ZEN of Windows 95 programming. Coriolis Group Books
13. Ivlev DD (1966) Theory of ideal plasticity (in Russian) Nauka Publishing House, Moskow
14. Künzi HP, Krelle W (1962) Nichtlineare programmierung. Springer–Verlag, Berlin Göttingen Heidelberg
15. Lawson CL, Hanson RJ (1974) Solving least squares problems. Prentice-Hall, Inc, Englewood Cliffs, N.J.
16. Merkin DR (1980) Introduction to mechanics of flexible ropes (in Russian) Nauka Publishing House, Moskow
17. Novozhilov VV (1953) Foundations of nonlinear theory of elasticity. Graylock Press, Rochester, N.Y.
18. Oden JT (1972) Finite elements in nonlinear continua. McGraw-Hill, New York
19. Ostashevsky AY (1973) Static analysis of systems with insufficient constraints In: Rainus G (ed) Basics of theory of instantaneously rigid systems (in Russian) LenZNIIEP, Leningrad: pp. 20–28
20. Panovko YaG, Gubanova. II (1965) Stability and oscillations of elastic systems; paradoxes, fallacies, and new concepts. Consultants Bureau, New York
21. Perelmuter AV (1969) Basics of theory of systems with unilateral constraints (in Russian) CINISA of the USSR State Committee for Construction, Moscow
22. Perelmuter AV (1969) Principles of analysis of cable-bar systems (in Russian) Stroyzdat, Moskow
23. Perelmuter AV (1974) On the influence of changes in stiffness on re-distribution of stresses in a redundant system (in Russian). Structural mechanics and analysis of structures, 5:64–67
24. Perelmuter AV (1978) On the convergence of a refinement of a working system (in Russian). Structural mechanics and analysis of structures, 5:76–77
25. Perelmuter AV (1995) On the estimation of survivability of load-carrying structures In: Metal structures: Works of the school of Professor N.S. Streletsky. (in Russian) Moskow state building uneversity publishing, Moskow, pp. 62–68.
26. Perelmuter AV, Slivker VI (1976) Peculiarities of algoritmization of the slope deflection method with additional constraints taken into account (in Russian) In: Rozin LA (ed) Finite elements method and structural mechanics: Proceedings of Leningrad Polytehnical Iinstitute, No 349 (in Russian). Leningrad, pp. 28–36
27. Perelmuter AV, Slivker VI (1979) On implementation of complex kinematical conditions in the slope deflection analysis of discrete systems (in Russian). In: Rozin LA (ed) Finite elements method and analysis of structures: Proceedings of Leningrad Polytehnical Iinstitute, No 363 (in Russian). Leningrad, pp. 26–39

28. Rabinovich IM (1950) Some points of theory of structures that include unilateral constraints (in Russian). Engineering bulletin, vol..У1 Publishing house of the USSR Academy of Science, Moscow–Leningrad
29. Reznikov RA (1971) Резников Р.А. Solution of problems of structural mechanics using digital computers (in Russian). – Stroyzdat Moskow
30. Robinson J, Hagenmacher GW (1971) The static force method as an eigenvalue problem. In: De Veubeke FB (ed) Proceeding IUTAM Symposium "Hige speed computing of elastic structures", Université de Liége, pp. 91–102
31. Rozin LA (1975) Bar systems treated as systems of finite elements (in Russian). SPbGTU Publishing House, Sankt-Petersburg
32. Rozin LA (1998) Problems of the theory of elasticity and numerical methods (in Russian). SPbGTU Publishing House, Sankt-Peters Rozin L.A. (1998) Problems of the theory of elasticity and numerical methods (in Russian). SPbGTU Publishing House, Sanct-Peterburg burg
33. Rzhanitsin AR (1982) Structural mechanics (in Russian). Visshaya shkola, Moscow
34. Shulkin YB (1984) Theory of elastic bar systems (in Russian). Nauka Publishing House, Moskow
35. Strang G (1976) Linear algebra and itf applications. Academic press, New York San Francisco London
36. Vinogradov AI (1957) On some properties of systems with given configurations of axes (in Russian). In: Rabinovich IM (ed) Research in structural theory, Issue 7.—. Stroyzdat Moskow, pp. 365–371.
37. Volmir AS (1956) Flexible plates and shells (in Russian) Gostekhizdat, Moskow
38. Vovkushevsky AV, Shoikhet BA (1981) Analysis of massive water engineering structures taking into account the opening of seams (in Russian). Energhiya, Moskow
39. Zukhovitsky SI, Avdeeva LI (1967) Linear and convex programming (in Russian). Nauka Publishing House, Moskow

4 Finite Element Models

A nonsense wearing a uniform is much more convincing than one totally naked. The very fact that a theory is expressed mathematically... somehow makes us treat it as serious.

J. Schwartz

4.1 Notes on a Structural Discretization

Long before the contemporary concept of discretization used in the finite element method (FEM) appeared, so-called physical discretization (or *physical lumping* in English literature) methods were in active use. There, to solve a continual problem, a system with a finite number of degrees of freedom used to be introduced on the basis of certain physical considerations. For example, recall a method by B.N.Zhemochkin [32] popular as long as thirty years ago, intended for the analysis of beams and slabs lying on an elastic semi-plane and elastic semi-space respectively. Particularly, mechanical bar models used to be constructed, equivalent to continual systems in a certain sense (for example, by accumulated energy). It may seem that the days of those theories are gone, and most researchers have realized that thy physical lumping is much weaker (and more naive) than the contemporary finite element analysis, now that the latter has all attributes of a "serious theory" based on advanced mathematics and theory of continua.

Though, there are some considerations that make us keep the methods of physical lumping ready at hand. The main of those is that the discretization of this kind is well adjusted to the intuitive thinking of many engineers who tend to build analogies and understand results of the finite element analysis better when involving those analogies. Also, the physical lumping can be used in the opposite direction, to represent a truly discrete structure as a continual one by means of "smearing".

Operations characteristic for the physical lumping such as the replacement of distributed masses with concentrated ones, the replacement of a solid elastic foundation with a system of closely spaced springs, are usually performed by guesswork. But there are examples of the physical lumping that lead to the same equations as FEM does, though there are not so many cases where it is feasible. The latter fact must be realized perfectly well in order not to fall victim to an unjustified intuition.

4.2 Basic Concepts of the Finite Element Analysis

In the finite element analysis, a continual area Ω, occupied by a structure that has an infinite number of degrees of freedom, is approximated by a discrete model consisting of a set of sub-areas (finite elements) having a finite number of degrees of freedom and interconnected only in nodal points.

Every finite element Ω_e ($e = 1,\ldots, m$) is characterized by the following properties:

- dimension of the space used (one-dimensional, two-dimensional, three-dimensional);
- geometrical shape, which is, most often, one of simplest geometrical figures (a line segment, a triangle, a rectangle, a tetrahedron etc.);
- set of nodes located at lines/surfaces of elements' contact (as a rule, though not always) which are common for adjacent elements;
- set of external and internal degrees of freedom used, most often referred to nodes (though not necessarily to nodes), such as displacements, slopes, etc. [1];
- system of approximant functions within the Ω_e region to define approximate expressions for displacement components, and their relation to the degrees of freedom of the finite element unambiguously determined by values of the components when known in the nodes for all degrees of freedom (we mean classical finite elements based on approximations of displacement fields);
- physical law that establishes a relationship between stresses and strains;
- definition of an energy space the elements of which are to be approximated by finite elements of this particular type. This energy space definition is equivalent to a reference to a class of problems for which this finite element type is applicable (FE of a plane stress plate, FE of a Kirchhoff – Love plate, FE of a Reissner plate lying on an elastic two-parametric foundation, a Timoshenko bar for spatial problems etc.);
- and the last (but not the least!), a list of limitations and recommendations regarding the applicability of a particular finite element.

One of most important properties of a finite element model is the maximum diameter of its elements

$$h = \max_e \left(\sup_{x,y\in\Omega_e} |x-y| \right), \tag{4.1}$$

which is often connected with error estimates of the method. In other words, h is a

---

[1] *Internal* degrees of freedom of a finite element are those whose respective generalized displacements do not figure in the expressions of the deformation energy of any other finite elements of the system different from this particular element. Generalized displacements that correspond to *external* degrees of freedom are included in the expressions of the deformation potential of both their respective element and some other (usually, topologically adjacent) elements.

least diameter of a sphere that can contain any finite element of the design model. Also, usually it is assumed that, as the diameter tends to zero infinitely, i.e. $h \to 0$, the following regularity condition holds: each finite element is capable of containing a sphere of radius $\rho \le Ch$ where constant C does not depend on h. This keeps one from using so-called *"needle"* elements (too much stretched rectangles, triangles with very small angles etc.).

As a rule, the approximant functions of FEM are polynomials or piecewise polynomials (a sub-area method), though there are elements with rational (so-called isoparametric elements), trigonometric, logarithmic and other approximations of the displacement field. The choice of the degrees of freedom of an element and its respective approximant functions determines completely the convergence rate and the error estimate of FEM.

If we fix all parameters of a finite element design model except for the diameter of the finite elements h, then we can imagine how we can get a sequence of approximate solutions $\mathbf{u}_h$ by changing this diameter. The convergence of FEM refers to this sequence's tending to the accurate solution $\mathbf{u}^*$ of the problem as $h \to 0$.

The above stated concept of a finite element modeling of a problem is based on what is called the application of *h-elements* in the literature on the subject. The relative simplicity of this representation enables one to solve the problem in a simple and efficient way. In recent years, some software systems (such as **NASTRAN** or **StressCheck**) sport another type of finite elements, so-called *p-elements*. Unlike *h*-elements, *p*-elements are capable of representing the curvature of a surface and peculiarities of the stress fields directly at simpler meshes containing less elements. The accuracy of the analysis is controlled by the *p* level (order of the polynomial) assigned to each element: the higher the *p* level, the better the accuracy. Technically, an *h*-element can be treated as a special case of a *p*-element where *p* is a fixed number defined by the software developers. When using *h*-elements, the accuracy of the result and the computational time grows as the number of elements increases. When using the *p*-elements, the same phenomenon takes places as the "*p* order" grows.

One of the primary motives for the use of *p*-elements is the case of areas of a complicated stress field where the geometry and external factors must be represented very accurately. These conditions include rounded corners in the design model instead of sharp ones, and distributed loads in place of concentrated ones. A more detailed model yields more detailed results. With *p*-elements, the model is adjusted by increasing the *p* level of the elements chosen either by the user manually or by the software by default until the answer satisfies a certain convergence criterion without changing the grid.

In spite of a lot of positive features possessed by *p*-elements, *h*-elements also have their advantages. Undoubtedly, the demand that *p*-elements should completely replace *h*-elements in all analyses, as well as the statement that *p*-elements are not finite elements, has no reasonable ground at all. The analysis of a global behavior of a structure can be often done better with *h*-elements. Also, those better represent solutions where there are jumps of stresses, for example, in ribbed plates, and in nonlinear problems. Finally, *h*-elements token an elaborated time-proved technology. To process models consisting of h-elements, there is a variety of mesh generators, preprocessing and post-processing tools.

A combined use of these methods where both h-elements and p-elements are involved provides advantages for complex analyses. The possibility to combine h-elements and p-elements in a single design model is advantageous for most structures which should be investigated roughly with h-elements to obtain a general answer, while some particular areas of interest can be investigated with p-elements to provide a higher accuracy in critical points.

As the computational time grows with the increase of the p level, a good modeling should involve both h-elements and p-elements to provide for more accurate solutions of large problems without losses that can happen due to the use of p-elements only. The strategy of solution that combines a detailed investigation and a rough general estimation, uses both rough and fine meshes, or applies p-elements to selected areas, is sometimes called a *global/local analysis*.

Further we will deal with the concept of h-elements only, because it is this one that has become most popular in a majority of commonly applicable software.

Thus, the h-element analysis means a process of construction of a sequence of approximate solution for a problem by the finite element method where each pair of successive steps of the sequence are different only in their size of the integration area mesh cell. *Types* of finite elements involved in different steps of this process are not modified from one step to another.

When a software manual refers to finite element types, it usually means a reference to a number or another identifier (such as a name) of the respective section of the *finite element library* used by the software. At the beginning of this section we have already mentioned the set of parameters and other features that define the finite element type. Those features include the set of degrees of freedom of a finite element referred to its nodes. This important characteristic enables us to select a class of finite elements, among others, with all their nodal degrees of freedom having the sense of linear displacements and, possibly, slopes of the nodes.

Confining ourselves to this class of finite elements, we define a node of a finite element model mechanically as a *perfectly rigid body of vanishing sizes*. More generally, a node means a geometrical point of the area occupied by a finite element where one or more degrees of freedom are located. In this connection, the respective class of finite elements can be called a class of finite elements with *rigid nodes*[2].

Finite elements with non-rigid nodes are also widely applicable. Let us mention some examples such as a fully consistent rectangular Bogner – Fox – Schmit finite element [7] with four degrees of freedom in each node for a flexural plate, a triangular element with six degrees of freedom in the node for the same problem (Cowper – Kosko – Lindberg – Olson [10]). In elements of this type, the generalized degrees of freedom are not only values of the deflection function in the nodes

[2] We do not claim this term to be linguistically perfect. Though, we'd like to emphasize the necessity of an explicit definition of the rigid-node finite element class. In the literature, elements with non-rigid nodes are sometimes entitled "*finite elements of higher order of accuracy*".

and its first derivatives with respect to the coordinates (mechanically, they are exactly linear displacements and slopes of the nodes), but also second derivatives of the same function.

The usual motivation for introducing additional degrees of freedom is the intention to eliminate breaks in stress fields (eliminate the inconsistency) by increasing the order of approximation of the desirable functions. The increased order of approximation is also useful for improving asymptotic convergence estimates at the same finite element meshes. But this profit is not gained for free, and it will be paid for by more degrees of freedom in the system. Also, another advantage of rigid-node finite elements will be lost: the physical obviousness of the nodal degrees of freedom. A practical engineer understands intuitively and feels better about the rigid-node finite elements because for him it is an imaginary bridge between structural mechanics of bar systems and the finite element method.

There is another upsetting circumstance in connection with finite elements of higher order of accuracy which is not often paid a proper attention to. The matter is that with rigid-node finite elements the linear transformation of coordinates (a transition from a local system of axes to a global one and vice versa) is done in a conventional way. Meanwhile, there can be problems with the finite elements of higher order of accuracy at this point, especially, if the involved nodal degrees of freedom include incomplete higher derivatives.

Let us illustrate this fact with a rectangular Bogner – Schmit finite element for a flexural plate. For this element (a pretty effective one, by the way), the following set of degrees of freedom in each node is adopted:

$$w\,, \qquad \theta_\xi = \frac{\partial w}{\partial \eta}\,, \qquad \theta_\eta = -\frac{\partial w}{\partial \xi}\,, \qquad \frac{\partial^2 w}{\partial \xi \partial \eta}\,, \tag{4.2}$$

where w is the deflection function of the plate, (ξ,η,ζ) is a local coordinate system (at the element).

Let the transition from the local coordinate system to the global coordinate system (X,Y,Z) be performed by the rotation by angle φ about axis ζ as shown in Fig. 4.1.

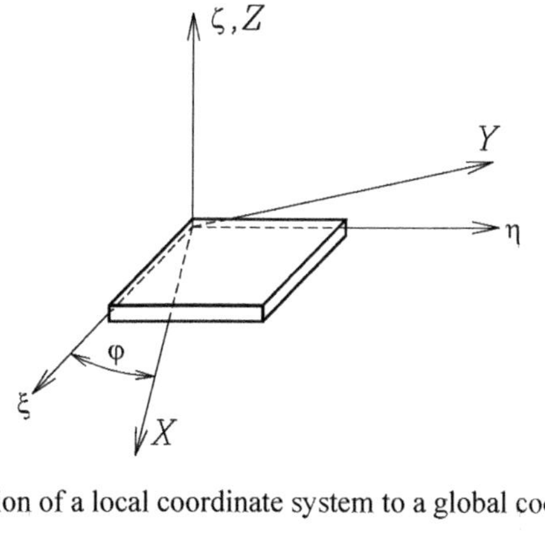

Fig. 4.1. A transformation of a local coordinate system to a global coordinate system

This transition is expressed analytically as a set of linear transformations

$$x = \xi \cos\varphi + \eta \sin\varphi \,, \qquad y = -\,\xi \sin\varphi + \eta \cos\varphi \,, \tag{4.3}$$

and it may be easily proved that the fourth degree of freedom in set (4.2) will be

$$\frac{\partial^2 w}{\partial\xi\partial\eta} = \frac{1}{2}\frac{\partial^2 w}{\partial x\partial x}\sin 2\varphi + \frac{\partial^2 w}{\partial x\partial y}\cos 2\varphi - \frac{1}{2}\frac{\partial^2 w}{\partial y\partial y}\sin 2\varphi \,. \tag{4.4}$$

Therefrom one can see that if angle φ is not a multiple of $\pi/2$ then the transition from the global coordinates to the local ones is possible only if the set of degrees of freedom of each node includes all second-order derivatives, and the only mixed derivative is not enough for the coordinate transformation.

Apparently, the circumstances noticed above together can explain the fact that the absolute majority of contemporary software developments of the industrial purpose confine their users to finite element libraries containing rigid-node elements only.

Meanwhile, there are situations where this limitation is too vexatious. For example, in the analysis of systems that include thin-walled bars of open profiles, the essential role in the general deformation belongs to the warping (deviation of points of bar cross-sections from the initial planes of these cross-sections). To allow for this, one has to (?) introduce an additional seventh (warping) degree of freedom in nodes. In the previous sentence, the interrogation mark is intentionally added after the phrase "one has to", because, as it follows from results of Section 8.8, there is a possibility to build a model of a thin-walled bar without involving any but rigid nodes. But that is only a peculiar trick that does not guarantee one against using non-rigid-node models in more complicated designs.

Though, we envision that in prospect, competitive commercial software products will be equipped with a wider range of finite element types available to the user, so that the limitation of using rigid-node elements only will be eventually eliminated with time.

4.3 Modeling of Bar Systems

All commercial software systems based on the finite element method feature the capability of analyzing bar design models. In these models, the shape functions intended to determine displacements within an element by displacements of its nodes present *accurate* solutions (for constant-section bars), and it is this feature that makes us distinguish the said systems from a number of others where the shape functions are only approximate representations of the displacement field.

In this connection we would like to note that P. Tong [29] has proved an interesting theorem regarding one-dimensional problems. Let the approximant functions (shape functions) used for building a finite element model on the basis of the Lagrangian minimization be functions that satisfy the respective homogeneous differential equation at each finite element. Also, let each finite element have two nodes, and let each node have m degrees of freedom where $2m$ is the order of the differential equation that governs the problem. Then, independently from the mesh parameters (the number and relative sizes of finite elements) and the form of the partial solution (caused by the load), the finite-element solution u^h for displacements in the mesh nodes will be the same as the exact solution. In other words, u^h interpolates the exact solution u in the following sense:

$$\frac{d^j u^h}{dx^j}(x_i) = \frac{d^j u}{dx^j}(x_i) \qquad (j = 1,\dots, m-1)\,, \tag{4.5}$$

where x_i is the coordinate of any nodal point.

The use of bar-type elements follows the common strategy of the construction and investigation of systems consisting of elements with a finite number of freedom degrees [21]. Software developments should provide the use of such elements to ensure the universality of design and analysis of various structural objects.

It should be remembered that though the statement of the FEM convergence problem is often senseless with respect to bar-type finite elements, it is not always the case. For example, when Hermitian polynomials are used as shape functions, being exact solutions of the static problem in the case of a constant bar rigidity along its length, they can be approximate solutions (violate the applicability conditions of Tong's theorem) in other conditions, say, at a variable rigidity in the static analysis or in a dynamic/buckling analysis. The same regards bars on elastic foundations. Meantime, there are FEM implementations that use Puzirevsky–Krylov functions (an oscillating bar), Kornoukhov functions (a compressed/tensioned bar), Belous–Chudnovsky functions (an oscillating compressed bar) as shape approximants. In these cases, the solution is exact for an oscillating bar too, which is analyzed with taking into account the influence of the longitudinal force on the moments. Therefore, the analyst must be aware what shape functions are operated by the software he uses.

Sometimes the problem of "small length" of a bar finite element arises. It means the following: the analyst remembers that textbooks on strength of materials or structural mechanics define the bar as an object with its one size (length) much more than the other two that describe its cross-section. But in a particular design, a bar, being only an element of the model, can be very short. The typical example is a bar simulation of stiffening ribs of a plate, and the lengths of those bars are defined by a pretty fine division of the plate into finite elements. It may seem that an explicit violation of the convention what is bar and what is not arises in this case. Actually, there is no violation here because the assumption of oblongness has been necessary only to validate the respective differential equation[3]. As for the method of solution, when we use a pretty fine division of the bar (or the integration interval) into segments, it by no means influences the form of

---

[3] Now it seems curious, but in middle sixties engineers were out in the field about the understanding of the finite element method. Feelings ran high between adherents of "physical lumping" and "mathematical discretization". One of smashing arguments of the lumping guys was based exactly on considerations of this kind.

the equation. It is certainly true with both the "exact" and approximate shape functions.

Discussing the refinement of a design model by making the mesh finer, the author of [2] says "*...this refinement leads to a by-effect due to the violation of principal assumption of the FEM theory of relationships that form the basis for shape functions and stiffness matrices of bar elements: l >> b. This causes singularity effects to arise, and the solution tends to diverge...*».

The very fact is noticed correctly, but its interpretation is not right: the matter is not that the relationships are violated (we discussed it above) but that we have a badly conditioned problem considered by Bruce Irons in detail [14]. He states, for example, that a known case of bad conditionality is a cantilever beam loaded by a force at its free end (Fig. 4.2). The full stiffness which determines the displacement under the force is equal to $G = 3EJ/L^3$, while the initial stiffness of one segment of the model is $K = 12EJ/(L/n)^3$. If the number of elements is $n = 50$, it is 5×10^5 times greater than G.

The main reason for the bad conditionality is that a solution-distorting effect is introduced in the analysis via large numbers, and it can blur more essential aspects of the system's behavior. The beam bends along a smooth curve, but if its elements are too short, errors in the calculated deflections may make this curve oscillate, and this will lead to more serious errors at big values of the rigidity of particular segments. For example, the deflection of one node may correspond to a greater deformation energy than that of all other deflections taken together.

It is interesting to notice another fact: a change of the used measurement units can cause the conditionality number of the governing equation system to change, and thus deteriorate the analysis accuracy in cases when this number grows. One may think the common scaling by changing meters to millimeters must not affect the computation. But, in fact, this change leaves the units of rotations/slopes intact (they were and still are radians), therefore for a bar element the stiffness matrix components EA/L that correspond to its linear displacements and $6EJ/L^2$ corresponding to its slopes change differently, and this leads to different relations between the maximum and minimum eigenvalues of the matrix, that is, the different conditionality number. Of course, it is convenient for a user to be able to choose measurement units in his software environment, and most existing CAD programs feature capabilities of setting up the system of units desired by the user. But it would be very useful too, if commercial software products could suggest measures to guarantee against computational troubles due to an ineffectual choice of data dimensions.

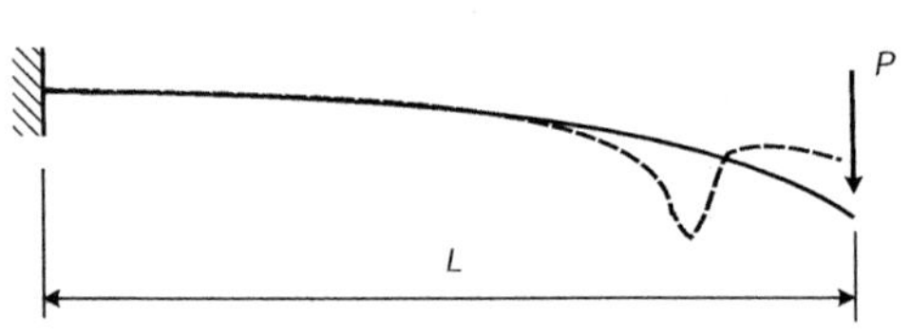

Fig. 4.2. A distortion of the deflection shape

The way out of the situation is pretty simple and almost evident. It is an automatic use of dimensionless parameters in the course of formation of the governing equation system. It is astonishing: as far as we know, no software system of the number of widely popular ones suggests a procedure of transition to dimensionless unknown variables. We do not see any particular difficulties in the implementation of this procedure. The only problem to be solved algorithmically by the developers of a software system is the choice of some characteristic linear dimension for any particular model which will determine the scaling factor for calculating dimensionless linear displacements.

We can indicate another method, maybe more preferable due to a simplicity of the algorithm, to perform the transition to dimensionless parameters in the system of governing equations of a problem (linear algebraic equations). With this purpose, the equation system of type (3.8), first written as $\mathbf{Ku} = \mathbf{p}$, will be primarily (before its solution) converted to an equivalent form using a linear transformation with matrix $\mathbf{\Lambda}$

$$\mathbf{\Lambda}^{-1}\mathbf{K}\mathbf{\Lambda}^{-1}\mathbf{\Lambda}\mathbf{u} = \mathbf{\Lambda}^{-1}\mathbf{p},$$

or $\mathbf{K}^{\circ}\mathbf{u}^{\circ} = \mathbf{p}^{\circ}$, where $\mathbf{u}^{\circ} = \mathbf{\Lambda}\mathbf{u}$, $\mathbf{p}^{\circ} = \mathbf{\Lambda}^{-1}\mathbf{p}$, $\mathbf{K}^{\circ} = \mathbf{\Lambda}^{-1}\mathbf{K}\mathbf{\Lambda}^{-1}$. It is convenient to take for matrix $\mathbf{\Lambda}$ a diagonal matrix composed of square roots of diagonal elements of the original stiffness matrix $\mathbf{K}$, that is, assume

$$\mathbf{\Lambda} = \left\| \begin{bmatrix} \sqrt{k_{11}} & & 0 \\ & \ddots & \\ 0 & & \sqrt{k_{nn}} \end{bmatrix} \right\| .$$

It is evident that all diagonal elements of matrix $\mathbf{K}^{\circ}$ scaled in this way become equal to one[4], while its other elements can be calculated by a simple rule

$$k_{ij}^{\circ} = \frac{k_{ij}}{\sqrt{k_{ii}}\sqrt{k_{jj}}} .$$

Let us return to the issue of dividing a bar into finite elements along its length. The necessity of dividing the bar into fine finite elements can be caused by the purpose of a nonlinear analysis (a deformed geometry analysis). The simplest statement of the problem includes a combined bending and compression when one has to take into account additional bending moments that arise in the bar's cross-section due to the longitudinal force causing a moment at lateral deflections of the bar's axis. This problem can be solved accurately (in a closed form) for bars of a constant cross-section with a given longitudinal force. But in statically indeterminate bar systems there can be problems due to the necessity to re-determine the longitudinal forces. Of course, there are systems where the bar flexural rigidity

[4] Obviously, this method of scaling is applicable only to matrices whose main diagonal contains strictly positive elements only. Positive definite matrices satisfy this requirement automatically.

change in the course of tension/compression causes only a slight redistribution of the axial stresses. Then there is no necessity to recalculate, but systems of this kind cannot always be easily recognized (those are called quasi-mono-cyclic according to a classification by N.V. Kornoukhov [17]).

We think it is reasonable to give here a useful classification of types of analysis of compressed/flexural and tensioned/flexural bar systems suggested by Kornoukhov in his day ([17], p. 80) and, as we believe, undeservingly forgotten.

If the set of longitudinal forces in bars of a system is known beforehand, before doing any computation (this is a rare though feasible case) then the deformed geometry analysis can be performed in one step using appropriate transcendental functions that depend on the longitudinal forces as parameters. This type of analysis was entitled *mono-cyclic* by Kornoukhov.

If the set of longitudinal forces in bars of the system can be determined with a sufficient accuracy without a deformed shape analysis (for example, by a primary linear analysis of the system) and then used in a one-step repeated analysis of the same system in its deformed geometry with no subsequent recalculation of the longitudinal forces, then this analysis, according to Kornoukhov, is called *quasi-mono-cyclic*.

Finally, if it is necessary to recalculate the system in its deformed geometry for many times, re-determining the longitudinal forces at each step, then we have a *poly-cyclic* analysis.

After the computer-aided FEM analysis has become commonly recognized, another approach has grown in popularity: to solve a problem as geometrically nonlinear one, dividing the bars themselves into finite elements, just as well as plates, shells and solid bodies included in a system. It is in full accordance with general principles of algorithms of geometrically nonlinear finite-element analysis. But one does not have to get rid of all old things only because they've been known for a long time. And new things are not necessarily better than old ones. Developers that understand this simple rule want to provide their users with multiple choices rather than confine them to strictly limited capabilities. We believe strongly that the latter attitude is a disrespect to the users, and we are not afraid to be misunderstood.

An analyst should understand clearly what options he can really choose between when using one set of finite elements or another. For example, the **SCAD** system features a bar finite element of type 10 for which one can specify a value of the longitudinal force, and it will be accounted for by the use of special functions[5] — and users are applauding.

But one can use also a bar element of type 310 for which one does not have to specify the longitudinal force (it will be determined by a geometrically nonlinear analysis) — more applause[6].

---

[5] In Russian engineering literature those are usually called Kornoukhov functions.

[6] A software development will be received with real thunders of applause when it performs a triple somersault at the users' request: a poly-cyclic analysis of a bar system with an automatic choice of longitudinal forces in the system's bars when transiting from step to step of the analysis.

In this case, the longitudinal strains of the bar's fibers are calculated accounting for nonlinear terms by the formula

$$\varepsilon_x = \frac{du}{dx} + \frac{1}{2}\left[\left(\frac{du}{dx}\right)^2 + \left(\frac{dv}{dx}\right)^2 + \left(\frac{dw}{dx}\right)^2\right] - y\frac{d^2v}{dx^2} - z\frac{d^2w}{dx^2}, \tag{4.6}$$

and displacements u, v and w are approximated by cubic polynomials, i.e. the solution is sought in the classical FEM style. One must remember to use the equilibrium equations for the deformed state because geometrical equations (4.6) correspond to the second floor of geometrical nonlinearity! The meshing of the bar in its span must be such as to provide a satisfactory accuracy of the solution. Usually, it suffices to have 4 to 6 segments in the bar's span.

If the analyst nevertheless has doubts about the needed number of segments the bar is divided into, the following estimation can be helpful. Coefficients of the stiffness matrix of a particular segment of the bar are related proportionally to the transcendental Kornoukhov functions

$$\alpha = \frac{\nu}{\mathrm{tg}\nu}\,\frac{\mathrm{tg}\nu - \nu}{2\mathrm{tg}\frac{\nu}{2} - \nu}, \qquad \beta = \frac{\nu}{\sin\nu}\,\frac{\nu - \sin\nu}{2\mathrm{tg}\frac{\nu}{2} - \nu}, \qquad \gamma = \frac{\nu^3}{2\mathrm{tg}\frac{\nu}{2} - \nu}, \qquad \delta = \alpha + \beta,$$

with their argument ν related to longitudinal force N, being $\nu^2 = l^2N / EI$ where l is the length of the bar's segment. At small values of ν (such as those less than $\pi/2$), these functions can be approximated pretty accurately by linear aggregates $C_1 + C_2\nu^2$ where C_1 and C_2 are constants that depend on the form of one of the four Kornoukhov functions. The correctness of this representation at relatively small values of ν (say, at $\nu \le \pi/2$) can be proved formally, by expanding these functions into Taylor – Maclaurin series in the vicinity of zero. In order not to bother the reader with extra mathematics, let's refer to diagrams of these functions (Fig. 4.3). It is easy to understand that after replacing the Kornoukhov functions with the linear constructions presented above the stiffness matrix $\mathbf{K}$ of the bar's segment will have the form $\mathbf{K} = \mathbf{K}_0 + \nu^2\mathbf{K}_G$ where $\mathbf{K}_G$ is exactly the geometrical stiffness matrix that participates in the geometrically nonlinear analysis. Hence a practical recommendation to check the sufficiency of a division of a compressed bar into segments: the length of a particular segment must be such that the inequality $\nu \le \pi/2$ should hold at it.

There is another peculiarity in the use of results of analysis of a bar system divided into fine elements. It is the fact that some normative recommendations regarding structural design involve the ratio of the cross-section's sizes to the bar's length explicitly or, more often, implicitly, and thus have a sense only for particular relations between those parameters. Therefore, at the stage of using the results, one must clearly distinguish between a bar finite element that represents a small segment of a bar, and the whole bar to which the recommendations apply. Without distinguishing those, concepts like bar flexibility criteria, oblique cross-section design, validation of local stability of webs, legs and flanges of steel beams etc. may prove senseless.

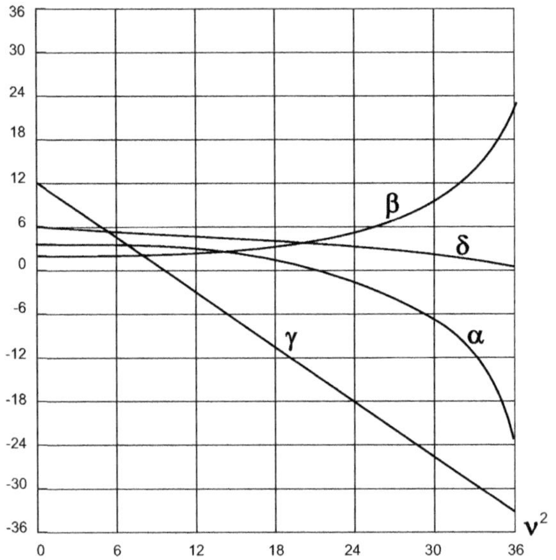

Fig. 4.3. Diagrams of the Kornoukhov functions

The implementation of this distinguishing has lead naturally to the concept of a *structural element*. It may be, for example, a continuous chain of finite elements of equal cross-sections with equally oriented principal axes of inertia, without hinges in their contact nodes.

This description refers to a prismatic bar of a constant section. Of course, it can be different (for example, the torsion of the bar or its inconstant section could be taken into account), but it is the very fact of the existence of such a concept that is important. Using it, we can operate, for example, a parameter such as the bar's length understood as the length of the structural element (not a part the structure is made of!). Apparently, the concept of a structural element may apply to some fragments of a finite-element plate or shell model if there are characteristics of the structural element which should refer to this element as an integral object and cannot be handled at the level of a finite element[7].

To conclude this section, let us return once more to the allowance for the longitudinal force in a bar. One often makes a peculiar mistake here: when considering pre-stressed structures such as prefabricated concrete systems assembled from parts pre-stressed during fabrication, one specifies the pre-stress as the longitudinal (parametric) force N. But the parametric force N must be **external** with respect to the element, while the mentioned pre-stress is a part of the internal system of forces self-balanced in the element (tension in reinforcement bars is compensated

[7] Note that a "structural element" is a design concept (in the sense that it participates with its geometrical parameters in structural design analyses intended to provide the compliance with standard regulations) rather than just a descriptive term such as an "item" in the **MicroFE** software environment.

by compression in concrete). In order to make sure it is a mistake to specify this force as a parametric longitudinal force, one should just ask himself a question: what is the sign of this force? The issue of the allowance for the pre-stressing is discussed in more detail further below.

4.4 Finite Element Grid Modeling

The theory of the finite element method deals very much with the problem of convergence, i.e. an asymptotic behavior of estimates of accuracy of an obtained approximate solution as the finite element mesh becomes infinitely dense. A number of important convergence theorems have been stated. For example, for consistent elements it has been determined that if (k–1) is the degree of a polynomial used within a finite element for approximating its displacements, and an elliptic boundary-value problem of order $2m$ is solved approximately by having found a displacement solution u^h, then the error of energy in comparison with the exact solution u will be

$$\mathrm{E}(u-u^h,\, u-u^h) \le Ch^{2(k-m)}\,\|u\|^2_k, \tag{4.7}$$

where h is the diameter of the finite element mesh defined by formula (4.1). Estimate (4.7) contains the norm $\|u\|_k$ in its right part, a root-mean-square of kth derivative of the desirable function u.

The following error estimates take place for s^{th} derivatives:

$$\|\, u-u^h \,\|_s \le Ch^{k-s}\,\|\, u \,\|_k\,, \qquad \text{if } s > 2m-k\,; \tag{4.8.a}$$

$$\|\, u-u^h \,\|_s \le Ch^{2(k-m)}\,\|\, u \,\|_k\,, \quad \text{if } s \le 2m-k\,. \tag{4.8.b}$$

Similar estimates for inconsistent elements have been obtained in a series of papers by I.D. Evzerov and V.S. Karpilovsky (see [11, 15] for examples).

Table 4.1 gives reference data on the convergence rate for some types of finite elements, such as those used in the **SCAD** computational software. These data are pretty typical, and the reference to a particular software development is of little importance here.

The data listed in the table enable one to assign an approximate density of the finite element mesh on the basis of, say, the following peculiar considerations: [30]: "*...let us just note that the convergence takes place assuming some natural limitations of initial data and the meshing of the area are satisfied, and the error of the determination of stresses/strains is of the order ch/L where c denotes a constant depending on the area's shape; h is the mesh spacing; L is a characteristic size of the area. This estimate should guide the definition of the mesh spacing, depending on the desired accuracy. For example, if you want the accuracy of the approximate solution to be 5%, choose the spacing of the mesh as some 1/20 of its characteristic size...*", i.e. a characteristic two-dimensional spot would include about 400 nodes.

Table 4.1.

Element's entitlement	Power of the convergence rate estimates by:			
	Displacements	stresses	moments	shear forces
Rectangular plate element	2	—	1	—
Triangular plate element	2	—	1	0
Quadrangular (4 to 8 nodes) plate element	2	—	1	0
Rectangular element of plane elasticity	2	1	—	—
Triangular element of plane elasticity	2	1	—	—
Quadrangular (4 to 8 nodes) element of plane elasticity	2	1	—	—
Quadrangular (4 to 12 nodes) element of plane elasticity	2	1	—	—
Parallelepiped	2	1	—	—
Tetrahedron	2	1	—	—
Triangular prism	2	1	—	—
Spatial isoparametric six-node element	2	1	—	—
Spatial isoparametric eight-node element	2	1	—	—
Spatial isoparametric twelve-node element	2	1	—	—
Rectangular shell element	2	1	1	0
Triangular shell element	2	1	1	0
Quadrangular shell element	2	1	1	0
Quadrangular (4 to 8 nodes) shell finite element	2	1	1	0

Mathematically proved convergence conditions of the finite element method, such as those formulated at the beginning of this section, may be supplemented by some considerations of mechanical nature to provide a possibility for an informal analysis. Recall that the solution of a mechanical problem is equivalent to the minimization of the full potential energy of the respective system (a Lagrange functional), and the approximation of the displacement field by a finite set of predefined functions bounds the system's capability of deforming arbitrarily, i.e. it can be treated as an imposition of some constraints. If finite elements are inconsistent, then on their boundaries there can be some displacements impossible in the continual analytic model (such as mutual rotations of plates) that correspond to the absence of some of the constraints.

As the number of finite elements grows and their size decreases, the total number of degrees of freedom increases, thus the influence of the imposed nodal constraints reduces. It is this process that provides the method's convergence for consistent finite elements under certain conditions. On the other hand, the same process leads to less displacements on the inter-element boundaries between inconsistent elements which can be interpreted as a kind of the restoring of previously removed constraints. Thus, the convergence of inconsistent elements can take place only in the case when positive tendencies of overcoming the imposed constraints prevail over the negative tendency of imposing new constraints on the inter-element boundaries.

4.5 On Practical Convergence

One should remember that the afore-said estimates of the convergence rate are oriented at asymptotic properties of the solution, while a practical analyst is interested with the extent of proximity of an approximate solution obtained at a particular finite element mesh. Of course, in most cases the asymptotic convergence is accompanied by an appropriate "practical convergence". The latter term will mean the possibility of obtaining an acceptable accuracy with a comparatively rough meshing. But this rule has its exceptions too. In this connection we'd like to cite from a great mathematician and physicist A. Poincare (quoted from [6]):

"*... of two series where their general terms are $1000^n/n!$ and $n!/1000^n$, mathematicians would name the first one to be convergent... because the millionth term is much less than $999{,}999^{th}$, whilst the second term would be treated as divergent for its general term may grow unlimitedly. Astronomers, on the contrary, would take the first series for a divergent one because its first 1,000 terms go increasing; and the second series would be called by them convergent because its first 1,000 terms go decreasing and this decrease is quick enough at the beginning.*" And then follows a breathtaking conclusion: "*Both views are valid: the first one in theoretical investigations, and the second one in computational applications.*".

One should also take into account the fact that theoretical estimates of the convergence rate are usually built on the basis of some pessimistic assumptions about the displacement field which do not necessarily take place in every particular problem. Also, in the course of proving convergence theorems chains of more and more strong inequalities are used, and finally the estimates like (4.7) and (4.8) contain unknown constants. In this connection, practically one often uses the analysis of a sequence of solutions on finite element meshes that thicken gradually. Then, one makes a decision on the acceptability of the level of details that has been chosen after the results of this comparative analysis.

Certainly, it is hardly reasonable to solve a whole big problem on gradually thickening meshes, but this analysis makes sense, obviously, for characteristic fragments of design models. The empirical fact of the solution's stability as the mesh is being thickened is a pretty convincing argument "pro" this approach.

All that has been said must not be treated as an appeal to a pure empiricism. Theoretical investigations of convergence are very important, and their results can be used in practice. But there are some serious problems stated above which an analyst should be sure to take into account. One of the first problems is that a satisfactory practical convergence of displacements does not necessarily imply a good convergence of internal stresses which the analyst is interested with. Those are determined by the differentiation of the displacements, and the differentiation is an incorrect operation in the sense that a slight variation of a function's value may cause a substantial variation of the function's derivative.

Thus, validation of practical convergence must be oriented at investigating results important for a particular problem to be solved. For example, the spacing of a mesh needed to obtain an acceptable accuracy in shell problems may depend on a number of reasons. It is often the case that a small thickness of the shell limits the area of bending moments to an edge zone where the moments change substantially. The membrane stresses can be calculated with a high accuracy even on a rough mesh, while the variation of the bending moments require an extensively fine mesh for being described properly. Also, there is a difficulty of comparing the stresses obtained on meshes of different densities: the stresses are often determined in centers of finite elements, and a certain effort is needed to compare stresses in different points.

When using some types of finite elements (such as triangular elements with the linear displacement approximation for plane elasticity), the stress fields are piecewise constant functions, and the areas of their constancy coincide with the mesh's triangles. Values of the stresses determined with these elements vary intensively from one element to another, therefore usually one applies an averaging of the stresses by elements of a star and refers those to a nodal point. The comparison of such stress fields is encumbered with the averaging operation.

Organizing the validation of the practical convergence, one should remember that the problem being solved may have unpleasant peculiarities due to an incorrect idealization of the structure. A typical example is the idealization of a load as a concentrated force (practically unfeasible situation) which may cause such features of the problem's solution as infinitely big values (a logarithmic singularity in the expressions of bending moments in a plate under a concentrated force) or high gradients of the stress field.

Thus, the practical convergence check should be performed with examples similar to the problems of interest but having accurate solutions and known unpleasant singularities. In this case the interpretation of the testing results becomes more informative. Some problems of this kind are discussed in the next section.

4.6 Convergence Validation for some Models

Below, a comparative analysis of a simply supported square plate, loaded by a uniformly distributed force throughout its area, is given.

The analyses were performed with four finite element meshes: 4x4, 8x8, 16x16 and 24x24. Isolines of all results of the comparative analysis were built (Fig. 4.4) to estimate the general stress and strain distribution.

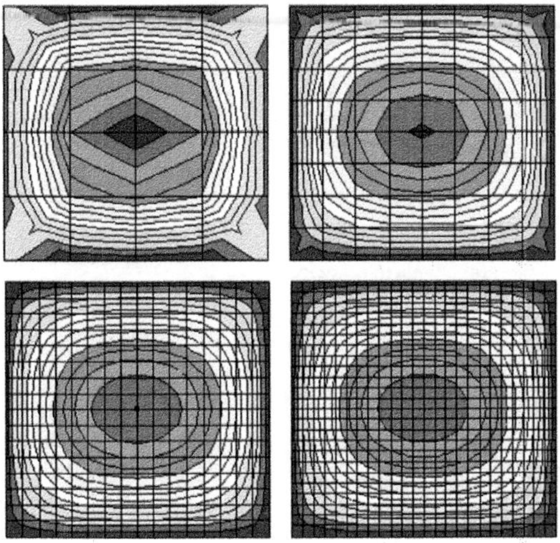

Fig. 4.4. Isofields of bending moments obtained with different design models and finite element meshes

The variation of deflections, moments and shear forces was under investigation throughout the plate's area and in points where their extreme values were achieved. The latter note is important because it is the extreme values that are questionable in cases where no convergence is warranted such as the shear forces (see Table 4.1). Generally, the results of the finite element analysis behave much better throughout the field.

Fig. 4.5 shows results regarding displacements, bending moments and shear forces obtained with finite elements of different types on the aforesaid meshes.

As one can see in Fig. 4.5, the practical convergence takes place for the deflections and bending moments with the use of finite elements of different types. For the shear forces, one of the element types gives values noticeably different from those obtained by the use of other finite elements. We should note that this element (a quadrangular element having 4 to 8 nodes) was used in an arrangement where it was connected to four nodes only, though it was possible to introduce more nodes on its sides. Checkup calculations of this system showed that the accuracy of the results improved significantly and became close to that obtained with meshes twice as dense. For example, for a mesh of 8×8 elements the deflection was 0.01701, the bending moment was 0.0442, and the shear force was 0.278.

In another series of numerical experiments, when the same plate was loaded by a concentrated force, results shown in Fig. 4.6 prove to be less optimistic. Here the practical convergence rate of the moments slows down, and that of the shear does even more significantly. The latter's values were taken in a point located at a quarter of the plate's thickness from its center. Apparently, one should not take into consideration values of the shear forces in points so close to the location of the concentrated force.

(*a*)

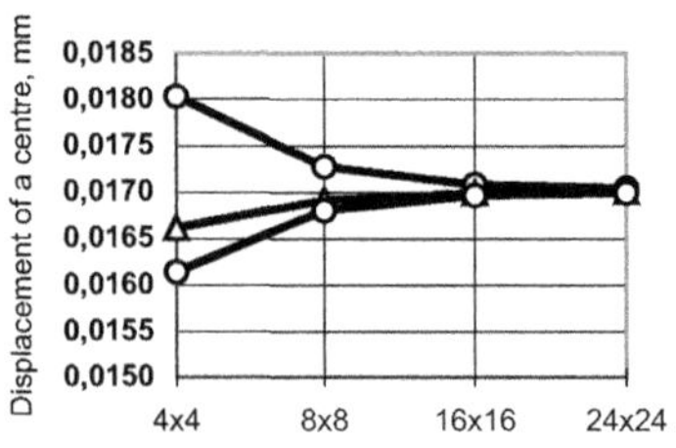

(*b*)

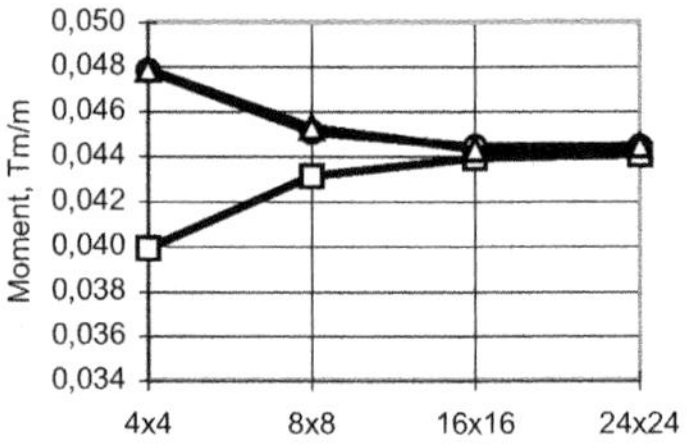

(*c*)

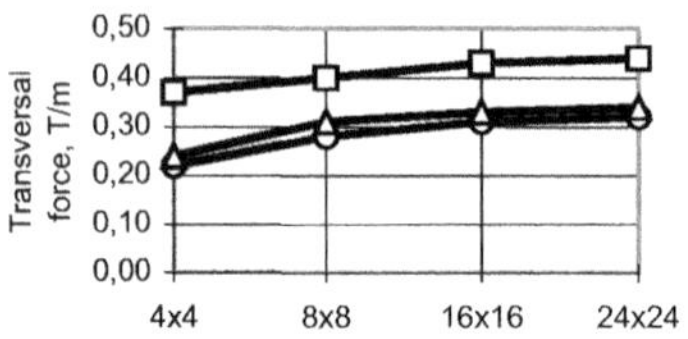

Fig. 4.5. The convergence of results for a distributed loading: (*a*) by deflections, (*b*) by moments, (*c*) by shear forces. Circles denote results for **SCAD** finite elements of type 11 or 41; squares show types 20 or 50; triangles show types 12 or 42

a)

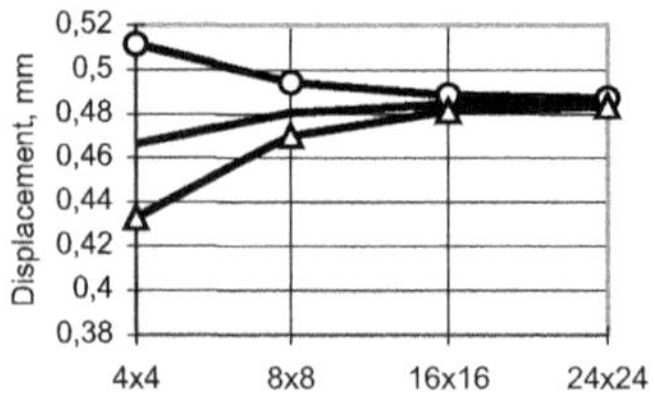

b)

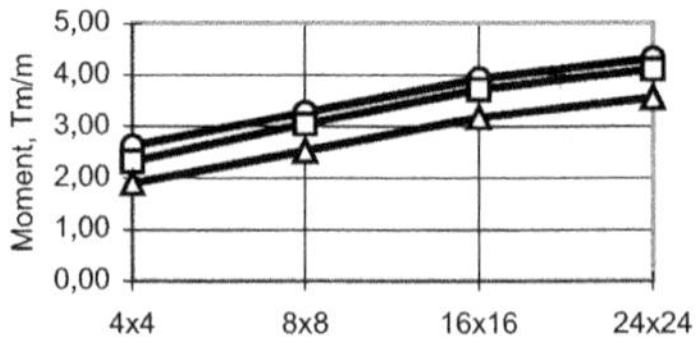

c)

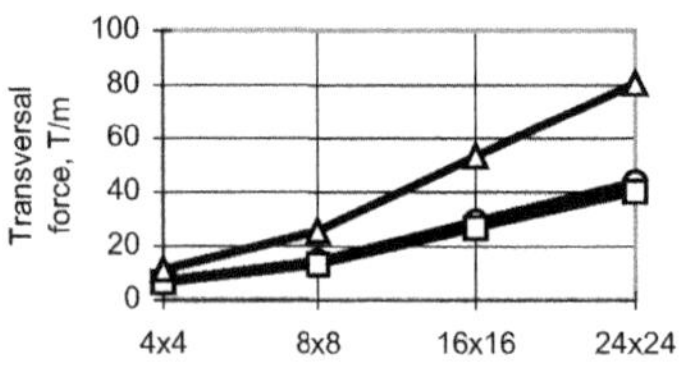

Fig. 4.6. The convergence of results for a concentrated force:

(*a*) by deflections, (*b*) by moments, (*c*) by shear forces. Circles denote results for **SCAD** finite elements of type 11 or 41; squares show types 20 or 50; triangles show types 12 or 42.

To conclude this section, we should notice that the practical convergence validated on a solely experimental basis in a series of kindred calculations does not necessarily correlates with the theoretical convergence (just take a look at the results regarding the shear forces!). This is due to the fact that each time a particular problem is under consideration, where unpleasant singularities of the desirable stress field may or may not reveal themselves, and the general case where the convergence is not guaranteed might not take place. We'd like to make a pretty long quotation here:

"*An interesting problem arises in connection with the question of completeness of a system of coordinate functions in methods like that by Galyorkin. Recall that even we begin with a complete system of functions, actually we end up choosing only a few from the system. On the other hand, one can easily make sure that any finite system of functions can be treated as a part of some complete system. But this by no means implies that finite systems of functions chosen to serve as coordinate ones are all the same from the standpoint of completeness.*

The system selected must have the property of a ***practical completeness with respect to a problem being solved*** (our boldface — A.P. and V.S.), *i.e. one must be sure (or at least one must reasonably hope) that the solution can be approximated with an acceptable accuracy ...*"[33].

This issue is discussed in more detail in sections that follow. Here we would like to note that most publications, unfortunately, seem to be too prudish to discuss the behavior of a solution with respect to shear forces. A pretty typical example of this is the monograph [4] that gives a lot of comparisons of the practical convergence rate but only for displacements and normal stresses (which means moments too).

Satisfactory results concerning the shear can be obtained only by the use of specially constructed elements that take into account the lateral shear strain. Fig. 4.7 illustrates it visually (the dash line shows accurate values).

One should note that the quicker convergence of results for some finite element types is paid for by a noticeably bigger computation time. Therefore one should compare the loss of time caused by the use of those elements and the computation time spent for the solution of the same problem on a finer mesh with elements of another type. Apparently, in relatively small problems where the total computation time is small, those comparisons can be of little sense.

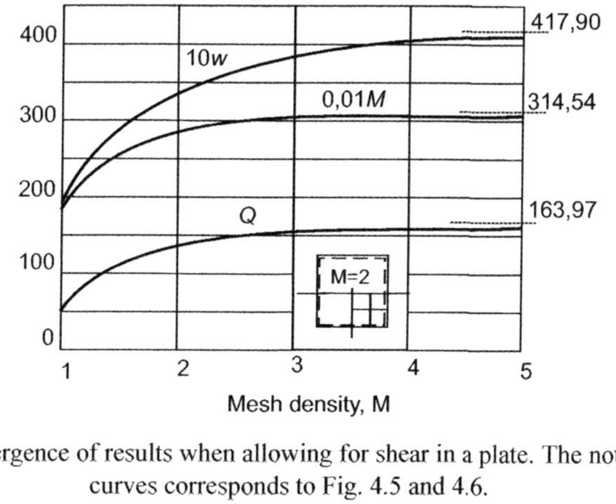

Fig. 4.7. Convergence of results when allowing for shear in a plate. The notation of the curves corresponds to Fig. 4.5 and 4.6.

4.7 Richardson Extrapolation

If analyses have been made on two geometrically similar meshes of different densities, then one can build an expression on the basis of estimates like (4.7) and (4.8) to describe a prognostic solution with its accuracy higher than that of a more detailed analysis. The basis for this prognosis is a Richardson extrapolation known in mathematics [18]. Without diving deep in mathematical aspects of Richardson extrapolation foundations, we'd like to outline its main idea.

Let a solution X_{2h} be known, for example, one determined on a finite element mesh with its characteristic size $2h$, and a solution X_h found on a mesh with the spacing h which has been derived from the previous mesh by splitting the latter's elements sides into halves. Let's assume that the shape of the area is the same in both problems, and the geometries of the finite element meshes are similar, so finite elements in the problems differ only by their scale not the shape. Suppose that the sequence of solutions on meshes becoming twice denser at each step $\{X_{2h}, X_h, X_{h/2}, \ldots\}$ is monotone and converges to an exact solution X_* (to be particular, let's assume that the convergence from below takes place so that $X_* > X_h$ for any h). Then an estimate of an approximate solution's accuracy is supposed to take place:

$$X_* - X_h = C_1 h^\alpha + C_2 h^{\alpha+1} + C_3 h^{\alpha+2} + \ldots \quad , \tag{4.9}$$

at any arbitrary characteristic size of the mesh h. Here we assume that the error of the approximate solution X_h can be represented as a power series with its principal term of the order h^α. As we have assumed that expansion (4.9) holds for any arbitrary diameter of the mesh h, in particular, we can present a similar estimate for a doubled mesh substituting $2h$ instead of h in (4.9)

$$X_* - X_{2h} = 2^\alpha C_1 h^\alpha + 2^{\alpha+1} C_2 h^{\alpha+1} + 2^{\alpha+2} C_3 h^{\alpha+2} + \ldots \quad . \tag{4.10}$$

The idea by Richardson is to try to construct one more solution out of two approximate (mesh) solutions using linear combinations. This solution, denoted as X_R here and further, is supposed to have a higher order of accuracy than those of the two original solutions. In other words, we assume

$$X_R = a_1 X_h + a_2 X_{2h} \tag{4.11}$$

with so far unknown coefficients a_1 and a_2.

Multiplying (4.9) by 2^α and subtracting (4.10) from what we obtain, we cancel the principal term in the expansion of the solution's error into the series. It results in the following estimate:

$$X_* - (2^\alpha X_h - X_{2h}) / (2^\alpha - 1) = -(C_2 h^{\alpha+1} + 3C_3 h^{\alpha+2} + \ldots)\, 2^\alpha / (2^\alpha - 1) \, . \tag{4.12}$$

It is clear now that assuming $a_1 = 2^\alpha/(2^\alpha - 1)$ and $a_2 = -1/(2^\alpha - 1)$, for the solution extrapolated according to Richardson we will have the order of the error estimate higher by at least one than that in (4.9). Thus, the principal term in the expansion of the error of the extrapolated solution becomes one of the order $h^{\alpha+1}$

$$X_* - X_R = -(C_2 h^{\alpha+1} + 3C_3 h^{\alpha+2} + \ldots) 2^\alpha / (2^\alpha - 1) \, . \tag{4.13}$$

We can derive an extrapolation on the basis of information obtained from a sequence of three mesh solutions such as X_{2h}, X_h и $X_{h/2}$ with their mesh diameters $2h$, h and $h/2$, respectively. To see this, apply (4.12) to the couple of meshes h and $h/2$ and have

$$X_* - (2^\alpha X_{h/2} - X_h) / (2^\alpha - 1) = -(C_2 h^{\alpha+1}/2 + 3C_3 h^{\alpha+2}/4 + \dots) / (2^\alpha - 1) . \qquad (4.14)$$

Now multiply (4.14) by $2^{\alpha+1}$ and subtract (4.12) from what has been obtained, and thus derive the following result:

$$\begin{aligned}(2^{\alpha+1}-1)X_* - (2^{2\alpha+1}X_{h/2} - 2^\alpha 3X_h + X_{2h}) / (2^\alpha - 1) = \\ = (3C_3 h^{\alpha+2} + \dots) 2^{\alpha-1} / (2^\alpha - 1) ,\end{aligned} \qquad (4.15)$$

from where one can see that the Richardson extrapolation by three sequential solutions should be performed as follows:

$$X_R = (2^\alpha X_{h/2} - 2^\alpha 3X_h + 2X_{2h}) / [(2^\alpha - 1)(2^{\alpha+1} - 1)] . \qquad (4.16)$$

The main obstacle for the practical extrapolation of this kind is that, as a rule, the order of α in estimate (4.9) is not known beforehand, especially when we deal with quantities like stresses (forces) in particular points of a structure.

Though, if we know solutions on three consecutive meshes, then the desirable parameter 2^α can be found by subtracting equality (4.14) from (4.12). After simple transformations we have

$$2^\alpha = (X_h - X_{2h})/(X_{h/2} - X_h) - (2 + 2^\alpha)\, C_2 h^{\alpha+1}/2^{2\alpha+1} - \dots \qquad (4.17)$$

and neglect the remainder in (4.17) to derive an approximate estimate

$$2^\alpha \approx (X_h - X_{2h})/(X_{h/2} - X_h) . \qquad (4.18)$$

It is easy to see that the convergence from below ($X_h < X_*$ for any h) makes the remainder in (4.17) negative, and the approximation of 2^α according to (4.18) is too small, otherwise it is too big.

In order to judge about the capabilities of the extrapolation, let's consider an example analysis of a clamped ferroconcrete slab. See the schematic in Fig. 4.8.

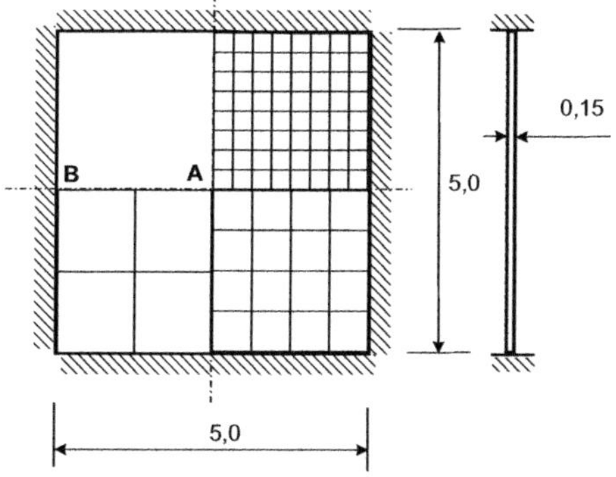

Fig. 4.8. A schematic of the example problem

The elasticity modulus is $E = 880$ kN/cm^2, Poisson ratio is $\nu = 0$. The slab is loaded by a distributed force $q = 50$ kN/m^2. Results of the analysis that involved three meshes of different densities are given in Table 4.2. The last column of Table 4.2 gives Richardson-extrapolated values of the desirable functions, and the calculations were done by (4.18) and (4.16). For the purpose of comparison, the same table gives exact values of the same functions calculated by formulae from [28].

The most interesting thing here is the Richardson extrapolation of values of the shear forces which have had the highest error. The value of the shear force 112.83 was found using solutions obtained on the sequence of meshes 4×4, 8×8 and 16×16, with its error being 1.1% instead of error 10.3% on the mesh 16×16. As one can see from the data given here, the Richardson extrapolation decreases the error of the approximate numerical solution significantly.

Table 4.2.

Result of analysis	**Mesh 4×4**	**Mesh 8×8**	**Mesh 16×16**	**Exact solution**	**X_R by (4.16)**
Deflection in point A, [mm]	17.25	16.34	16.07	16.00	15.96
Moment in point A, [t-m/m]	26.46	23.41	22.69	22.01	22.47
Moment in point B, [t-m/m]	59.20	63.72	64.95	64.43	65.43
Shear force in point B, [t/m]	66.63	88.60	100.12	111.61	112.83

The extrapolation technique has been discussed a lot of times and validated numerically [27], but the experience of its practical application is comparatively poor. Apparently, the matter is that the absolute majority of commercial CAD systems are not capable of performing the operation (4.11) or (4.16) which must be specially prepared by additional determination of $X_{(2h)}$ values in nodes of a denser mesh. Then the Richardson extrapolation can be done easily. Of course, one could calculate by formulae (4.11) or (4.16) manually, performing a selective estimation of most valuable results, but there is hardly an analyst who would attempt such a boring and complicated thing.

Though, in spite of all difficulties of the software implementation, we think it's probable (and promising too!) that the interest of developers of commercial software to the extrapolation procedure of a finite element solution refinement will grow. This is one of few issues of the software development that remains little addressed to by researchers and promises a real breakthrough. Moreover, theoretical investigations of the recent decade on extrapolation have outpaced the developments of software manufacturers and thus let us hope that practically important bilateral estimates will be obtained for grid methods, particularly for the finite element method [24].

It is necessary to notice that the Richardson extrapolation can be well-founded only when the shape of elements remains the same as the mesh becomes finer. If it is not the case, then one may find oneself in the situation where there is no monotone convergence. The illustration for this is the solution of the Kirsch problem of tension of a plate weakened by a round orifice, taken from [4]. The solution was found on meshes built from different numbers of elements along the radius (parameter K) and along the arc (parameter N). Fig. 4.9 shows how error ε is not monotonic at fixed K and variable N. This occurs due to a variable ratio of the finite element sides.

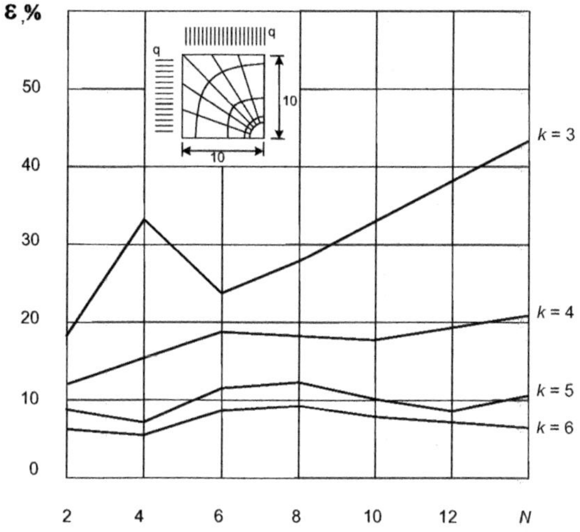

Fig. 4.9. Error of solution of the Kirsch problem

This phenomenon can be encountered almost each time when the finite element mesh is based on the polar coordinate system. Then the fining of the mesh by its opening angle can result in the emergence of "needle" elements. Two tendencies may start to fight each other: the refinement of the solution on a finer mesh, and the fall of the accuracy due to degenerated element shapes. The result of this fight is unpredictable.

4.8 Circumventing Singularities

In the vicinity of singular points — where a drastic stress concentration occurs — the use of finite elements (as well as other discretization techniques) is usually encumbered, especially in the representation of the stress field. One has to dramatically fine the finite element mesh, thus increasing the dimension of the analysis.

Also, the fining of the mesh may or may not lead to a proper result, and this fact makes us perform a deeper analysis of the situation. One of most popular judgments is the following: a concentrated force is an abstraction that does not exist in reality. If it had been actually implemented, then it would pass through a structure of any strength and ran away to infinity without being resisted. So it happens that this idealization creates an artificial difficulty which may be fought heroically only to find that this heroism is of very little practical sense. One had better remember how the force represented as a concentrated one in the model is implemented in the actual structure, then all issues of how to provide the convergence of the finite-element solution to an exact one might be resolved.

Let's consider the said problem by the example of the analysis of a plate on an elastic foundation, the plate being loaded by a concentrated force. This example is a frequent idealization used in the design of bedplates (the concentrated force simulates the load caused by a column), aerodrome and road pavements (here the concentrated force imitates a wheel's pressure) and other practically important objects.

There have been other alternatives proposed instead of the replacement of a concentrated force by a distributed load applied to a small area:

- use an exact solution for a concentrated force that acts in an infinite body, with the use of displacements of the respective fragment's boundaries in the finite element analysis;
- combine a two-dimensional idealization of an object with refinements that follow from a three-dimensional analysis.

The second option seems more practical because its implementation does not require the respective software to know all exact solutions for concentrated forces applied to infinite bodies. Let's show how to do it by a pretty simple example of a round plate simply supported along its contour, having radius a and cylindrical rigidity $D = (Eh^3/12)/(1-\nu^2)$, and loaded by a concentrated force P in its center.

This problem has an exact solution based on conventional assumptions of the theory of plates. It is expressed by the following formulae for deflections w, radial moments M_r and shear forces Q_r:

$$w = \frac{Pa^2}{16\pi D}\left[\frac{3+\nu}{1+\nu}\left(1-\rho^2\right)+2\rho^2\ln\rho\right], \tag{4.19}$$

$$M_r = \frac{P}{4\pi}(1+\nu)\ln\rho\,, \tag{4.20}$$

$$Q_r = -\frac{P}{2\pi a\rho}\,. \tag{4.21}$$

As $\rho^2\ln\rho \to 0$ when $\rho \to 0$, the deflection function does not have any singularities in the center where the relative radius $\rho = r/a$ is zero, while the moment function has a logarithmic singularity. As we approach this point, the values of the moment and the shear tend to infinity (Fig. 4.10), that evidences an insufficient correctness of the analytic model.

Let's extract a fragment from the structure with its boundaries defined by the relative radius $\rho = 0.2$. As one can see from Fig. 4.10, the internal stresses begin to grow dramatically at $\rho \leq 0.05$, thus the fragment's size is such that it lies beyond the interval in which the solution is questionable and should be validated.

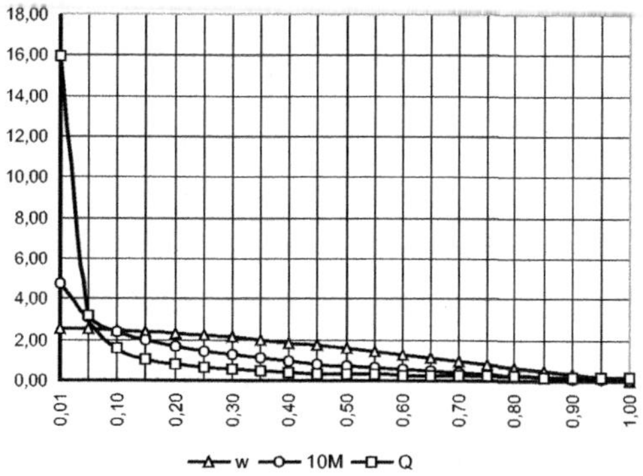

Fig. 4.10. Deflections and stresses in a round plate loaded by a concentrated force in its center

The analysis of the fragment has been performed for a model of spatial elasticity under reactions caused by the removed part of the plate. The schematic is shown in Fig. 4.11. These reactions, applied to the contour of the fragment, form a balanced system of forces. In the analysis the fragment was restrained against its displacement as a rigid body so that no static indeterminacy arose in its supports. Reactions in the respective constraints were practically zero, which served as a validation of accuracy of specification of the loads applied to the fragment.

It should be noted that a concentrated force is not the only type of singularities encountered in practical designs. The idealization of a slab supported by a wall in the form of restrained nodal displacements in cases when the support does not end at the edge of the slab results in a singularity too (Fig. 4.12, *a*). Here, the reactions of the constraints at the ends of the supporting strip tend to infinity. Actually, one can treat this situation as the loading of the slab via a perfectly rigid strip die used to transfer the reaction along the AB line.

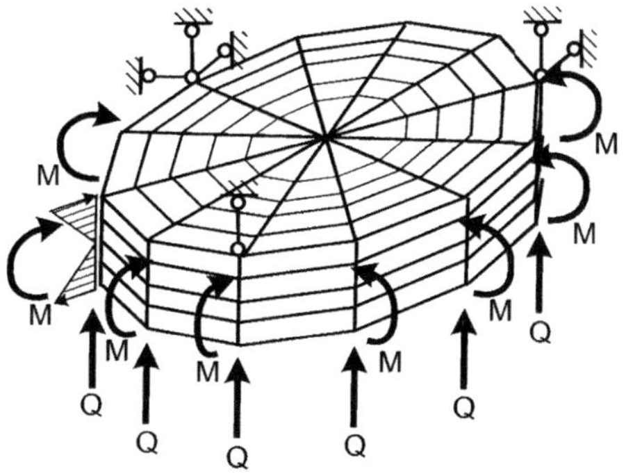

Fig. 4.11. Design model of the fragment

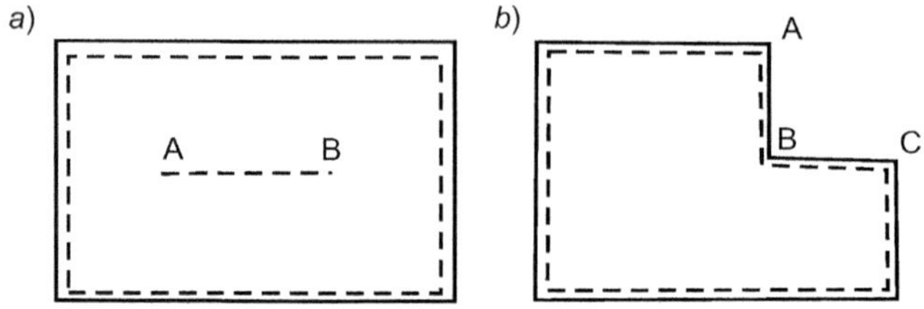

Fig. 4.12. Singularities in discontinuous supports along lines

A similar situation appears in the vicinity of a reentrant angle of an L-shaped plate supported along its edge (Fig. 4.12, *b*). In both cases the origin of the singularities is the same: the strip constraint along the AB line (same as the CB line in Fig. 4.12, *b*) ends up sharply, and at once the plate that could not deform along the said lines begins to participate in deformation at the continuations of the lines. Thus the curvature in the direction of the "cut" constraint suffers a discontinuity, and a singularity appears.

The reactions calculated in points A and B increase as the mesh becomes finer, as it usually happens in a singular point. To circumvent this situation, one can rebuild the model in two ways:

- extract a fragment and build another model within it, such as a spatial elasticity model, just as we have done in the problem with the concentrated force;
- use a more realistic model to represent the response of the wall that supports the slab, for example, allow for its compliance (compressibility) in the vertical direction.

Results of the solution of the problem shown in Fig. 4.12, *a*, where the compliance of the wall along the support line AB is allowed for by introducing realistic elastic constraints, are shown in Fig. 4.13. This figure makes it clear that the real model has no drastic bend in points A and B. As in the vicinity of this singularity significant shear forces appear (formally they tend to infinity) while the bending moments are small, one can expect that the change of the curvature will be caused by the shear. This fosters one to use the Reissner model of flexural plates instead of that by Kirchhoff, in spite of the plate's being thin.

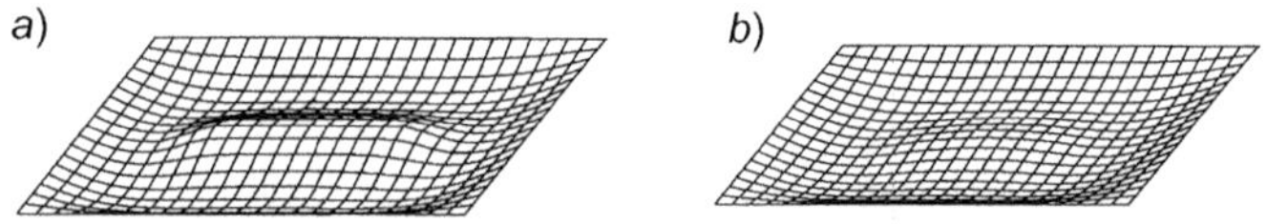

Fig. 4.13. Displacements in a finite element model of the problem from Fig. 4.12, *a*: (*a*) for a simple support; (*b*) for an elastic support.

A local modeling by the use of spatial finite elements is also feasible. The result of such modeling is presented in Fig. 4.14 that shows the diagram of reactions along the AB line (from the symmetry axis to the support line break) calculated with the spatial elasticity model (solid line) and the thin plate model (dash line).

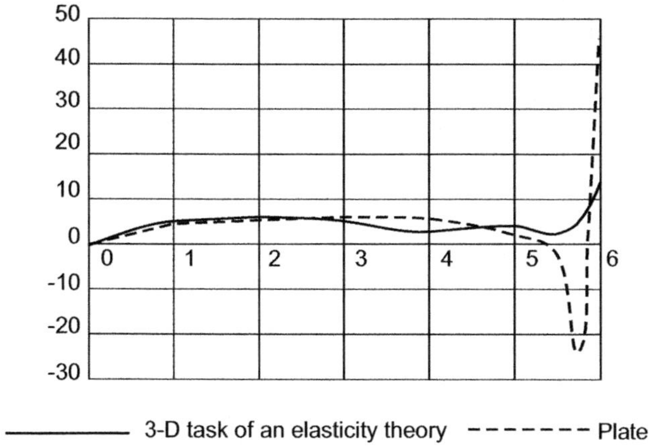

Fig. 4.14. Diagram of reactions along the AB line

Regarding singular points at boundaries of contact zones, a similar problem was considered in the notable book [12] where they analyzed in detail the problem of a concentrated reaction *P* appearing in the bending of bars supported by an infinitely rigid foundation (see problem 56, pp.128–134).

One might have concluded from what has been said that the existence of complicated situations with singular solutions originated in the essence of a problem being solved should orient us at the prevention or circumvention of such situations, or at the use of some "over-sophisticated" techniques of analysis.

Though, a software developer should remember that in a great many situations this "over-sophistication" is introduced only to overcome pathologies that may arise in principle, and nothing more. Therefore the neglect of some delicate issues is an unquestionably feasible thing. The obligatory requirement is that the software user must be warned of it, or, what is far better, he must be recommended what to do to avoid such pathologies (for example, it should be said that a concentrated force must be represented as one distributed on some small spot, and the result of the analysis will be the moment at the boundary of this spot).

4.9 Finite Element Mesh Generation

Contemporary personal computers enable one to solve problems with as many nodes as thousands or tens of thousands. Naturally, nobody can think of making the appropriate finite element mesh "manually". Most software products available today include various kinds of automatic mesh generators. The quality of such generators is determined by their list of implemented functionalities. The number of such may amount to tens in powerful computational systems such as **ANSYS**. These include, for example, an automatic matching of already constructed meshes on surfaces to be connected, creation of a spatial three-dimensional mesh from an available surface one, reconstruction of a mesh according to results of an analysis, and a lot of others.

The convenience of using triangles as elementary areas occupied by separate finite elements is due to their facilitating the description of boundaries of pretty complex shapes. Also, the triangular grid is useful to make elements finer and denser in areas where high stress gradients are anticipated, for example, in stress concentration locations such as indents, re-entrant angles etc. The triangles fit well for an automatic generation of a finite element mesh, and most available mesh generators included in advanced computational software are based on the area triangulation. These advantages of triangular finite elements make them so widely popular in CAD software systems.

For triangular elements, efficient mathematics has been developed to represent shape functions of the elements in barycentric coordinates that enable one to easily obtain analytical expressions for the stiffness matrix components of the elements by simple formulae of exact integration of polynomials in triangular areas [19]. Though, certain difficulties arise during the construction of the vector of loads referred to nodes. The advantages of the barycentric coordinates take place only when integrating through the whole area Δ occupied by a finite element, and most loads upon a structure are usually defined on spots not related directly to the finite element grid.

There's a necessity to integrate along the area Ω formed by the intersection of the triangle Δ and the loaded spot Ψ (Fig. 4.15), and if $\Delta \neq \Psi$, then the barycentric coordinates are of no advantage. That's the reason why most CAD programs suppose that the loaded spot covers the respective finite element completely, i.e. $\Delta = \Psi$, or another approximate (often pretty rough) technique is used to refer loads to nodes of a triangular mesh.

Publication [13] describes in detail a special algorithm that enables one to form components of the load vector on the basis of independent information of the load and the position of a finite element.

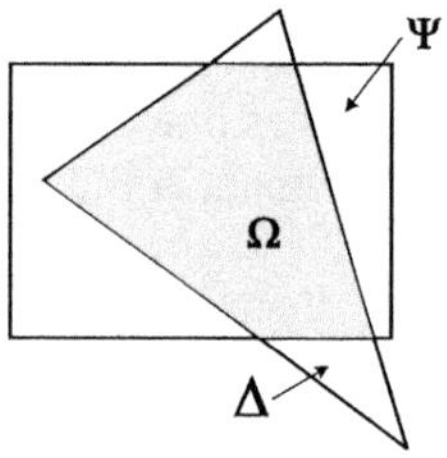

Fig. 4.15. An incompletely loaded triangular element.

Though, one should note that there is a tendency to avoid using triangular and quadrangular pyramidal elements because they usually yield slightly less accurate solutions comparing to those obtained with rectangular and hexahedral elements. In this connection, the above mentioned **ANSYS** mesh generator is capable of making ordered or structured meshes from spatial hexahedrons (bricks) and building irregular meshes of quadrangular elements on arbitrary surfaces. Problems of mesh fining/coarsening in various parts of the domain where triangular elements are traditionally considered a must, can be resolved using techniques shown in Fig. 4.16.

As for the approximation of a complicated boundary, as a rule, one cannot utterly avoid using triangular elements. The **SCAD** system includes a function of merging triangular finite elements into quadrangles in all places where it is possible.

Anyway, it is useful that a software have a capability of automatic building of load vectors for triangular or quadrangular elements on the basis of independently specified information about the meshing of the domain and loaded spots. The user will also appreciate the capability of specifying a variable intensity of the load along the spot, at least in the form of a linear law.

An important feature of a finite element mesh is its smoothness (a gradual variation of the elements' size throughout the mesh) because a drastic variation of the characteristic element size will introduce a noticeable disturbance in the stress and strain distribution. It is especially clear with stresses represented by level curves. Good mesh generators are capable of controlling the densification of the mesh (at a point, at a line, to the center or perimeter of a domain etc.), and these capabilities let one maintain a certain degree of smoothness of the mesh. An example of a smooth variation of the mesh taken from a report by Shevchuk [23] is shown in Fig. 4.17.

The density of the mesh in automatic generators is controlled by such properties as the maximum distance between nodes and the allowable minimum angle at the vertex of a triangle (to be specified). The effect of the latter parameter can be seen in Fig. 4.18 taken from the same paper. Also, another controlling parameter is used — element's side increment rate. When the curvature of a contour increases locally, algorithms of automatic mesh densification are involved.

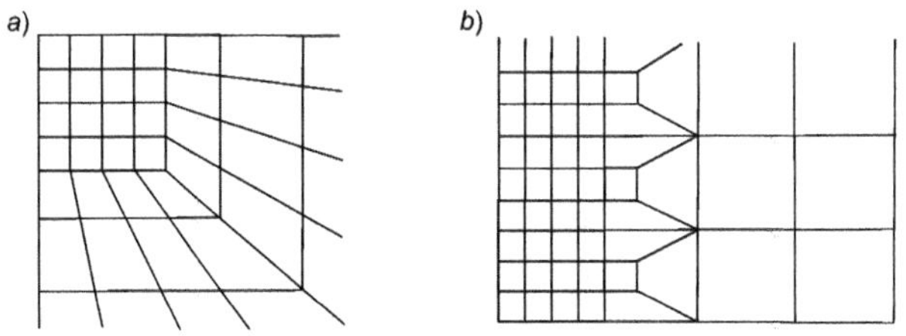

Fig. 4.16. Fining of quadrangular elements

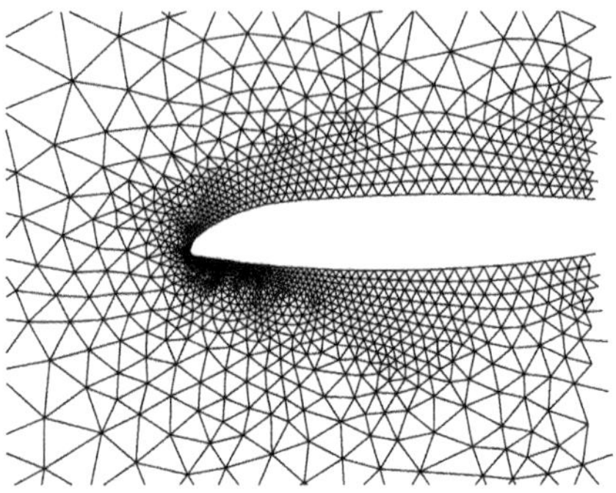

Fig. 4.17. Coarsening of the mesh in the direction from the interior contour

If the software is not capable of smoothing, then one should keep to the following empirical rule to properly perform it manually: do not increase the mesh spacing more than twice at one "front" of densification, and those "fronts" should not be placed closer than in two or three rows of elements. Also, one should pay attention to symmetries and regularities.

Sometimes the concept of typical structural elements can be efficient [16]. Each element of this kind performs the division of a canonic domain into finite elements using one or more parameters that define the shape of elements and the spacing of the mesh.

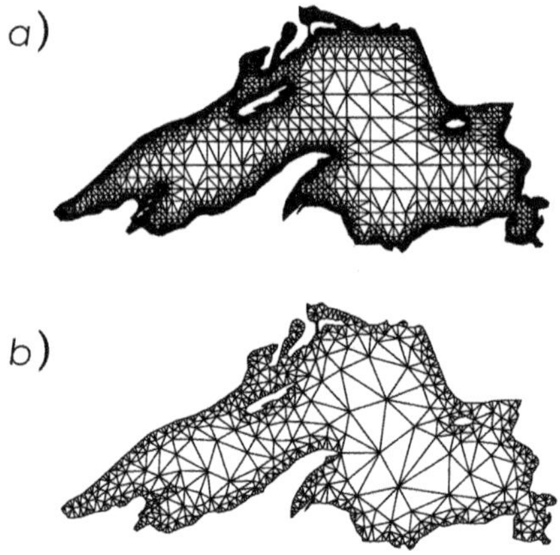

Fig. 4.18. Automatic triangulation of a domain: (*a*) at the minimum angle 15°; (*b*) at the minimum angle 25°.

4.10 Using Hybrid Finite Elements

Many authors insist on advantages of so-called hybrid models of the finite element method where the displacement field is determined only at the boundaries of the elements, and all other parameters are to be found on the basis of an assumed stress distribution. These models use the principle of minimal complementary energy (Castigliano principle) or the conditions of stationarity of a mixed Reissner-type functional.

Several variants of these elements are known. Particularly, such are used in the **MicroFE** software development [22]. A modification of the hybrid method used in **MicroFE** implements ideas by the creators of the method Pian and Tong [20] who both developed and entitled it. Practically it means that this method deviates from the classical finite element technique only in immanent properties of particular finite elements (stiffness matrices and load vectors). As for the solution of discrete problems, this variation of the hybrid method is again the old good slope deflection method. Note that generally the theoretical proof of convergence for these elements is a hard problem to solve, though this proof has been obtained for particular classes of problems. We must do justice to these elements meaning that in many numerical experiments the hybrid finite elements demonstrate excellent results, especially on meshes of a medium size. Though, the statement that in any application the hybrid elements would exceed the classical finite elements based on the displacement field approximation, is too strong or just untrue. It is not very hard to find opposite examples where the classical finite elements outperform their hybrid counterparts.

There's another circumstance that makes the hybrid elements a surrogate rather than a fully functional replacement of the classical elements. The matter is that in spectral analysis problems (such as a frequency-mode analysis of structures) the classical consistent finite elements guarantee the sign of the error which is very important for the analysis (see Section 1.6 concerning majorant and minorant models). At the same time, the hybrid finite elements do not provide a unilateral convergence in these problems.

Meaning to satisfy the user, we would say that the preferable choice for a professional software, in order to take into account all that was said, seems to be a library full of both classic finite elements and various types of hybrid finite elements (and more!). The obtrusion of solely hybrid elements on the users is the same as a diet consisting of caviar only. The result is known[8].

The authors of papers [22] state that the principal advantage of their development is the capability of refining the stress field even on pretty coarse mesh models. Also, they claim the successful resolution of the "shear locking" issue.

[8] There's an immensely popular Russian "Eastern" movie (as opposite to Western movies) called "White Sun of the Desert" taking place in Central Asia in 20ths. One of its characters, an old grave tsarist customs officer living near the Caspian sea, is known for his eating caviar almost solely for the lack of bread and abundance of fish in those parts. Unfortunately, he is killed in the movie. Hence the harmful effect of a dreary diet. A popular sentence from this movie says "East is a delicate matter". — *Translator's note.*

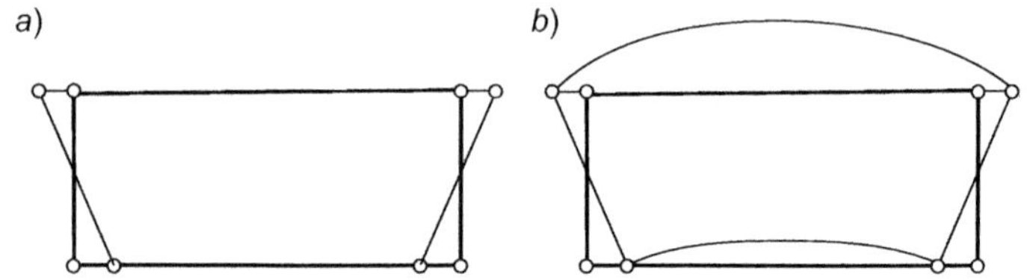

Fig. 4.19. To the explanation of the false shear effect

"Cramping", "shear locking", "false shear effect" [4] are different names for the phenomenon when an element of a flexural Reissner plate[9] with the integration by the thickness does not reproduce the expected Kirchhoff solution as the plate's thickness decreases. Elements of the Reissner plate with a small thickness/length ratio appear too stiff to be used for simulating a simple bending. Thus the possible shear mode of deformation (Fig. 4.19, *a*) somehow replaces the actual mode of deformation of the simple bending (Fig. 4.19, *b*). The shear mode itself presumes that the strain compatibility conditions hold strictly within a finite element, and this does not correspond to the Kirchhoff assumption of straight normals.

In order to eliminate the false shear effect, they use artificial techniques of reducing the numerical integration accuracy, actually reducing the number of degrees of freedom for stresses. Though, one should be very careful here because an over-mitigation of the conditions expressed in stresses may result in a degenerate stiffness matrix.

We should say that the false shear effect grows as the relative thickness of the plate decreases, thus it demonstrates an example of a special class of singularities in the finite element analysis where there can be a degeneration in the very formulation of the problem. This results in the existence of so-called *degenerating* problems. Without being able to discuss this topic here in more detail, we suggest that the reader refer to some publications of disciples of L.A.Rosin's school [5, 8, 9]. Particularly, these papers state that the hybrid method is really efficient in the solution of degenerating problems. Though, together with the hybrid method, the same results in the sense of accuracy can be obtained by the use of other techniques that do not step out of the line of the slope deflection method in the sense of the discretization procedure. In this connection we would like to notice a *moment scheme of the finite element method* (A.S. Sakharov, [4]) and a peculiar *double approximation* technique (V.P. Boldichev [8]). The possibility of eliminating the false shear effect follows also from a circle of ideas of how to build optimum approximations for the slope deflection method [26] demonstrated by a one-dimensional problem of the Timoshenko beam bending solved by a *method of coupled approximations* [25]. Results of the listed papers show clearly that the capabilities of the displacement-based discretization in degenerating

[9] Sometimes the theory of bending of medium thickness plates that allows for shear and, in some versions, for normal stresses σ_{zz} in sections parallel to the median surface of the plate, is called a *Timoshenko – Mindlin theory* [31].

problems are not confined to the standard finite element procedure and are by no means less efficient than the hybrid method.

Many practical examples demonstrate a good quality of solutions obtained with hybrid elements on relatively coarse meshes [22]. Though, the difference between the accuracies of the stress solutions is smoothed pretty fast, which can be seen by the example of a problem (Fig. 4.20) taken from the manual [1]. In this model, the thickness of the plate and its ribs is 0.1 m; the load of intensity 10 kN/m is applied to lines along the ribs.

Table 4.3 shows a comparison of the stresses for three solutions: one obtained by summation of three terms of a trigonometric series in [1], one obtained by the use of hybrid elements in the **MicroFe** environment, and one obtained by the conventional finite element method in the **SCAD** environment.

As one can see from this table, the difference does not exceed 3–5% which is quite satisfactory by accuracy. As for the computation time and disk storage needed, there's nothing even to talk about.

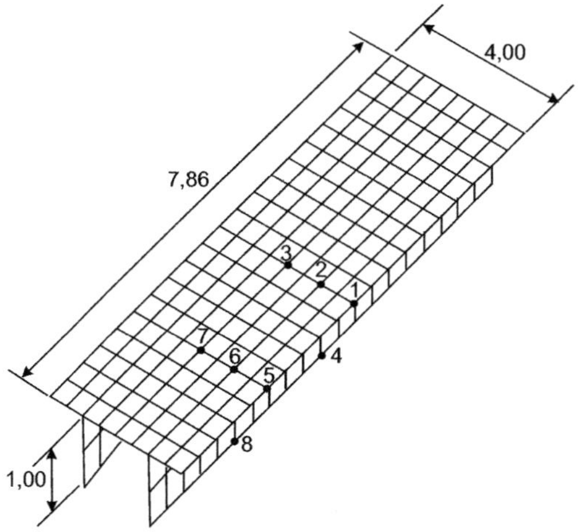

Fig. 4.20. A schematic of a finite element analysis

Table 4.3.

Point	Stresses according to:		
	paper [1]	**MicroFe**	**SCAD**
1	471.7	456.6	446.7
2	522.6	550.8	536.2
3	487.9	484.4	469.7
4	2642.9	2633.4	2564.2
5	344.8	344.8	368.4
6	405.5	405.5	436.9
7	359.8	359.8	362.2
8	1959.9	1959.8	2064.5

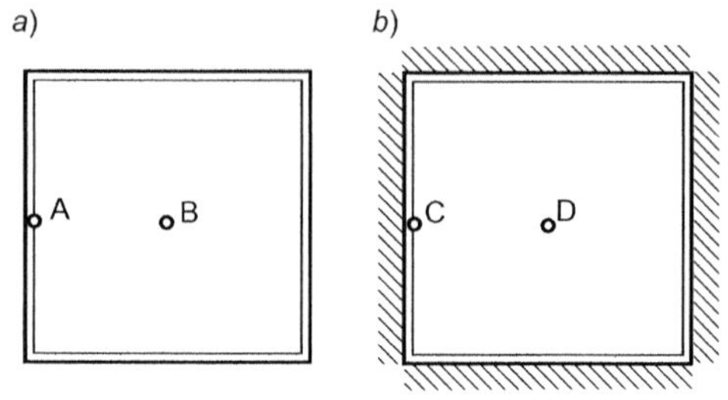

Fig. 4.21. A square plate: (*a*) simply supported, (*b*) clamped

Table 4.4 contains results of analyses of a square plate, either simply supported (Fig. 4.21, *a*) or clamped along its contour (Fig. 4.21, *b*), obtained by Allman [3] by the use of hybrid elements and re-calculated by the software, in comparison with an analytical solution taken from monograph [28]. Values of the nodal moments obtained by **SCAD** are given there after a simple averaging by elements of the plate adjacent to the respective node.

Also, this table demonstrates that the refinement of the stress distribution claimed for hybrid finite elements has been pretty quickly outperformed by a moderate finite element mesh.

Table 4.4.

Result	Mesh	Numerical values:		
		From [28]	From [1]	**SCAD**
$w_B/(pL^4D)$	4×4	0.00406	0.00386	0.00426
	8×8	0.00406	0.00401	0.00411
$w_D/(pL^4D)$	4×4	0.00126	0.00132	0.00136
	8×8	0.00126	0.00128	0.00129
V_A/pL	4×4	−0.4200	−0.2850	0.2273
	8×8	−0.4200	−0.3400	0.2344
$M_{B,x}/pL^2$	4×4	0.0479	0.0515	0.0387
	8×8	0.0479	0.0487	0.0468
$M_{B,y}/pL^2$	4×4	0.0479	0.0504	0.0387
	8×8	0.0479	0.0482	0.0348
$M_{C,x}/pL^2$	4×4	−0.0513	−0.0473	0.0456
	8×8	−0.0513	−0.0495	0.0492
$M_{D,x}/pL^2$	4×4	0.0231	0,0314	0.0204
	8×8	0.0231	0.0250	0.0221
$M_{D,y}/pL^2$	4×4	0.0231	0.0316	0.0204
	8×8	0.0231	0.0247	0.0221

To conclude this section, we would like to mention some general issues of the comparison between various versions of the finite element method. Nowadays, when contemporary personal computers with a high performance and immense disk storage resources are widely used, such feature of the FEM implementation as the computation time becomes little significant (especially when taking into account the data given in Fig. 2.2). And only for very big problems it may be essential to take into consideration the needed disk storage volume.

Therefore, the possibility of using very coarse meshes ceases to be a determinative feature, and other properties of finite elements step forward, such as their universality, mathematical strictness and proved convergence, the capability of a simple detection and localization of pathologies etc. In this connection it seems to us that in many cases the classical FE analysis is quite competitive as compared to hybrid elements. Attempts of giving a boost to hybrid models and get rid of classical ones are inspired rather by marketing ambitions of some software developers than by their sincere aspiration to create an absolutely best product which customers will love. Once more, our claim to developers is that they should enable the user to make his own choice in every particular case by providing him with both the standard FEM technique based on a classical slope deflection method and a hybrid technique. To do justice, we'd like to say that it is exactly the way some software developers, such as the authors of the **GTSTRUDL** software, follow.

References

1. Aleksandrov AV, Lashchennikov BJa, Shaposhnikov NN (1983) Structural analysis. Thin-walled space constructions (in Russian). Stroyzdat, Moskow
2. Alekseenko BG (1999) On boundaries of applicability of FEM for analysis of bar systems operated in corrosive media (in Russian) In: Proceeding of the second Belarusian congress on theoretical and applied mechanics "Mechanics-99. «Strinko», Minsk
3. Allman DJ (1984) A compatible triangular element including vertex rotations for plane elasticity analysis. Computer and Structures 12:1-8.
4. Altenbach J, Sacharov AS (eds) (1982) Die Methode der Finite Elemente in der Festkőrpermechanik. VEB Fachbuchverlag, Leipzig
5. Bayenkhayev AV (1985) On the development of peculiar mixed algorithms of the finite element method (in Russian) In: Rozin LA (ed) Finite elements method and analysis of structures: Proceedings of Leningrad polytehnical institute, vol. 405. Leningrad, pp. 41–48
6. Blekhman II, Myshkis AD, Panovko JaG (1983) Mechanics and applied mathematics: Logic and peculiarities of mathematical applications (in Russian). Nauka Publishing House, Moskow
7. Bogner FK, Fox RL, Schmit LA (1966) The generation of interelement compatible stiffness and mass matrices by the use of interpolations formulas. In: Proc. of the Conference of Matrix methods in Struct. Mech. pp. 397–443
8. Boldychev VP (1979) A double approximation of the slope in the analysis of plates of medium thickness (in Russian) In: The bulletin VNIIG, vol.133. "Energija", Leningrad, pp. 68–74
9. Boldychev VP (1985) On connections between various algorithms of the finite element method in the solution of degenerating problems (in Russian) In: Rozin LA (ed) Finite

elements method and analysis of structures: Proceedings of Leningrad polytehnical institute, vol. 405 (in Russian). Leningrad, pp. 32–41

10. Cowper GR, Kosko E, Lindberg GM, Olson MD (1969) Static and dynamic applications of a high-precision triangular plate bending element. AIAA Journal, 7:1957–1965
11. Evzerov ID (1983) Estimates of displacement error caused by the use of inconsistent finite elements (in Russian) In: Numerical methods of continuum Vol. 14, 5:24-31
12. Feodosi'ev VI (1967) Selected problems and questions of strength of materials (In Russian), Nauka Publishing House, Moscow
13. Galasova KP, Slivker VI (1988) On the reduction of external loads to nodes of triangular finite elements (in Russian). Structural mechanics and analysis of structures, 6:64-68
14. Irons BM (1967) A criterion of rounding in solutions by the direct method of rigidities AIAA Journal, vol.3, 7:107-113
15. Karpilovky VS (1982) Use and construction of some finite element types for problems of structural mechanics (in Russian). Ph.D, thesis, The Kiev transport university
16. Khriakov AD (1986) A.D. Geometrical discretization of plane systems (in Russian). Structural mechanics and analysis of structures, 3:47–50
17. Kornoukhov NV (1963) Strength and stability of bar systems: Selected writings on structural mechanics (in Russian). Academy of Science of Ukraine, Kiev
18. Marchouk GI, Shaidurov VV (1979) Improvement of accuracy in solutions of difference schemes (in Russian). Mir, Moskow
19. Peano A (1974) Hierarchies of conforming finite element for plane elasticity and plate bending . Computer and Mathematics, 3–4:211–224
20. Pian T, Tong P (1969) Basis of finite element methods for solid continua. Int. J. Num. Meth. Eng. 1:3–28
21. Rozin L.A. (1975) Bar systems treated as systems of finite elements (in Russian). SPbGTU Publishing House, Sankt-Petersburg
22. Semenov VA., Semenov PYu (2001) Finite rotation shell element with three rotation parameters. In: pp. 158-159 (ful text in CD-ROM). In: ECCM-2001. 2-nd European conference on computational mechanics. Abstracts, vol.1/. "Vesalius" Publisher, Crakow
23. Shewchuk JR (1997) Delaunay refinement Mesh Generation: Ph.D. thesis, Carnegie Mellon University
24. Shvarzman BS (1992) An extrapolation method for finding numerical solutions with a given accuracy (in Russian). In: Structural mechanics and analysis of structures. Proc. of St. Petersburg State Technical University, St.-Petersburg, pp.157-162
25. Slivker AV (1986) Analysis of beams of medium thickness using coupled approximations for the deflection and the slope (in Russian). In: Bulletin of *VNIIG*. Collected scientific papers, Vol. 191.— Energhiya, Leningrad, pp. 96–103
26. Slivker AV (1987) Optimal approximations of displacements in the finite element method (in Russian)., Structural mechanics and analysis of structures, 2:27–30
27. Strang G, Fix GJ (1973) An analysis of the finite element method. Prentice-Hall Inc., Endlewood Cliffs, N.J
28. Timoshenko SP, Voinovsky-Kriger C (1959) Theory of plates and shells, McGraw-Hill, New York Toronto London
29. Tong P (1969). Exact solution of certain problems by finite element method. AIAA Journal, 1:178–180

30. Vovkushevsky AV, Shoikhet BA (1981) Analysis of massive water engineering structures taking into account the opening of seams (in Russian). Energhiya, Moskow
31. Washizu K (1975) Variational methods in elasticity and plasticity. Pergamon Press, Oxford New York Toronto Sydney Paris Frankfurt
32. Zhemochkin BN, Sinitsyn AP (1947) Practical methods of analysis of ground beams and slabs on elastic foundations (in Russian). Stroyizdat, Moskow
33. Zienkiewicz OC (1977) The finite element method in engineering science. McGraw-Hill, London
34. Zienkiewicz OC, Irons BM, Scott FC, Campbell JS (1971) Three dimensionals stress analysis. In: De Veubeke FB (ed) Proceeding IUTAM Symposium "Hige speed computing of elastic structures", Université de Liége, pp. 293–305

5 Mistakes and Pitfalls, Special Techniques to Build Finite Element Models

In ancient times every man knew both the purpose and the design of their devices: a hammer, a bow, an arrow. An ever progressing differentiation of labor abated gradually this individual knowledge, so in the modern industrial society there is a clear distinction between those who maintain devices (a workman, a technician) or use them (a man using an elevator, watching TV, driving a car) and those who know the design of the devices.

S. Lem.

5.1 Fragmentation

What we mean by fragmentation is an extraction of some part of a structure from it in order to consequently include only this extracted part, called a *fragment*, in a design model. Particularly, we above used this technique in Section 4.8. Now we are going to look at it from a bit more general point of view. We will consider the following two-phase procedure:

- Neglecting local peculiarities of a design solution, build a rough first-iteration design model of a full structure to enable us to estimate the stress and strain distribution of the object as a whole, and perform the calculation.
- Select a fragment of the structure that contains a singularity of our interest. Apply forces directly working on this part of the structure to the fragment selected. The full boundary of the fragment Γ will be a combination of two parts: $\Gamma = \Gamma_1 \cup \Gamma_2$ where Γ_1 is a part of the fragment's boundary (possibly a zero one) that is at the same time the boundary of the full design model while Γ_2 is a part of the fragment's boundary that appears after its being extracted from the whole structure and is not a boundary of the structure.

Boundary conditions at the boundary Γ_1 inherit the respective boundary conditions of the whole structure at the same part of the boundary. As for the boundary conditions at Γ_2, there are three feasible options: imposition of static boundary conditions at Γ_2, imposition of kinematical boundary conditions at Γ_2, imposition of mixed boundary conditions at Γ_2.

If static boundary conditions are imposed on Γ_2, it means that this boundary of the fragment is subjected to forces of interaction between the fragment and the removed part of the full structure as if those were external forces. So, these are reactions obtained by removing the rest of the structure. If kinematical conditions are imposed on Γ_2, then this boundary is restrained against all displacements, and external influences on the fragment are supplemented by displacements specified at Γ_2 with their values obtained previously by the analysis of the rough design model. With mixed boundary conditions specified, the boundary of the fragment Γ_2 is subjected partially to force and partially to kinematical external influences[1].

Further, the fragment will be analyzed using a more detailed design model, and of the solution thus obtained only a part will be used which relates to points located at some distance from the boundary of the fragment.

This approach is concordant with the practice of choosing a series of design models to analyze various peculiarities of a structure's behavior. But it requires a bit of intuition and experience in order to exclude the danger of an irreducible error that appears in the analysis of a rough model. To facilitate the decision making for an analyst, below we present an analysis of possible error sources.

First let's consider the case of purely static boundary conditions imposed on the Γ_2 boundary. We will indicate any quantities that regard the accurate solution of the problem with the $*$ subscript, quantities related to the solution of the full problem on a rough mesh with the H subscript, and the same quantities on a fine mesh with the h subscript.

The boundary conditions at the Γ_2 boundary needed for the analysis of the fragment are forces of interaction between the fragment and the removed part of the structure (to simplify the discussion, let's assume the plane elasticity). That is, the following conditions will define the normal p^n and tangential p^t external forces at Γ_2:

$$p^n(s) = \sigma^{ij} n_i n_j , \qquad p^t(s) = \sigma^{ij} n_i t_j , \tag{5.1}$$

where σ^{ij} are components of the stress tensor in points of the Γ_2 boundary, n_i are components of the external normal unit vector at the Γ_2 boundary (the one that is external with respect to the fragment), t_j are components of the tangential unit vector at Γ_2 (here we assume a conventional summation by repeated indexes).

If the $\boldsymbol{\sigma}$ tensor in (5.1) is stress tensor $\boldsymbol{\sigma}_*$ of the accurate solution of the full problem, then the analysis of the fragment by any possible technique will have its error strictly equal to that introduced by only the employed technique. In our case

---

[1] The determination of interaction forces between a fragment and the rest of the structure (especially their specification as external loads upon the fragment) is a laborious process that can tire a user out. In our opinion, an elegant solution is implemented in the **SCAD** system. The developers of it have made sure to see about interests of users who really need the fragmentation implemented. With this purpose, **SCAD** provides an analyst with a particular service implemented in the *"Loads from structure's fragment"* mode. Switching this mode on will automate the calculation of forces working on a selected fragment from the rest of the structure and thus will render these forces external.

it will be the error caused by the finite element method on a fine mesh of diameter h. Though, tensor $\boldsymbol{\sigma}_*$ is unknown to us, so instead we have to use some kind of its approximation in (5.1) in order to impose static boundary conditions on Γ_2. This approximation will be the solution of the full problem on a coarse mesh H, that is, in (5.1) we assume $\boldsymbol{\sigma} = \boldsymbol{\sigma}_H$. As a result, even the statement of the fragment's analysis problem contains an irreducible error as "noise" of the static boundary conditions on Γ_2

$$\Delta p^n(s) = (\sigma_*^{ij} - \sigma_H^{ij})\, n_i n_j \,, \qquad \Delta p^t(s) = (\sigma_*^{ij} - \sigma_H^{ij}) n_i t_j \,. \tag{5.2}$$

As the fragment has been extracted along lines of dissection of the full structure (this is exactly Γ_2) where no high gradient of the stress field is anticipated, thus the accuracy of the solution around Γ_2 is pretty good on the coarse mesh, noise (5.2) in the boundary conditions will be inessential and will depend little on the solution obtained for the fragment with the fine mesh h. This is true, too, for that part of the fragment where there is a singularity in the solution of a particular interest for the analyst and at a sufficient distance from Γ_2, if the noise in (5.2) is a system of forces locally self-balanced at fragments of the Γ_2 boundary.

In order to simplify the consideration, let's confine ourselves to the case when Γ_2 is a closed contour, and the fragment extracted is not restrained from translation/rotation displacements.

As contour stress $\boldsymbol{p}_*$, and stress $\boldsymbol{p}_H$ also, balances all external load imposed on the selected fragment, this implies that components of "noise" contour forces $\Delta p^n(s)$ and $\Delta p^t(s)$ satisfy homogeneous equilibrium equations. In other words, the noise at the contour of the selected fragment is self-balanced. But it does not suffice to have a noise in the boundary conditions self-balanced at the Γ_2 contour in general. We would like to prove its being self-balanced locally too.

The theory of the finite element method [32] states that the sign of the stress error alternates within the mesh spacing H (by the way, it implies the existence of singular stress points called Burlow points which will be discussed in Section 5.2). If now we expand the noise stress (5.2) into a Fourier series along the Γ_2 contour, it's easy to find that the principal term of this expansion will be a harmonic with the half-wave length about H. And it is this harmonic that is a self-balanced load with the local self-balance area $2H$ long.

It is known that locally applied self-balanced loads introduce an addition to the solution of a problem that subsides at a distance from their localization area. That's what the Saint-Venant principle states, and for systems where this principle holds[2] the localization error will subside quickly with the distance from the location of the self-balanced forces. These forces include also the locally self-balanced part of the noise in the boundary conditions at Γ_2. The smaller the local fragment of Γ_2 with its respective noises $\Delta p^n(s)$ and $\Delta p^t(s)$ self-balanced, the smaller the area of their sensible influence on the stress and strain distribution along the frag-

---

[2] There are systems where the Saint-Venant principle does not hold. In this regard we recommend [10], p. 62, and an interesting paper [16] dedicated to foundations and applicability of the Saint-Venant principle.

ment. Actually it means that the noise of the boundary conditions does not play the main part in the error estimate $\|\boldsymbol{\sigma}_* - \boldsymbol{\sigma}_h\|$ of the solution at the fragment in a certain norm.

The above said means that in the analysis of a fragment there exists a boundary area adjacent to Γ_2 where the stress distribution reacts sensibly to errors in the stressed state of the whole structure determined on a coarse mesh. At a sufficient distance from this area, say, at the distance δ from the Γ_2 boundary, the effect of this error can be neglected (Fig. 5.1).

In order to estimate the dimensions of the sensitivity area and the subsidence rate of the solution error caused by the noise in the boundary conditions[3], let's consider a model problem of finding the stress distribution in a round fragment of radius R under radial q_r and tangential q_t contour stresses

$$q_r = q\sin(k\theta)\,, \qquad q_t = q\cos(k\theta)\,, \tag{5.3}$$

where k is an even positive number, θ is an angular coordinate.

It is easy to check that forces (5.3) are self-balanced both at the whole contour Γ_2 and locally within intervals $-\pi/2k \le \theta \le 3\pi/2k$, that is, the general vector and moment of these forces are zero. We may assume that the boundary load (5.3) corresponds to noise forces at Γ_2 obtained on a coarse mesh with its characteristic spacing

$$H = \pi R/k\,. \tag{5.4}$$

Let's use a known solution from the elasticity theory [23] to find the stress distribution in the fragment. If we use the condition of finiteness of the displacements at the center of the plate to exclude terms containing r (current radial coordinate) raised to negative powers, then the stress tensor components in the polar coordinate system (r, θ) will be expressed as:

$$\begin{aligned}
\sigma_r &= q\,[-k(k-1)Ar^{k-2} - (k-2)(k+1)Cr^k]\sin(k\theta),\\
\sigma_\theta &= q\,[\,k(k-1)Ar^{k-2} + (k+2)(k+1)Cr^k]\sin(k\theta),\\
\tau_{r\theta} &= -q\,[k(k-1)Ar^{k-2} + k(k+1)Cr^k]\cos(k\theta).
\end{aligned} \tag{5.5}$$

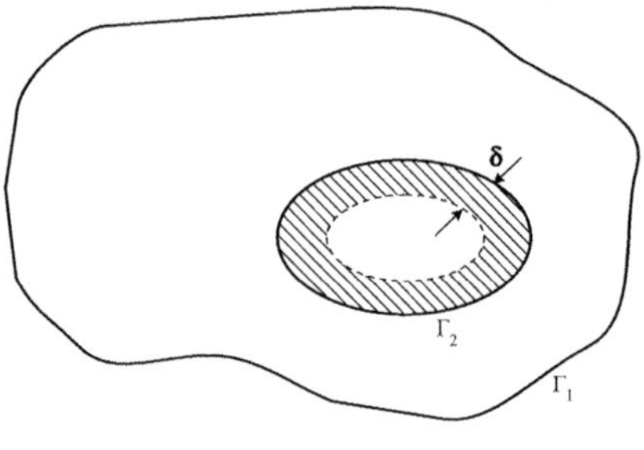

Fig. 5.1. A full structure, a fragment, and an area of sensitivity

[3] An estimate of the subsidence rate of the stress caused by a local self-balanced system of forces for various types of structural mechanics and elasticity problems is given in [16].

Integration constants A and C will be found from the conditions at the contour

$$\text{At } r = R: \qquad \sigma_r = q_r, \qquad \tau_{r\theta} = q_t \,. \tag{5.6}$$

The result is $A = 1/[k(k-1)R^{k-2}]$ and $C = 0$, which gives the following after being substituted to (5.5)

$$\sigma_r = -q\rho^{k-2}\sin(k\theta)\,, \quad \sigma_\theta = q\rho^{k-2}\sin(k\theta)\,, \quad \tau_{r\theta} = -q\rho^{k-2}\cos(k\theta)\,, \quad \rho = r/R\,. \tag{5.7}$$

Let's introduce a relative depth of the sensitivity area β:

$$\beta = \delta/H = k\delta\,/\pi R\,, \tag{5.8}$$

Then at the boundary of the sensitivity area (at $r = R - \delta$) the fall of the stress with respect to the contour Γ_2 points will be defined by a general parameter η that describes the stress subsidence at a distance from the Γ_2 boundary:

$$\eta = \sigma_r(r)/\sigma_r(R) = \sigma_\theta(r)/\sigma_\theta(R) = \tau_{r\theta}(r)/\tau_{r\theta}(R) = (1 - \beta\pi/k)^{k-2}\,. \tag{5.9}$$

For example, assuming $\beta = 1$ and $k = 12$, we have $\eta = 0.048$ from formula (5.9). By doubling the sensitivity area, i.e. assuming $\beta = 2$, we obtain $\eta = 0.0006$. Hence a conclusion that the sensitivity area equal to one or two cells of the H mesh is sufficient even for a pretty coarse mesh (at $k = 12$ we have $H \approx R/4$).

Let's now discuss the case when Γ_2 is subjected to kinematical boundary conditions. Again, the accurate formulation of these boundary conditions is

$$u(s) = u_*(s)\,, \qquad v(s) = v_*(s)\,. \tag{5.10}$$

But when we use approximations of u_* and v_* instead of the functions themselves, those taken from the solution of the full problem on the coarse mesh, we have a noise in the kinematical boundary conditions:

$$\Delta u(s) = u_*(s) - u_H(s)\,, \qquad \Delta v(s) = v_*(s) - v_H(s)\,. \tag{5.11}$$

Here and (5.10) mean the components of the vector of displacements of the Γ_2 boundary by u and v. Kinematical noise (5.11) causes, in its turn, a noise in the stress distribution at Γ_2. By assuming that this secondary noise of the stress distribution at Γ_2 is locally self-balanced, again we can appeal to the Saint-Venant principle.

It should be clear that all considerations and statements given above are qualitative and by no means form a strict mathematical proof of the sensitivity area determination technique during a structure's fragment extraction. Nonetheless one may assume tentatively that outside an area adjacent to Γ_2 with its width about $H \ldots 2H$ the effect of the error of the full solution with the coarse mesh on the fragment's analysis results is not essential and this error can be neglected.

Assuming that a stress convergence estimate of order α takes place for the full structure analysis, that is

$$\|\sigma_* - \sigma_H\| \le CH^\alpha, \qquad \|\sigma_* - \sigma_h\| \le Ch^\alpha, \tag{5.12}$$

we have the following for the analysis of the fragment with rough boundary conditions obtained from the coarse mesh solution, outside the sensitivity area,

$$\|\sigma_* - \sigma_h\| \le \eta C H^\alpha + C h^\alpha. \tag{5.13}$$

A conscientious application of the fragmentation technique requires one to understand well all sources of errors introduced by the fragmentation.

In the course of the fragmentation technique application, it may turn out at a joint of the fragment with the main structure that the fragment's nodes do not have their counterparts on the other side of the joint (Fig. 5.2). Also, the situation of inconsistency between the meshes on two- and three-dimensional finite elements located at different sides of a joint may arise. Thus there arises an issue of the use of special finite elements to match joints at the boundary between a fragment and its main structure. These finite elements must be capable of adding more nodes at sides of rectangles.

Generally, additional matching nodes can be introduced using the poly-constraint technique as shown in Fig. 2.21 — those poly-constraints again! An important role can be played, seemingly, by a mesh smoothing optimization (a gradual change of its elements' sizes) which was addressed in [26].

It should be noted that after a fragment is extracted from the structure the former might prove geometrically changeable. If its changeability is a consequence of the fragment's internal build (an *internal changeability*) then it is not possible to impose purely static boundary conditions on Γ_2, and we have to use kinematical or mixed conditions that keep the fragment geometrically unchangeable. If the fragment is internally unchangeable but capable of moving as a rigid body then it must be restrained against movement in a statically determinate way.

As all external loads working on the fragment are self-balanced (our discussion is confined to the case when the Γ_1 boundary either does not exist or is subjected to static boundary conditions), then the reactions in the additionally imposed constraints must be zero if only the restraint is statically determinate. This statement is true only in theory because it does not allow for the inevitable numerical solution error. The accumulation of the rounding errors results in small but nonzero reactions instead of zero ones. This shows that in a real numerical representation things look like there is a small nonzero force applied to the fragment.

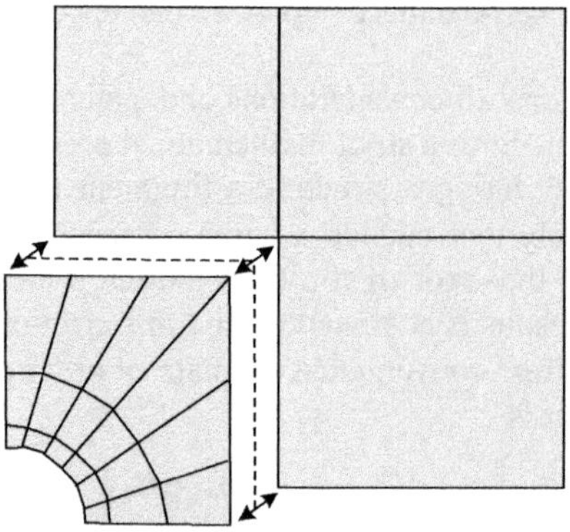

Fig. 5.2. A joint design

As the mentioned force is resisted by conventionally imposed constraints, values of the respective reactions appear to be dependent on the way those constraints are imposed. Note that the same problem arises not only with the fragmentation but in other cases too, when one has to analyze *flying* structures: this is a professional slang for the analysis of structures with no external constraints imposed under the action of self-balanced external loads.

If such constraints are placed as far from one another as possible, then the reactions will be small and won't disturb the stress distribution essentially. But, for example, if one restrains a flexural plate in one point against two rotations and one displacement, which is theoretically feasible, then the disturbance may prove substantial and the computational results in the vicinity of such node will be distorted.

A paper by S.A. Alekseev [3] deals with the issue of the best statically determinate fixation of a body in the sense of minimization of the respective reaction values. It demonstrates that for a plane case (a disk) the best fixation is such that the triangle formed by the constraint lines is a maximum possible one of those circumscribed around the given disk (Fig. 5.3, *a*). This corresponds to the determinant of the disk equilibrium equation system with respect to the support reactions $\Delta = 1.673d$ that defines the fixation quality. The structure in Fig.5.3, *b* has its determinant $\Delta = d$, and the one in Fig.5.3, *c* has the determinant $\Delta = 1.414d$. For a spatial body the maximum possible value of the determinant is achieved in the case when a maximum possible circumscribed parallelepiped can be built on the directions of the constraints.

All that's been said above regards also any case of the fixation of a system that possesses rigid-body degrees of freedom.

It should be noted that the maximization of the determinant yields a most stable equilibrium equation system with respect to the reactions in the imposed constraints, theoretically. But the system itself is of a trifle order, so the main error is caused by disturbances accumulated during its formation rather than its solution. Therefore the issue of minimizing exactly this error is no less (or even more) important, and this one may prove really noticeable in problems with a great deal of unknowns.

Being somewhat previous, we'd like to notice it is reasonable to fix the nodes that result in a least compliant structure of those restrained externally. The background for this recommendation and some details are given in Section 5.5.

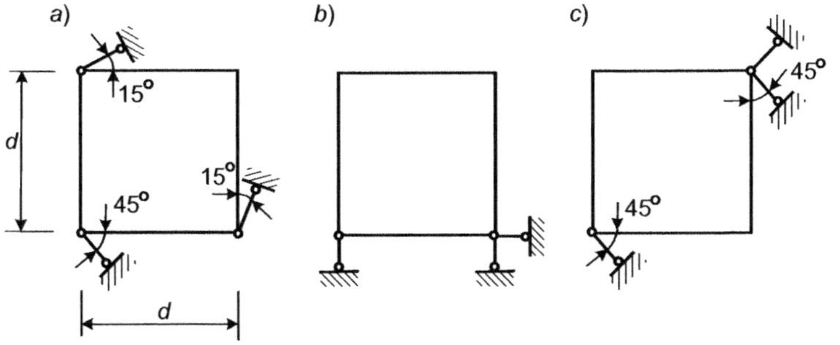

Fig. 5.3. Imposition of constraints to parry the changeability

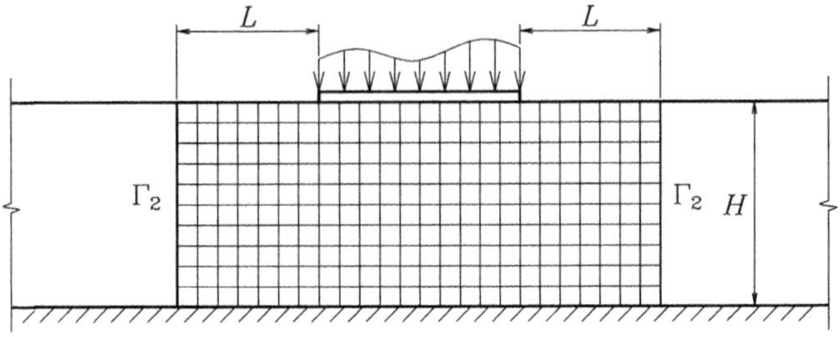

Fig. 5.4. A fragmentation to replace an infinite elastic layer by a finite one

Note that in engineering practice the fragmentation technique is used not only to extract a part of a structure to study its behavior with a much denser mesh. It is often the case that the fragmentation proves to be a useful technique to pass from infinite regions occupied by a full model to bounded regions. Typical examples include the analysis of beam/slab constructions lying on an elastic bed where a finite-thickness elastic layer is assumed as a mechanical model of the elastic foundation (Fig.5.4).

As a rule, an analyst is interested with the stress distribution of the structure itself while that in the elastic layer is a less meaningful information for the purpose of understanding the structure's behavior. In order to employ the finite element analysis software, one has to limit the plane dimensions of the layer neglecting the response of a part of the elastic foundation located sufficiently far from the structure's edge. And the very first question that arises immediately is what distance is sufficiently far from the structure's edge in the elastic layer in order to be able to neglect the remote part of the layer in the analysis.

An empiric rule known to experienced analysts states that it is quite enough to supplement the design model with an elastic layer at each side of the structure at the distance of $L = H \ldots 1.5H$ where H is the height of the bulk being compressed (Fig. 5.4).

An estimate of the distorting effect of this fragmentation on the stress distribution of a structure can be obtained by comparing two solutions with different boundary conditions along a vertical section of the elastic layer (which is exactly the Γ_2 boundary of the fragment).

In the first case, the boundary conditions along the vertical section of the layer are assumed to be static with zero external forces along Γ_2. In the second case, a full fixation of Γ_2 against all displacements is assumed. If you like, these two models of a fragment can be treated as a majorant and minorant model discussed earlier in Section 1.6.

5.2 Building Continuous Stress Fields with FEM

The finite element approach that employs displacements causes a number of issues with the calculation of stresses. A local calculation of the stresses on each element gives stress fields discontinuous at inter-element boundaries (Fig. 5.5). In order to get nodal stress values, one has to use a method of averaging the stress contribution of particular elements to reduce it to their common node (an averaging by the element star of a node)[4].

It is apparently useful to know the extent of inconsistence of the stress field in nodes. To obtain this data, one can evaluate Δ by the formula (5.14) that follows in each node, where the maximum are to be taken over all elements adjoining i^{th} node. The summation is performed over the same elements

$$\Delta = \frac{\left| \max_i \sigma_{(i)} - \min_i \sigma_{(i)} \right|}{\sum_i \left| \sigma_{(i)} \right|} . \tag{5.14}$$

An estimate of the Δ value average over the whole system (or some specific sub-area) is a useful additional information that enables one to assume a proper attitude to results yielded by computation.

The issue of displaying a discontinuous stress fields and its properties can be resolved best if the software is capable of employing several techniques of result demonstration. One of most advantageous systems is **ADINA** which can display (and print) the following results:

- discontinuous values (**None**);
- values averaged in nodes (**Average**);
- a maximum value of those calculated on all elements around a node (**Maximum**);
- a minimum value of those calculated on all elements around a node (**Minimum**);
- an absolute difference of the maximum and minimum values (**Difference**);
- a relative difference of the maximum and minimum values (**Error**).

The solution of a problem shown in Fig. 5.6 as stress isofields in accordance with the above listed modes is demonstrated by Fig. 5.7.

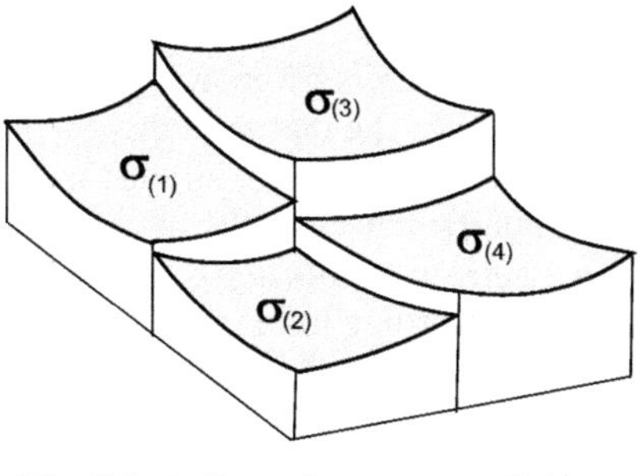

Fig. 5.5. A discontinuous stress field

[4] An *element star of a node* is a region consisting of finite elements that adjoin this node. The node itself is called a *center of the star*.

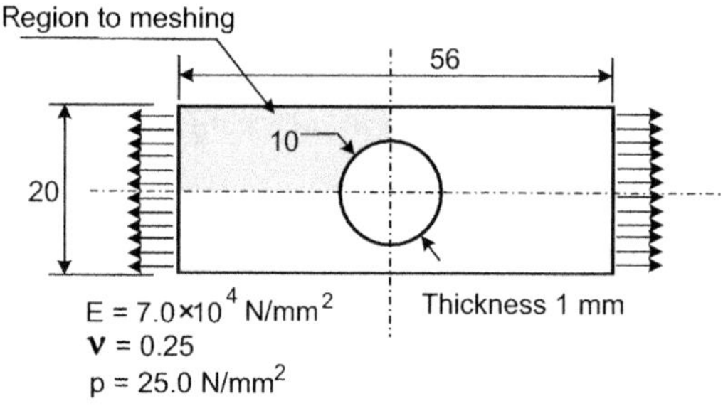

Fig. 5.6. An illustrative problem

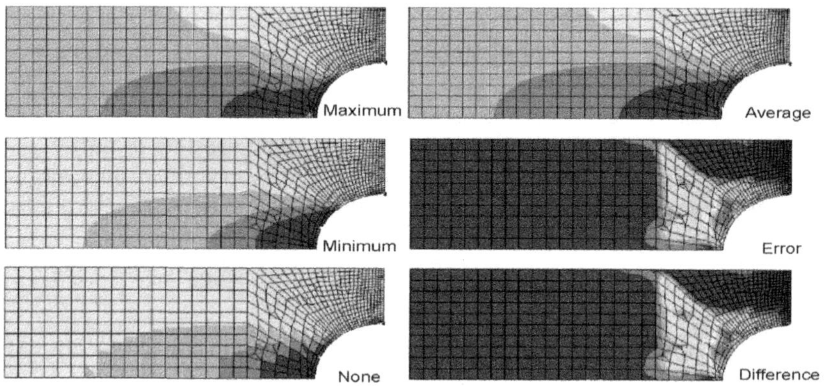

Fig. 5.7. A variety of ways to represent stress fields

The averaging of stress fields should be employed with care when one deals with constrained nodes. For example, in a rectangular plate completely restrained from displacements along the ABCDEF contour and some internal line BE, and loaded inside the BCDE rectangle, there can be no moments at ABEF. But the averaging of moments at the BE line causes a distribution shown in Fig. 5.8 to appear.

As the stress determination issue is of great importance for analysts, it makes sense to discuss it in more detail. The matter is that the principal disadvantage of FEM in its slope deflection form continues, strange as it seems, its main advantage. Indeed, the most important peculiarity of a variational formulation of a problem that employs a Lagrangian functional is weakened requirements to the smoothness of the solution. According to terms by J. Descloux [8], a symmetric or *semi-weak* formulation of a generalized solution is involved. On one hand, this extension of the solution's domain helps build the space of comparable displacement functions, on the other hand, the requirements prove too "weak" for the stresses because the differentiation operation needed to obtain the stresses by displacements breaks the continuity of the stress field components.

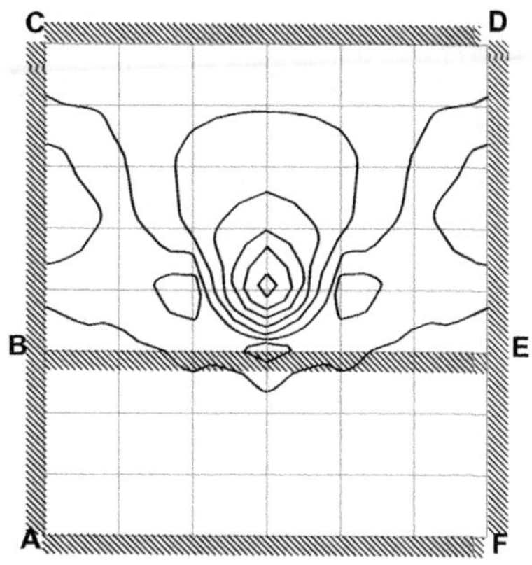

Fig. 5.8. An error caused by averaging

Let $\mathbf{u}_h$ be a vector field of displacements obtained by minimization of the Lagrangian over some finite-dimensional subset of the space of kinematically admissible displacements[5]. Let $\mathbf{Q}$ be an operator that maps displacements $\mathbf{u}$ to stresses $\boldsymbol{\sigma}$

$$\boldsymbol{\sigma} = \mathbf{Q}\mathbf{u} \Rightarrow \boldsymbol{\sigma}_h = \mathbf{Q}\mathbf{u}_h \tag{5.15}$$

In many cases an engineer possesses a prior information regarding the smoothness of the stress in the classical accurate solution of a problem. If this information is contradictory to the observed/expected smoothness $\boldsymbol{\sigma}_h$, then the engineer wishes to make this solution "look better" by mapping it to a smoother set

$$\boldsymbol{\sigma}_{h*} = \mathbf{P}\boldsymbol{\sigma}_h\,. \tag{5.16}$$

Thus the essence of the problem is how to build a smoothing operator $\mathbf{P}$ to project a discontinuous solution $\boldsymbol{\sigma}_h$ onto a surface the smoothness of which is matched to the available prior information. This mapping is expected to give a higher accuracy of the smoothed stress field $\boldsymbol{\sigma}_{h*}$ than that of $\boldsymbol{\sigma}_h$.

Along with the above said averaging operator, there are known attempts of solving the same problem with difference formulae of higher order of accuracy employed to determine nodal stress values. See papers [6, 9] for examples. Though, this way is not algorithmic because it's not clear how to use it on irregular meshes.

---

[5] The concept of kinematically admissible displacements needs some circumstantiation. In scientific and educational books one can find a definition of kinematically admissible displacements in a strictly mechanical sense as those not violating constraints imposed on a system. One should realize that in the formal mathematical sense this definition implies two requirements to be fulfilled:

- the displacements must satisfy kinematical boundary conditions;
- the displacements must be sufficiently smooth functions.

As different definitions of a problem solution (*strong*, *weak*, *semi-weak solution* [8]) require a different smoothness of the displacements, the very set of kinematically admissible displacements changes together with the definition of the problem's solution.

For some time people had expectations of an idea by Burlow mentioned in a known book by H.Strang and J.Fix [32]. According to the idea, the $\boldsymbol{\sigma}_h$ solution is to be determined by formulae (5.15) but in some *optimal points* different from nodes in which the stress error is guaranteed to be minimal. This idea has not become really popular as well because of difficulties that appear when searching for the optimal points. Also, as a rule users are interested with stresses in nodal points of a mesh or in other particular points the position of which can be specified explicitly.

A peculiar method for obtaining continuous stress fields was suggested and implemented by Loubignac who published his co-authored proposal in [17]. According to the consideration by Loubignac, a particular form of the **P** operator that maps discontinuous stresses $\boldsymbol{\sigma}_h$ into a field of continuous stresses $\boldsymbol{\sigma}_{h*}$, is unimportant and can be assumed as simple as possible, including an averaging operator. That's actually the way Loubignac follows. Two essential points of this method follow:

- the distribution of the stresses in the finite elements is described by the same shape functions $\mathbf{U}(X)$ which are used to approximate the displacements;
- the difference of the stress fields $\boldsymbol{\sigma}_0 = \boldsymbol{\sigma}_{h*} - \boldsymbol{\sigma}_h$ is treated as a system of prestress.

If $\boldsymbol{\sigma}_0(X)$ is a pre-stress field, then the reduction of this peculiar distributed load to nodes will give the main governing equation in the form [40]

$$\mathbf{Ku} + \int_{\Omega} \mathbf{B}^{\mathrm{T}} (\boldsymbol{\sigma}_{h*} - \boldsymbol{\sigma}_h)\, d\Omega = \mathbf{p}\,, \tag{5.17}$$

where $\mathbf{B}^{\mathrm{T}}$ is a transposed matrix of connection between the nodal displacements and strains. This system of equations is solved by Loubignac in an iterational way, assigning the (k) index to values calculated at k^{th} iteration.

$$\mathbf{Ku}^{(k)} + \int_{\Omega} \mathbf{B}^{\mathrm{T}} (\mathbf{P}\boldsymbol{\sigma}_h^{(k-1)} - \boldsymbol{\sigma}_h^{(k-1)})\, d\Omega = \mathbf{p}\,. \tag{5.18}$$

By introducing a material elasticity matrix **C** we obtain [40]

$$\boldsymbol{\sigma}_h^{(k-1)} = \mathbf{CB}\,\mathbf{u}^{(k-1)}, \qquad \mathbf{K} = \int_{\Omega} \mathbf{B}^{\mathrm{T}}\,\mathbf{CB}\, d\Omega\,, \tag{5.19}$$

and the iteration process (5.18) acquires its final form

$$\mathbf{K}(\mathbf{u}^{(k)} - \mathbf{u}^{(k-1)}) + \int_{\grave{U}} \mathbf{B}^{\mathrm{T}}\,\mathbf{PCBu}^{(k-1)}\, d\Omega = \mathbf{p}\,. \tag{5.20}$$

The iterations are performed until the difference $(\mathbf{u}^{(k)} - \mathbf{u}^{(k-1)})$ becomes sufficiently small. Final continuous stress fields in each finite element will have the form

$$\boldsymbol{\sigma}_e(X) = \mathbf{U}(X)_e\,(\boldsymbol{\sigma}_{\mathrm{av}})_e, \tag{5.21}$$

where $(\boldsymbol{\sigma}_{\mathrm{av}})$ is a vector of nodal stresses obtained by averaging over the star of each node of an element.

This algorithm was tried by Loubignac with a problem of a cantilever beam under a load distributed according to a square parabolic law at its free end, and with a problem of a thick-walled tube under an internal pressure. Computational results for these problems show a substantial improvement of the accuracy after five iterations of the stresses.

Fig. 5.9, *a* shows an example from [17] to illustrate the results of the described analysis. A cantilever deep beam is under consideration, with its end loaded by a parabolically distributed force. Fig. 5.9, *b* shows values of the stresses in point A. The solutions were obtained with different finite element meshes. It turned out that even a pretty coarse mesh yielded good results close to the theoretical values.

Unfortunately, the method by Loubignac still does not have a theoretical foundation, and the conditions of the iteration convergence are unclear. Also, as Loubignac himself notices, there can be algorithmic difficulties with problems where stress fields are discontinuous. Finally, the iterative nature of the Loubignac's procedure makes it poorly applicable to a problem with multiple load distributions specified because the computational time is approximately proportional to the number of those.

An interesting idea how to obtain nodal stress values belongs to A.V. Vovkushevsky [38]. According to his suggestion, a *star* of elements of a node is taken into consideration together with this node where one wants to know the stresses[6]. Let k be the number of the star's nodes including its center.

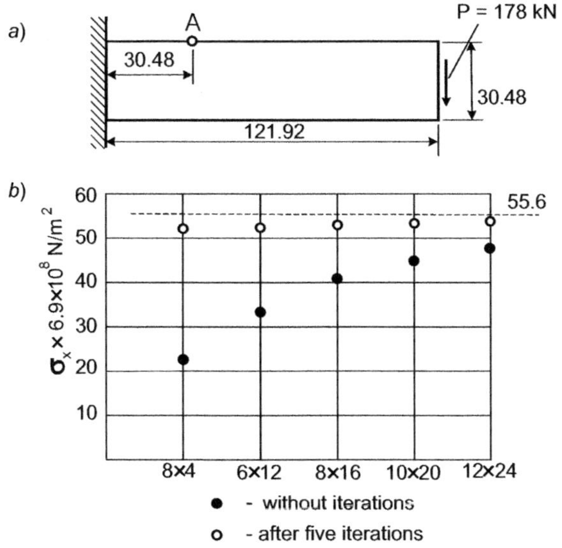

Fig. 5.9. Analysis of a deep cantilever beam

[6] The *star of a node* means the set of finite elements incident to the node.

To be particular, let's discuss a plate bending problem. Within the star a new approximation of displacements w is introduced, generally not the same as that used to determine nodal displacement values. According to this new approximation we assume

$$w=\sum_{i=0}^{s}\sum_{j=0}^{s}a_{ij}f_i(x)f_j(y), \tag{5.22}$$

where $f_i(x)$ and $f_j(y)$ are polynomials of power i and j with respect to x and y, a_{ij} are coefficients to be found. In total the expansion (5.22) contains $(s+1)^2$ unknown coefficients a_{ij}.

We demand that in every node of the star the approximations (5.22) satisfy the generalized displacements obtained by the finite element analysis. Each node thus provides N equations where N is the number of degrees of freedom of the respective node.

So, for the finite elements with solid-node elements ($N = 3$) and for m^{th} node of the star we'll have the following:

$$\left.\begin{aligned}&\sum_{i=0}^{s}\sum_{j=0}^{s}a_{ij}f_i'(x_m)f_j(y_m)+\theta_{ym}=0,\\&\sum_{i=0}^{s}\sum_{j=0}^{s}a_{ij}f_i'(x_m)f_j(y_m)-\theta_{xm}=0,\\&\sum_{i=0}^{s}\sum_{j=0}^{s}a_{ij}f_i(x_m)f_j(y_m)-w_m=0,\end{aligned}\right\}\quad(m=1,\dots,k), \tag{5.23}$$

where w_m, θ_{xm}, θ_{ym} are deflections and slopes of mth node. Let's choose the s number so that the number of conditions of the type (5.23) be not less than the number of coefficients a_{ij} to be determined in the representation (5.22), that is

$$(s+1)^2\le N\,k\,. \tag{5.24}$$

The equation system (5.23) is redundant if the strict inequality holds in (5.24), therefore its solution can be found by, say, the least squares method. By the way, it's reasonable to primarily refine this system to some extent by putting the star's center into a privileged position with respect to the other nodes. This can be done by introducing a weight factor $p > 1$ into equations of the system (5.23) that correspond to the center of the star. Having determined the coefficients a_{ij}, we can find the stresses at the center of the star by differentiating (5.22) and using the physical relationships.

It should be mentioned that Vovkushevsky himself applied his idea to the plane elasticity. An attempt of using the same technique for the analysis of a plate bending encounters a difficulty: though the final equation system of the least squares method has a low order, but it is poorly conditioned to a great extent and thus very sensitive even to smallest disturbances of quantities w_m, θ_{xm}, θ_{ym} ($m = 1,\dots, k$). Another drawback of Vovkushevsky's method is its orientation at strictly continuous stress fields. Also, there are algorithmic issues that arise with coupling finite elements of different types.

In publications a so-called *conjugate approximation method* by Oden – Reddy [21] is very popular. Its technique (without much interpretation) can be found even in educational literature [27]. A detailed explanation of the Oden – Reddy method is found in [20] though that explanation is much complicated by bulky index transformations and inessential details that encumber the reader's understanding.

After a careful look, the following interpretation can be given to the Oden – Reddy technique by relating it to the variational formulation of the problem of stress determination by a known displacement field. The stress field is to be sought as an expansion by the same basis functions used to minimize the Lagrangian. Further, we compose a functional of the following form that will be referred to as an Oden functional (though Oden himself does not treat his method as a variational one):

$$\mathrm{O}(\boldsymbol{\sigma}) = \frac{1}{2}\int_{\Omega}[\,\sigma^{ij}\sigma_{ij} - \sigma_{kl}C^{ijkl}(u_{i,j} + u_{j,i})]\,d\Omega\,, \tag{5.25}$$

where C^{ijkl} are components of the elasticity tensor that relate the stress tensor $\boldsymbol{\sigma}$ with the deformation tensor $\boldsymbol{\varepsilon}$, $\varepsilon_{ij} = (u_{i,j} + u_{j,i})/2$. The stationarity conditions for the Oden functional $\mathrm{O}(\boldsymbol{\sigma})$ lead to the physical elasticity equations

$$\sigma^{ij} = C^{ijkl}\varepsilon_{kl}\,. \tag{5.26}$$

Let the set of functions φ_k ($k = 1,..., n$) form a system of basis functions used to minimize the Lagrangian thus finding the displacements by a finite element analysis. By expanding the stresses into finite sums by the same basis functions and substituting the expansions to functional (5.25), we transform the functional into a quadratic form. The conditions of its stationarity give us a system of equations to find coefficients of the stress expansion by the system of basis functions $\{\varphi_k\}$.

The conjugate approximation method by Oden, though it yields continuous stress fields, has a number of drawbacks:

- a system of linear algebraic equations needs to be composed and solved in order to determine the stress expansion coefficients by functions $\{\varphi_k\}$;
- the static boundary conditions still cannot be satisfied by the Oden technique accurately;
- the approximation of the stress field is bound to the displacement approximation and does not provide the possibility to use other coordinate functions;
- the variational interpretation of the Oden method shows that the governing equations of the method follow from the condition of stationarity of functional $\mathrm{O}(\sigma)$ that does not have an explicit sense of energy which can be immediately seen by considering the dimensionality of quantities in the integrand of (5.25);
- the method is oriented at obtaining continuous stress fields and does not help allow for discontinuities such as those caused by jumps of elasticity constants in two- or three-dimensional problems.

Finally, a *method of two functionals* suggested in [29] should be noted. In a sense (in its procedure) this method is close to the Oden – Reddy method because it requires the formation and solution of an additional equation system as well. Therefore the first drawback of the above list of those of the Oden – Reddy conjugate approximation technique is inherent in the two functionals method too. Though, drawbacks listed in other items of the list are eliminated thereby. Without going into theoretical details of the background of the two functionals method, we will state its main idea in brief.

Just as in the conjugate approximation by Oden, the approximate solution of a problem by the two functionals method is sought in two stages. At the first stage, the minimization of the Lagrangian gives a conventional finite element solution $\mathbf{u}^h$ that describes the displacements. Then, a functional depending on stress components is built

$$\mathsf{D}(\boldsymbol{\sigma}) = \mathsf{E}_\sigma(\boldsymbol{\sigma}) - \Pi_\sigma(\boldsymbol{\sigma}) , \tag{5.27}$$

where $\mathsf{E}_\sigma(\boldsymbol{\sigma})$ is the complementary energy functional, $\Pi_\sigma(\boldsymbol{\sigma})$ is the potential of reactive forces

$$\mathsf{E}_\sigma(\boldsymbol{\sigma}) = \frac{1}{2}\int_\Omega D_{ijkl}\sigma^{ij}\sigma^{kl}d\Omega \,, \quad \Pi_\sigma(\boldsymbol{\sigma}) = -\int_\Omega u_i\sigma^{ij}_{,j}d\Omega + \int_{\Gamma_u}\bar{u}_i\sigma^{ij}n_j d\Gamma \,. \tag{5.28}$$

The notation follows:

D_{ijkl} are components of the elastic compliance tensor inverse to the C^{ijkl} tensor;
Γ_u is a part of the boundary of the Ω area where kinematical boundary conditions are specified as given displacements $\bar{u}_i$;
n_i are components of an external normal ort to the Γ_u boundary.

Having replaced displacements u_i by their approximations u_i^h obtained at the first stage in the functional $\mathsf{D}(\boldsymbol{\sigma})$, we have a variational form of the second stage of the problem's solution. This stage is intended to find the stresses only. At this point we are searching so-called *statically semi-admissible* stress fields rather than any arbitrary fields for the minimum of the functional $\mathsf{D}(\boldsymbol{\sigma})$. The statically semi-admissible stresses are those satisfying static boundary conditions and conditions of discontinuity known a priori, that is, before the problem is solved. The basis functions for the purpose of the minimization of $\mathsf{D}(\boldsymbol{\sigma})$ can be the same functions $\{\varphi_k\}$, just as in the conjugate approximation method. Though, a different system of coordinate functions can be used to allow for peculiarities of a problem. Not being able to go too deep into details of the two functionals method, we refer the reader to the paper [29] for particularities and examples.

It should be noted that so far the existing software developments do not employ the conjugate approximation method by Oden, nor the two functionals method. So there remains a hope for future software implementations. And the authors do not doubt to say that a team of developers first to risk an additional effort and release a marketable computational software of a higher stress field determination accuracy will have a considerable advantage on this hard market.

5.3 Mistakes and Traps in Coupling Elements of Different Dimensionality

It is only a rare case that a whole structure can be represented by elements of the same type (such as plate elements). More frequently the same design model may include bars and shells, and other types of elements.

Contemporary software allows almost any combination of finite elements to be employed, the elements being of various types, dimensionalities, sizes and geometric shapes, and numerical stiffness properties. This flexibility is a great boon for an experienced user which enables him to exhibit his inventiveness in order to expand the applicability of the software and provides him with a tool to crack hard nuts not even realized by the developers of the product. On the other hand, this software flexibility is a dynamite, sometimes explicit, sometimes hidden thus especially insidious[7]. Anyway, from the standpoint of the mechanical problem formulation and the equivalency of the finite element model to the respective real structure, the correctness of a design model relies on the user. See Fig. 2.1 in this regard.

This section deals with some traps that await the user in connection with structural members of different dimensions combined in a design model, the members being described mathematically in totally different ways. The situations like this may arise, say, when a design model includes both one-dimensional (bars) and two-dimensional (plates and/or slabs) members, though it is not the only source of the danger.

Table 5.1 lists possible options of combining parts of a structure that contain elements of different dimensionality. The "–" sign designates dangerous combinations of structural members from the standpoint of correctness of a design model.

As will be shown below, this merging of a variety of different elements in their common design model requires a careful attention, especially in places where the said elements of different dimensionality are joined.

Table 5.1.

Dimensionality	Elements		1	2	3	4	5
One-dimensional	Bars	1	+	–	–	–	–
Two-dimensional	Slabs	2		+	–	–	–
	Plates	3			+	–	–
	Shells	4				+	–
Three-dimensional	Bulk elements	5					+

[7] An expert in programming languages and compiler development would call input languages of CAD software weak typing languages. It is known that the weak typing causes one to make hard-to-find mistakes. In the English-language tradition of programming, such are called in a zoologic manner — bugs. One can easily imagine holes and gaps in design models where these filthy insects can creep in.

The common rule is: when joining elements of different dimensionality, one should not rely on formal computational tools provided by finite element software. As a rule, one needs to perform a more detailed analysis of joints between finite elements of different dimensionality to keep the design model correct.

Let's discuss some situations of those listed in Table 5.1 in more detail.

5.3.1 Bars + Plates

Let's take as an example of combined slab and bar elements the analysis of a spatial framing together with a slab foundation[8].

The introduction to this book mentions that lately users tend to further complicate their design models. In particular, the analysis of frameworks of buildings together with their foundation structures has become popular in recent years. A separate analysis of a framing and, say, a foundation slab of a building looks like an anachronism, now that powerful design software tools are so widely popular and easily available. In the community of structural designers such an analysis may be even treated as a professionally indecent behavior. Meantime, not every captain who sets sail to the sea of structural design knows how to take a smooth water and escape all reefs on the way, and a reliable pilot is not always available. At the same time no "pilot chart" (which means just "a software reference manual") available to the authors contains warnings of the danger.

So, let us consider a discrete design model combining plate/slab finite elements and bars rigidly attached to the slab. The fact that the slab lies on an elastic bed is of no importance for the problem being discussed. The finite element mesh is formed so that the bars of the structure's framework hit the nodes of the meshed slab. If no additional measures are taken, the model described above will provide the consistency of both vertical displacements of the slab/framework (in the perpendicular direction to the plate) and the respective slopes in nodes where the plate and bar elements join one another. Though, bending moments in cross-sections of the columns adjoining the slab, calculated with this model, have nothing to do with the real distribution of internal stresses. This being the case, the distribution of the internal stresses in other elements of the structure's framework will be affected too.

Indeed, let us imagine how the mesh of the slab is being denser, and the user convinced of the convergence of the employed plate finite element by the developers of his software, expects a further refining of the accuracy of the analysis. Though, if the software works all right then, starting from a certain density of the mesh, the further division of the mesh should result in permanently decreasing absolute values of the bending moments in the bars in places of the latter's attachment to the slab. When passing to a limit, as the maximum size of the mesh cell tends to zero, these bending moments should tend to zero too. This means that this design model provides a hinged rather than stiff attachment of the framework ele-

[8] In structural mechanics, the terms "slab" or "plate" mean the same: "a plate capable of bending", "a flexural plate". We will use "plate" to be brief.

ments to the slab. The fact that the user gets some formal nonzero numerical values of the bending moments on a specific mesh evidences only a computational error of the finite element method. And there is no reason at all to take the discretization error for a credible numerical result.

Actually, with a fixed mesh put on the slab the user employs a design model where elements of the framework are, as it seems, fixed elastically by external constraints to prohibit the slopes of the respective bar cross-sections. The stiffness coefficients of these external "clock springs" depend on the diameter of the slab's mesh and tend to zero together with the maximum size of the mesh cells.

As can be seen, the bending moment in a bar of this model, independent of the mesh spacing, is transferred to the slab as a concentrated moment in a node of the mesh (the moment is concentrated because the bar is one-dimensional). On the other hand, the slab under the concentrated bending moment acquires an infinite slope in the moment's plane at its point of application. More exactly, the expression of the slope includes a logarithmic singularity. Thus the slab does not resist to a concentrated slope and does not restrain the framework elements.

In order to be particular, let's refer to an accurate analysis of a simply supported round plate loaded with an external concentrated moment M at its center as shown in Fig. 5.10.

Denoting the plate's radius as R, its flexural rigidity as D, its material's Poisson ratio as ν, the current radial and angular polar coordinates as ρ and φ, we have the following expression for the plate deflection w [35]

$$w = \frac{MR}{8\pi D}\Big[\frac{1+\nu}{3+\nu}(\rho^3-\rho) - 2\rho\ \ln\rho\Big]\cos\varphi\,, \tag{5.29}$$

and the slope θ of the normal to the plate's median surface in the plane of the M application (i.e. at $\varphi = 0$) will be

$$\theta = \left.\frac{\partial w}{\partial \rho}\right|_{\varphi=0} = \frac{MR}{8\pi D}\Big[\frac{1+\nu}{3+\nu}(3\rho^2-1) - 2 - 2\ln\rho\Big]\,, \tag{5.30}$$

hence the said logarithmic singularity (at $\rho \to 0$) that makes the slope angle θ infinite in the point of application of the concentrated moment[9].

In order to amend the model, one should allow for the design of the joint between the framework and the slab. If the columns of the framework are attached to the slab via plinths, then the latter can be treated as perfectly rigid bodies not changing their sizes whatever modifications are made to the finite element mesh.

---

[9] The book [18] gives more solutions for plates under concentrated bending moments. In all cases the expression of the slope of the normal to the median surface of a plate at the application point of the moment contains a singular term that makes the slope infinite. By the way, the same book demonstrates that this is not a sole property of the thin plate bending theory (Kirchhoff – Love plates). There's also a singularity in similar solutions obtained with the theory of medium-thickness plates that allows for lateral shear (Reissner plates).

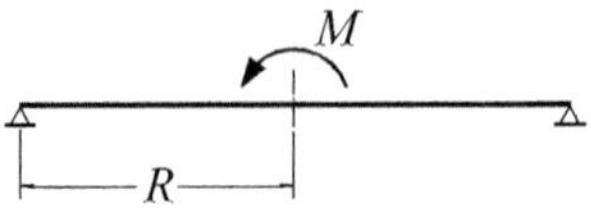

Fig. 5.10. A load upon a round slab

The size of the rigid body itself can be assigned on the basis of the plinth's size ($a{\times}b$), allowing for the application of its pressure to the slab at the angle of 45° (Fig. 5.11) which conforms to the common construction practice.

In this model the concentrated moment from the framework's columns is transferred to the slab via a stiff spacer the sizes of which $a+h$ and $b+h$ do not depend on the spacing of the finite element mesh, nor the discretization error respectively. Though the discretization error will affect the numerical values of the bending moments in the columns, it won't cause the design model to become so fatally different from the real structure as it used to be in the model with no intermediate rigid body. When no plinths are used, one may use the sizes of the column's cross-section as those of the rigid body, taking into account the application of the pressure to the slab at 45 degrees as shown in Fig. 5.11.

In order to give a confirmation of all said above and to a try test example, let's discuss results of analysis of a square plate clamped along its contour with a single column fixed at the center of the plate.

An external concentrated force P is applied to the free top end of the column. It is directed along the global axis Z (Fig. 5.12). In this case the bending moment in the bottom cross-section of the column is a constant value independent of the finite element mesh spacing because the system is statically determinate with respect to the column. Though, the same effect manifests itself by a horizontal displacement of the column's top that grows infinitely as the mesh is becoming denser because of the growing slope in the central node of the plate.

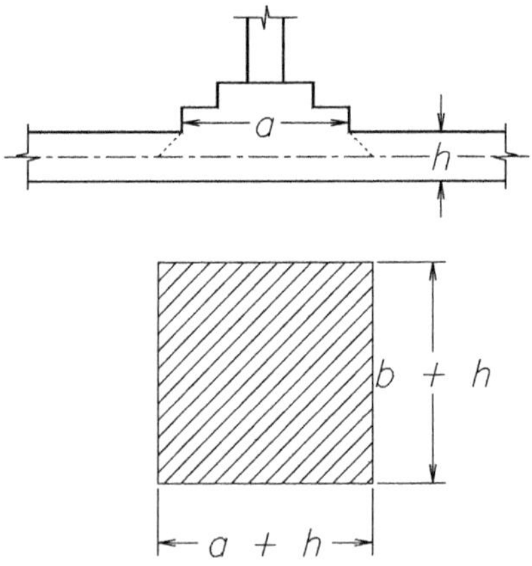

Fig. 5.11. An attachment of a column by a plinth: a and b are plane sizes of the plinth's foot

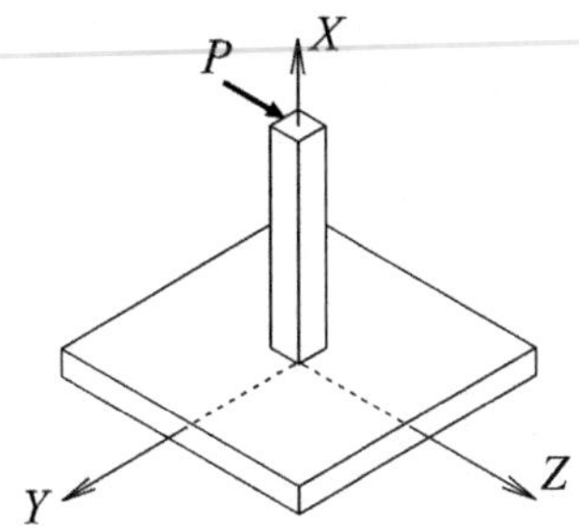

Fig. 5.12. A column fixed at a slab

The second line of Table 5.2 gives results of calculation of the displacement w_n of the column's free end in the direction of the Z vs. the finite element mesh size ($n \times n$) on a quarter of the plate's plane projection.

The following source data are used by the analysis:

- the plate thickness $h = 0.5$;
- the full plane dimensions of the plate 10.0×10.0;
- the size of the cross-section of the column 0.5×0.5;
- mechanical constants of the column and slab material $E = 3 \cdot 10^7$, $\nu = 0.25$.

Table 5.2.

$n \times n$	2×2	4×4	8×8	16×16	32×32	64×64
$w_n \cdot 10^4$	11.826	11.996	12.162	12.326	12.492	12.659
$\underline{w}_n \cdot 10^4$	–	11.209	11.194	11.180	11.172	11.169

It is easy to see that each doubling of the mesh causes the deflection w_n to grow by almost the same value $\Delta \cong 0.167 \cdot 10^{-4}$, or in other words, the column's deflection grows linearly with $\log_2 n$. More exactly,

$$w_n = w_2 + \Delta \cdot \log_2(n-1),$$

wherefrom the infinite growth of the slope of the column's bottom cross-section (as the finite element mesh is becoming denser) follows.

Let's now see how the result of the analysis will change if a perfectly rigid body of a plinth is included therein. Taking the designations from Fig. 5.11, we assume $a + h = b + h = 2.50$.

The third row of Table. 5.2 presents displacements $\underline{w}_n$ of the column's free end obtained with the plate model that includes the perfectly rigid body of the sizes indicated above. Even a careless eye looking at these solutions will immediately notice a quick convergence of the solution as the mesh condenses.

Taking a careful look, one may notice a strange circumstance that accompanies the design model with a rigid body. Indeed, though the convergence of $\underline{w}_n$ takes place, the quantity is decreasing. As $\underline{w}_n$ is the work of external forces (up to a constant coefficient), all principles of the finite element analysis tell us that the value of $\underline{w}_n$ should have grown as the mesh becomes denser. The mechanics of the process includes the removal of more and more constraints from the discrete system as the mesh condenses, that is, the system is acquiring more degrees of freedom. Though, this explanation is true only if fully consistent finite elements are employed. With inconsistent (or *non-conforming* as they are sometimes referred to) elements, these considerations are of no use. On one hand, the condensation of the mesh increases the compliance of the system, on the other hand, it imposes new constraints on it by closing gaps between the system's inconsistent elements that appear in the course of its deformation. Any of the two tendencies (stiffening of the system or its becoming more compliant as the mesh condenses) may prevail in a particular problem. It is hard to predict which one it will be.

Note also the fact that actually a *discrete perfectly rigid body* participates in the model rather than a continuous one.

To clear up the question what we mean by a discrete perfectly rigid body, let's take a look at Fig. 5.13 wherein a quarter of the plate's plane projection is shown together with the rigid body and a mesh of finite elements 8×8. The conditions of the perfect rigidity hold along the 2–3–4 line only in the mesh nodes lying on the boundary rather than on the whole continuous line. Accounting for the symmetry, the following relationships can be set up for the nodal displacements:

$$u_1 = u_2 = 0, \qquad u_3 = 0.625\theta_y, \qquad u_4 = u_5 = u_6 = 1.25\theta_y,$$

where u_i is a vertical displacement (along an imaginable X axis) of ith node shown in Fig. 5.13, θ_y is an angle of rotation of the rigid body about the Y axis.

For a continuous rigid body these conditions would look like:

$$u(z) = z\theta_y$$

for all values of z within the boundaries of the body.

So, a discrete perfectly rigid body restrains respective displacements of only specific nodes of this body.

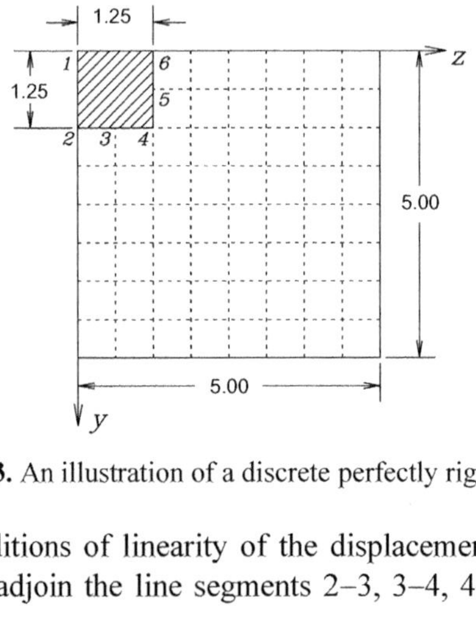

Fig. 5.13. An illustration of a discrete perfectly rigid body

Generally, the conditions of linearity of the displacements are not obliged to hold at elements that adjoin the line segments 2–3, 3–4, 4–5 and 5–6. Then, in intermediate points between the nodes the conditions of consistency of vertical displacements u of the rigid body and its adjacent finite elements will be violated. But in this case the condensation of the mesh, while taking off the system additional constraints in the plate, imposes new constraints between the rigid body and the plate (by eliminating allowed displacement gaps). Though, it can be proved for an Adini – Clough finite element [1] employed by us that there is going to be no gap in the u displacements between the rigid body and finite elements adjoining it. The respective consideration is given in Subsection **5.3.4** of this section. But with other types of plate finite elements this inconsistency may well arise. It can be explained by a difference of the behavior of the discrete and continuous models of perfectly rigid bodies.

Here we should emphasize that a conscientious use of perfect rigid body models is possible only when one understands quite clearly the difference between a continuous and a discrete rigid body.

It should be noted that in both cases the computation was performed by the **GTSTRUDL** software with the finite element type BPR (Bending Plate Rectangle), a rectangular finite element to describe plate bending. This is a popular among users and well-reputed inconsistent finite element having three degrees of freedom per node, an Adini – Clough element. Though, the particular choice of a plate bending element is not critical for this problem. The same conclusion will be made for any finite element type that ensures the convergence.

Finally, we'd like to draw the reader's attention to the circumstance related to a joint analysis of a spatial framework and its slab foundation. Assuming columns of the framework to be orthogonal, we find that the slab does not resists a column's torsion as well as displacements v, w of the column's bottom in the direction of Y, Z respectively because the plate finite elements just do not have appropriate degrees of freedom to restrain those linear displacements and torsions.

If there are no other external horizontal constraints in the model (those in the plane Y, Z), then the mechanical system turns out to be unsecured and geometrically unstable, and the software will have to response to this circumstance somehow. The type of this response depends on the forethought and inventiveness of the software developers. For example, an inelaborate abort of the analysis is quite feasible. But the best response boasted of by most advanced computational software developments is an attempt to correct the user politely by adding external constraints. J.Alger [4] even constructed a special abbreviation DWIMNIS to designate this logic of operation: "*Do what I mean, not what I say*". The locations for an automatic installation of external constraints depend on the attitude adopted by the developers. It goes without saying that a software must inform its user of corrections made to a design model on its own responsibility by generating appropriate warnings.

An even more tricky bug is the absence of constraints that keep the columns from torsion because in a spatial framework no geometrical instability would arise due to it, and the respective software might lack mechanisms for detecting the errors of this kind. This circumstance should be kept in mind when building a design model of a complex combined structure.

One of feasible solutions to join plate and bar elements can be an imposition of external constraints on the rotations of nodes in the plane of a plate. The engineer must be confident of the impossibility for such rotations to appear.

Generally, the problem of constructing reliable algorithms for kinematical analysis of mechanical systems is non-trivial. It is worth discussing separately and in sufficient detail. We will discuss the matter when an opportunity occurs. Some considerations on the subject are given in Section 3.2. Here we have just touched the topic in connection with a particular example where the problem reveals itself.

5.3.2 Bars + Plane Stress

To analyze a high-rise building including both a framework and stiffening diaphragms under horizontal loads, one has to include heterogeneous elements in its design model. Some peculiarities of design models appearing when a stiffening diaphragm is simulated by a Timoshenko bar and the framework by conventional Bernoulli – Euler bars are discussed later in Section 5.4. Here we are concentrating on a situation when the relative sizes of the stiffening diaphragm do not allow classifying it as a bar element even though the shear be taken into account. In this case the diaphragm is treated as a deep beam, and its behavior is described by the plane elasticity theory.

So, let the design model of a structure include a *plate* and a framework made of bars, for example, as shown in Fig. 5.14[10].

If one takes an utterly formal attitude towards the formation of a discrete design model and describes the plate by a set of plane stress finite elements while the bar framework by one-dimensional elements without any additional measures taken, then both the bending moments and longitudinal/lateral stresses in crossbars that adjoin nodes at the boundary between the framework and the diaphragm (nodes 1 and 2 in Fig. 5.14) will be totally unrelated to the realities of life.

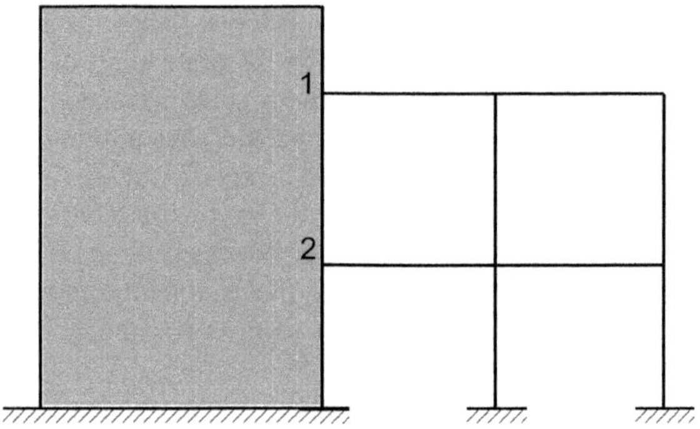

Fig. 5.14. A junction between a framework and a plate

[10] Here a plate will mean a two-dimensional structural element that experiences a plane stress state.

Let's handle the bending moments first. Let the model of the diaphragm use simplest finite elements with two degrees of freedom in each node (linear displacements *u* and *v* in two mutually orthogonal directions). Such finite elements do not resist to a rotation of an adjoining node because the elements just lack the appropriate degrees of freedom. Therefore these finite elements do not transfer any moments to the adjoining nodes. As the software will require all equilibrium conditions to be satisfied, including the equilibrium of each node by moments, the crossbars that join these nodes rigidly will have to transfer zero moments to the nodes. Thus the bending moments in the crossbars in these nodes will be equal to zero. It corresponds to a model of hinged attachment of the crossbars to the diaphragm, and it does not satisfy the requirement of a user oriented at the analysis of crossbars clamped in the diaphragm.

In order to avoid such distortion of the design model, sometimes users introduce external constraints into crossbar-to-diaphragm junction nodes to restrain the latter from rotation. But this technique can hardly be approved because therewith we introduce a distortion to the model again. This time it is an increase of the total stiffness of the model structure because there are no constraints like this in the real structure.

The interesting thing is that in spite of this problem's being known to all the users for a long time, discussions on the subject continue arising again and again. Among recommendations being invented one can hear even an appeal to employ the moment elasticity theory to simulate the diaphragm's behavior. As we will show later, the problem can be solved with simpler models and the behavior of a real structure can be adequately described only by considering structural solutions of joints between the crossbars and the diaphragms and allowing for these solutions in the design model.

Among propagated recommendations, which we cannot uphold for reasons stated below, there is one to use plane stress finite elements having additional degrees of freedom with the meaning of nodal rotations[11]. The finite elements of this particular type are quite feasible and are one of options to build finite element models of higher order of displacement/stress approximation. That's why the text below must not be treated as a negative opinion towards those finite element types. We'd like to draw the reader's attention to the fact that an attempt to circumvent the problem stated above is fruitless, moreover, dangerous because makes a user think this is a solution while actually it is not. Again the gap for this new species of "bugs" is just masked not eliminated (it is the user who may miss the gap, but a "bug" wouldn't!) by the discretization error: it is true that the bending moments in the crossbars are nonzero.

Just as in the problem of the plate-to-bar coupling, the main issue is not a drawback of the discrete model itself. The matter is that the original mathematical

[11] For example, the user manual of the **STARK_ES** software belonging to the **MicroFE** family says the following when describing plane stress finite element types: "*Making use of three degrees of freedom in a node provides an improved accuracy of the analysis and helps match the degrees of freedom in combined systems (such as a deep beam and a bar)*".

formulation of the problem which implies a pointwise joint between a plane stressed diaphragm and a one-dimensional bar, when solved accurately, yields zero moments at the junctions of the bars and the diaphragm. To prove this statement it suffices to make sure that the respective rotation component ω_z of the plane elasticity solution is singular in the vicinity of the concentrated moment's application point, that is, tends to infinity.

To see this, let's consider a round plate of the radius R restrained from any displacements at its exterior contour as shown in Fig. 5.15. Let a perfectly rigid disk of the radius c be embedded into the plate at its center, and let a torsional moment M be applied to it.

Denoting radial and tangential displacements of the plate's points as u and v, it's easy to see that the components of the displacement vector in any point outside the rigid disk are defined by the expressions

$$u = 0\,, \qquad v = \frac{M}{4\pi G h}\,(1/\rho - \rho/R^2)\,, \tag{5.31}$$

where ρ is the current radial coordinate, G is the shear modulus, h is the thickness of the plate.

As one can see, all equilibrium equations hold together with the boundary conditions in this case.

Denoting the rotation angle (slope) of the rigid disk as ω_z, we have the following

$$\omega_z = v(c)\,/\,c = \frac{M}{4\pi G h}\,(1/c^2 - 1/R^2)\,, \tag{5.32}$$

wherefrom one can see that the slope ω_z tends to infinity as the size of the disk tends to zero.

Let's return for a moment to the recommendation of the **STARK_ES** software user manual, quoted in the footnote[11], that regards the matching of the degrees of freedom of bars and plane stress finite elements (apparently authors of the recommendation understand this matching as equating slopes and linear displacements of all elements adjoining a bar node).

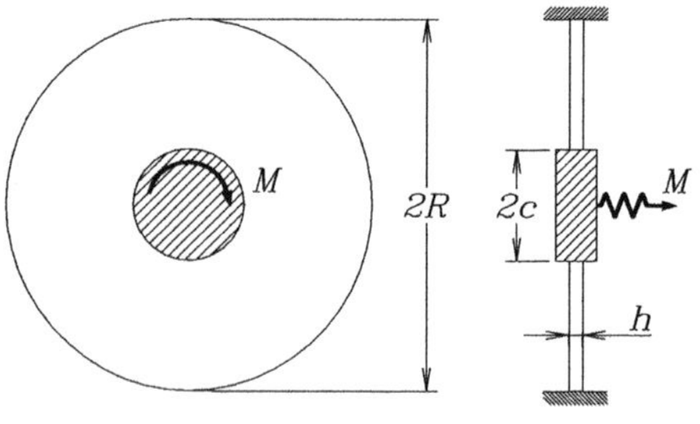

Fig. 5.15. A round plate loaded with a torsional moment

We should note that this is exactly the way one must not follow. Slopes of cross-sections of bars and slope components ω_z are totally different things, and equating these two can only make a design model entangled. An analyst should realize clearly that the introduction of additional parameters ω_z into a finite element model may be justified and useful for the purpose of reducing the discretization error on the same finite element mesh. But there is no reason at all to treat these parameters as slopes of the nodes to which the bars adjoin.

Another argument sometimes given "pro" the introduction of additional parameters ω_z is related to the problem of coupling a plane stress plate and a bar along the whole length of the bar rather than pointwise. Such problems arise when one borders plates along their whole or partial boundary by explicit ribs simulated by bar elements. The conventional finite element analysis technique provides the consistence of the bar's and its adjoining plate's displacements in their nodes only, while in intermediate points (between the nodes) one cannot avoid discontinuities of the bar's and plate elements' displacements in the perpendicular direction to the orthogonal bar axis. The introduction of parameters ω_z enables one to get rid of this inconsistence fully or at least partially depending on approximations used to approximate the plate's displacement fields. It's true all right, but let us see what price is paid for this restored consistence of the displacement.

Let's consider a plane elasticity problem posed in an area having a rectilinear fragment of its boundary Γ provided with a bar along the whole length of the boundary.

We introduce a Cartesian coordinate system (X,Y) so that the X axis match the longitudinal axis of the bar and denoting displacements of the bar's cross-sections along the X and Y axes as $u°(x)$ and $v°(x)$ respectively. We should impose the following boundary conditions on the Γ boundary, or, exactly, interface conditions

$$u(x,0) = u°(x)\,, \quad v(x,0) = v°(x)\,, \tag{5.33}$$

where $u(x,y)$ and $v(x,y)$ are usual components of the displacement vector along the X and Y axis respectively. Also, if we try to identify the bar cross-section slopes $\theta_z = dv°/dx$ with rotation components ω_z at Γ, then we'll have an additional condition to be satisfied at the Γ boundary.

To be particular, let's discuss an introduction of the rotational component ω_z as an averaged slope in the vicinity of a point of the elastic medium[12], that is

$$\omega_z = (\partial v/\partial x - \partial u/\partial y)/2\,. \tag{5.34}$$

In this case the said additional condition at Γ is reduced to a requirement that there be no shear γ_{xy}. By equating the expressions of $\theta_z(x)$ and $\omega_z(x,0)$ we obtain

$$\gamma_{xy}(x,0) = (\partial v/\partial x + \partial u/\partial y)|_{y=0} = 0\,, \tag{5.35}$$

---

[12] An additional degree of freedom of a plane elasticity finite element having the sense of a rotational component ω_z (a nodal slope) can be introduced in a variety of ways. The paper [34] gives a critical review and analysis of suggestions of various kinds regarding the introduction of rotational degrees of freedom into finite element nodes.

thus at the Γ boundary there are three conditions instead of two implied by the biharmonic equation. This renders the mathematical formulation of the problem incorrect. It can be easily imagined at the mechanical level, too. Indeed, condition (5.35) leads to the prohibition of tangential stresses τ_{xy} along the Γ boundary on the basis of physical relations for an isotropic elastic body. It means that the body can slide freely along the Γ boundary. On the other hand, it follows from (5.33) that all displacement components must acquire certain values at the boundary. Thus the two conditions contradict each other.

The mentioned considerations awaken a suspicion. Therefore the described technique can hardly be recommended for a wide practical employment without a careful (theoretical) study of conditions of its safe application. To be fair, we should note that published results of test computations, indeed, demonstrate a reduction of the discretization error after the degrees of freedom discussed above have been introduced into plane stress finite elements. But those tests are based on classical examples and do not comprise, say, issues of coupling plane stress solutions and bars.

At this point the authors deem it necessary to explicate their argumentation. The mater is that the above said thesis about an incorrect equation of a so-called "angular degree of freedom" ω_z of plane stress finite elements to slopes θ_z of elements of other types adjoining the same node (particularly bar elements) caused a longstanding discussion. Arguments of our opponents can be summarized by a statement that even if this equation introduces some distortion to the final solution of the problem, the chances are that it is of local nature.

Before answering the question on its merits, we should admit it is we who bear the blame, not for the essence of the matter but for the beginning of the discussion. Thereby we have broken a principle of persuasiveness formulated brilliantly by Romans: «*Don't speak so that you could be understood; speak so that you could not be misunderstood*» [13].

We should note it is not our intention to criticize any particular way of introducing an additional degree of freedom into a plane stressed plate. It is not the method of the ω_z parameter introduction which is vicious, it is the very principle of its identification with the θ_z parameter. Whatever sophisticated technique were applied to introduce ω_z into a design model, there would be no reason at all to state the equality

$$\omega_z = \theta_z \,. \qquad (???)$$

From our point of view, this very equality appears only because of an interpretation of the ω_z parameter as a nodal slope. Then the following logic, or rather a wordplay and not a harmless one, causes a confusion of concepts: once the ω_z parameter acquires the meaning of a nodal slope, and a bar element comes to this same node having it cross-section slope θ_z, these two quantities should be the same. But the matter is that those are totally different quantities! That's why it has

[13] We deem it appropriate to refer to a book by Y.G. Panovko [22] wherefrom we have taken this sentence.

been said "*...the true meaning of a term is found in its application rather than in its definition*" [39] [14].

In order for our statement to be absolutely clear and holeproof against any further verbal acrobatics, let's point out a totally transparent example and provide it with strict mathematics: there's nothing more convincing than the mathematical formalism.

So, let us consider the finding of a stress and strain distribution in a plate that occupies an area Ω with a boundary Γ in the (X,Y) plane. Let the plate be bordered by a Timoshenko bar along a part of its boundary Γ_1 which we assume to run along the X axis only. We emphasize the use of a Timoshenko bar rather than a Bernoulli one. It is the case where contradictions caused by equality (???) reveal themselves most evidently.

Let the plate is restrained from all displacements at the other part of its boundary $\Gamma_2 = \Gamma - \Gamma_1$. We assume the x coordinate of the longitudinal axis of the bordering bar varies between 0 and l. It is easy to see that the variational formulation of the problem can be reduced to the minimization of the Lagrangian L. Here we don't neglect to present it in its complete form, with a conventional notation:

$$\mathsf{L} = \frac{E}{2(1-\nu^2)} \int_\Omega \left[u_{,x}^2 + v_{,x}^2 + v_{,x}^2 + 2\nu u_{,x} v_{,x} + \frac{1-\nu}{2}(u_{,x} + v_{,x})^2\right] d\Omega \; +$$

$$+ \frac{1}{2}\int_0^l [EI\,\theta_{z,x}^2 + GF(v_{,x}^\circ - \theta_z)^2 + EA(u_{,x}^\circ)^2]dx - \Pi,$$

where the first integral represents a deformation energy accumulated in the plate itself, the second one the energy accumulated in the bar, and Π is a drop of potential of external forces [15]. Apparently, the Lagrangian L is a functional of three independently varied functions

$$\mathsf{L} = \mathsf{L}(u, v, \theta_z),$$

as $v^\circ(x) = v(x,0)$ and $u^\circ(x) = u(x,0)$, so $u^\circ(x)$ and $v^\circ(x)$ are not independently varied functions. The minimum of the functional L is sought in the set of functions $u(x,y)$, $v(x,y)$, $\theta_z(x)$ that satisfy main (kinematical) boundary conditions

$$u = 0\,,\, v = 0 \;\in \Gamma_2,$$

and have generalized square-summable first derivatives. That's all required by the variational principle of the full potential energy minimum. The Lagrange variational principle does not establish any predefined relation between the bar section

[14] It is another demonstrative example of an "anything goes" attitude towards the terminology of mechanics which Y.G. Panovko treated as totally unacceptable [22]. Indeed, "*You name a boat the way it will float*".

[15] To simplify the notation, here we denote derivatives with respect to certain coordinate by putting this coordinate in the subscript after a comma. For example, $u_{,x} = \partial u/\partial x$. Note also that the integrand of the Lagrangian L includes F that means the section's shearing area different from the geometrical area A by a section shape factor.

slopes $\theta_z(x)$ and the $u(x,y)$ and $v(x,y)$ displacements of the plate. If now we introduce an expression for the "slope" ω_z in a point (x, y) of the plate, it will have a form like $\omega_z(x,y) = A(u, v)$ anyway, where the A operator will define the method of this introduction. Now it's clear that the relationship (???) is just an imposition of an additional constraint on the system, the constraint being

$$A(u, v)\Big|_{y=0} = \theta_z(x),$$

that introduces by no means justified distortion into the original statement of the problem whatever form the A operator might have. It is apparent that the independence of the θ_z function on functions u and v is lost owing to the condition (???). By the way, this distortion makes the system stiffer, just as does any additional constraint. It implies that the distorted finite element solution is further by the energetic norm from the accurate solution of the problem than the similar finite element solution without the additional constraint (???). In other words, we have the following inequality

$$\mathsf{E} \geq \mathsf{E}_h \geq \mathsf{E}_h^+ ,$$

where E is the energy of the system corresponding to the accurate solution, E_h is that of the usual finite element solution, E_h^+ is the energy of the finite element solution disturbed by the additional constraint (???). Apparently, these estimates hold only for fully consistent finite elements. Also, these estimates imply that the external influence on the system consists of forces only.

These being the facts, there's no need to discuss the second statement of our opponents, that is, a local effect of the disturbance introduced by the constraint into the solution. Is it really a worthy occupation, to overcome difficulties created only to be overcome?

There are other problems, too, the solution of which with finite elements of this kind risks a distortion of results and/or deterioration of the computation accuracy. Thus, in plane elasticity the derivatives of the displacements are discontinuous, as it is known, when the domain of a problem includes two or more regions of different elastic properties (in structural design, such different materials may be encountered when a brick wall is to be erected on a concrete wall panel) along junctions of the inhomogeneous materials. Equating the ω_z parameters of all adjacent finite elements in nodes of such a line forces an elimination of some jumps in the displacement derivatives. In other words, the finite element solution is bound to be extra smooth to comply with a predefined condition. This is an opposite case, in a sense, to the inconsistency situation, and it can be entitled similarly: an *over-consistency*. Users of computational software do care about the answer to the question how this over-consistency will affect the accuracy of the analysis.

We cannot fix the attention of the reader on the issues of applicability of the finite elements in question any longer. After all, those must be a subject of special theoretical research publications written for other readers and by other authors, as we believe, who can call themselves experts in theoretical foundations of the finite element method more deservedly. Here we'd like to confine ourselves to a single example where the ω_z rotational components are identified formally with slopes θ_z

of respective nodes. With this purpose, let's consider an analysis of a square plate shown in Fig.5.16.

Let's assume that the plate is bordered along its upper boundary (at $y = 0$) with a bar perfectly rigid against bending and torsion/compression. So, in this case $v°(x) = x\theta$ where θ is the slope of the bar in the (X,Y) plane. By restraining one of nodes of the plate's upper boundary (say, one in the coordinate origin) from a horizontal displacement, we have $u°(x) = 0$. The following source data are used in the analysis:

- plate thickness $h = 1{,}0$;
- dimensions of the plate in the (X,Y) plane 16.0×16.0;
- material constants $E = 3 \cdot 10^5$, $\nu = 0{,}25$;
- applied load $P = 1000$.

Results of a numerical experiment conducted with this problem are given in Table 5.3 vs. a finite element mesh ($n\times n$) and a design model employed:

- The "A" item corresponds to a straightforward model consisting of conventional rectangular elements with a bilinear displacement approximation, that is, finite elements having two degrees of freedom in each node.
- The "B" item describes the same data obtained with finite elements having three degrees of freedom per node. Due to the merging of the elastic medium's upper border nodes with the bar nodes, the said nodes have the same slopes equal to the common slope θ_n of the rigid bar. Also, zero slopes are specified along the lower boundary of the area.
- Finally, the "C" item uses the same finite elements having three degrees of freedom per node, but the slopes of the bar and those of the plane elasticity rectangular finite elements are assumed to be mutually independent variables. At the lower boundary of the rectangular area, the boundary conditions do not include the nodal slopes.

For each of the three models, Table 5.3 gives the θ_n slope of the rigid bar and values of the respective reactive forces H_n arising in the external constraint that prohibits the horizontal displacement of the bar.

As one can see from the table, the error of the analysis in an energetic metrics (the θ slope is a work of external forces up to a constant factor) shown by the three listed models is almost the same.

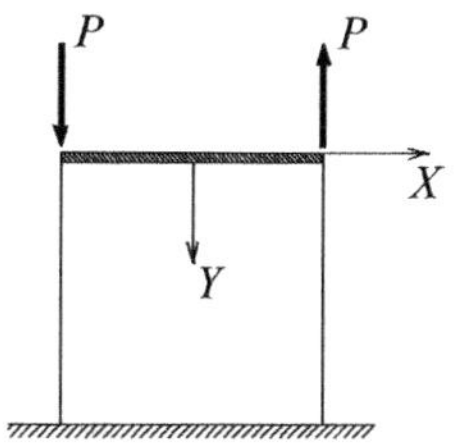

Fig. 5.16. A square plate bordered with a bar

Table 5.3.

Model	$n \times n$	2x2	4x4	8x8	16x16	32x32	64x64
"A"	$\theta_n \cdot 10^4$	11.979	13.314	13.785	13.929	13.972	13.985
	H_n	915	890	878	874	873	872
"B"	$\theta_n \cdot 10^4$	12.111	13.302	13.754	13.908	13.960	13.978
	H_n	863	877	875	874	873	872
"C"	$\theta_n \cdot 10^4$	14.371	14.123	14.033	14.006	13.996	13.994
	H_n	961	886	872	872	872	872

Though, a tendency is nonetheless noticeable. The listed results show that $\mathsf{E}_\mathrm{C} > \mathsf{E}_\mathrm{B}$, which is the same as $\mathsf{E}_h \geq \mathsf{E}_h^+$ stated before but uses a different notation. Moreover, starting from the 4×4 mesh even $\mathsf{E}_\mathrm{A} > \mathsf{E}_\mathrm{B}$ in spite of the fact that the number of degrees of freedom employed by the "B" model is greater than that of the "A" model by half.

Note that the analysis using the conventional elements (the "A" model) was performed by the **GTSTRUDL** software while the three-freedom-degree analysis used the **MicroFE** software, version 6.13. Finite elements [28] employed by **MicroFE** use the third nodal degree of freedom according to ideas originated in a paper by Allman [5].

Also, it should be said that there is no immediate capability of specifying perfectly rigid bodies in the **MicroFE** environment, unlike **GTSTRUDL**, therefore the horizontal bar was assigned a finite rigidity of a great magnitude. Namely, the following values were used in the analysis: $EI = 3 \cdot 10^{12}$, $EA = 3 \cdot 10^{15}$.

At the same time, the tangential stress in a nearby area of the horizontal bar falls expectedly when the "B" model is used. It can be seen clearly in Fig.5.17 that shows isolines of the tangential stress built by the **MicroFE** program at the 8×8 finite element mesh.

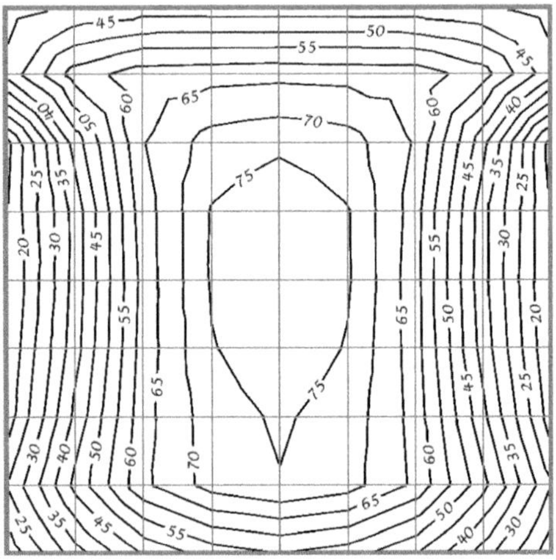

Fig. 5.17. Tangential stress isolines for the "B" model

By the way, these isolines show a suspicious lack of equilibrium in the said region: the area of the tangential stress diagram along the horizontal cross-section of the plate in the immediate vicinity of the bar is much less than the similar area along the cross-sections $y = 4$, $y = 6$ and $y = 8$ though all of those must be equal to the magnitude of the reactive force H. With the "C" model, one does not observe this effect.

Now let's discuss longitudinal and lateral stresses in bar elements adjoining a plate. It can be proved that the action of a concentrated force on a plane stressed/strained body causes the displacement of this force's application point to be singular.

For example, consider the same problem of a round plate restrained from all displacements along its contour (Fig. 5.15) but subjected to a force P applied to the disk and directed along a radius. The respective u displacement of the disk will be [18]

$$u = \frac{P}{8\pi Gh}\left[(3-\nu)\ln(c/R) - \frac{(1+\nu)^2\left(c^2/R^2-1\right)}{(3-\nu)\left(c^2/R^2+1\right)}\right], \tag{5.36}$$

thus there is a logarithmic singularity at $c \to 0$.

Using the same logic with a bending moment, we come to a conclusion that the accurate solution of the pointwise bar-to-plane coupling problem (i.e. the matching in a single node) will yield a zero force transferred to the bar. Again, nonzero values of the longitudinal and lateral stresses in bars formally obtained using a discrete model are caused by the discretization error solely. The design model itself (not to be confused with its discrete counterpart) yields both zero moments and stresses in the bars.

The same conclusion can be made after a careful analysis of numerical experiments on the condensation of the mesh in the case of a concentrated force applied to the plate. To do it, let's return to the problem shown in Fig.5.16. This time let us assume that the boundary $y = 0$ is not reinforced by any bar, and one vertical force P is applied directly to the plate, instead of two, in the point (0,0).

Table 5.4 presents results of the computation where v_n is the displacement of the P force application point in the direction of Y. It's clear that the displacement tends to infinity as the finite element mesh condenses, exactly as expected.

Table 5.4.

$n \times n$	2×2	4×4	8×8	16×16	32×32	64×64
$v_n \cdot 10^4$	55.278	68.282	82.665	97.296	111.989	126.695

So, the general conclusion is: joining bar elements and plane elasticity finite elements in points results in an incorrect statement of the problem.

What corrections should be made to the original mechanical model or the mathematical statement of the problem (which is the same thing) in order to describe the real structure behavior adequately? As it was stated before, to do that one should consider the design of a joint between a bar and a diaphragm in sufficient detail.

For example, let a steel I-beam be embedded partly in a brick wall as shown in Fig.5.18, *a*. Then it suffices to add a one-dimensional bar element that penetrates into the plane stressed wall by an appropriate part of its length, to the respective design model and into its discrete counterpart, as shown in Fig.5.18, *b*.

Another design can be suggested for a monolithic joint between a ferroconcrete wall panel and a spandrel beam of a building — see Fig.5.19, *a*. Here one can take into account actual dimensions of the respective cross-section: the height of the beam, with a perfectly rigid body as high as the cross-section of the beam placed along the wall's border as shown in Fig.5.19, *b*. This rigid body complies with the Bernoulli's plane section assumption for the beam itself. The assumption states that the beam's section must remain flat and must not change its size after its deformation.

Certainly the two ways of building design models given here by no means comprise the wide scope of possible situations. In every particular case an analyst should take into account peculiarities of his structure's design rather than some fancied and questionable lockstep patterns.

Also, it should be noted that the "joint area extension" technique as in Fig.5.18 can be used to match other types of elements of different dimensionality. Fig.5.20 shows an example wherein a cylindrical shell of a wall and a spherical shell of a cover of a nuclear reactor containment are matched with a massive ring. It should be noted that in places of interpenetration of the structures some perturbation of rigidity properties arises due to a summation of the shell's and the massive body's rigidities in parts that occupy the same region of space. This fact should be taken into account when assigning the rigidity constants.

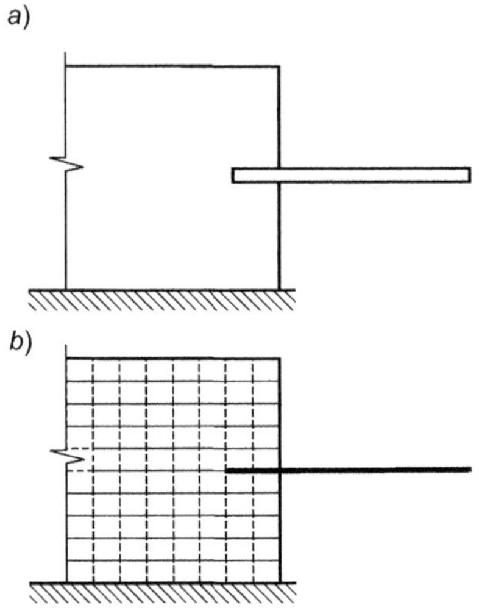

Fig. 5.18. A cantilever clamped in a wall

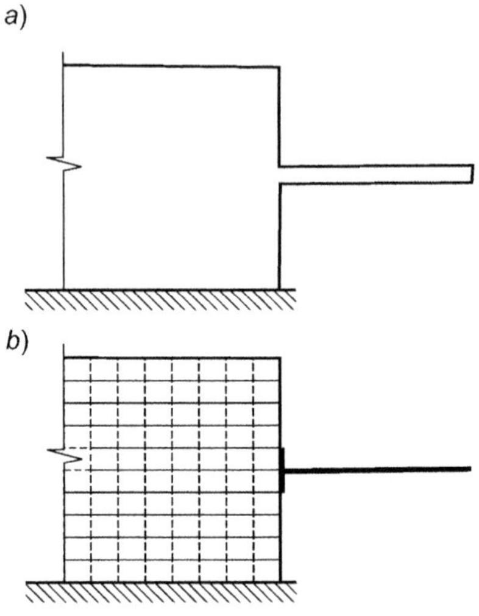

Fig. 5.19. A monolithic joint

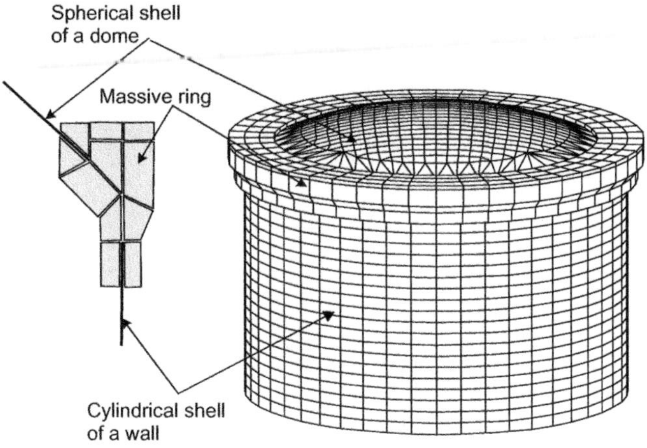

Fig. 5.20. Coupling a reactor containment's shell and cover with a massive ring.

5.3.3 Bars + Massive Elements

Problems where both bars and massive elements are involved to simulate structures include, for example, a joint analysis of a building's framework and massive foundations under columns. Issues of forming correct design models are similar to those described in the previous section, and we won't give any more details here. Just as in the case of coupling bars and plane stress elements, a formal nodal coupling is not applicable. The design model should be corrected so as to allow for particularities of the joint between the bar and the bulk body. Further details are omitted.

5.3.4 Plane Stress + Flexural Plates (Coupling Shell Elements)

The analysis of spatial plate structures, folded shells, and smooth shells approximated by piecewise-linear surfaces requires a coupling of plane stress elements and flat flexural members joint along their common boundary at an angle α (Fig. 5.21). Sometimes every finite element that participates in a discrete model can be a plane one and a flexural plate at the same time (such one is often called a plane shell element). This kind of elements has five degrees of freedom in each node.

Let us introduce a local Cartesian right-oriented coordinate system (ξ, η, ζ) referred to a particular plane element so that axes ξ, η belong to the median surface of the element and axis ζ is orthogonal to the former two axes.

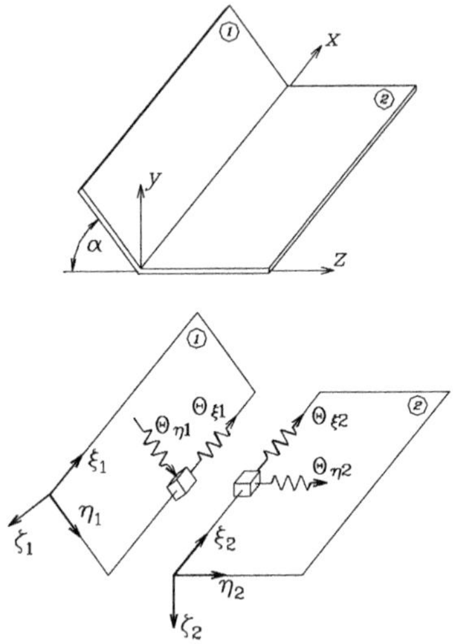

Fig. 5.21. Coupling plane shell elements at an angle

Let u, v, w be linear displacements of points of the median surface of the shell with respect to the local coordinate system, θ_ξ, θ_η be slopes of the normals to the shell's median surface about respective axes ξ, η, and

$$\theta_\xi = \frac{\partial w}{\partial \eta}, \qquad \theta_\eta = -\frac{\partial w}{\partial \xi}. \tag{5.37}$$

These 5 parameters — u, v, w, θ_ξ, θ_η (more exactly, values of these functions in nodal points of finite elements) — correspond to five degrees of freedom of each plane shell element node.

Let (X, Y, Z) be a global coordinate Cartesian system which we assume to be an orthogonal Cartesian right-oriented system. Introduce a matrix of direction cosines $\mathbf{\Lambda}$ that defines the mutual orientation of the global and local coordinate systems. We have:

$\mathbf{\Lambda}$ =

ξ	η	ζ	
$\lambda_{x\xi}$	$\lambda_{x\eta}$	$\lambda_{x\zeta}$	X
$\lambda_{y\xi}$	$\lambda_{y\eta}$	$\lambda_{y\zeta}$	Y
$\lambda_{z\xi}$	$\lambda_{z\eta}$	$\lambda_{z\zeta}$	Z

.

Now the components of the full nodal slope vector $\boldsymbol{\theta}$ of a particular finite element with respect to the global coordinate system (θ_x, θ_y, θ_z) can be expressed via its components referred to the local coordinate system (θ_ξ, θ_η, 0):

$$\theta_x = \lambda_{x\xi}\theta_\xi + \lambda_{x\eta}\theta_\eta + \lambda_{x\zeta}0\,,$$

$$\theta_y = \lambda_{y\xi}\theta_\xi + \lambda_{y\eta}\theta_\eta + \lambda_{y\zeta}0\,,$$

$$\theta_z = \lambda_{z\xi}\theta_\xi + \lambda_{z\eta}\theta_\eta + \lambda_{z\zeta}0\,.$$

If the α angle at which two adjacent shell planes (Fig. 5.21) cross is neither 0 nor π, then any node belonging to the intersection of the planes will acquire three linearly independent slopes, i.e. three rotational degrees of freedom. As can be seen in Fig. 5.21, two slope couples ($\theta_{\xi 1}$, $\theta_{\eta 1}$) and ($\theta_{\xi 2}$, $\theta_{\eta 2}$) are not coplanar (do not lie in the same plane) therefore form three linearly independent rotations in the three-dimensional space.

Thus, a node belonging to the boundary will have all six degrees of freedom. Now suppose a couple of planes intersect at the α angle equal to π. Then the node at the joint of the planes will lose one of three rotational degrees of freedom because all four slopes ($\theta_{\xi 1}$, $\theta_{\eta 1}$) and ($\theta_{\xi 2}$, $\theta_{\eta 2}$) are coplanar and do not form components of a rotation vector orthogonal to this plane.

It is here where the first issue of coupling plane shell elements arises. Some nodes acquire six degrees of freedom while the others only five. Some most advanced software developments can handle this situation by providing each node automatically only with degrees of freedom allowed by all finite elements adjoining the node. In other cases the user has to control this process manually by imposing external constraints on nodal slopes not resisted by adjoining elements.

Many developments employ a traditional principle of providing a design model with a node typical for this particular type of design. The source data for software use a concept of *structural type* (*scheme type*) that means a set of degrees of freedom of a typical node. For example, the "*spatial truss*" structural type conforms to a node having three linear displacement degrees of freedom in three non-coplanar directions. The "*plane frame*" structural type conforms to nodes with two non-collinear displacements and a slope with the latter's vector orthogonal to the linear displacement plane.

The structural type concept does make sense. It is useful, for example, to track some kinds of errors in the source data. If a structure of, say, the "*plane frame*" structural type happens to include a flexural plate finite element, then the respective software has a chance to spit out every uncomplimentary opinion that the developers have about the level of expertise of such a user.

We believe that one should follow the concept down-the-line. Why only five structural types are provided by software to include in the source data for an analysis? Why kill the fantasy of users beforehand?

Suppose every node is provided with its own local coordinate system — it's a simple but useful idea now implemented by many programs. Let's confine ourselves to solid nodes (see Section 4.2 for details in this regard). We can assign up to six degrees of freedom to each node corresponding to linear u_α, u_β, u_γ and angular θ_α, θ_β, θ_γ displacements referred to the local coordinate system (α, β, γ). Then a "structural type" can be defined using the table:

u_α	u_β	u_γ	θ_α	θ_β	θ_γ

,

where in the lower row the user may put marks (digits, ticks, pluses to the developer's liking) to denote the assigning of the respective degree of freedom to a standard node. Obviously, in total $2^6-1=63$ "structural types" can be defined in this way, and the user has plenty of room to practice his fantasy. We exclude a degenerate case of a node having no degrees of freedom at all.

So what is better: the "*structural type*" concept, or providing a node with degrees of freedom imposed by its adjoining elements? We prefer the answer "both" because an extra degree of freedom for the user (not a node!) provides a better flexibility of operation. By the way, the option of empty fields in the table above can be employed by the user to refuse the "structural type" definition and commit the responsibility for assigning nodal degrees of freedom to the program.

Let us now return to our problem of coupling plane shell elements, supposing that somehow we have contrived to solve the problem of matching the nodal degrees of freedom by their number. For the sake of simplicity, assume that two intersecting planes are orthogonal, i.e. the α angle is equal to $\pi/2$ in Fig. 5.21. Then a fragment of the intersection line of the two planes located between two nearest nodes of the line will belong to two finite element simultaneously, and thus it will acquire inter-nodal displacements distributed by a certain law in one element (say, a linear law when the boundary moves in the element's plane) and by a different one in the other element (say, a polynomial law if the boundary moves out of the element's plane). It's clear enough this situation will result in discontinuity of the displacements. In other words, flexural plate finite elements and plane stress plate finite elements that intersect (at an angle other than π) will be inconsistent. And the very first question immediately and naturally asked by the user goes like this: how dangerous is this inconsistency from the standpoint of violation of the finite element solution convergence?

A qualitative description of the effect of the said inconsistency on the finite element solution convergence can be obtained from the following simple consideration.

Let us introduce a global coordinate system (X, Y, Z) in our problem where two adjoining planes intersect, as shown in Fig. 5.21. Let $u_1(x, y)$, $v_1(x, y)$, $w_1(x, y)$ be the displacements of points of the plane 1 in the directions of axes X, Y, Z, respectively, and $u_2(z, x)$, $v_2(z, x)$, $w_2(z, x)$ the displacements of points of the plane 2 in the same directions. Then, the coupling conditions at the common boundary of the two planes can be stated as

$$u_1(x,0) = u_2(0,x),\ \ v_1(x,0) = v_2(0,x),$$
$$w_1(x,0) = w_2(0,x),\ \ \theta_{1x}(x,0) = \theta_{2x}(0,x), \tag{5.38}$$

where $\theta_{1x}(x,0) = \dfrac{\partial w_1}{\partial y}(x,0), \qquad \theta_{2x}(0,x) = -\dfrac{\partial v_2}{\partial z}(0,x)$.

And that's all folks! Any additional coupling conditions not following directly from relations (5.38) lead to a mathematically incorrect problem statement.

It is clear the displacement discontinuities that bother us so much are due to a violation of the second and third coupling conditions (5.38). It suffices to choose one of them, say, the second one which holds in nodal points belonging to the two planes intersection line and does not between the nodes. We mark coordinates and displacement components referring to two adjacent nodes that belong to the said line of intersection of the two planes, with indices i and i+1. Then for all $x \in [x_i, x_{i+1}]$, assuming

$$l = x_{i+1} - x_i, \qquad \xi = (x - x_i) / l ,$$

we have

$$v_1(x, 0) = v_i\, \mathrm{L}_1(\xi) + v_{i+1}\, \mathrm{L}_2(\xi) ,$$
$$v_2(0,x) = v_i\, \mathrm{H}_1(\xi) + \theta_{i,z}\, l\, \mathrm{H}_2(\xi) + v_{i+1}\, \mathrm{H}_3(\xi) + \theta_{i+1,z}\, l\, \mathrm{H}_4(\xi) , \tag{5.39}$$

where an additional notation is included:

$\mathrm{L}_1(\xi) = 1 - \xi$, $\mathrm{L}_2(\xi) = \xi$ are linear functions,

$\mathrm{H}_1(\xi) = 2\xi^3 - 3\xi^2 + 1$, $\mathrm{H}_2(\xi) = \xi^3 - 2\xi^2 + \xi$,

$\mathrm{H}_3(\xi) = -2\xi^3 + 3\xi^2$, $\mathrm{H}_4(\xi) = \xi^3 - \xi^2$ are Hermitian polynomials,

$\theta_{i,z}$, $\theta_{i+1,z}$ are rotations/slopes of respective nodes i and i+1 about the Z axis,

v_i, v_{i+1} are linear displacements of nodes i and i+1 in the Y axis direction.

Note that the expression of $v_2(0,x)$ in (5.39) holds for rectangular Adini – Clough elements.

As the mesh becomes denser, the distance l between adjacent nodes tends to zero, and the nodal slopes θ_{iz} and $\theta_{i+1,z}$ can be replaced by difference relations becoming more accurate with less l. Taking into account the rule of signs for the slopes, we can write

$$\theta_{i,z} = \theta_{i+1,z} = (v_{i+1} - v_i)/l . \tag{5.40}$$

Now replacing slopes in (5.39) by the difference relations from (5.40), we have $v_1(x,0) = v_2(0,x)$, that is, discontinuities in the displacements are disappearing as the spacing of the mesh approaches zero.

Software developers have to attempt solving two problems at the same time:

(1) match the number of degrees of freedom of nodes both belonging and not belonging to the same plane together with all adjoining plane shell elements;
(2) eliminate the inconsistency of displacements of elements joined at an angle other than zero or π.

With this purpose, some programs suggest plane shell finite elements with a sixth degree of freedom, and this one is treated geometrically as a rotation of the node in the plane of a finite element. The inconsistency of the displacements of these elements cannot be fully eliminated, but the convergence of the finite element solution is ensured by the conditions of piecewise testing. Though, the software actually requires six coupling conditions to be satisfied along the intersection line of the two planes, instead of the four from (5.38), and two more coupling conditions violate the mathematical correctness of the problem. To illustrate this, let's return to our example of an orthogonal intersection of two planes.

Introducing a rotation component ω_{1z} for the plane 1 (say, as an average rotation in the vicinity of the point in question, i.e. a node), and demanding that ω_{1z} be equal to the slope θ_{2z} of the normal to the median surface in the plane 2, along the common boundary of the two planes, we have:

$$\frac{1}{2}\left[\frac{\partial v_1}{\partial x}(x,0) - \frac{\partial u_1}{\partial y}(x,0)\right] = \frac{\partial v_2}{\partial x}(0,x)\,, \tag{5.41}$$

wherefrom, taking into account (5.38), we come up with a condition of no shear in the plane 1 where it joins the plane 2, that is

$$\gamma_{1xy}(x,0) = \frac{\partial v_1}{\partial x}(x,0) + \frac{\partial u_1}{\partial y}(x,0) = 0\,. \tag{5.42}$$

In the same way we obtain the second additional condition along the junction line of the two planes:

$$\gamma_{2zx}(0,x) = \frac{\partial u_2}{\partial z}(0,x) + \frac{\partial w_2}{\partial x}(0,x) = 0\,. \tag{5.43}$$

It should be clear that relations (5.42) and (5.43) are conditions of a mutual slip of the two planes along the X coordinate axis, that is, they contradict the first coupling condition of (5.38) interpreted mechanically as conditions of "adhesion" of the two planes.

So we repeat once more: there is nothing bad about introducing additional sixth degree of freedom, but the user should understand its meaning clearly and not identify it formally with the slope of the respective node. It is of particular importance for the formulation of boundary conditions and conditions of coupling of different plates of a folded shell intersecting at an angle.

5.4 A Paradox of Coupling Bernoulli and Timoshenko Bars in the Same Model

The previous section deals with some issues that may arise in the analysis of combined systems built of finite elements of different dimensionality. To continue this discussion, let us demonstrate here that the sources of dangerous mistakes are not limited to the difference of finite element dimensionality.

A combination of elements of the same dimensionality but based on different theoretical assumptions in a single design also requires an analyst to be careful about the model of his structure. Combined mechanical models of this kind may provoke incorrect computational results.

An illustration of this statement is the problem that follows. It is simple in its formulation, yet yields a result paradoxical at the first glance.

Suppose we need to analyze a single-span multistory (40 stories) frame (a so-called "étagère") shown in Fig. 5.22.

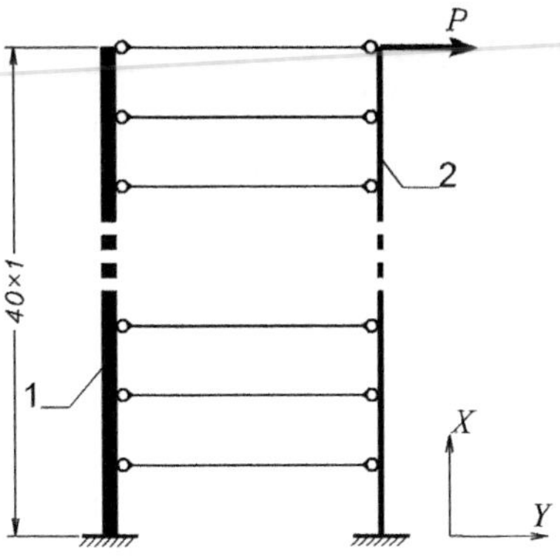

Fig. 5.22. An étagère design

Assume both columns of the frame to be made of the same material and to have rectangular sections of an equal width b (the lateral size to the plane of the frame) but a different height h (the size in the plane of the frame). Let the height h_1 of the stronger column (No.1 in Fig. 5.22) be to the full height of the frame l as 1 to 5. It is known that for a ratio like this between the section height and beam length the classical beam bending theory requires a correction to be made to take into account the shear which affects much the general strain distribution in the member.

Commonly, these circumstances force an analyst's decision of allowing for the shear. In other words, the column 1 of the model will be described by the Timoshenko bar bending theory. Next, let the section height h_2 of the column 2 is one third of h_1. As it makes $h_2/l = 1/15$, it is reasonable to simulate the column 2 by a classical Bernoulli – Euler flexural bar based on the plane sections assumption and ignoration of the shear. As for crossbars attached to the columns, we will assume them to be nondeformable longitudinally, that is, perfectly rigid against tension/compression.

For a rectangular section of the dimensions ($b \times h$), it is easy to find a relation between the flexural EI and shear GF rigidities of the bar as a function of the principal dimensionless parameter h/l and the bar length l. Here and below I is the moment of inertia of the bar's section, F is a shearing area that allows for a so-called section shape factor, the latter being 6/5 for a rectangular section. This gives

$$I = \frac{bh^3}{12}, \qquad F = \frac{5}{6} bh .$$

Seeing that the elasticity modulus of the material E and the shear rigidity G are related as $G = E/(2+2\nu)$ where ν is the Poisson ratio, and assuming $\nu = 0.25$ to be particular, we have

$$EI = l^2 GF \left(\frac{h}{2l}\right)^2. \tag{5.44}$$

The analysis of the frame shown in Fig. 5.22 is by no means a hard task and can be performed by any software available to a designer, which can handle both Timoshenko bars and classical Bernoulli – Euler bars. Let's take the following rigidity property for an example, a conventional one and specified in conventional units of measurement. We assume

$$GF_1 = 0.25 \cdot 10^9,$$

then

$$EI_1 = (40)^2 \times 0.25 \cdot 10^9 \times (1/5)^2/4 = 4.0 \cdot 10^9, \quad EI_2 = EI_1(h_2/h_1)^3 = 1.481 \cdot 10^8.$$

It is natural to expect that a main part of the load will be resisted by the stronger column, that is, the column 1 because its flexural rigidity EI_1 is 27 times greater than that of the column 2, EI_2. Unexpectedly (at first blush), the results of the computation show that the shear force in the lower cross-section of the column 1 is $Q_1(0) = 0.444P$ while the column 2 is affected by all the other part of the load, that is, $Q_2(0) = 0.556P$. This distribution of the shear forces $Q_1(0) < Q_2(0)$ contradicts (quite reasonably!) a common intuition of an engineer who would expect a directly opposite result.

In order to understand the origination of this effect, imagine an idealized model close enough to that shown in Fig. 5.22, namely: suppose both columns are joined along their whole height by continuous links to provide the full consistency of the lateral displacements of the columns 1 and 2: displacements $v(x)$ in the direction of the Y axis. The boundary conditions in the places of fixation follow:

$$\text{column 1:} \qquad v(0) = 0\,,\ \theta_1(0) = 0\,,$$

$$\text{column 2:} \qquad v(0) = 0\,,\ v'(0) = 0\,, \tag{5.45}$$

where $\theta_1(x)$ denotes the slope of the column 1 cross-section located at the distance x vertically from the place of fixation. For a Timoshenko bar, the shear angle $\gamma_1(x)$ is defined as a difference between the slope of the tangent to the curved bar axis $v'(x)$ and the slope $\theta_1(x)$ of the cross-section of the bar:

$$\gamma_1(x) = v'(x) - \theta_1(x)\,, \tag{5.46}$$

and the shear force $Q_1(x)$ is proportional to the shear force with a factor equal to the shear rigidity of the bar GF_1, that is,

$$Q_1(x) = GF_1\gamma_1(x)\,. \tag{5.47}$$

For our idealized scheme of connection between the two bars, the boundary conditions (5.45) and relations (5.46), (5.47) yield immediately $Q_1(0) = 0$. Then the equilibrium equations of the whole frame in projections onto the Y axis give that $Q_2(0) = P$.

So, in the idealized model consisting of two columns and continuously distributed links between them all the load in the place of fixation is resisted by the Bernoulli bar solely. This conclusion is true **independently from a relation between the flexural rigidities of the two columns, and from a relative (but finite which is essential) value of the shear rigidity of the column 1!**

The original problem of the forty-story frame also includes links that provide the consistency of the displacements of the two columns but those are specified in a discrete set of points rather than distributed continuously. Therefore the shear force in the fixation point of the frame's column 2 is somewhat less than the full load P. Note that the manifestation of this effect in a framed structure depends essentially both on the frequency of the cross links (the frame's crossbars in our case) and relative values of the flexural and shear rigidities of the columns, respectively.

For generality, let's consider the behavior of the idealized model in more detail supposing both columns are simulated by Timoshenko bars. The same problem with the column 2 modeled by a Bernoulli bar will be derived as a particular case of a more general Timoshenko columns problem by passing to a limit.

The simplest way to obtain a governing system of differential equations for the problem in question is to use the Lagrange variational principle. With this purpose, let us write out an expression of the full potential energy of the system — a Lagrangian functional L. It yields

$$\mathsf{L}(v, \theta_1, \theta_2) = \frac{1}{2}\int_0^l [EI_1(\theta_1')^2 + EI_2(\theta_2')^2 +$$

$$+ GF_1(v' - \theta_1)^2 + GF_2(v' - \theta_2)^2]dx - Pv(l)\,, \tag{5.48}$$

where the first two terms in the integrand define the flexural deformation energy accumulated in the bars while the next two define the energy of the shear. Here, as before, the subscript 1 denotes quantities related to the column 1 and the subscript 2 denotes those of the column 2. Functional arguments in parentheses in the left part of equation (5.48) are there to emphasize the independence of these parameters as the functional L is varied.

Necessary conditions of the system's full potential energy minimum, $\delta\mathsf{L} = 0$, is to be sought in a set of independently varied functions that satisfy the main (kinematical) boundary conditions

$$v(0) = 0\,, \qquad \theta_1(0) = 0\,, \qquad \theta_2(0) = 0\,. \tag{5.49}$$

Equating the first variation of the functional L to zero, we have the following after employing the formula of integration by parts:

$$-\int_0^l \{[GF_1(v' - \theta_1)' + GF_2(v' - \theta_2)']\delta v + [EI_1\theta_1'' + GF_1(v' - \theta_1)]\delta\theta_1 +$$

$$+ [EI_2\theta_2'' + GF_2(v' - \theta_2)]\delta\theta_2\}dx + [EI_1\theta_1'\delta\theta_1]\Big|_0^l + [EI_2\theta_2'\delta\theta_2]\Big|_0^l +$$

$$+ \{[GF_1(v' - \theta_1) + GF_2(v' - \theta_2)]\delta v\}\Big|_0^l - P\,\delta v(l) = 0, \tag{5.50}$$

wherefrom we derive the Euler equations for the functional L taking into account the main boundary conditions (5.49)

$$GF_1(v'' - \theta_1') + GF_2(v'' - \theta_2') = 0\,,$$
$$EI_1\theta_1'' + GF_1(v' - \theta_1) = 0\,,$$
$$EI_2\theta_2'' + GF_2(v' - \theta_2) = 0 \tag{5.51}$$

and natural boundary conditions

$$\theta_1'(l) = 0\,, \quad \theta_2'(l) = 0\,, \quad GF_1[v'(l) - \theta_1(l)] + GF_2[v'(l) - \theta_2(l)] = P\,. \tag{5.52}$$

The first two boundary conditions in (5.52) demand that the bending moments at the free ends of the columns be zero, whereas the third one in (5.52) ensures the equality of the total shear force at the ends of the columns to the applied load P.

By integrating the first equation in (5.51) once, we obtain the following after considering the third boundary condition of (5.52)

$$GF_1(v' - \theta_1) + GF_2(v' - \theta_2) = P\,, \tag{5.53}$$

which evidences the equality of the total shear force in the two columns to the magnitude of the applied load P at any value of the independent coordinate x.

Let's introduce the following dimensionless parameters:

$$k_1 = \frac{EI_1}{l^2 GF_1}\,, \qquad k_2 = \frac{EI_2}{l^2 GF_2}\,, \qquad t = \frac{GF_2}{GF_1}\,, \qquad \lambda = \frac{EI_1}{EI_2}\,, \tag{5.54}$$

together with a relative deflection $u = v/l$ and a relative coordinate $\xi = x/l$.

Using the dimensionless parameters, the equation (5.53) and the last two equations in (5.51) can be re-written as (from this point on in this section we denote the differentiation with respect to the dimensionless coordinate ξ by a stroke)

$$u' - \theta_1 + t(u' - \theta_2) = \frac{P}{GF_1}\,, \quad u' - \theta_1 + k_1\,\theta_1'' = 0\,, \quad u' - \theta_2 + k_2\,\theta_2'' = 0\,. \tag{5.55}$$

From (5.55) it follows immediately that

$$k_1\,\theta_1'' + t\,k_2\,\theta_2'' = -\frac{P}{GF_1}\,, \tag{5.56}$$

that gives the following after an integration

$$k_1\,\theta_1 + t\,k_2\,\theta_2 = -\frac{P\xi^2}{2GF_1} + C_1 + C_2\xi\,.$$

Using the boundary conditions $\theta_1(0) = 0$, $\theta_2(0) = 0$, $\theta_1'(1) = 0$, $\theta_2'(1) = 0$, we find the integration constants $C_1 = 0$, $C_2 = P/GF_1$ and thus

$$k_1\,\theta_1 + t\,k_2\,\theta_2 = \frac{P(2\xi - \xi^2)}{2GF_1}\,. \tag{5.57}$$

By excluding u' from the last two equations of the system (5.55) and taking into account (5.56), we derive

$$\theta_1 = \theta_2 - (1+t)\, k_2\, \theta_2'' - \frac{P}{GF_1}\,. \tag{5.58}$$

Substituting the expression of θ_1 from (5.58) to (5.57), we obtain the following second-order differential equation with respect to only one unknown function θ_2

$$\theta_2'' - r^2\theta_2 = \frac{P(-2k_1^2 - 2\xi + \xi^2)}{2aGF_1}\,, \tag{5.59}$$

where we use the notation

$$a = (1+t)k_1k_2\,, \qquad r^2 = \frac{k_1 + tk_2}{a}\,. \tag{5.60}$$

A direct substitution helps check that the general solution of the differential equation (5.59) can be represented as

$$\theta_2 = C_3 \operatorname{sh} r\xi + C_4 \operatorname{ch} r\xi + \frac{P(2b + 2\xi - \xi^2)}{2ar^2GF_1}\,, \tag{5.61}$$

where

$$b = k_1 - 1/r^2\,.$$

To determine the integration constants C_3 and C_4, we use the boundary conditions $\theta_2(0) = 0$ and $\theta_2'(1) = 0$. It yields

$$C_4 = -\frac{Pb}{ar^2GF_1}\,, \qquad C_3 = -C_4 \operatorname{th} r\,.$$

A known formula of hyperbolic trigonometry

$$\operatorname{ch} r(1-\xi) = \operatorname{ch} r \operatorname{ch} r\xi - \operatorname{sh} r \operatorname{sh} r\xi$$

helps derive a final expression of the function θ_2, and now it is

$$\theta_2 = \frac{P}{2ar^2GF_1}\left[2b + 2\xi - \xi^2 - 2b\frac{\operatorname{ch} r(1-\xi)}{\operatorname{ch} r}\right], \tag{5.63}$$

wherefrom a sequential double differentiation derives

$$\theta_2'' = -\frac{P}{ar^2GF_1}\left[1 + br^2\frac{\operatorname{ch} r(1-\xi)}{\operatorname{ch} r}\right]. \tag{5.64}$$

Let's now determine the shear force in the column 2. As $Q_2(\xi) = GF_2(u' - \theta_2)$, the consideration of the last equation in (5.55) gives

$$Q_2(\xi) = \frac{Ptk_2}{ar^2} [1 + br^2 \frac{\operatorname{ch} r(1-\xi)}{\operatorname{ch} r}] .$$

This formula can be rewritten in a different form a bit more convenient for reviewing and analyzing. Note that from (5.54), (5.60) and (5.62) these formulae follow

$$k_1 = \lambda t k_2 , \quad a = \lambda t(1+t)\, k_2^2 , \quad r^2 = \frac{1+\lambda}{\lambda(1+t)k_2} ,$$

$$ar^2 = (1+\lambda)t\, k_2 , \quad br^2 = \frac{(1+\lambda)t-(1+t)}{1+t} .$$

Substituting those to the formula of $Q_2(\xi)$, we transform the latter to the form

$$Q_2(\xi) = \frac{P}{(1+\lambda)} [1 + \frac{(1+\lambda)t-(1+t)}{1+t} \cdot \frac{\operatorname{ch} r(1-\xi)}{\operatorname{ch} r}] . \tag{5.65}$$

In particular, in the fixation point the shear force $Q_2(0)$ is

$$Q_2(0) = P\,\frac{t}{1+t} = P\,\frac{GF_2}{GF_1+GF_2} , \tag{5.66}$$

and seeing that $Q_1 + Q_2 = P$ as can be seen by considering the equilibrium in projection onto the Y axis, an unexpected conclusion can be made: when using two Timoshenko columns, the **shear forces in the columns near the fixation points are distributed proportionally to the shear rigidities** of the respective bars, that is

$$\frac{Q_1(0)}{Q_2(0)} = \frac{GF_1}{GF_2} , \tag{5.67}$$

thus the distribution of these shear forces **does not depend on the relation between the flexural rigidities of the columns.** It is worth emphasizing that this conclusion is true only for the root cross-section $\xi = 0$ but not for every cross-section of the columns.

Now let's find out the behavior of the $Q_2(\xi)$ function by passing to a limit case when the column 2 conforms to the classical theory without allowing for the shear. To pass to the limit, we should assume

$$GF_2 \to \infty , \text{ hence } t \to \infty \text{ and } k_2 \to 0 .$$

As can be seen, at $GF_2 \to \infty$ the physical relationships $Q_2(x) = GF_2\gamma_2(x)$ imply $\gamma_2 \to 0$ and next $v'(x) - \theta_2(x) \to 0$, which is in full accordance with the classical bar theory. In this context the r^2 parameter can be represented conveniently as

$$r^2 = \frac{(1+\lambda)t}{(1+t)k_1} , \tag{5.68}$$

which helps us avoid the indeterminacy in the product $t\,k_2$. At $t \to \infty$ we have

$$r^2 \to (1+\lambda)/k_1, \tag{5.69}$$

i.e. r^2 tends to a finite value. In the limit, we can derive from formula (5.65)

$$\eta(\xi) = \frac{Q_2(\xi)}{P} = \frac{1}{1+\lambda}\left[1 + \lambda\,\frac{\operatorname{ch} r(1-\xi)}{\operatorname{ch} r}\right]. \tag{5.70}$$

The consideration of formula (5.70) shows that the relative shear force $\eta(\xi)$ along the column 2 near the fixation point tends to grow quickly which is characteristic for edge effects. In the original problem of the forty-story frame, the parameters are assigned such values that the edge effect area is long enough to cover crossbars of lower stories.

In the counterpart of this problem that uses our idealized model, the following parameters are used:

$$k_1 = 4.0\cdot 10^9 / (40^2 \times 0.25\cdot 10^9) = 0.01; \quad \lambda = 27; \quad r = 52.92.$$

A plot of the $\eta(\xi)$ function built by these parameters is shown in Fig. 5.23 where the abscissa axis tracks the ξ coordinate, and the ordinate axis gives values of η.

Let's see how the computational results change after the column 2 is assigned $GF_2 = GF_1 \times h_2/h_1 = 0.833\cdot 10^8$. Now the analysis yields

$$Q_1(0) = 0.808P, \quad Q_2(0) = 0.192P,$$

which both corresponds to intuitive expectations of engineers and helps avoid possible mistakes in structural solutions related to the transfer of the shear force to the foundations of the two columns.

It is not incurious to study a general limiting case of the solution (5.65) when the flexural rigidities of both columns tend to infinity while the shear rigidities remain finite.

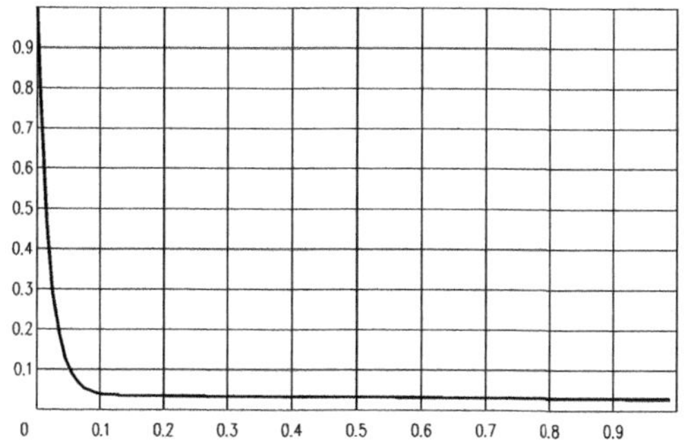

Fig. 5.23. The $\eta(\xi)$ function plot

With this purpose in mind, we assume

$$EI_1 \to \infty, \qquad EI_2 \to \infty .$$

Let the relation between the respective rigidities is kept the same in the course of passing to the limit, that is,

$$\lambda = EI_1 / EI_2 = \text{Const}, \qquad t = GF_2/GF_1 = \text{Const}.$$

In our case $r \to 0$, therefore

$$\frac{\text{ch}\, r(1-\xi)}{\text{ch}\, r} \to 1 .$$

By substituting this to formula (5.65) we transform the latter to its limit form

$$Q_2(\xi) = P\,\frac{t}{1+t} = P\,\frac{GF_2}{GF_1 + GF_2},$$

and it becomes clear that the distribution of the shear forces is the same along the whole length of the columns and proportional to their shear rigidities. Certainly this last result can be derived directly by more elementary means. Though, it is always reasonable to pass to a limit in a formula in order to make sure our previous conclusions are right even in those limit cases.

5.5 Approximating Geometric Shapes and Fixations

Together with an approximation of the displacement field with a selected system of finite elements, sometimes one makes other approximations too, required by the necessity of dealing with a discrete finite element model. Those include a replacement of the geometrical shape of a structure by a similar one such as a polygon, or that of a smooth shell median surface by some similar polyhedron.

A classic by J.Strang and J.Fix [32] dedicates a separate section to this problem where a number of delicate mathematical issues are elicited. The principal conclusion is pretty simple to formulate in mechanical terms: when a plane region is approximated by a curvilinear polygon, there is a *boundary layer* two elements thick in the vicinity of the boundary where the derivative error is high, but this singular behavior subsides quickly along the normal to the boundary. Here one can see "*... a modification of the Saint-Venant principle that regards geometry to a greater extent than boundary data as it usually does*".

The approximation of a system's geometry may cause the boundary conditions to change together with the modification of the shape because those should be reformulated for a boundary of another shape. Here one can easily find oneself in a trap due to the fact that *passing to the limit of the contour's shape* does not necessarily causes *passing to the limit of kinematical properties*. It is evidenced by a well-known Sapondjan paradox for a simply supported polygonal plate [22].

The paradox deals with a simply supported plate having the shape of a regular polygon inscribed in a circle of the radius R, with a uniformly distributed load q

applied to it. The essence of the paradox is that when the number of sides of the polygon increases, the k coefficient in the formula of the maximum deflection $w_{max} = kqR^4/D$ becomes 0.04688 instead of the correct value 0.0637 (the Poisson ratio is assumed to be $\nu = 0.3$). For a plate clamped along its contour, passing to the limit yields a correct result $k = 0.01563$ (see Fig. 5.24).

The explanation for this paradox given in a book by Y.G. Panovko [22] is based on a consideration of edge conditions for a polygonal plate. If the plate is simply supported, the normal to its median surface in any point of its contour (except for an angular one!) can rotate only around the respective rectilinear boundary (Fig. 5.25). That is, the a–a normal rotates around n–n, the b–b normal around m–m etc. But then the normal in an angular point cannot rotate at all. Those form a kind of stiff "paling" that becomes denser as the number of sides grows, and changes the conditions of the simple (hinged) support to something different, something between an embedding and a simple support.

It should be noted that the above said does not necessarily hold if the computation is performed by FEM. In this case no Sapondjan paradox will manifest itself if the edge conditions of the discrete model are formulated accurately, with distorting the area's boundary geometry but without distorting the edge condition formulation itself.

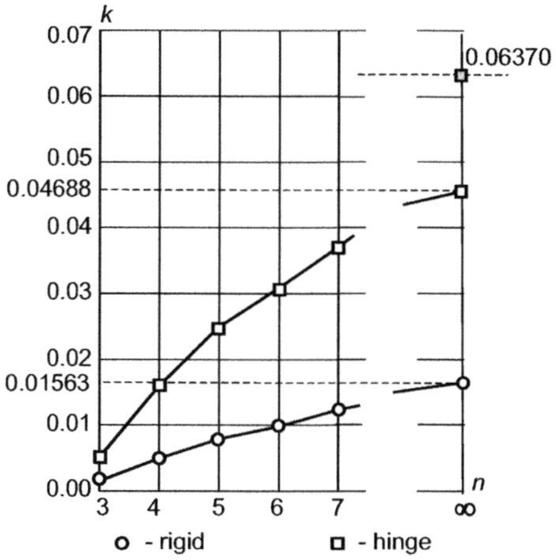

Fig. 5.24. Sapondjan paradox

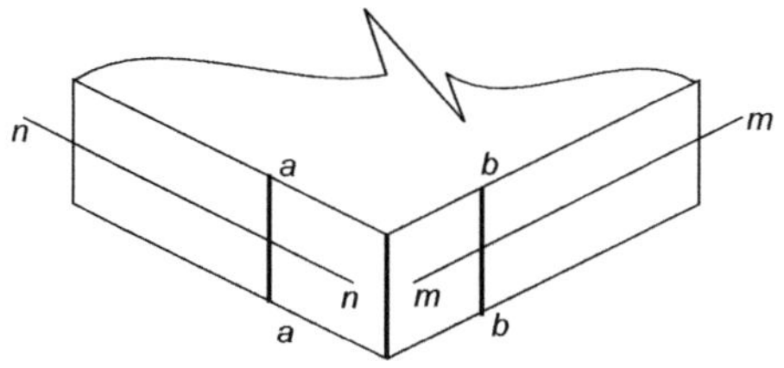

Fig. 5.25. Boundary conditions in an angle

To illustrate the point, let's give results of a finite element analysis of a simply supported round plate of the radius R under a uniformly distributed load q. In its finite element model the plate is approximated by a regular n-sided polygon inscribed in a circle. Due to a cyclic symmetry found in this problem, it suffices to include only one sector with the opening angle $\varphi = 2\pi/n$ in the discrete model.

Assuming $n = 12, 18, 24, 30$ sequentially, we will divide the radius into k parts where $k = n/6$. Fig. 5.26 gives an example of a finite element mesh at $n = 18$. The following fixation of nodes against displacements is assumed:

- in node 1: $\theta_x = 0, \theta_y = 0$;
- in nodes 2, 4: $\theta_x = 0$;
- in node 6: $\theta_x = 0, w = 0$;
- in nodes 3, 5: $\theta_\xi = 0$;
- in node 7: $\theta_\xi = 0, w = 0$,

where θ_x, θ_y, θ_ξ are nodal slopes with respect to the X, Y and ξ axes, respectively.

It's clear that the boundary conditions stated above are in full correspondence with the simple support of the round plate without any distortions introduced.

The analysis performed by the **GTSTRUDL** software gives values of the deflection in the center of the plate (in a dimensionless form) shown in Table 5.5 that follows.

The plate's triangular finite element used in the analysis was a BPH-type one, and the quadrangular element was a BPHQ. These are hybrid flexural plate finite elements (representing the Kirchhoff – Love theory). As one can see in Table 5.5, finite element solutions for inscribed polygons converge to the round plate solution.

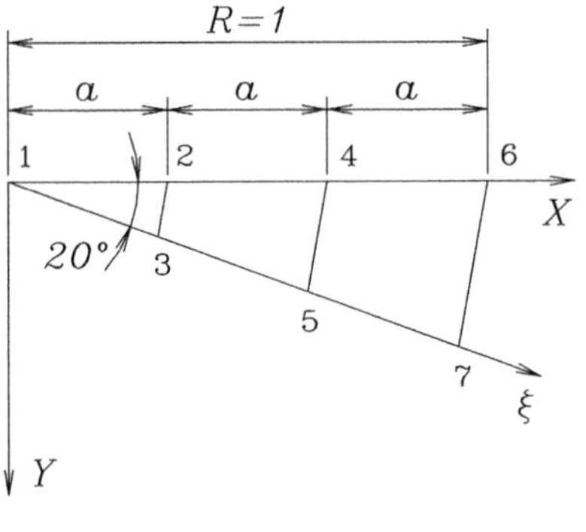

Fig. 5.26. A finite element mesh for an 18-sided polygon

Table 5.5.

Accurate solution for a round plate	Accurate solution for an n-gon, $n \to \infty$	$n = 12$	$n = 18$	$n = 24$	$n = 30$
0.0637	0.04688	0.06439	0.06380	0.06375	0.06373

The following should be noted in this regard. It is an error introduced by a distortion of boundary conditions rather than that caused by an approximation of the region's geometry which is of particular importance for the finite element analysis in general.

Another special thing to discuss is a simulation of the geometry of shell structures. There are two feasible approaches:

- employ plane finite elements thus replacing the true surface of a shell by a polyhedron of a close shape;
- make use of curvilinear finite elements possessing the same curvature as the surface of a shell in question.

In the former case the shell finite element is essentially a combination of a flexural plate finite element and a plane stress plate element. The curvature of the shell's surface is replaced by discrete jumps of the normal direction at lines of junctions between finite elements which introduces a certain amount of inaccuracy into the solution.

In the latter case the curvature varies smoothly but another difficulty arises. Displacements of an element as a whole rigid body can hardly be approximated by a class of commonly used shape functions, and it may cause spurious stresses to appear in the model. The thinner the shell, the bigger the effect of this error. Very often one chooses not to employ the Kirchhoff assumption in order to reduce this error. In this case simpler approximations of the displacements can be used. Curvilinear finite elements are used most often to analyze cylindrical shells where they facilitate the specification of source data to a great extent — it suffices to specify the radius of the cylinder's generatrix. With more complex surfaces where values of the curvature must be specified at each finite element and boundaries of the elements are usually principal curvature lines, it may become a problem. Of course, from the user's standpoint this problem is solved easily by selecting an automatic finite element mesh generation, but it becomes really cumbersome when elements are employed not included in options of the respective mesh generator.

5.6 Computational Error and Ways to Dispose of it

Errors of any kind of calculation, not counting mere blunders, are usually caused by inaccurate source data, errors of an employed numerical technique, and rounding errors. In connection with calculations performed by computers the latter reason is of particular importance. First, an analysis of this kind involves very long chains of calculations, and errors of particular operations may well accumulate if the analyst failed to choose a good technique of solution. This accumulation may prove a real problem so that starting from some point one will not deal with an expected result spoiled to a certain extent by "noise" of the errors. Rather, one will deal with a pure noise spoiled to some extent by traces of the expected result. The issue of the rounding error effect is closely related to that of the conditionality of governing equation systems discussed before.

A paradoxical phenomenon may arise when solving a problem with an unstable method, in the computational sense, or/and with poorly conditioned design models: making the finite element mesh denser in order to improve the accuracy of the solution increases the number of arithmetical operations and causes the rounding error to grow, thus deteriorating the solution eventually. To detect this trap, one is recommended to employ an artificial technique of solving a simple problem and increasing its dimensionality until it reaches the scale of a really complex model. If an accurate solution is known for a test problem, the detection of abnormalities in its results becomes trivial.

Though, the problem of the computational error in question is of much wider scope. We deem it useful to discuss the problem from a little bit different standpoint.

We begin our analysis with a consideration of a perfectly trivial technique to demonstrate the effect of the sequence of calculations on the accuracy of its final result.

Let the task be to determine the moment of inertia I of an annular cross-section with its exterior diameter D and its internal one d. As is known, the quantity in question can be found from the formula

$$I = \frac{\pi}{64}(D^4 - d^4)\,. \tag{5.71}$$

For a very thin ring with the ratio of its diameters

$$\eta = d/D\,, \tag{5.72}$$

nearly equal to one, the direct calculation of the moment of inertia I by formula (5.71) may result in a noticeable error in the final result of the computation. Here we mean a computation using a calculator capable of keeping only a small number of significant digits. Therefore an expert in computing will perform some transformations of formula (5.71) before starting the calculation itself, meaning to suppress the major computational error introduced by the loss of accuracy in intermediate arithmetic. From the standpoint of the computational mathematics this error is caused by a limited machine word length used to represent a floating-point number. Introducing a small quantity of α

$$\alpha = 1 - \eta\,, \tag{5.73}$$

we re-write the annular section's moment of inertia (up to a constant factor $\pi D^4/64$ inessential for our consideration) in the following two forms:

$$I_1 = 1 - \eta^4 \qquad \text{and} \qquad I_2 = \alpha(1+\eta)(1+\eta^2)\,. \tag{5.74}$$

It is obvious that the two forms of the moment of inertia expressions are absolutely equivalent in the mathematical sense. Though, the direct use of the expressions for the calculation differs in the order (sequence) of computing.

It is easy to notice that the first expression (the I_1 formula) suggests exactly the same order of calculations as formula (5.71).

The expression of I_2 is particularly interesting because it yields the following useful estimate immediately:

$$I_2 < 4\alpha\,, \tag{5.75}$$

and the estimate gets closer with smaller α.

Here we present a short program enabling one to run a computational experiment in order to compare the accuracy of the I_1 and I_2 calculation:

```
#include <iostream.h>
typedef     float     real;
void        main(void)
{
    const   real      one = (real)1.0;
    const   real      zero= (real)0.0;
    real    alf, eta, J1, J2, r1, r2, r3, r4;
    while(1)
    {
            cin>>alf;               if (alf <= 0) return;
            eta = one - alf;
            r1 = eta * eta;     r2 = r1*r1;      J1 = one - r2;
            r3 = one + eta; r4 = one + r2;  J2 = alf*r3*r4;
            cout<<"J1="<<J1<<" J2="<<J2<<endl;
    }
}
```

Table 5.6 lists results of calculations performed by this routine for some values of the α parameter. The short floating-point computer arithmetic was used (the **float** format for such programming languages as C and C++).

Table 5.6.

α	I_1	I_2
10^{-5}	4.005×10^{-5}	4.000×10^{-5}
10^{-6}	4.053×10^{-6}	4.000×10^{-6}
10^{-7}	4.768×10^{-7}	4.000×10^{-7}

In a personal 32-bit-architecture computer, the **float** format allocates 23 bits for the significant part of a floating-point number (so-called *mantissa*) which is equivalent approximately to six decimal digits. Actually, the analysis of errors caused by rounding in the floating-point arithmetic is more complicated to an extent. It depends on such additional attributes of this arithmetic as normalization, whether there is rounding or the rounding is blocked etc. A discussion in whatever detail could be sufficient to explicate this topic does not follow the main line of our book and would draw the reader's attention to a discussion of undoubtedly interesting but inessential particularities of "trees behind which one cannot see the wood". We refer a reader who is interested in details to special publications on the subject where answers are given to a lot of questions related to the rounding error analysis in the floating-point arithmetic. In particular, a pretty elementary though by no means lacking strictness discussion on this subject can be found in a book by J. Forsyte and C. Moler which seems to have become a classic judging from the number of references to it [11].

It is not too hard to understand the reason for the fall of accuracy of the I_1 quantity as the α parameter decreases.

With this purpose in mind, let's introduce a small number denoted as ε_f, one depending on a particular computer implementation. As the computer word length representing a **float**-format number is limited, the machine instruction of adding any positive number $\varepsilon < \varepsilon_f$ to one is insensitive to the value of ε. In other words, in the machine representation the following relationships take place:

$$1.0F + \varepsilon = 1.0F, \;\; 1.0F + \varepsilon_f > 1.0F \qquad (5.76)$$

for any positive $\varepsilon < \varepsilon_f$. The designation 1.0F taken from the C/C++ programming languages points out the use of the **float** format to represent the number one.

The ε_f value that depends on the hardware and software environment is equal to the following, in our case:

$$\varepsilon_f = 1.2\times 10^{-7}. \qquad (5.77)$$

Therefore, at small α comparable with ε_f in their order of smallness, the calculation of I_1, being a subtraction of two very close values, throws away an essential part of information regarding the numerical results because it makes use of computer storage devices limited in their floating-point representation capabilities[16].

A critical reader may have two following objections against the example given above by the authors:

- first, Table 5.6 itself evidences that any noticeable loss of accuracy of the computation can be observed only at values of the α parameter so small that they are totally impractical and do not correspond to any reasonable ratio of an annular section's diameters. In real problems the effect of the accuracy loss (in this particular example) won't be essential to any extent.
- second, if even we assume there can be a substantial loss of accuracy of a computation because of an ineffectual choice of the calculation sequence, what does the software user have to do with it? Let this circumstance bother the software developers: they're educated enough to save the user from the trouble of this kind.

---

[16] Solicitous readers who have an access to a C compiler such as Visual C++, can find the ε_f number in the <float.h> header file where there is a preprocessor directive:

```
#define FLT_EPSILON 1.192092896e-07F /*smallest such that
  1.0+FLT_EPSILON != 1.0*/
```

As for the first one of the objections above (supposedly advanced by readers), we answer that we did not have in mind to construct a practically applicable example. Our objective was to demonstrate the very possibility that a source of errors may be a wrong arrangement of calculations. And this objective required a most simplified example rather than a practical one because a complicated practical problem would not be so illustrative. Yes, it is really the case that with a reasonable ratio of the ring's diameters the error of the calculation performed in any order will be a mere trifle. But small errors have an unpleasant tendency to accumulate in solutions of problems of a high dimensionality. As the saying goes, "light gains make heavy purses".

Regarding the second one of the predicted objections to our example by a hypothetic critical reader, we agree immediately and implicitly that it is the software developer who is responsible for the order of calculations. Though, as will be shown below, a computational engineer must not adopt a passive attitude to a CAD software he uses, or, saying it in the language of wrestling, he must not get on all fours and wrestle on his knees. A really experienced and highly qualified user employs a software **actively**, forcing it to do **ultimate** things which even the software's developer might not think it could be capable of doing! Examples of this kind are found in this book too.

On the contrary, nobody seems to blame a manufacturer of hammers for a customer's using a purchased hammer to drive nails with its handle. The only thing the customer could blame the manufacturer for is that the former was not provided with a manual where the whole technology of nailing would be explicated in all possible detail. We omit a pathological case of a user who cannot read. Also, it is highly probable that once the problem is stated, an inventor will appear who will design a hammer absolutely prohibiting any contact of its handle with a nail. That's a red-hot example of a feedback between software users and developers desired by everybody.

Now let's start constructing an example where the user can affect the sequence of calculations, and this sequence will influence the error of the solution. Frankly speaking, this example has been notorious among software developers and experienced analysts for a long time as a sort of folklore. Nonetheless, the professional experience of communicating with both tells the authors that people tend to step on the same rake again and again. Maybe it happens all the time because, to the knowledge of the authors, so far there have been published no reasonable and convincing explanation of the respective effects[17].

So, let us consider a model of a conventional cantilever beam composed of n segments of different length and loaded by a concentrated force P at its free end.

[17] Here authors find it possible to quote from their own words said ten years ago on the question being discussed: "*This fact detected experimentally does not have a convincing interpretation so far. Yet we argue that the following hypothesis of a more general scope may prove credible: the sequence of unknowns' exclusion least sensitive to the loss of accuracy is the one wherein at each step of the exclusion a variable chosen to be excluded corresponds to the least diagonal element*" [12].

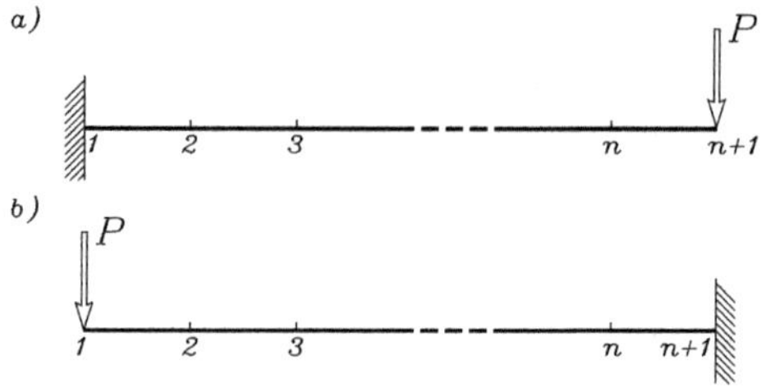

Fig. 5.27. Two systems of numbering of unknowns: (*a*) a "bad" one (*b*) a "good" one

One can think of two possible sequences of the node numbering, both resulting in a minimal band width of the structure's stiffness matrix: one with the beginning at the fixation point (Fig. 5.27, *a*), and another with the beginning at the free end (Fig. 5.27, *b*).

Assuming the length of a segment of the beam to be equal to one, we have the full length of the beam $l = n$. Also, assuming $EI = 1.0\times10^9$, $P = 3$, we must obtain the deflection w^* at the end of the cantilever equal to

$$w^* = \frac{Pl^3}{3EI} = \left(\frac{n}{1000}\right)^3. \tag{5.78}$$

Table 5.7 below gives results of the deflection calculation by the **SCAD, GTSTRUDL** and **ROBOT** software at different values of the number of segments n. The analysis with **SCAD** used the following settings:

- the *Decomposition accuracy* property was assigned a maximum possible value 10^{-15};
- the *Solution check* option was disabled;
- the matrix profile optimization mode was enabled.

The first setting means that all numbers less than 10^{-15} are treated as zeros. That is, a residual rigidity which may arise because of an "incomplete exclusion" due to a rounding error (see below) is limited by this value.

Table 5.7.

Result	n	5 000	10 000	15 000	20 000
By (5.78)	w^*	125	1000	3375	8000
SCAD	Scheme *a*	125	997	591	605
	Scheme *b*	125	1000	3375	8000
GTSTRUDL	Scheme *a*	125	1056	2121	1427
	Scheme *b*	125	971	2975	5772
ROBOT	Scheme *a*	125	976	2869	1610
	Scheme *b*	125	1000	3375	8000

The second setting means that by default the system will perform an automatic accuracy check of the solution of the governing equation system. This check consists of the analysis of the structure subjected to an artificial load such that all linear displacements are equal to one and all angular ones are zero (a test by A.V. Gorbovets). If this option is enabled, then the solution will be interrupted and a message will be output to the log file: "*Error of matrix decomposition*".

As the **GTSTRUDL** software employs a grouped unknowns exclusion technique that introduces an additional computational error, results of the analysis performed by **GTSTRUDL** are not so illustrative and peculiar as those by **SCAD**[18]. Therefore our further consideration below will be based on the results of the numerical experiment performed by the **SCAD** software.

As one can see in the results given in Table 5.7, the *a* scheme causes a more intensive accumulation of the computational error than the *b* scheme. What is the reason for this?

In order to clear up reasons for a high sensitivity of the *a* scheme to the accumulation of the computational error, let us express the computation itself in terms and concepts of mechanics. Recall the physical sense of the Gaussian exclusion procedure for solving linear algebraic systems which is employed to implement the slope deflection method. The matter is that a sequence of the node numbering specified by the user is an implicit instruction for the program how to number the unknowns in the principal system of the slope deflection method. If we exclude from the further consideration the nodal displacements along the beam (inessential for the current analysis), then two unknown displacements will remain in each node: a linear deflection in the perpendicular direction to the beam's axis, and a slope of the node. As commonly known, the principal system of the slope deflection method is formed by constraining displacements of the original system's nodes. Then for this principal system a stiffness matrix **K** and a load vector **p** are formed. We can get a hint as to how to explain the effect in question if we consider the Gaussian exclusion procedure for solving the linear algebraic equation system

$$\mathbf{K}\mathbf{u} = \mathbf{p}\,, \tag{5.79}$$

mechanically (see Appendix A) as a sequential removal of constraints imposed on the original system during the formation of the principal system of the slope deflection method.

Seeing that the Gaussian exclusion is a numerical procedure performed by means of machine arithmetic, small errors of computation are unavoidable, and those cause the constraints to be as if removed incompletely. To simplify the consideration, let's assume that these errors are concentrated in elements of the matrix's main diagonal. It seems that these computational "tails" keep some sort of springs of very small stiffness in places of removed constraints. These remaining springs, in their turn, increase the error of the next Gaussian exclusion step thus picking up the stiffness of new "spurious" springs in the next node etc. As a result, actually after removing all imposed constraints we have a distorted system with a

[18] To the credit of the **GTSTRUDL** software, it does not remain a passive observer of the solution losing its accuracy. When this insidious computation job is running, **GTSTRUDL** pricks up its ears and starts bombarding the user with storm warnings.

set of springs that appeared because of the computational error accumulation, rather than the unchanged original one. The algebraic aspect of this is a completion of a so-called direct pass of the Gaussian exclusion of unknowns.

In the *a* numbering scheme the stiffness of the springs grows from the fixation to the free end while in the *b* scheme in the inverse direction. It should be said that the sign of the error (the stiffness of the springs) is an undetermined value, therefore one can obtain the deflection of the cantilever's end either less or greater than the true one with different sets of the input data.

If we considered a little bit more complicated conditions of the beam's loading, such as a distributed load along the whole beam, then the error would be accumulated in the load vector as well as in the matrix itself. That is, additional "spurious" external forces would appear, increasing in their intensity from lower node numbers to higher ones.

Now it becomes clear (at least mechanically) how the *b* scheme can be more noise-immune to the error accumulation in the calculation of the cantilever beam's free end deflection.

The problem being discussed is of key importance, while utterly ignored by the scientific and educational literature on the subject of structural design. Therefore, in order to further improve the reader's understanding of the situation above, let us discuss a simplest possible problem of tension/compression of a bar shown in Fig. 5.28 by a force $P = 1$ applied to the free end of the bar.

It's obvious that the accurate solution of this problem will yield the longitudinal displacement u^* at the free end of the bar

$$u^* = Pl/EF\,. \tag{5.80}$$

Further, a subscript *a* and *b* will distinguish the **K** stiffness matrix or the **p** load vector depending on what node numbering scheme those refer to.

The stiffness matrices and load vectors of the *a* and *b* schemes will look, respectively,

$$\mathbf{K}_a = \frac{EF}{l/n}\begin{bmatrix} 2 & -1 & & & \mathbf{0} \\ -1 & 2 & -1 & & \\ & \ddots & \ddots & \ddots & \\ & & -1 & 2 & -1 \\ \mathbf{0} & & & -1 & 1 \end{bmatrix}, \qquad \mathbf{p}_a = P\begin{bmatrix} 0 \\ 0 \\ \vdots \\ 0 \\ 1 \end{bmatrix}, \tag{5.81, a}$$

$$\mathbf{K}_b = \frac{EF}{l/n}\begin{bmatrix} 1 & -1 & & & \mathbf{0} \\ -1 & 2 & -1 & & \\ & \ddots & \ddots & \ddots & \\ & & -1 & 2 & -1 \\ \mathbf{0} & & & -1 & 2 \end{bmatrix}, \qquad \mathbf{p}_b = P\begin{bmatrix} 1 \\ 0 \\ \vdots \\ 0 \\ 0 \end{bmatrix} \tag{5.81, b}$$

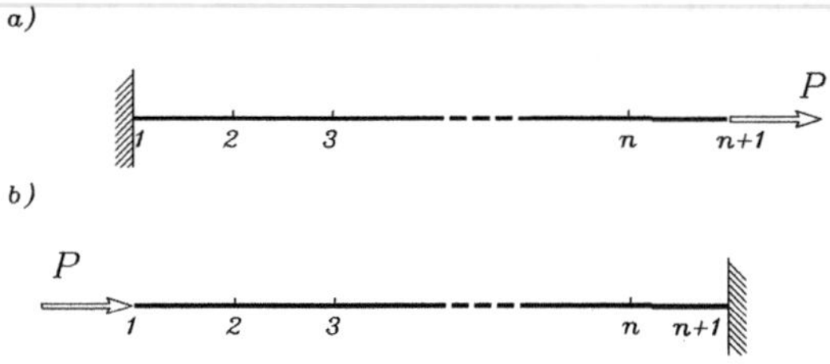

Fig. 5.28. A lengthwise loaded bar with different numbering of unknowns

Consider the equation system (5.79) after a transformation of the direct pass of the Gaussian exclusion procedure. It is equivalent to a decomposition of the equation system matrix into triangular cofactors, a so-called **LU** decomposition [31] where **L** is a lower triangular matrix containing units in its main diagonal, and **U** is an upper triangular matrix

$$\mathbf{K} = \mathbf{LU}\,, \qquad \mathbf{Uu} - \mathbf{L}^{-1}\mathbf{p} = \mathbf{0}\,. \tag{5.82}$$

In particular, for the matrices and vectors in (5.81, *a*) and (5.81, *b*) we have:

$$\mathbf{U}_a = \frac{EF}{l/n}\begin{bmatrix} \frac{2}{1} & -1 & & & \mathbf{0} \\ 0 & \frac{3}{2} & -1 & & \\ & \ddots & \ddots & \ddots & \\ & & 0 & \frac{n}{n-1} & -1 \\ \mathbf{0} & & & 0 & \frac{1}{n} \end{bmatrix}, \qquad \mathbf{L}_a^{-1}\mathbf{p}_a = P\begin{bmatrix} 0 \\ 0 \\ \vdots \\ 0 \\ 1 \end{bmatrix}, \tag{5.83, a}$$

$$\mathbf{U}_b = \frac{EF}{l/n}\begin{bmatrix} 1 & -1 & & & \mathbf{0} \\ 0 & 1 & -1 & & \\ & \ddots & \ddots & \ddots & \\ & & 0 & 1 & -1 \\ \mathbf{0} & & & 0 & 1 \end{bmatrix}, \qquad \mathbf{L}_b^{-1}\mathbf{p}_b = P\begin{bmatrix} 1 \\ 1 \\ \vdots \\ 1 \\ 1 \end{bmatrix}. \tag{5.83, b}$$

Now performing the reverse pass of the Gaussian exclusion procedure, we have the displacement of the *P* force application point in the cases *a* and *b*, respectively.

$$u_a = u_{n+1} = \frac{P\,l/n}{EF}\,\frac{1}{1/n} = \frac{Pl}{EF} = u^*,$$

$$u_b = u_1 = \frac{P\,l/n}{EF}\,n = \frac{Pl}{EF} = u^*, \tag{5.84}$$

which is quite expected.

Evidently, both cases have yielded the accurate result only because we performed all Gaussian procedure steps with designations rather than numbers, therefore did not lose the accuracy in intermediate calculations. But computer programs deal with numbers not letters.

Now let's see what happens if the Gaussian procedure is performed with floating-point numbers. It is easy to make sure that our *b* numbering case does not cause the error to accumulate. But the *a* case is a different thing.

Here we present a short program allowing to simulate the process of the equation system (5.79) solution with its matrix and load vector from (5.81, *a*):

```
#include    <iostream.h>
typedef     float    real;
void        main(void)
{
    const   real    one = (real)1.0;
    const   real    two = (real)2.0;
    real    a, r, Za;
    long int n, i;
    while(1)
    {
            cin>>n;         if (n <= 0) return;
            a = two;
            for (i=3; i<n+1; i++)      // i – No. of node
            {
                    r = one / a;       a = two – r;
            }
            r = one / a;               a = one – r;
            Za = (one / a) / (real)n;
            cout<<" Za = "<<Za<<endl;
    }
}
```

Results of a numerical experiment performed with this program at different values of *n* are given in Table 5.8. Notice the fact that these calculations are performed with values represented as computer single floating-point numbers (the **float** format).

Table 5.8.

N	1000	2000	4000	5000	10000
u^*/u_a	0.9999	0.9804	0.8364	0.6691	0.3345

To perform the same calculations with double accuracy computer numbers, modify the second line of the program above by re-defining the **real** type as

```
typedef    double    real;
```

In this latter case a noticeable accumulation of the computational error begins at a much higher value of n, about 80,000,000, but the common tendency takes place all the same.

So, everything that was said makes us conclude that the user possesses some (though limited) tool to regulate the computational error to some extent.

Generally, to minimize the error affected by the sequence of the node numbering it is reasonable to number the nodes beginning from a most flexible part of the structure and moving towards places of its fixation.

Keeping to this rule, one can expect that a computational noise in the form of fictitious additional springs and/or additional external forces will grow towards less compliant part of a structure, thus expectedly distorting as little as possible the general behavior of the mechanical system. We realize quite well that the bold-faced recommendation formulated above is pretty nebulous and is not a clear algorithmic instruction. The development of such algorithm(s) can be done on the basis of strict mathematical estimates by experts in numerical analysis and computing. But this is a subject of other publications (and a concern of other authors).

By the way, from what was said above we can derive the following recommendation for the structural design software developers — so far as we know this recommendation is not yet implemented anywhere in the software. It would be really useful if a CAD program suggested a possibility (at a user's request or independently from such) to perform calculations over stiffness matrices both in the direct and in the inverse node numbering. A comparison of these two solutions would be a basis for a judgment of the computational error in the hands of an experienced user.

It should be said that in many cases the inverse numbering does not require a re-construction of the stiffness matrix and load vector(s). It suffices only to organize another procedure of solving the linear algebraic system with the Gaussian exclusion to be performed in the descending order of the unknowns. A similar idea appears in a frontal method by Irons, the second option of the front movement being an inverse pass through elements.

Though, one must take into account that the inverse unknown exclusion is not always possible without re-constructing the stiffness matrix.

As can be seen, there are three principal types of structures of the equation system's (5.79) stiffness matrix **K**. The latter does not change its structure (Fig. 5.29) in the course of the Gaussian exclusion: ribbon, ladder, or tower (profiled).

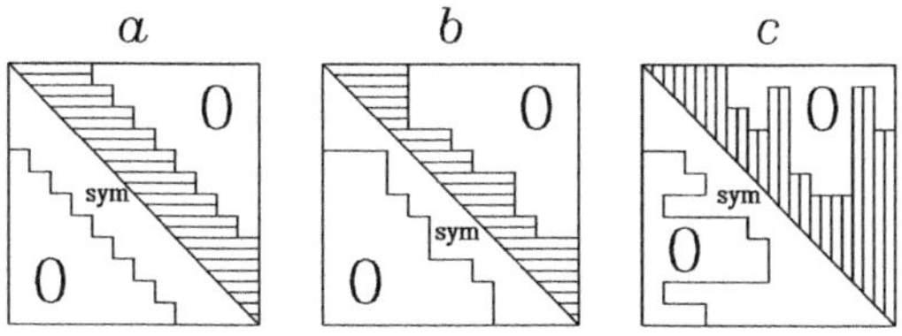

Fig. 5.29. Structure of stiffness matrices: (*a*) a ribbon-type matrix; (*b*) a ladder-type matrix; (*c*) a tower-type matrix

The ribbon structure can be treated as a particular case of a ladder with equal steps, that is, when the horizontal and vertical sizes of its steps are equal: both its tread and its rise.

It is easy to notice how the inverse exclusion scheme maintains the matrix's original structure in the first two cases (a ribbon and a ladder), while in the case of a tower this important condition does not hold. For example, if the topology of a structure resembles a bicycle wheel with spokes, then the central node (where all spokes meet) should be the last one in order. In this case the stiffness matrix structure will be tower-like with a minimum number of initial zero elements within each particular tower. If the exclusion of unknowns starts with this node, then the rest of the matrix will be completely filled with nonzero elements.

It seems that the most flexible option is a possibility of an explicit specification of the unknowns exclusion sequence by the user (a full or partial specification of groups of unknowns ordered in the sequence of their exclusion). Also, it's attractive to be able to charge the system with the procedure of unknowns ordering, or to specify some additional conditions for the ordering procedure. It goes without saying that the software should make a decision as for the order of the unknowns exclusion involving criteria such as the computational resource saving or any other requested by the user. Some of these recommendations are implemented in the **SCAD** software. For example, it suggests the capability of specifying direction cosines of a vector along which the node numbering should be performed, or choosing one of available options of the matrix profile optimization.

5.6.1 Notes on the Super-Element Application

Another possibility for suppressing the computational error is the employment of super-elements in cases where such are repeated many times in the same structure of interest. In the above stated problem of the bar tension/compression, one could form a super-element consisting of two nearby segments of the bar, then form a second-level super-element by merging two first-level super-elements and so on. Observing the sequence of the computation thus performed and paying attention to the exclusion of unknown displacements of a super-element's internal nodes (the procedure is called a static condensation), one notices that there is no accumulation of the error in this process.

Once we touch the issue of the super-element employment by structural analysis software, we'd like to discuss it in more detail. The idea of super-elements is far-back, and its first implementations on the tide of over-optimistic expectations of super-element advantages were made with veteran electronic tube computers[19]. Somehow, in the course of time the practical interest of software developers to super-elements faded away. We think it did due to a fantastic growth of the computer performance both in the speed of calculation and in the available memory

[19] We cannot say for sure who was the first to advance this idea and who is the author of the "super-element" term itself.

resources. Many developers think it's a sufficient compensation for not using super-elements. But we don't!

The employment of super-element models seems to us useful in a few aspects:

- to reduce the amount of input data;
- to dramatically reduce the software's demand of computational resources (time and memory);
- as an efficient tool of fighting the accuracy loss in matrices of a high order, particularly eliminating the effect of the node numbering on the computational error accumulation;
- as a ready-made tool for static condensation of selected nodes and their allotted degrees of freedom.

The last capability is very much useful, for example, to subsequently allow for a nonlinear behavior of a structure where all the nonlinearity is concentrated in selected nodes in the directions of selected degrees of freedom.

Certainly these advantages of the super-element modeling really prove out only if a system includes a good deal of repeated equivalent super-elements the number of internal nodes in which is substantially greater than that of external nodes, and the super-elements are implemented in a software with a deep nesting allowed, thus forming an extended hierarchy of super-elements as *element types*. It's also obvious that a good super-element model implementation should impose no limitations on external influences (pretty different even in particular instances of the same super-element type), nor on their mutual spatial position which means particularly that super-elements of the same type should allow different orientations with respect to the global coordinate system.

There is an idea that the super-element modeling is too complicated to be understood and used correctly by ordinary users. We answer to this opinion with words by Allen Holub[20] said in connection with a similar idea regarding the full-featured C++ application in the software development: "*it's the most ridiculous excuse for a poor implementation of a project that we have ever heard*". By the way, the analogy with C++ is intentional: the idea of super-elements as types defined by the user rather than taken from a predefined finite element library is a counterpart of the C++ concept of classes. This analogy is even deeper than it may seem at first because, on one hand, there is an inheritance and hierarchy of classes in C++, and on the other hand, there is a hierarchy of super-element types in FEA.

5.6.2 Notes on a Software Testing

All that was said above should make one pay a special attention to the procedure of testing a software that one is going to purchase. Before purchasing that, one is

---

20 "*Average users are too stupid to do everything right; actually they are not even interested in learning how to work right, and it is going to be a hard job to teach them. So we do not intend to do either thing. We'll just make our product a bit more stupid*" [14].

recommended to solve a few known "bad" problems such as a cantilever beam where nodes are numbered from the fixation to the free end.

A technique that is a must for testing is a check for an agreement of the analysis results obtained with a symmetric and asymmetric division of an area into finite elements (Fig. 5.30). Only if the results agree well enough, one can be sure that results of a practical analysis will be trustworthy in cases where no symmetric division is possible such as irregular boundaries, stress concentration etc.

An efficient check of performance of a program and its employed finite elements is a *patch test* [15] suggested by Bruce Irons in 1975 but not having become really well-known to users. The idea of the test is that an extremely chaotic design model should be used intentionally to solve a "good" problem where an analytic solution is known.

A typical example can be a rectangular plate in conditions of a uniform compression, or spherical/cylindrical bending (Fig. 5.31). A chaotic arrangement of the finite elements that corresponds to the entitlement of the test (resembling patches on a clothing) can reveal peculiarities of a design model that could remain concealed if a regular model were used.

An example of such testing is shown in Fig. 5.32 demonstrating isofields of displacements obtained with regular and patch meshes using the **SCAD** software.

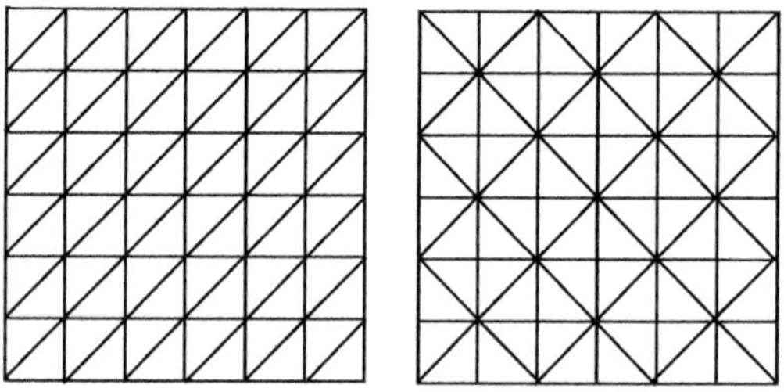

Fig. 5.30. Two ways of meshing an area

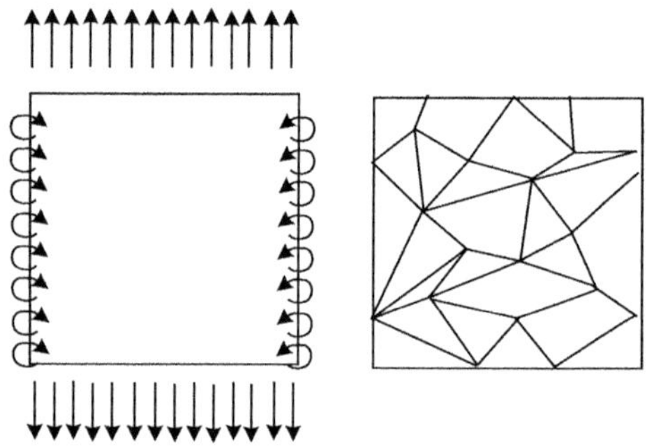

Fig. 5.31. A patch test: a scheme and a design model

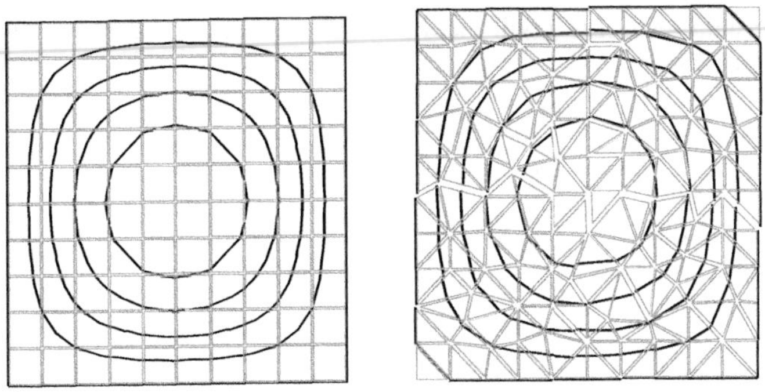

Fig. 5.32. Displacement field isolines for bending

5.7 Step-by-Step Procedure

It does not seem casual that a step-by-step procedure was clearly outlined as a technique for analyzing nonlinear structures in the paper [36], the same one where the finite element method was presented formally for the first time. As it always happens, after that a great deal of publications appeared where priorities were argued or revised. In particular, it was indicated that a similar idea (a method of successive loading) had been presented by V.Z. Vlasov [24] before. Or, even earlier purely mathematical results by L. Euler were referred to. From our viewpoint, all this argument is inessential when discussing practical applications of the step-by-step procedure in computer-aided design of structures. Any contemporary analyst is convinced of the step-by-step procedure being an integral part of the nonlinear finite element analysis.

General definitions of the step-by-step procedure are commonly known (see [33], for example). We think it's important to point out its key features that define the accuracy and the possibility to check solutions obtained and help understand how it really works.

First of all, we should note that the very transition from a system of nonlinear equilibrium equations

$$f_i\,(x_1, x_2, \dots , x_n) = 0 \qquad (i = 1, 2, \dots , n) \tag{5.85}$$

to equations with a parameter

$$F_i\,(x_1, x_2, \dots , x_n, t) = 0 \quad (i = 1, 2, \dots , n) \tag{5.86}$$

and then to a chain of linearized equations of the step-by-step procedure, depends on how the time-like t parameter is introduced into the system, the parameter's increments being used to transit from one step to another.

If the parameter is chosen so that the solution of the (5.86) system is known at some $t = t_0$, and at $t = t^*$ the system (5.86) is equivalent to (5.85), then the differentiation with respect to t will yield a system of linear differential equations with variable coefficients

$$\sum_{j=1}^{n} \frac{\partial F_i}{\partial x_j} \cdot \frac{dx_j}{dt} + \frac{\partial F_i}{\partial t} = 0 \qquad (i = 1, 2, \dots, n), \tag{5.87}$$

and initial conditions known at $t = t_0$. Solving the initial-value Cauchy problem for the equation system (5.87), we obtain an integral curve

$$x_1 = x_1(t), \quad x_2 = x_2(t), \; \dots \; , x_n = x_n(t), \tag{5.88}$$

which gives a solution of the original problem (5.85) at $t = t^*$.

There are known a few versions of the step-by-step procedure. In essence, they differ only in the way the t parameter is introduced and/or numerical techniques of the initial-value problem stated above. All of those have been tested practically and have their scope of application. For example, the **SCAD** system suggests the following options of the step-by-step procedure:

- simple step-by-step procedure;
- step-by-step refining;
- step-by-step iterations.

In the first case, at each step a linearized problem is solved and then the next step of loading starts, implying that the obtained solution is accurate enough. Actually, this procedure corresponds to a simplest method of solving a system of differential equations that builds a Eulerian polygonal line (Fig. 5.33, *a*) rather than an integral curve (5.88). The accuracy of the nonlinear solution, i.e. a deviation of the Eulerian polygon is not controlled: the error is thought to be small due to a small step of integration.

The second option includes a check of the residual at each step and an iterational refinement of the next-step loading by allowing for the residual of the equilibrium equations. The iterations are performed with a constant value of the linearized stiffness matrix (Fig. 5.33, *b*), calculated at the beginning of the step.

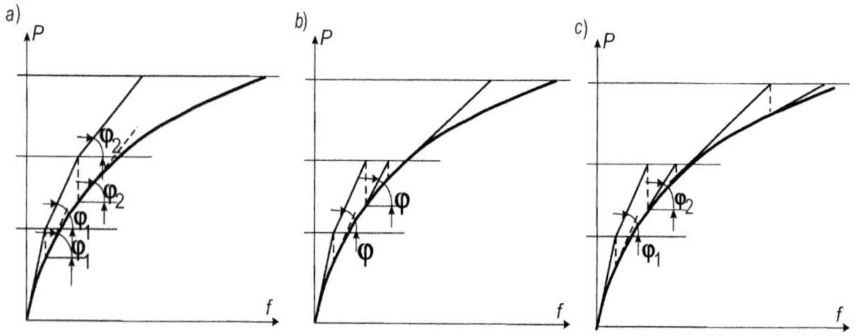

Fig. 5.33. A graphical illustration of the step-by-step procedure

Finally, the third option includes an iterational refinement of the solution at each step, also with a correction of the linearized stiffness matrix at each iteration (Fig. 5.33, *c*).

It should be noted that we can obtain an illustrative information regarding the loaded system's behavior by varying the method of introducing the t parameter. Assuming that this parameter is present as a factor at all terms in equations (5.85) which describe the external loads directly, we have that its monotone increasing from $t_0 = 0$ to $t^* = 1$ will correspond to a proportional growth of the loads, and the integral curve (5.88) will describe the behavior of the generalized coordinates in the course of a direct active loading. In conservative (nonlinearly elastic systems) this simplified idea of a load application must not affect the result — the very definition of a conservative system prescribes that its resulting state be independent of the sequence in which the loads are applied. Though, this theoretical statement does not allow for the fact that equations are solved numerically with an error, and this error is capable of accumulating, and its value appears dependent on the modeling of the loading sequence.

One may fancy also other ways of applying and alternating loads, more complicated and close to the reality. By varying those one can run a series of "mathematical experiments" regarding the analysis of a structure's behavior in different modes of loading.

For example, the **SCAD** implements such programme of investigations by providing an option to study the history of a combined load. This is performed by specifying a sequence of action of particular load groups (basic loadings). The beginning of action of a next load group coincides with the end of application of the previous group (a full level achieved), that is, the new basic loading is a continuation of the previous one. In particular, the use of a previously modeled basic loading is possible with a negative intensity factor thus allowing to investigate the full "loading/unloading" cycle. This technique enables one to estimate the accuracy of the analysis by a discrepancy Δf between the initial and final states of the system (Fig. 5.34).

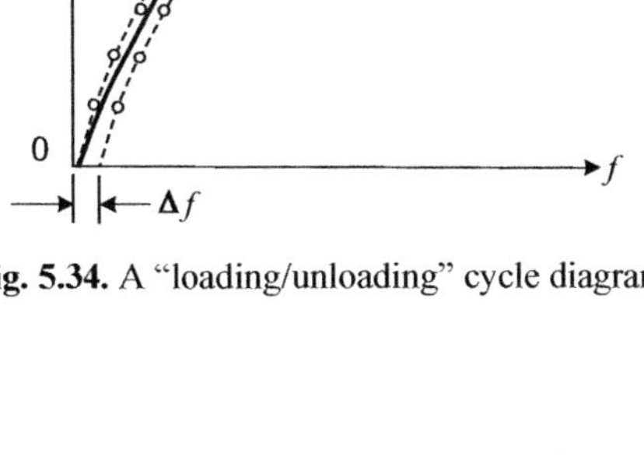

Fig. 5.34. A "loading/unloading" cycle diagram

The comparison of a systems behavior in the "loading/unloading" cycle will sometimes give one a chance to notice a peculiar loop in the diagram that reflects the dependence of a characteristic displacement on the intensity of the load. This loop may evidence a buckling in the system which has not been detected at the selected step of the load subdivision. Thereby the system may have transited from its stable equilibrium at the *ON* branch to a vicinity of a secondary stable equilibrium state at the *MA* branch, and all equilibrium quality control techniques (such as negative coefficients on the main diagonal of the stiffness matrix transformed by the direct Gaussian exclusion) thus failed.

A loop at the equilibrium state curve indicates the presence of a lower critical value (the *M* point) the determination of which is very often the goal of the unloading process investigation. In this regard there's a peculiar quotation (see [37], p. 830):

> "***Let's decrease*** *the unloading parameter λ and observe the behavior of solutions that determine secondary states. That value of the λ parameter at which the secondary states of the body cease being stable, we call a lower critical value*...».

In mathematics the nonlinear equation system solution technique described above is entitled a method of solution extension by parameter. It is treated as a method of building a set of solutions of nonlinear equations that contain a parameter. A simplest example of such a set is a curve in a multi-dimensional space where the coordinates are unknowns and the parameter. One of creators of the method, D.Davidenko, noticed that the solution extension parameter can be any of the unknowns, not necessarily the parameter itself. Paper [13] proves that the best computational properties are ensured by the extension parameter chosen as the length of the solution set curve.

It should be noted that the feasibility of multiple solutions, being a characteristic feature of nonlinear problems, and an uncritical attitude towards this phenomenon have been often the reasons for an utter denial of the finite element approach's applicability in problems with a strong nonlinearity [19].

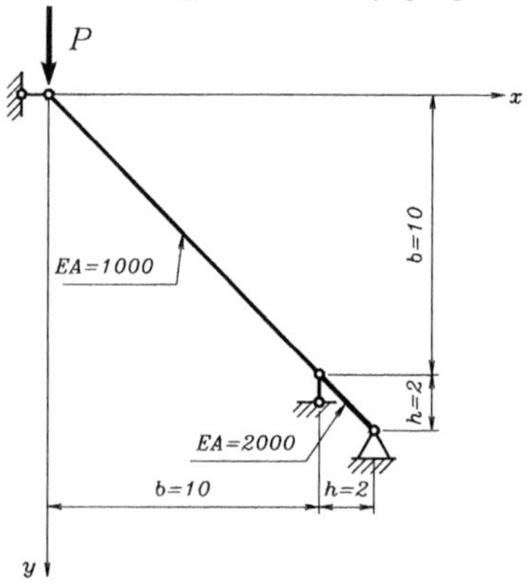

Fig. 5.35. An example of an analysis of a strongly nonlinear system

A detailed analysis of an "anti-example" (Fig. 5.35) from the paper [19] was performed in [7] using the **UAI/NASTRAN** software. It was noted there that this nonlinear problem could not be solved with standard settings of the software: those proved to be too "timid", and as soon as a slightest trace of the lack of convergence was detected the computation was terminated.

Thus one has to set up an algorithm that would provide a convergent solution in all points of the integral curve. Here's a list of principal settings to define the course of the iterational solution of this problem:

- method of solution — the arc length method;
- method of iteration (to be selected automatically: depending on the convergence rate, either the σ-ε curve tangent method or the unbalanced load iteration method is chosen);
- the increment of the load (to be selected automatically at each iteration depending on the solution's convergence);
- convergence criteria (by error/residual of the energy, load, or displacements);
- results of the finite element analysis (Fig. 5.36).

A peculiarity of the step-by-step procedure and its attractive difference from other nonlinear solution techniques, is that a system linearized in the vicinity of a loading can be analyzed by conventional methods of linear structural mechanics. It makes sense to estimate the stability coefficient (apparently, this estimate will describe a possibility of the load intensity to grow from the achieved level) or natural frequencies/modes of this linearized system.

The said peculiarities of the step-by-step procedure characterize it as a tool for analysis of the structure's properties, not just the way to obtain a solution. In some cases these require the engineer to use a pretty fine division of the load into steps. Generally, especially for geometrically nonlinear problems where the behavior of a system is "smooth" enough, it is more reasonable to use a combination of pretty large loading steps with iterations at each of such steps.

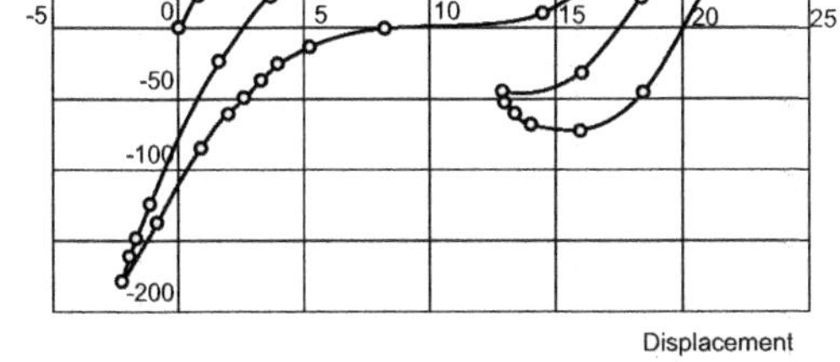

Fig. 5.36. A solution of the problem in Fig. 5.35

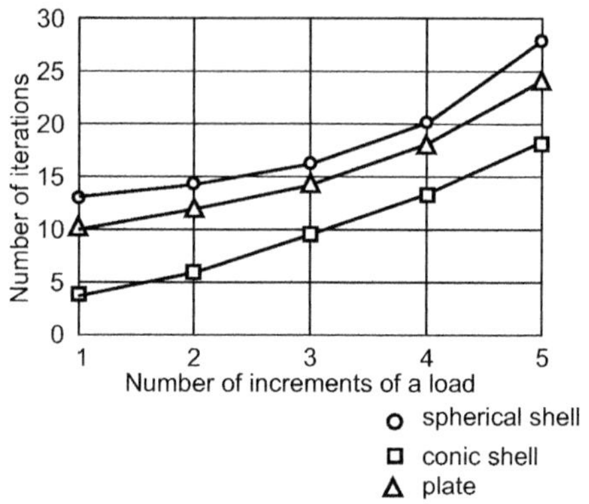

Fig. 5.37. A dependence of the number of iterations on the step length

Fig. 5.37 shows a dependence of the total number of iterations on the length of steps into which the load is divided. The dependence has been obtained from solutions of geometrically nonlinear problems [25]. A fine division of the load may prove reasonable in physically nonlinear problems where rigidity properties of finite elements can change drastically when entering the plastic phase of deformation, and also in systems having unilateral constraints where similar drastic changes are related to the activation/deactivation of the constraints.

It should be noted that the investigation of the behavior of a nonlinear system may require the previous design model to be revised. In particular, the finite element mesh may need to be made finer. For example, when analyzing some shells [30] one found that the growth of the load (especially in the post-critical phase) caused peculiar regions of the shell to appear where the deformed shape of the median surface is not just smooth but has an intensively variable profile. A good approximation of such a shape may require more than a conventional scheme of finite element meshing commonly used in the linear analysis. The experience of practical calculations shows that trustworthy results, including those of an ultimate load design, can be obtained only with 3-4 or more finite elements fitting in a half-wave of a bending shape (or a buckling shape).

References

1. Adini A, Clough RW (1960) Analysis of Plate Bending by the Finite Element Method: (Report submitted to the National Science Foundation, Grant G7337)
2. Aleksandrov AV, Laschenikov BY, Shaposhnikov NN, Smirnov VA (1976) Methods for analysis of bar systems, plates and shells using computers. Part 1.(in Russian). Stroyizdat, Moscow
3. Alekseev SA (1977) Quality of a statically determinate fixation of nodes and solids (in Russian) In: Alekseev SA, Novozhilov VV (eds) Analysis of spatial structures, vol. XVII.— Stroyzdat, Moskow, pp.131–140.
4. Alger J (1998) C++ for real programmers. Academic press, Boston

5. Allman DJ (1984) A compatible triangular element including vertex rotations for plane elasticity analysis. Computer and Structures, 12:1-8
6. Bereš E (1980) High accuracy interpolation of stresses. Acta techn.Acad. sci. hung, Vol.91, 4:257-263
7. Danilin AN, Zuyev NN, Snegovsky DV, Shalashilin VI (2000) On a use of finite element method to solve geometrically nonlinear problems (in Russian). In: CAD and graphics, 2:60–64
8. Descloux J (1973) Méthode des éléments finis (in France). Ecole polytechnique Federale de Lausanne, Lausanne, Suisse
9. Fan Yuan-xun, Wang Szing (1979) On the application of an interpolation matrix for computation of stresses in finite elements. In: Numer. Meth. Geomech.: Proc. 3rd Int. Conf., Aachen, vol. 4, Rotterdam, pp. 1273–1280,
10. Feodosi'ev VI (1969) Ten lectures-conversations on strength of materials (in Russian). Nauka Publishing House, Moskow
11. Forsythe GE, Moler CB (1967) Computer solution of linear algebraic systems. Prentice-Hill Inc., Englwood Cliffs, N.J.
12. Gorodetsky AS, Perelmuter AV, Slivker VI (1989) An intelligent computational analysis software. A prognosis of new capabilities (in Russian). In: Computer-aided design systems for structural engineering. Issue 6. Budivelnyk, Kiev, pp. 43–56.
13. Grigolyuk EI, Shalashilin VI (1991) Problems of Nonlinear Deformation. Dordrecht et al. Kluwer
14. Holub AI (1995) Enough Rope to Shoot Yourself in the Foot. Rules for C and C++ McGraw-Hill
15. Irons BM (1975) The superpatch theorem and other proposition relating to the patch tests. In: Proceedings of the 5th Canadian Congress of Applied Mechanics, Fredericton, pp. 651–652.
16. Janelidze GY, Panovko YG (1950) The Saint-Venant principle and its application in the plate and shell theory (in Russian). In: Umansky AA (ed) Analysis of spatial structures. Issue 1. Stroyzdat, Moskow, pp. 329-342.
17. Loubignac G, Cantin G, Touzot G (1977) Continuous Stress Fields in Finite Element Analysis. AIAA Journal, vol.15, 11:1645–1647
18. Lukasiewicz S (1979) Local loads in plates and shells. Noordhoff international publishing, Leyden
19. Nazarov D (2000) A review of modern finite element analysis programs (in Russian). In: CAD and graphics, 2:52–55
20. Oden JT (1972) Finite elements in nonlinear continua. McGraw-Hill, New York.
21. Oden JT, Reddy JN (1973) Note on an approximate method for computing consistent conjugate stresses in elastic finite elements. Int. J. Numer. Meth. Eng, 1:55–61
22. Panovko YG (1965) Mechanics of deformable solids: Modern concepts, mistakes and paradoxes (in Russian). Nauka Publishing House, Moskow
23. Papkovich PF (1979) Theory of elasticity (in Russian). Oborongiz, Leningrad
24. Petrov VV (1959) On the analysis of shallow shells with finite deflections (in Russian). In: Proceedings of high schools. Structural engineering, 1:15-18
25. Sakharov AS, Altenbakh I (1982). Finite element method in mechanics of solids (in Russian). Vyscha shkola, Kiev
26. Schiermeier JE, Housner JM, Ranson JB, Aminpour MA, Stroud WJ (1996) The application of interface element to dissimilar meshes in global/local analysis. In:MSC 1996 World Users' Conference Proceedings

27. Segerlind LJ (1984) Applied finite element analysis, 2nd ed. Wiley, New York
28. Semionov VA, Semionov PY (1998) Higher-accuracy finite elements for analysis of complex spatial systems (in Russian). Proc. XVI Intl. Conf. "Mathematical modeling in solid mechanics. Finite element and boundary element methods", June 23-26, 1998. Vol.1. St. Petersburg, pp. 68-69
29. Slivker VI (1982) Ritz' method in problems of elasticity, based on a successive minimization of two functionals (in Russian). Izvestiya Akademii Nauk SSSR, Mekhanica Tverdogo Tela, Vol.17, 2:57-65. Translation: Mechanics of Solids (1982), vol.17, 2:47-53.
30. Srinivasan RS, Bobby W (1976) Buckling and postbuckling behavior of shallow shells. AIAA Journal, vol.14, 3:289-290
31. Strang G (1976) Linear algebra and its applications. Academic press, New York San Francisco London
32. Strang G, Fix GJ (1973) An analysis of the finite element method. Prentice-Hall Inc., Englewood Cliffs, N.J.
33. Stricklin JA, Haisler WE, Von Riesemann WA (1973) Evaluation of solution procedures for material and/or geometrically nonlinear structural analysis. AIAA Journal, vol.11, 3:292-299
34. Svojsky FM (1986) Inconsistent finite elements for analysis of plate spatial structures (in Russian). In: Application problems of strength and plasticity. Publishing house of Gorky University,Gorky, pp. 59-70
35. Timoshenko SP, Voinovsky-Kriger C (1959) Theory of plates and shells, McGraw-Hill, New York Toronto London
36. Turner MJ, Clough RW, Martin HC, Topp LJ (1956) Stiffness and Deflection Analysis of Complex Structures. Journal of the Aeronautical Sciences, 9:805–823
37. Volmir AS (1963) Stability of elastic systems (in Russian). Nauka Publishing House, Moskow
38. Vovkushevsky AV (1979) On computing of stress in finite element analysis of elasticity problems (in Russian). Bulletin of *VNIIG*, Vol.133. Energhiya, Leningrad, pp.18-22.
39. Young LC (1969) Lectures on the calculus of variations and optimal control theory. W.B. saunders Company, Philadelphia, London, Toronto
40. Zienkiewicz O (1977) The finite element method in engineering science. McGraw-Hill, London.

6 Estimating and Interpreting Results

"... at a general assembly of members of the British Royal Society of Naval Engineers, when a report regarding newly built Majestic naval ships was discussed, presented by Sir W. White who was a chief naval architect at the time, Admiral Lord Charles Beresford said: "We will drown on those ships, and Sir W. White will then explain why we have done so".

A.N. Krylov

6.1 An Estimation Issue

A complex structure can be designed either by a strict calculation or by good fortune, or by implementing a traditional project. When manual methods of design were common, the strict calculation used to be hardly possible. The good fortune is not a thing to rely upon, it seems. So one used to adopt traditional approaches the reliability of which had been tried by a long lasting experience. But now the situation is quite different.

Though presently available computational capabilities are huge, it does not mean every analysis performed by computers using a specialized software is good enough. The behavior of a complex structure is complex: this trivial idea is not always realized clearly, hence a variety of errors. It appears that a good analysis requires more than a high-end computer program available: it requires a highly qualified analyst too. The employment of a complex software system with an insufficient understanding of its theoretical basics and methods of solution, nor with an elaborate informal estimation of results, often leads to just an illusion of a firm analytic foundation under the adopted design decisions. In such case it is a really good fortune that one needs to avoid a big trouble.

Apparently, at the present time the following issues can be treated as obstacles to a good analysis of a complicated responsible structure:

- an insufficient understanding of principal ideas of a performed computer analysis by an analyst, an insufficient level of expertise of a user;
- an incomplete information regarding features of the software employed (such as properties of the finite elements involved in the analysis);
- limitations to the time of computation;

- a lack of demonstrativeness of results obtained, when critical data may be masked by a multitude of unimportant particulars.

The issue of the user's expertise was addressed for many a time in specialized discussions. Sometimes even an idea of "*user certification*" arose, though no good solution has been found yet. The absolute number of expert users grows gradually, but their relative number remains comparatively small, as it seems. The only hope is that the problem can be helped by introducing special techniques into the contemporary software to facilitate the analysis thus making it better performable by moderate-skill users too. Some of the techniques are discussed below.

At an international conference on stability/buckling held in 2001 and dedicated to the 70th anniversary of Professor Johann Arboč (Director of W. Coiter Institute in Delft) it was said that by the present time a new engineering specialty had appeared and started to develop intensively: a **reviewer of computational analysis results** which is in great demand among structural design institutions[1]. The demand of an opinion by an expert who receives ready-made analysis results and produces an interpretation indicates another aspect of the problem: it is not just a backward translation from a finite-element model into the language of a traditional "continuous" description, it is also an expert review of differences between the two models which (every)one needs.

6.2 What Analysis Results are Needed

The conventional set of results to be found (displacements, forces, stress fields) has become so traditional that it is never subject to any discussion. But this piece of information is by no means enough for one to digest all the features of a solution obtained. Also, methods of representation of such results are often too bulky for a clear understanding of the peculiarities of a problem solved.

Results of static/dynamic analysis of a complex system represented in numbers are huge arrays of data practically incomprehensible by a human analyst. The capability of printing only selected results available in most CAD environments is of little help too because the analyst does not always know what results are going to be critical.

A graphical representation of the results as plots and isofields is much more illustrative: it compacts the volume of information and makes it visual rather than textual. By the way, most software systems are capable of displaying graphical isofields only for chosen components of the stress and strain distribution in a structure. But other graphical representations were known long before the modern computational facilities appeared. Those even got their special names such as *isostates*, *isochromes*, *isoclines*, and *isopachics*.

We should remind that isostates are lines, in their every point tangential to one of the principal stresses in the same point of space. Isochromes are lines of equal maximum tangential stresses. Isoclines are curves across points of the area where

[1] This refers to a report by Professor V.A. Zarutsky

the directions of the principal stresses are the same. Finally, an isopachic is a curve in every point of which the sum of the principal stresses is the same.

Ideally, a software should be able to display isofields (and/or isolines) of those stress/strain magnitudes the functional dependence of which on components of the strain/stress tensors is formulated by the user at the latter's will.

But it might not be enough for a good analysis too, because the graphical information for a whole structure might prove insufficiently legible (Fig. 6.1). Its fragmentation restores the legibility but causes a new problem: how to find a fragment in which the calculated values of the user's interest are located. This problem cannot be solved in a trivial way for a structure of any noticeable complexity.

The way out of the situation can be a technique employed by the **SCAD** software and based on a color indication control. There is a capability of the color indication of an isofield's parts that belong to a certain range of values while all the other levels are turned off (Fig.6.2). The location of "critical" values obtained by the analysis thus becomes obvious. An even more accurate localization of such critical places is provided by a so-called *magnitude map* because it operates a "transparent" structure (Fig. 6.3).

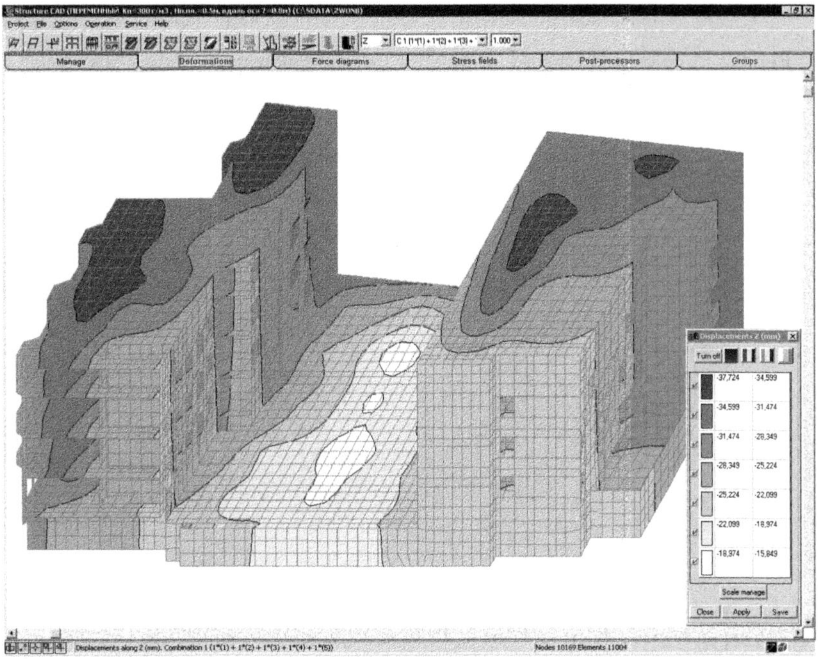

Fig. 6.1. A color graphical representation of information (bending moment fields) — a complete picture of the data

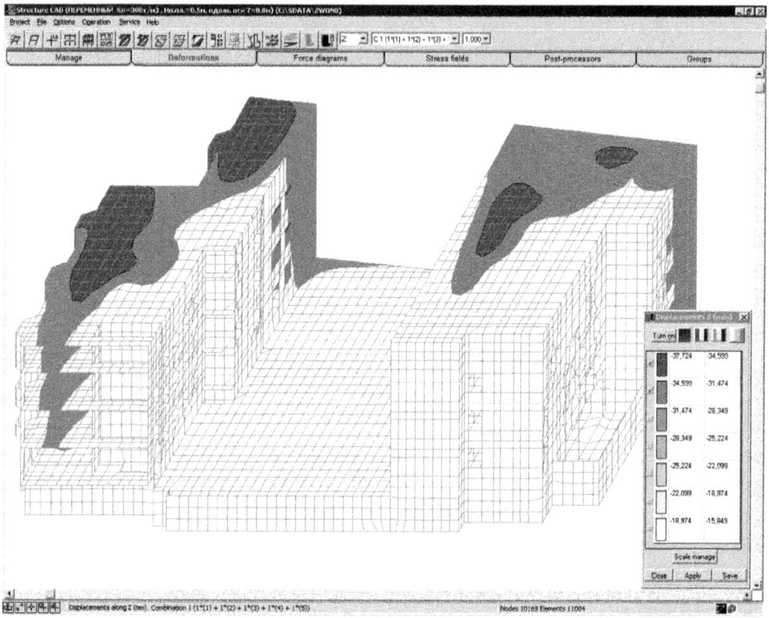

Fig. 6.2. A color indication of information (bending moment fields) — a selection of top levels;

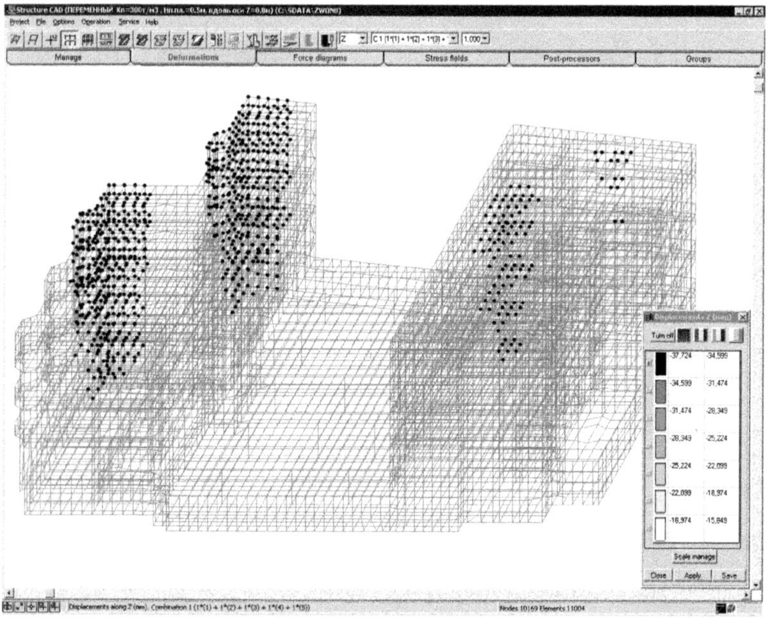

Fig. 6.3. A color indication of information (bending moment fields): viewing a top-level localization map

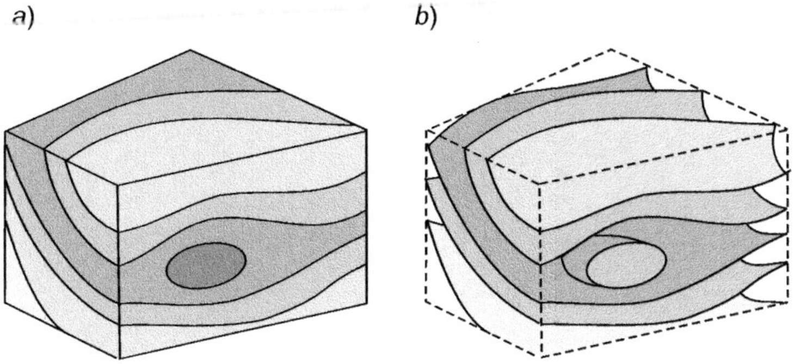

Fig. 6.4. An example of iso-surfaces

It should be noted that the isofield representation is well-illustrative only for two-dimensional objects. In analyses where spatial bodies are present, only isofields on external boundary surfaces are visible (Fig. 6.4, *a*), and isofields on user-specified cross-sections can be built also. Though, the procedure of constructing a cross-section is pretty toilsome, and what is worse, does not give a clear idea what happens in the whole structure generally. In this regard, a capability of displaying a system of iso-surfaces available in some software environments (such as **ANSYS**) is noteworthy (Fig. 6.4, *b*).

An ineffectual visualization of analysis results may render useless any advantage achieved by the analysis itself. Developers of software tools usually address this aspect carefully, providing the user with a number of visualization options (Fig.6.5). Together with isofields (Fig. 6.1, 6.2) other representations of the analysis results (displacements, forces, stresses) can be available, such as so-called maps wherein the user can paint finite elements or nodes with colors corresponding to certain ranges of values of his choice, or put special color markers on finite elements (Fig.6.3). There is a pretty useful "probe" function available in some software environments that enables one to have isofields digitized in any points selected by the mouse pointer (Fig.6.5, *d*).

It is convenient when the color scale of isofields or maps can be controlled by the user as the **SCAD** system suggests: in this case one can refine away unimportant result ranges by choosing not to color them. Also, the user can control the number and size of intervals between the levels on his own which improves its capability of reviewing the results because this implements a command like "*Display an area where the X result (stress, displacements) ranges between* X_1 *and* X_2".

In spite of all conveniences of the graphical result representation, numerical magnitudes are also extremely important, not only for post-processing purposes but also to have a clearer idea of the data represented graphically. One of most advantageous choices, to our opinion, is a capability of presenting the analysis data both graphically and numerically at the same time which is provided by some software tools.

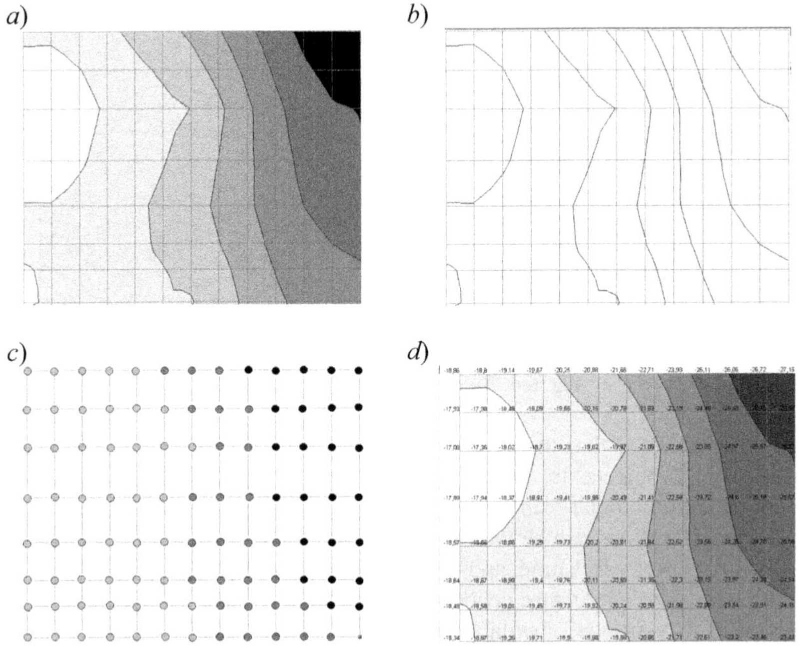

Fig. 6.5. Examples of various methods of the resulting information representation in the **SCAD** environment: (*a*) an isofield; (*b*) isolines; (*c*) a map; (d) probed points digitized

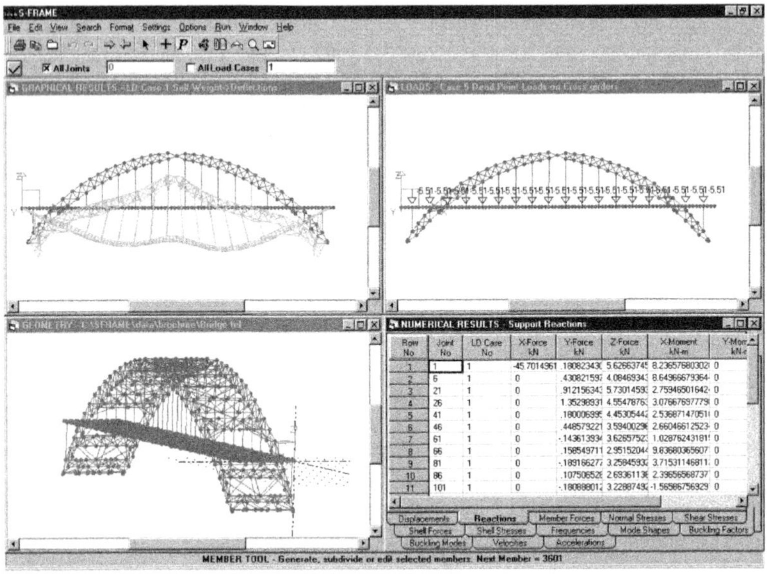

Fig. 6.6. A simultaneous representation of results in different forms

An example from the **S-FRAME** environment is presented in Fig. 6.6 where the computer screen displays graphical views of a model, a scheme of load application, the deformed model, a table of numerical data obtained by the analysis, all at the same time.

Though, the capability of finding and documenting needed results still does not resolve the issue of estimating those.

It is a kind of tradition to represent results of an analysis (displacements, stresses etc.) as "final", forgetting all the errors present in the source data and a possible inaccuracy of the computation itself. It is important to remind that a solution is approximate, especially in nonlinear problems where accurate solutions are hardly available. But a user should be introduced to this information effectually, otherwise a confusion may arise. A curious example of this is a reaction of users to values of a residual presented in results of a nonlinear analysis for equilibrium equations corresponding to the obtained approximate solution. This data that appeared in a newer release of the **SCAD** software (it was not displayed in preceding versions) has caused complaints because some users think that version worked accurately and this one works with errors.

The computational error is accumulated also when solving linear problems of a high dimensionality. Therefore some software systems provide an estimation of the stiffness matrix decomposition accuracy. This estimation operates a particular stiffness matrix of a structure in question, and is performed by solving a special auxiliary problem that has a known accurate solution. One of such auxiliary problems is the case when all linear displacements are equal to one and all nodal slopes are zero (*a Gorbovets test*). Also, an inverse version of this test might prove useful: consider only nonzero slopes. The right part column for this problem is constructed quite easily by summing respective elements of the stiffness matrix, and the solution is performed by the direct/reverse pass through this column. The maximum deviation from the accurate solution estimates the error of computation and provides a basis for judging about the correctness of the design model. Unfortunately, poorly qualified users do not pay much attention to such estimate.

The tradition of an "accurate" analysis result representation contradicts the stochastic nature of source data stated above. Random source data lead to a random computational result, quite expectedly. Therefore an estimation of the computational error should be supplemented by an estimation of a possible dispersion of results due to an inaccuracy of the source data. This information is a good addition to the traditional set of results. Also, the authors' opinion is that the results should be supplemented also by an estimation of the solution's sensitivity to variations of principal parameters. This issue is addressed further below.

The problem of documenting analysis results is immensely complicated and is not yet solved adequately. Of course, any advanced computational software can provide its users with hard copies of any subset of the analysis results, including the capability of printing the whole set of data available. It is a tradition that such listings supplemented by printouts of source data are transferred to archives of a design institution and stored there as original design documentation. But any software lives its own life, it is corrected and updated, errors are detected and amended, new versions are released etc.

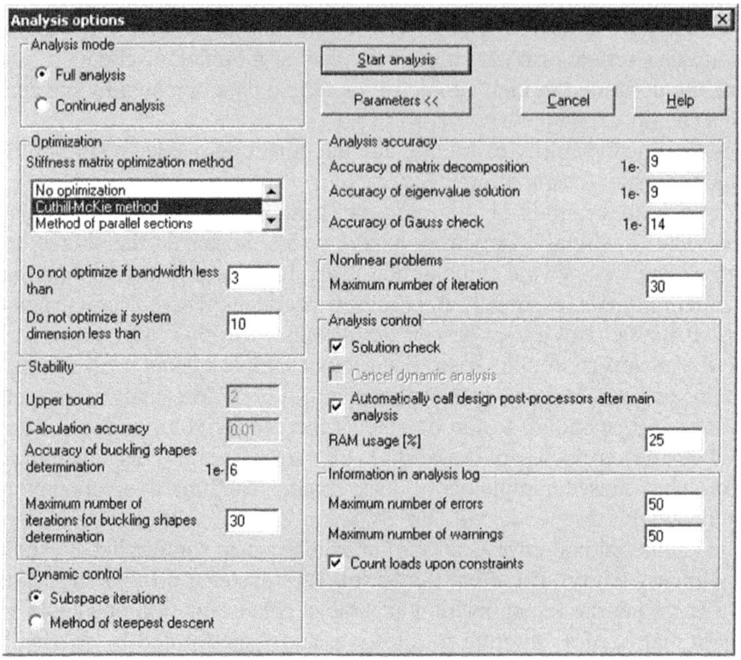

Fig. 6.7. SCAD system settings

Therefore there is no assurance that absolutely identical results would be obtained if a necessity arose to check the previous analysis results. This kind of a legal dead-end can be amended by storing a *really full set of information* on magnetic or optical data carriers, including the software release used, settings selected, sets of source data and results in internal formats. Also, it is necessary to store all the information of the settings that were used in the course of the solution, in order to be able to repeat the analysis and obtain perfectly identical results. This capability is provided by the **SCAD** software, for example, where settings (Fig. 6.7) are logged in a source data file during the analysis.

Though, not every software environment provides such a capability, and this is a drawback as far as the authors may judge.

6.3 General Validation

The book [1] states: "*The final type of the check performed after an engineering computation is a check of the* ***physical sense*** *of the result. There are good reasons not to confuse the physical sense with a* ***common sense*** *or a public opinion... Not anything said or done in the name of the common sense has been really sensible. It was the common sense that made people fagot witches, persecute Galileo, and* *mock at the steam engine.*" The same book also indicates the following obvious points of the physical sense check:

- whether the numerical results are practically feasible;
- whether the results are contradictory from the standpoint of their dimensions;
- an ultimate case check;
- a trend check (tendencies of variation);
- a law check;
- a completeness check (whether all essential factors have been taken into account).

This general list of checks may be extended or abridged, but one thing is clear: the general validation is a pretty important phase of the design work. The user can be told about such checks but cannot be forced to perform those. It reminds us of a story about a horse and water: the horse can be led to water but cannot be forced to drink. Therefore the presentment below is just an attempt to lead the horse to the water.

After obtaining analysis results, their common estimation is performed as a rule. Primarily, one should check most apparent expected properties of the solution (such as a symmetry of displacements in a symmetric structure under a symmetric load). One should not grudge one's effort for the testing of this kind. The whole experience of the computer-aided analysis evidences a stream of source data errors that never ends. Therefore one's confidence that the source data are free from errors costs very much. Sometimes special "trial" patterns of loading are created intentionally to yield predictable peculiarities of solutions.

It is natural that the estimation of results begins with that of messages generated by a system during the computation. These messages are recorded in an analysis log provided by most modern software tools to reflect both the sequence of computations performed and trouble that might have arisen on the way.

A typical sample log is given in Table 6.1. The following particulars should be emphasized:

- Messages are given regarding a detected geometrical changeability of the structure, and additional constraints are defined which the **SCAD** system has installed on its own to parry this changeability. The situation needs to be analyzed and the actions of the system estimated. In particular, it might be reasonable to eliminate the kinematical instability in another way more fitting for the problem in question, and solve the problem once more;
- The generated message of total external loads in 1st and 2nd patterns of loading is convenient to validate the source data specification because an analyst usually knows at least approximate estimates of these quantities;
- The log may include an information regarding negative results of the solution accuracy check (Table 6.1 does not present such) with an indication of a node where this check failed. If there is no serious singularity in the indicated node, one may try to increase the specified accuracy of the matrix decomposition or, on the contrary, reduce the solution check accuracy (the "Accuracy of Gauss check" setting).

Table 6.1.

```
ANALYSIS PROGRESS LOG
Wed Dec 22 07:55:17 1999
  Full analysis.
    File – "C:\SDATA\08.SPR",
07:55:18 — Inputting source data of the principal scheme
07:55:20 — Ordering the stiffness matrix of the principal scheme.
07:55:21 — Ordering the matrix with the Cuthill-McKie inverse algorithm.
( old profile 1.20416e+07, new profile 1.7203e+06).
07:55:21 — Renumbering degrees of freedom
07:55:22 — Forming the stiffness matrix of the principal scheme
07:55:31 — Decomposing the stiffness matrix of the principal scheme
07:56:15 — Geometrically changeable system. NODE 170
CONSTRAINT TYPE 2
07:56:15 — Geometrically changeable system. NODE 170
CONSTRAINT TYPE 1
07:56:15 — Accumulating loads of the principal scheme.
07:56:15 — Total external loads
              X    Y    Z    UX   UY   UZ
       1-     0    0    100  0    0    0
       2-     0    20   100  0    0    0
07:56:15 — Calculating displacements in the principal scheme.
07:56:19 — Work of external forces
       1 -  0.0204431
       2 -  0.0168025
07:56:19 — Checking the solution of the principal scheme.
07:56:21 — Calculating stresses in the principal scheme.
07:56:24
                    JOB COMPLETED
                   Time spent : 1.12 min.
```

The next stage of the general validation of the analysis results can be a review of the general deformation of the structure. Its visualization gives an idea of a good deal of peculiar features of the structure's behavior. Some of those can be predicted by the analyst (such as an expected symmetry) while others might draw his attention by their unexpectedness.

Anyway, such blunders as fixations incorrectly set or abrupt changes in rigidity properties (especially in a part of a structure) can be often detected by this kind of informal reviewing. A typical example is a situation when there are two nodes in the same point of the space (see point C in Fig. 6.8, *a*), and adjoining elements are attached to different nodes. In the original scheme this mistake may well be missed (though the **SCAD** software, for example, suggests a "*Show coincident nodes*" function for this purpose), but it will be immediately detected when browsing the deformed shape of the structure (Fig. 6.8, *b*).

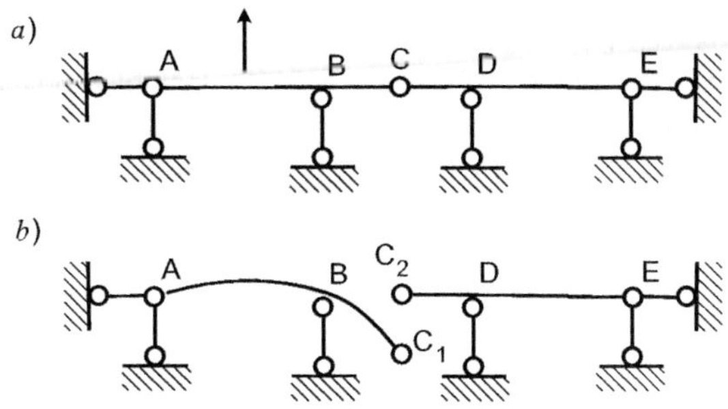

Fig. 6.8. A mistake of coincident nodes: (*a)* original geometry; (*b*) deformed shape

A similar response to spatially coincident nodes is implemented in the **GTSTRUDL** environment. This software generates a warning automatically (independently of the user's will) that includes a list of couples of nodes located at a smaller distance from each other than a specific small value. The user is suggested to make a decision whether to ignore the warning, to change the node vicinity criterion, or to alter the design somehow. The same response should be generated by software to a multiple connection of the same set of nodes by different finite elements. Sometimes a user places such parallel finite elements in the source data intentionally (the authors disapprove of the prohibition of models having coincident finite elements practiced by some software developments), but the same situation may be caused by a mistake in the source data. Therefore a response of a CAD program by means of a warning seems the most flexible choice.

Generally, searching for unexpected effects can often yield pretty useful results. Such a search enables one both to detect bugs in the source data and to critically treat the design model in question by refining and updating it. No general rules can be given for this process: things depend on the experience and intuition of the analyst. Though, the only general rule that is a must for an engineer is the readiness to perform this search every time in his design activities.

It is important to remember that there is no "inessential deviations" in the unexpectedness analysis: the fact of an unexplainable phenomenon itself is more important than its magnitude. As a physicist noticed nimbly, "*...if the clock is striking a thirteenth hour, it is not just the thirteenth strike that one counts wrong. It reflects upon the preceding twelve strikes being right, too*" [2].

6.4 Characteristic Displacement

One of most popular methods of analyzing the behavior of structures (especially nonlinear ones) involves a construction of a dependence between a certain components of the stress/strain distribution and the load's intensity. Particularly, it can be a curve of equilibrium states that establishes a relationship between a load intensity factor and some displacements that define the deformed shape of the structure. For a system having a single degree of freedom, this curve enables one to study the quality of the equilibrium because stable states of the system correspond to an ascending branch of the curve while unstable ones are described by a descending branch. It should be emphasized that this criterion is applicable only to systems with one degree of freedom. In more complex systems the "load–displacement" relation can be quite whimsical even in the region of stable equilibrium states. A typical example is shown in Fig. 6.9 where a "normal pressure–deflection" diagram is presented for four points of a cylindrical panel clamped along its contour. The source data follow:

- span $2c = 20$ cm;
- element $2l = 20$ cm;
- rise $f = 5$ cm;
- thickness $h = 0.1$ cm;
- elasticity modulus $E = 2.1\times10^6$ кг/ cm^2;
- Poisson ratio $\nu = 0.3$.

Plots of Fig. 6.9 show that even in the pre-critical state the deflections of points 1 and 4 do not change monotonously, unlike those of points 2 and 3.

There are a plethora of examples of characteristic displacements chosen to build equilibrium diagrams, and most often this choice has no logical ground under it. Sometimes one involves "intuitively obvious" considerations (such as a deflection in the center of a plate), sometimes one uses displacements that reflect a certain expected result most distinctly (for example, to illustrate a possible equilibrium bifurcation one considers the lateral deflection of a compressed cantilever's end). Certainly the analyst has the right to deal with any kinematical parameter of his choice, but it is better to use such parameters for which general properties of a "*displacement parameter*" vs. "*load parameter*" dependence curve have been strictly established. The load parameter should be, not necessarily but most often, a quantity that characterizes the magnitude of a load. A more sophisticated method of introducing such load parameter is the position of a load on a structure, or even the position of the structure itself in a fixed coordinate basis. An example of the latter load parameters can be encountered in problems related to a bridge span launching.

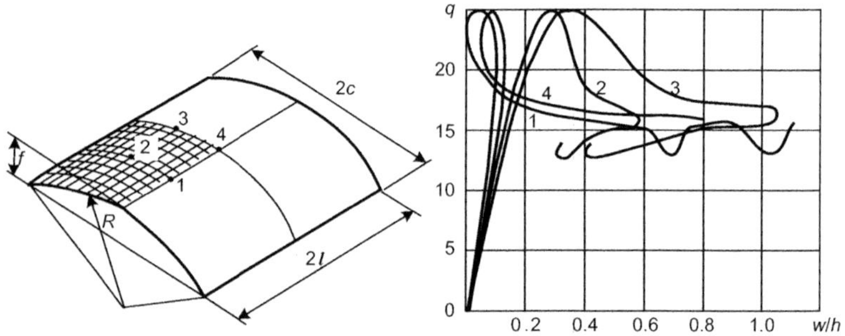

Fig. 6.9. Equilibrium diagrams

Let the applied load $P(\lambda) = P_0 + P_1(\lambda)$ be characterized by a load intensity parameter λ the value of which can be treated as a generalized force working upon the system. Then, in order to build a *characteristic curve* $\lambda = \lambda(q)$, one is recommended to employ a characteristic displacement q, i.e. a generalized displacement conjugate to the λ generalized force by energy in the sense that the product $\lambda\, dq$ yields an increment of the work of the generalized force λ at increments of the displacement dq. In this case, as proved in [3], the characteristic curve will possess the following important properties for a system loaded by conservative forces:

- at stable branches of equilibrium trajectories the inequality holds

$$dq \,/\, d\lambda \geq 0; \tag{6.1}$$

- when an arbitrary linear constraint is imposed on the system, the absolute value of the characteristic displacement cannot increase.

It is easy to give examples that show a violation of these properties if an arbitrary (non-characteristic) displacement is involved. For example, in a bar supported by a nonlinearly elastic bearing with its reaction proportional to the cube of the displacement (Fig. 6.10, *a*) the dependence between the slope angle φ and the force factor λ is such that only at $0 \leq \lambda \leq 7.483$ the derivative $d\varphi/d\lambda \geq 0$ (Fig. 6.10, *b*) though the equilibrium of the structure is stable at any λ.

Also, an arbitrary displacement may lack the second property immanent to characteristic displacements. A simplest example of such a non-characteristic displacement employed is a slope of the middle cross-section of a simply supported beam which is zero under a uniformly distributed load of any intensity. But once a constraint is added to one of the supports to prohibit its end slopes, this non-characteristic displacement immediately becomes nonzero, i.e. its absolute value grows.

The violation of the said condition explains also the fact that sometimes the deflection calculated in the center of a square plate under a uniformly distributed load proves to be *somewhat greater* than its accurate value (see [7] for examples) though the employed slope/deflection finite element technique does not permit a realization of every arbitrary deformed shapes thus imposing certain constraints upon the system.

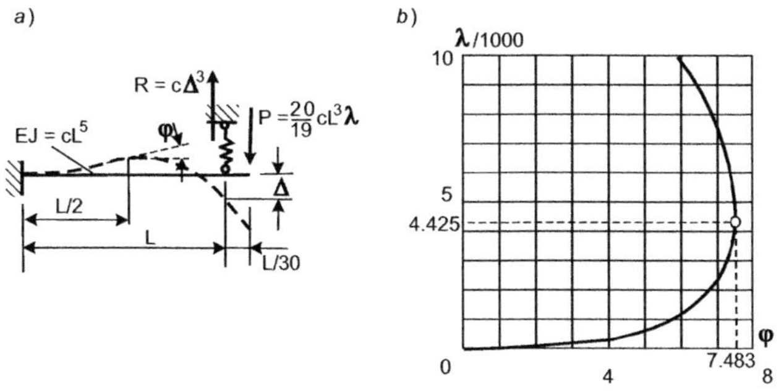

Fig. 6.10. A dependence between a slope and a load intensity

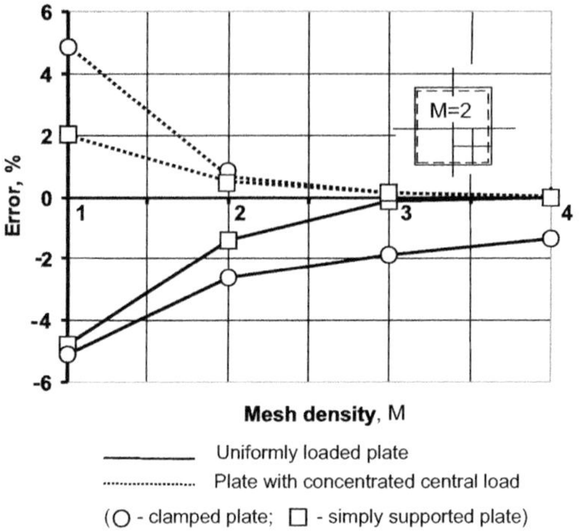

Fig. 6.11. Errors of a deflection's calculation at the center of a square plate

In the present case the characteristic displacement is the volume of a body bounded by the deflected and original surfaces, rather than the deflection at the center. This is the way how the effect of a system's "stiffening" in the finite element analysis may be blurred.

Under a concentrated force applied at the central point of the plate the deflection of the plate at this point is a characteristic displacement, and expectedly we have an approximation from below (Fig. 6.11).

It is noteworthy that the finite element method ensures an approximation from below for a characteristic displacement, as well as for other integral properties (such as a natural frequency), only when consistent elements are used. Inconsistent elements do not warrant the sign of the error.

6.5 Calculating Energy of Deformation

Earlier we have mentioned a possibility of employing a procedure of a finite element solution's internal convergence check by performing a series of computations with a sequence of ever denser meshes. The solution is to be found with a different level of detailing of the model (using globally nested meshes or those condensed in regions of high stress gradients, or other way).

As it is known, the generalized Pythagorean theorem [5] defines the square of the solution error's energetic norm as a difference of squares of the energetic norm of the accurate solution $\mathbf{u}^*$ of the original continual problem and the $\mathbf{u}^h$ solution of the discrete problem

$$\|\mathbf{u}^* - \mathbf{u}^h\|^2 = \|\mathbf{u}^*\|^2 - \|\mathbf{u}^h\|^2 , \tag{6.2}$$

so that $\|\mathbf{u}\| = \sqrt{\mathrm{E}(\mathbf{u})}$, where $\mathrm{E}(\mathbf{u})$ is a deformation energy accumulated by the mechanical system in its elastic displacements described by the vector **u**, and the superscript h characterizes a dimensionless value of the maximum finite element diameter in the current mesh.

If we have a sequence of solutions obtained with successively detailed meshes $h_1, h_2, \ldots, h_k$, then the relation (6.2) both yields an information of the internal convergence of the process as the mesh condenses and enables us to predict an effect taking place in this sequence as the finite element model is being further detailed. For this purpose one needs to know the system's deformation energy at its finite element solution, which can be calculated efficiently without taking additional resources of memory by means of an algorithm proposed in [4].

The calculation of the deformation energy is an option suggested by many software tools. Some of them are capable of building dependencies between variable parameters of a finite element model and the deformation energy obtained respectively. Thus the **StressCheck** system [6] employing a p-technique of the finite element analysis (see Section 3.2) can generate a plot of the deformation energy at a constant finite element mesh vs. the degree of the approximant polynomial varying from 1 to 8. The energy norm of the solution's error is estimated by the following inequality:

$$\left\| \mathbf{u}^* - \mathbf{u}_N \right\|^2 \leq \frac{C}{N^{2\beta}}, \tag{6.3}$$

where $\mathbf{u}_N$ is a finite element solution for a system possessing N degrees of freedom. The value of N changes with the change of the exponent p, and the constant C does not depend on N. The exponent β that determines the convergence rate is estimated by comparing respective solutions at different p.

Table 6.2 and Fig. 6.12 represent characteristic results of these kinds of estimations generated by the **StressCheck** software when solving test problems.

Table 6.2.

p	Number of degrees of freedom	Energy of deformation	Exponent β	Error estimate, %
1	65	1.2622	0.00	70.55
2	171	2.2141	0.74	34.49
3	283	2.4646	1.81	13.88
4	433	2.4991	1.48	7.41
5	621	2.5091	1.77	3.92
6	847	2.5121	2.37	1.88
7	1111	2.5127	2.49	0.96
8	1413	2.5129	2.49	0.53

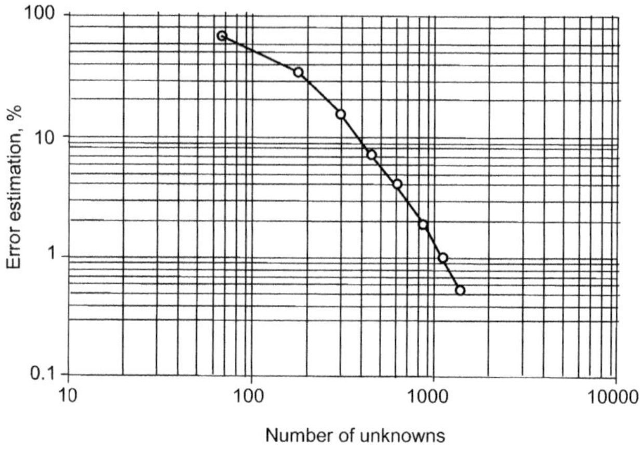

Fig. 6.12. An error estimate plot

6.6 Further Processing of Results

The determination of stresses/displacements provided by any finite element analysis program is not enough for most practical design applications. Some standards advance requirements to such deformation parameters as "tilt" or "twist" generally related to systems of a certain type (such as a column, rigid body etc.) having comparatively few degrees of freedom. These parameters are undefined for a general elastic system, and only some credible estimations can be made for such.

The concept of a tilt of a building can be involved when all nodes of the deformed shape are located along some oblique line defined by the following equations (the Z axis is assumed vertical):

$$u = a + k_x z; \qquad v = b + k_y z. \tag{6.4}$$

The tilt in the X direction is defined by the coefficient k_x, and that in the Y direction by the coefficient k_y, which together determine the tilt angles unambiguously:

$$\theta_x = \text{arctg}\,(k_x); \qquad \theta_y = \text{arctg}\,(k_y). \tag{6.5}$$

These coefficients can be found from the condition of a minimal sum of squares of deviations of the actual nodal displacements u_i and v_i ($i = 1,\ldots, m$) from the displacements u and v calculated by formulae (6.4)

$$\sum_{i=1}^{m}\left[\left(u_i - a - k_x z_i\right)^2 + \left(v_i - b - k_y z_i\right)^2\right] \Rightarrow \min\,, \tag{6.6}$$

where z_i is a coordinate of ith node, m is the total number of the nodes in question. Similarly, a common tilt or a common twist of a floor panel disk, a foundation slab etc. can be defined. Operations like these are performed by particular post-processors in the **SCAD** system, for example.

Another peculiar example is a problem of finding the magnitude and the application point of a resultant of all forces transferred to a foundation slab from its superstructure. These data are often needed to perform a primary estimation of the foundation slab dimensions and choose parameters for describing the subsoil.

For example, the **SCAD** system performs this work as follows:

- using the available option "*Calculate loads from a fragment of the model*", stresses N_i, M_{xi} and M_{yi} ($i = 1, \ldots, k$) are found in all nodes where the columns adjoin the foundation slab;
- the resultant of the vertical loads is calculated

$$R = \sum_{i=1}^{k} N_i \,; \tag{6.7}$$

- the coordinates of this load's application points are determined

$$X_0 = \left[\sum_{i=1}^{k}\left(N_i y_i + M_{yi}\cos\alpha_i + M_{xi}\sin\alpha_i\right)\right] \Big/ R\,;$$

$$Y_0 = \left[\sum_{i=1}^{k}\left(N_i x_i - M_{yi}\sin\alpha_i + M_{xi}\cos\alpha_i\right)\right] \Big/ R\,. \tag{6.8}$$

In the equations (6.8), x_i and y_i stand for the coordinates of ith column, and α_i for an angle between its axis and the X axis of the global coordinate system.

Though, any set of results generated by a certain software system is limited, and, rich as it was, there will always be a user demanding something extraordinary beyond any predefined capabilities. Therefore it seems reasonable to provide the user with the option of its own processing of results by his desirable formulae. To implement this, a software should include a special mode or a built-in calculator (Fig. 6.13) capable of operating the computational results.

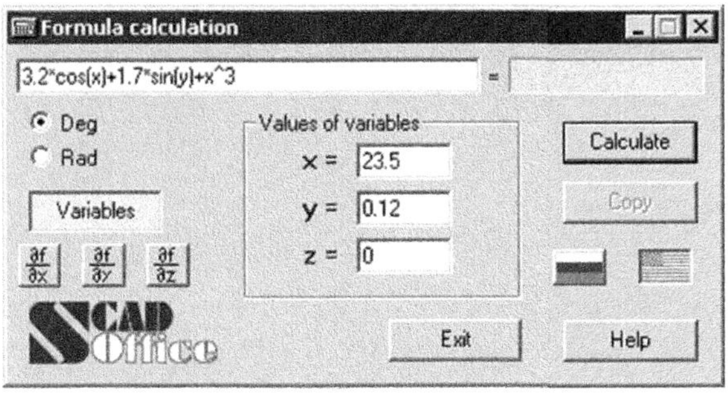

Fig. 6.13. A "Formula calculation" tool of the **SCAD** system

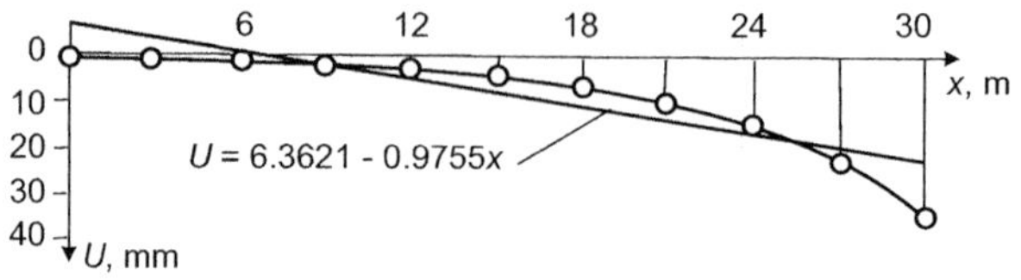

Fig. 6.14. Building a smoothing line

One of the options available for resolving this issue is to enable the user to transfer the analysis results to widely popular general data processing environments such as **EXCEL**. By employing functions of such software (sorting, calculation etc.) the user can perform almost any kind of post-analysis processing of the results. Fig. 6.14 shows an example of such post-processing where a smoothing line that approximates deflections has been built in the **EXCEL** environment by using the command "*Build a trend line*".

References

1. Dixon JR (1966) Design engineering: inventiveness, analysis and decission making. McGraw-Hill, New York Toronto London Sydney
2. Khurghin YI (1979) How to comprehend the incomprehensible (*In Russian*). Znaniye, Moscow
3. Perelmuter AV, Slivker VI (1987) A characteristic curve of an elastic system (*In Russian*). In: Research in structural theory, Issue XXV. Stroyizdat, Moscow, pp. 92-102.
4. Slivker VI (1991) Calculation of the deformation energy of a structure by the strength analysis software (*In Russian*). In: Structural mechanics and analysis of structures, 3:52-56.
5. Strang G, Fix GJ (1973) An analysis of the finite element method. Prentice-Hall Inc., Englewood Cliffs, N.J.
6. StressCheck.. The Stress Analysis Tool for Everyday Use. (1998) USER'S GUIDE. Engineering Software Research & Development, Inc., St. Louis, USA
7. Zienkiewicz OC, Taylor RL (2000) The finite element method. Vol.2. Solid mechanics. Butterworth-Heinemann, Oxford Auckland Boston Johannesburg Melbourne New Delhi

7 Uncertainty of Parameters

As the complexity of a system grows, our ability to formulate accurate and sensible statements regarding its behavior weakens to a certain threshold beyond which the accuracy and the sense become mutually exclusive things.

Zade L.A. Basics of a new approach to complex system analysis and decision making.— Moscow: *Znaniye*, 1974.

7.1 Key Sources of Uncertainty

A conventional engineering education is based on the principle of "accuracy". Engineers have been recognizing and perceiving the accuracy as one of their key principles ever since they were students. Any deviations are deemed undesirable and thus become as if non-existent in the psychological sense, though everyone realizes there can be no fault-free technologies, perfectly accurate measurements etc. In this connection, operating uncertain parameters requires one to readjust one's mind.

Nearly always, parameters of a design model manipulated by a researcher are inaccurate quantities. Often such parameters are essentially random variables, and when we assume one value of such or the other, we deal only with a realization of this randomness. Sometimes the value of a determinate parameter is simply unknown, or the available knowledge is extremely vague due to our limited capabilities. Just imagine how people performed the design of the lunar module in the "Apollo" program and what environmental properties were assumed for this design[1]. Finally, the uncertainty may arise due to an approximate modeling of a phenomenon (the typical example is a wind load modeled in a pretty rough way due to an approximate character of the aerodynamic flow simulation near a structural object).

---

[1] It is known that the Designer General of Soviet space systems S.P. Korolyov in his day "assigned" the lunar ground to be rigid and confirmed his decision by an appropriate order in writing. It should be noted here that the uncertainty of knowledge accompanies any kind of scientific research, not just the structural mechanics. That's how it is defined by a known British physicist H. Bondy [5]:

"*More than that, I regard it is an essential of any scientific theory to have room for putting in what one does not know yet* ".

Loads can be treated as most changeable elements of design models. Even the dead weight of a structure can change by ±10...15% according to the effectual standards, it being one of most stable kinds of loads. Other loads may vary in a much wider range both in their intensity or direction and in their position upon a structure. It was this peculiarity of the loads that initiated the development of methods of searching for a so-called "*worst possible loading*" i.e. such a combination of various external influence components that gives an extremum to a certain internal force, a stress in a cross-section, a nodal displacement etc.

From classics dedicated to foundations of methods of searching for worst possible positions of movable loads on an influence line (see [24], for example) to contemporary works on modern techniques of multi-criteria optimization solutions [11], this problem has been attracting a close attention of structural engineering experts. The more surprising is the circumstance that the consideration of variability of other design parameters has proved to be far less popular among the research community though its practical side is highly important.

Rigidity properties of members of a design model can be very much uncertain. This uncertainty may arise due to a natural spread of magnitudes of a parameter such as the elasticity modulus of concrete. It is a tradition to assume it using effectual design standards, and it is no less a tradition to ignore its potential variability of realizations, including its possible variations within an object and in time[2]. It may be caused by a pretty narrow range of variation of the elasticity modulus, but in many other cases there is a good reason to allow for the variability of the rigidity properties[3]. Particularly, such cases include the allowance for damage accumulated in a structure in the course of its operation. This damage should be taken into account in any analysis related to an existing structure condition survey and be assumed such as actually detected by the inspection. But not every element of an existing structure can be accessible by such inspection, therefore the rigidity properties of those elements are judged by the condition of other accessible parts of the structure. When dealing with responsible structures where knowing the rigidity properties is critical for the assessment of an existing structure's condition, various nondestructive tests are applied. Particularly efficient for this purpose are methods based on a simplest dynamical test when the rigidity properties are restored from the measured actual dynamical response of a structure. Though, the range of uncertainty can be pretty wide in any case.

An intensive variability, thus the uncertainty of values, is a characteristic feature of the natural foundation elasticity properties. This is partly due to an incomplete geological knowledge, partly to an approximate nature of the foundation models (Winclaire's, Pasternak's, semi-space etc.). Apparently, the best cases give the spread of the values about ±30 %, and sometimes much greater uncertainties

---

[2] The paper [8] presents data regarding the initial elasticity modulus of concrete the spread of which is ±25%.

[3] Another variable elastic parameter is the Poisson ratio. Measurements of elastic constants of fiberglass materials performed in the Institute of Mechanics of the Academy of Science of Ukraine have shown that the Poisson ratio varies by 10...40% as the elasticity modulus varies by 3...20%.

may take place. Results published in [28] that relate to a stress/strain analysis of frame structures on statistically heterogeneous elastic foundations are typical.

Dependencies of rated moments vs. ratios between the crossbar and the column rigidities were found by a numerical analysis of a frame shown in Fig.7.1, *a*, under various assumptions of statistical properties of the elastic foundation (see Fig.7.1, *b*).

The spread of the results determined by the variation coefficient, i.e. the ratio of the standard deviation σ to the average of the rated moment $\overline{M}$, depended noticeably on the form of a correlation matrix that defined a statistical relationship between random compliances of the foundation under different columns (Fig. 7.2). In other words, this matrix characterized a spatial heterogeneity of the rigidity properties of the foundation.

Three correlation matrices were involved: a matrix $\mathbf{K}_1$ equal to a unit one, which assumed a statistical independence of the elastic foundations under the different columns; and matrices $\mathbf{K}_2$ and $\mathbf{K}_3$ showing the increasing correlation between those:

$$\mathbf{K}_2 = \left\| \begin{bmatrix} 1.0 & 0.1 & & & \\ 0.1 & 1.0 & 0.1 & & \\ & 0.1 & 1.0 & 0.1 & \\ & & 0.1 & 1.0 & 0.1 \\ & & & 0.1 & 1.0 \end{bmatrix} \right\| \quad \mathbf{K}_3 = \left\| \begin{bmatrix} 1.0 & 0.3 & 0.2 & 0.1 & \\ 0.3 & 1.0 & 0.3 & 0.2 & 0.1 \\ 0.2 & 0.3 & 1.0 & 0.3 & 0.2 \\ 0.1 & 0.2 & 0.3 & 1.0 & 0.3 \\ & 0.1 & 0.2 & 0.3 & 1.0 \end{bmatrix} \right\|$$

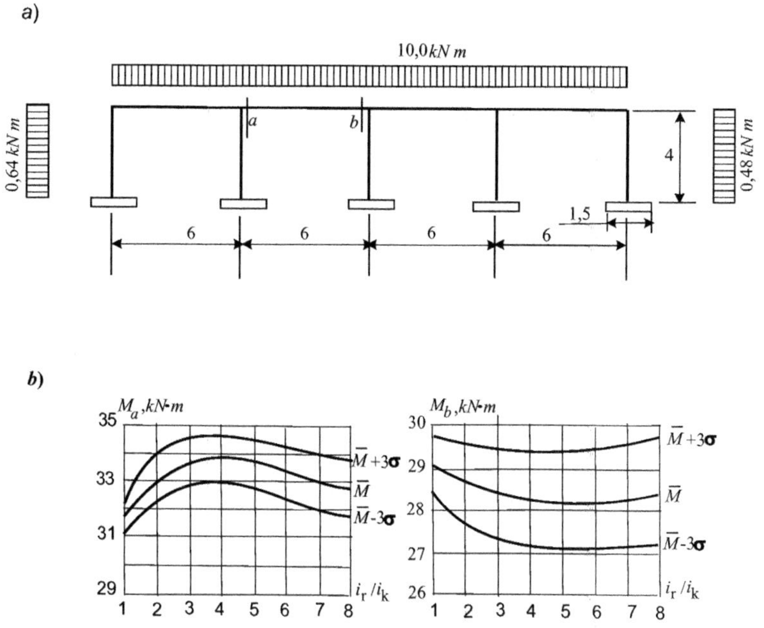

Fig. 7.1. A frame structure

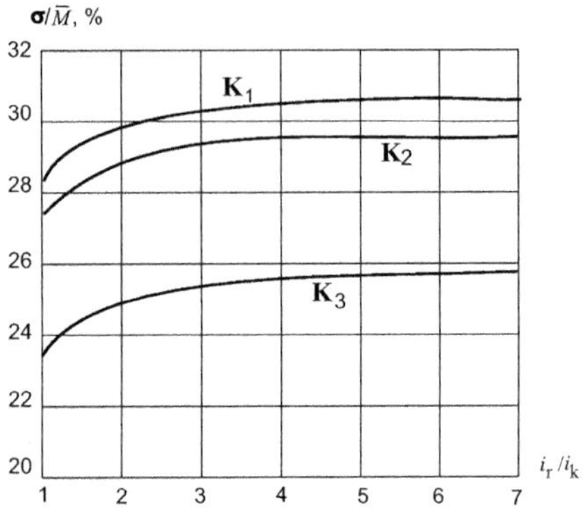

Fig. 7.2. Spread of the results vs. spatial correlation

It is a usual case in design models that an idealized joint between members in a node is assumed (such as a perfectly rigid or a hinged one) though the joint might have a certain finite elasticity. But the compliance of that is pretty hard to estimate, and the only way this estimation is feasible is by experiment. Very few reliable analytic methods are available presently for obtaining estimates of this kind. Therefore an engineer has to take the compliance of a nodal joint with a great degree of arbitrariness, by orienting himself at results of experimental investigations regarding "similar" structures.

It is a kind of tradition to assume certain structural solutions to be hinged or rigid, and such a tradition is relevant in the design practice when a designer has tools to implement the structural solution adopted. But the same tradition may prove impotent in conditions when an already existing structure is to be inspected and the structural solution cannot be adjusted to the model. A pretty typical approach to the solution of this problem is to make an expert estimation. An example of such from [17] is presented in Table 7.1.

Table 7.1.

Constraint type	Values of $m = R/r$					
	μ=1.0	μ=0.8	μ=0.6	μ=0.4	μ=0.2	μ=0.1
Perfectly rigid	1000.0	100.0	50.0	20.0	10.0	5.0
Very rigid	100.0	50.0	30.0	15.0	10.0	4.0
Pretty rigid	30.0	25.0	10.0	5.0	4.0	3.0
Not really rigid	10.0	5.0	4.0	3.0	2.0	1.0
Compliant	3.0	1.0	0.5	0.3	0.2	0.0
Pretty compliant	0.6	0.5	0.3	0.2	0.1	0.0
Very compliant	0.3	0.25	0.2	0.1	0.0	0.0
No constraint at all	50.0	0.1	0.1	0.1	0.2	0.3

The designations are:

R is a reaction of an isolated constraint caused by a unit strain;

r is its reaction within the structure caused by a unit displacement in the direction of the constraint;

μ is a level of credibility of the statement that regards the constraint type (a validity function).

When dealing with dynamical problems, one should allow for a possible variation of inertial properties. It may prove an uncommon thing for many analysts who treat the inertial mass only as a dead weight of a structure, and this one is not really variable. Though, nearly always a design model should take into account an associated mass determined by an equipment placed on the structure, stored materials, furniture and other weights that the current design codes treat as working loads. A load-carrying structure that swims in a liquid must be analyzed together with an associated mass of the liquid involved in the movement as the structure oscillates.

Unfortunately, it is a rare case when they publish an estimate of the source data spread effect on the results of an analysis, even in those few cases in which such estimation has been performed. That's the reason why every publication of this kind is so important. Here we present data taken from the paper [2] that considers two models of a heavy foundation intended for a turbine (Fig. 7.3): a bar model *A* and a model consisting of spatial eight-node elements *B*.

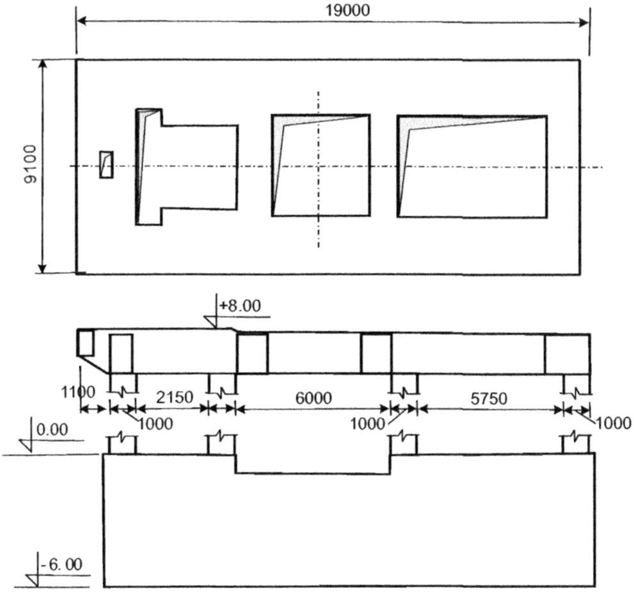

Fig. 7.3. A model of a foundation under a turbo-generator set

Various factors of uncertainty and their influence on the natural frequencies and oscillation mode shapes were compared by their importance. For quantities f being functions of uncertain parameters $X_1, \ldots, X_n$, a range estimate by each parameter was introduced:

$$S_i = [f(X_i)_{max} - f(X_i)_{min}] / [f(X_i)_{max} + f(X_i)_{min}], \quad (7.1)$$

and the total error was estimated as

$$S_f = \sqrt{\sum_i S_i^2}\,. \tag{7.2}$$

If vector values **V** are under consideration (such as oscillation modes), then a correlation-type criterion is used:

$$M_{Aj,Bk} = (\mathbf{V}_{A,j}\mathbf{V}_{B,k})^2 / [(\mathbf{V}_{A,j}\mathbf{V}_{A,j})(\mathbf{V}_{B,k}\mathbf{V}_{B,k})]\,, \tag{7.3}$$

where indexes A and B mean that the respective oscillation mode **V** is formed using two different models A and B, and the indexes j and k indicate the numbers of the mode shapes for which the estimation is performed. The criterion given above is used in the following estimate:

$$S_\mathrm{V} = [1 - M_{Aj,Bk}]/[1 + M_{Aj,Bk}]. \tag{7.4}$$

The natural frequencies were compared by the criterion (7.2), and the mode shapes by the criterion (7.4). Results of the analysis are presented in Table 7.2.

Table 7.2.

Comparison by:	S_f, % — in the frequency range:		S_V, % — by modes:	
	$f_1 \dots f_3$	$f_4 \dots f_{20}$	$\mathbf{V}_1 \dots \mathbf{V}_3$	$\mathbf{V}_4 \dots \mathbf{V}_{20}$
Group of uncertainties:				
(a) properties of the soil bed	43	49	3	40
(b) elasticity modulus and Poisson ratio of the structure	8	8	2	57
(c) design model	5	15	3	48
Total by all uncertainties	44	52	5	85

Expectedly, least credible results are those relating to higher natural frequencies and oscillation mode shapes. It is another evidence against the habit of using too much detailed models.

Another source of uncertainty includes parameters of a pre-stress specified in a structure. This fact was addressed in [20] where they showed how an inaccurate pre-stress in mast structures rated with the ±8% tolerance might result in a changed compliance of an anchor stay, which causes values of bending moments in some sections of the mast's shaft to vary almost twice (Fig. 7.4).

A pretty high uncertainty characterizes seismic analyses of structures [3]. Here there is a substantial spread of source data determined by the following sources of errors:

- the seismic disturbance itself is specified only approximately;
- the real bed's structure could be much different from a horizontal stratified model (it is hard to determine the position and thickness of the strata which also may vary within the site);
- a quasi-linear model of the soil behavior involved, and an uncertainty of its properties some of which can be noticeably variable from season to season.

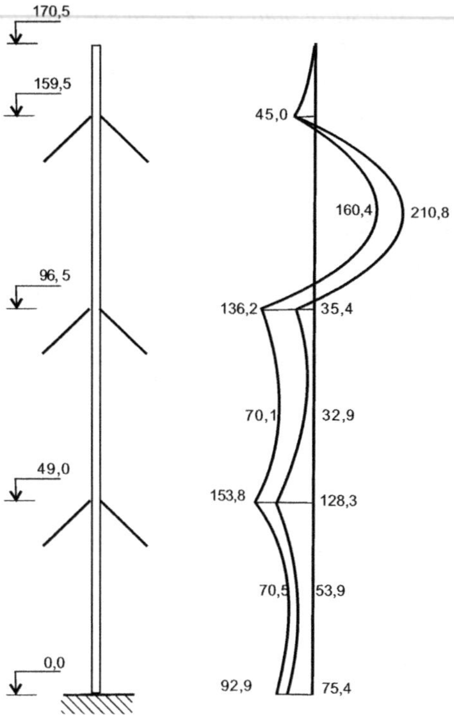

Fig. 7.4. An envelope of the moment curves in a mast's shaft

There are other uncertainties, too, which can result in a spread of the analysis results. There is the following practical approach to allow for those. First, we assume artificially that only one parameter should be varied, say, the shear modulus G of the soil. Further we assume it to vary between $G^*(1 + G_v)$ and $G^*/(1 - G_v)$ where G^* is its most probable estimated value and G_v is its coefficient of variation taken to be at least 0.5. These recommendations are formulated in U.S. nuclear power plant design standards [26].

A general recommendation can be that expressed in the book [4]:

"*Problems that contain hardly credible source data or fuzzy quantities should be paid special attention. In such problems, sometimes the credibility of the solution can be improved by an arbitrary variation of the said quantities within limits characteristic for a particular problem. Naturally, only results stable with respect to this variation should be treated as credible, and these results also show the degree of certainty of those quantities.*"

If a strong instability with respect to a certain parameter's variation is detected, one should try to understand its origin. Here another quotation from [4] suggests itself:

"*Spikes that might be encountered in the main line of some dependence are often confused with an error of an experiment. A typical anti-example follows. There are known average statistical data covering many years that regard the amount of greeting cards delivered by the postal service on the last day of April, June, July, August and September. Could these data be of any help to predict the result of December 31?"*

7.2 Methods of Sensitivity Analysis

To estimate the role of uncertain or variable parameters of a system (design variables), assuming their variations are relatively small, it may be very convenient to introduce coefficients of sensitivity, i.e. partial derivatives of some analysis result of our interest with respect to a design variable.

Let's use the equations in displacements (3.6) — (3.8) represented in the following form

$$\mathbf{K}(\mathbf{b})\mathbf{u} = \mathbf{p}_0 + \mathbf{p}_1(\mathbf{b}) - \mathbf{QF}(\mathbf{b})\mathbf{d}\,, \tag{7.5}$$

where the dependence of the stiffness matrix on the design variables **b** is specified explicitly. The latter usually include stiffness properties (cross-section dimensions), though other design variables can be involved, too, if we want to study the effect of variations of those.

As can be seen in (7.5), a part $\mathbf{p}_1$ of the load vector **p**

$$\mathbf{p} = \mathbf{p}_0 + \mathbf{p}_1(\mathbf{b})$$

can be also dependent on the design variables. If the latter are cross-sections of members, then the variable load can be the dead weight, for example. As for the vector of dislocations **d**, we will assume it to be independent of the design variables, though it is not really hard to allow for its dependence too.

Differentiating (7.5) with respect to some, say k^{th}, parameter and seeing that the displacements **u** are implicit functions of the design variables, we have

$$\mathbf{K}(\mathbf{b})\frac{\partial \mathbf{u}}{\partial b_k} = \frac{\partial \mathbf{p}_1}{\partial b_k} - \frac{\partial \mathbf{K}(\mathbf{b})}{\partial b_k}\mathbf{u} - \mathbf{Q}\frac{\partial \mathbf{F}(\mathbf{b})}{\partial b_k}\mathbf{d}\,. \tag{7.6}$$

Now using (3.7) and allowing for geometrical equations in (3.4), we rewrite (7.6) as

$$\mathbf{K}(\mathbf{b})\frac{\partial \mathbf{u}}{\partial b_k} = \frac{\partial \mathbf{p}_1}{\partial b_k} - \mathbf{Q}\frac{\partial \mathbf{F}(\mathbf{b})}{\partial b_k}\boldsymbol{\Delta}. \tag{7.7}$$

This relationship shows that the vector of partial derivatives $\partial\mathbf{u}/\partial b_k$, being the coefficients of sensitivity of the displacements with respect to a possible variation of the design variable b_k, can be found by solving the equation system with the same coefficient matrix as that of the static analysis itself.

If the direct Gaussian pass (or the Choletsky decomposition) has been performed in the course of the static analysis, and the triangular decomposition of the stiffness matrix according to (5.82) has been obtained and remembered, then it suffices to perform only a reverse substitution with the right parts (*pseudo-loads*) of the equations (7.7) to find $\partial\mathbf{u}/\partial b_k$. Evidently, this approach is much more efficient than a repeated solution of the equation system (7.7), though it takes memory resources to store the triangular decomposition of the matrix **K**.

The sensitivity of the stresses is determined by differentiating the equations (3.9) whence we have

$$\frac{\partial\mathbf{s}}{\partial b_k} = \frac{\partial\mathbf{F}}{\partial b_k}\left(\mathbf{Q}^{\mathrm{T}}\mathbf{u}+\mathbf{d}\right)+\mathbf{FQ}^{\mathrm{T}}\frac{\partial\mathbf{u}}{\partial b_k} = \frac{\partial\mathbf{F}}{\partial b_k}\boldsymbol{\Delta}+\mathbf{FQ}^{\mathrm{T}}\frac{\partial\mathbf{u}}{\partial b_k}\,. \tag{7.8}$$

After determining the vector $\partial\mathbf{u}/\partial b_k$ from (7.7), its substitution to (7.8) yields the solution:

$$\frac{\partial\mathbf{s}}{\partial b_k} = \left(\mathbf{I}-\mathbf{FQ}^{\mathrm{T}}\mathbf{K}^{-1}\mathbf{Q}\right)\frac{\partial\mathbf{F}}{\partial b_k}\boldsymbol{\Delta}+\mathbf{FQ}^{\mathrm{T}}\mathbf{K}^{-1}\frac{\partial\mathbf{p}_1}{\partial b_k}\,. \tag{7.9}$$

Note an auxiliary relationship, pretty useful for checking the calculation, which we can obtain from (7.9) by multiplying the latter by the matrix **Q** on the left. Taking (3.7) into account, this gives

$$\mathbf{Q}\frac{\partial\mathbf{s}}{\partial b_k} = \frac{\partial\mathbf{p}_1}{\partial b_k}\,. \tag{7.10}$$

Mechanically, the check equality (7.10) can be easily interpreted as an equilibrium between increments of external forces and increments of stresses in the system's members.

The approach described above that has yielded the equations (7.7) to (7.9) is called a *direct method of differentiation*. Also, another approach is possible, which is called a *method of conjugate variables* and involves an introduction of auxiliary unknowns

$$\boldsymbol{\theta} = (\mathbf{FQ}^{\mathrm{T}}\mathbf{K}^{-1})^{\mathrm{T}} = \mathbf{K}^{-1}\mathbf{QF}, \tag{7.11}$$

where the symmetry of the matrices **K** and **F** is used.

Multiplying (7.11) by **K** on the left, we obtain a system of equations

$$\mathbf{K}\boldsymbol{\theta} = \mathbf{QF}, \tag{7.12}$$

from which we can determine the conjugate variables $\boldsymbol{\theta}$, involving the same technique of accelerated calculation as described above. Then we substitute (7.12) to (7.9) which gives

$$\frac{\partial\mathbf{s}}{\partial b_k} = (\mathbf{I}-\boldsymbol{\theta}^{\mathrm{T}}\mathbf{Q})\frac{\partial\mathbf{F}}{\partial b_k}\boldsymbol{\Delta}+\boldsymbol{\theta}^{\mathrm{T}}\frac{\partial\mathbf{p}_1}{\partial b_k} = \frac{\partial\mathbf{F}}{\partial b_k}\boldsymbol{\Delta}+\boldsymbol{\theta}^{\mathrm{T}}\left[\frac{\partial\mathbf{p}_1}{\partial b_k}-\mathbf{Q}\frac{\partial\mathbf{F}}{\partial b_k}\boldsymbol{\Delta}\right]. \tag{7.13}$$

Note that the right parts of the equations (7.12) are nothing but nodal forces caused by unit internal strains of elements $\boldsymbol{\Delta}$, which fact can be easily derived from the first and third equations of the system (3.4). Thus the conjugate variables are displacements of nodes of the system caused by unit strains of its elements (strains of elements not related to a design variable being varied may be excluded from the consideration).

The book [13] gives a comparison of the approaches described above from the standpoint of a computational effort involved, and there they demonstrate that the conjugate variable method works almost always better than the direct differentiation technique, with a possible exception of rare cases when one estimates the sensitivity with respect to very few design variables and for one loading case only. The latter statement allows for the right part of the system (7.7) being dependent on the loading and the right part of (7.12) being an invariant of the system though a multi-column matrix.

The calculation of derivatives of the matrices $\partial \mathbf{K}/\partial b_k$ and/or $\partial \mathbf{F}/\partial b_k$ included in the formulae of the sensitivity coefficients is of certain difficulty. Of course, one can always use a finite-difference approximation such as that from [13]:

$$\frac{\partial \mathbf{K}(\mathbf{b})}{\partial b_k} \approx [\ \mathbf{K}(\mathbf{b} + \tau \mathbf{e}_k) - \mathbf{K}(\mathbf{b})\ ] / \tau \,, \tag{7.14}$$

where the vector $\mathbf{e}_k$ contains one in its kth position and zeros in all other positions, and τ is a small number that defines the magnitude of the disturbance b_k.

Though, it is better to use the formula for calculating the stiffness matrix components

$$\mathbf{K}(\mathbf{b}) = \mathbf{Q}\mathbf{F}(\mathbf{b})\mathbf{Q}^{\mathrm{T}}. \tag{7.15}$$

In truss structures where the components of the design variable vector $\mathbf{b}$ are areas of bar cross-sections, and in purely flexural systems where this vector consists of inertia moments, the stiffness matrix $\mathbf{K}(\mathbf{b})$ is a linear function of the vector $\mathbf{b}$ components, and it can be represented in the form

$$\mathbf{K} = \sum_k \mathbf{K}_k^0 b_k \,, \tag{7.16}$$

where $\mathbf{K}_k^0$ is a matrix of constant coefficients that defines the contribution of kth element. Then,

$$\frac{\partial \mathbf{K}(\mathbf{b})}{\partial b_k} = \mathbf{K}_k^0 \tag{7.17}$$

and remembering the matrices $\mathbf{K}_k^0$ during the composition of the global stiffness matrix can be of great help in the calculation of the sensitivity coefficients.

In the presentment above, the variable properties of the system were stiffness properties of its elements. Though, it is not the only choice of a design variable type. The coordinates of the system's nodes, $\mathbf{z}$, can be the variable design parameters just as well.

In a statically determinate case, one can take the equilibrium equations in the following form:

$$\mathbf{Q}(\mathbf{z})\, \mathbf{s} = \mathbf{p} \,, \tag{7.18}$$

and derive this by differentiating those with respect to the coordinate z_j

$$\mathbf{Q}\frac{\partial \mathbf{s}}{\partial z_j} = -\frac{\partial \mathbf{Q}}{\partial z_j}\mathbf{s} \tag{7.19}$$

under the assumption that the nodal load vector **p** does not depend on the coordinate **z**.

If the statically determinate system is not kinematically changeable, then **Q** will be a non-degenerate matrix, and the solution of the equation (7.19) will determine the sensitivity coefficients.

In a statically indeterminate case we can take the equations

$$\mathbf{Q}(\mathbf{z})\mathbf{F}(\mathbf{z})\mathbf{Q}(\mathbf{z})^{\mathrm{T}}\mathbf{u} = \mathbf{p} - \mathbf{Q}(\mathbf{z})\mathbf{F}(\mathbf{z})\mathbf{d}\,. \tag{7.20}$$

Under the same assumption that the nodal load is independent of the system's coordinate variations, we derive the following by differentiating the (7.20) relationships:

$$\mathbf{K}\frac{\partial \mathbf{u}}{\partial z_j} = -\left(\frac{\partial \mathbf{Q}}{\partial z_j}\mathbf{F} + \mathbf{Q}\frac{\partial \mathbf{F}}{\partial z_j}\right)(\mathbf{Q}^{\mathrm{T}}\mathbf{u} + \mathbf{d}) - \mathbf{Q}\mathbf{F}\frac{\partial \mathbf{Q}^{\mathrm{T}}}{\partial z_j}\mathbf{u}\,. \tag{7.21}$$

Differentiating the relations (3.9) gives us a connection between the sensitivity coefficients of the displacements and those of the stresses:

$$\frac{\partial \mathbf{s}}{\partial z_j} = \frac{\partial \mathbf{F}}{\partial z_j}(\mathbf{Q}^{\mathrm{T}}\mathbf{u} + \mathbf{d}) + \mathbf{F}\frac{\partial \mathbf{Q}}{\partial z_j}\mathbf{u} + \mathbf{F}\mathbf{Q}^{\mathrm{T}}\frac{\partial \mathbf{u}}{\partial z_j}\,. \tag{7.22}$$

Finding the displacement sensitivity coefficients from (7.21), we can then substitute those to the equations (7.22) and solve the problem posed. The substitution will give us this after some elementary transformations:

$$\frac{\partial \mathbf{s}}{\partial z_j} = \left(\mathbf{I} - \mathbf{F}\mathbf{Q}^{\mathrm{T}}\mathbf{K}^{-1}\mathbf{Q}\right)\left(\frac{\partial \mathbf{F}}{\partial z_j}\mathbf{D} + \mathbf{F}\frac{\partial \mathbf{Q}}{\partial z_j}\mathbf{u}\right) - \mathbf{F}\mathbf{Q}^{\mathrm{T}}\mathbf{K}^{-1}\frac{\partial \mathbf{Q}}{\partial z_j}\mathbf{F}\mathbf{D}\,. \tag{7.23}$$

Again, after multiplying (7.23) by the matrix **Q** on the left, we have a useful check equality

$$\mathbf{Q}\frac{\partial \mathbf{s}}{\partial z_j} = -\frac{\partial \mathbf{Q}}{\partial z_j}\mathbf{F}\boldsymbol{\Delta}\,. \tag{7.24}$$

The check equality (7.24) can be easily derived directly by differentiating the first equation of (3.4) and allowing for the load vector **p** being independent of the systems' node coordinates.

7.3 Sensitivity of Natural Oscillations

In a dynamical analysis of a system, the main things to determine are natural frequencies and modes of the system's oscillation. The respective system of governing equations follows:

$$\mathbf{K}(\mathbf{b})\mathbf{u} = \omega^2\mathbf{M}(\mathbf{b})\mathbf{u}\,, \tag{7.25}$$

where ω is a frequency that is, just as the oscillation mode $\mathbf{u}$, an implicit function of design variables, while the dependence of the stiffness matrix and mass matrix on those variables is stated explicitly in (7.25).

Now assuming $\lambda_k = \omega_k^2$ to be an eigenvalue that nulls the determinant of the matrix $\mathbf{K}(\mathbf{b}) - \lambda_k\mathbf{M}(\mathbf{b})$ and supposing that $\mathbf{u}_k$ is its respective vector of kth natural oscillation mode, we differentiate (7.25) with respect to one of components of the design variable vector $\mathbf{b}$ to obtain the following (for the purpose of briefness, we denote the varied component of the design variable vector as b with no indexes)

$$\left(\frac{\partial \mathbf{K}}{\partial b} - \lambda_k \frac{\partial \mathbf{M}}{\partial b}\right)\mathbf{u}_k - \frac{\partial \lambda_k}{\partial b}\mathbf{M}\mathbf{u}_k + (\mathbf{K} - \lambda_k\mathbf{M})\frac{\partial \mathbf{u}_k}{\partial b} = 0\,. \tag{7.26}$$

Seeing that for a natural oscillation mode it holds:

$$(\mathbf{K} - \lambda_k\mathbf{M})\,\mathbf{u}_k = 0, \tag{7.27}$$

and the matrices $\mathbf{K}$ and $\mathbf{M}$ are symmetrical, the multiplication of (7.26) by $\mathbf{u}_k^{\mathrm{T}}$ on the left eliminates the last term of the left part of the obtained relationship.

Let's assume that all eigenvectors are normalized by mass so that any two eigenvectors $\mathbf{u}_i$ and $\mathbf{u}_k$ are orthogonal by stiffness and orthonormal by mass:

$$\mathbf{u}_k^{\mathrm{T}}\,\mathbf{K}\,\mathbf{u}_i = \delta_{ik}\lambda_k\,, \qquad \mathbf{u}_k^{\mathrm{T}}\,\mathbf{M}\,\mathbf{u}_i = \delta_{ik}\,, \tag{7.28}$$

where δ_{ik} is a Kronecher's symbol equal to 1 at $i = k$ and zero otherwise.

With (7.27) and (7.28) taken into account, the relation (7.26) multiplied by $\mathbf{u}_k^{\mathrm{T}}$ on the left yields an expression to calculate the eigenvalue sensitivity coefficient

$$\frac{\partial \lambda_k}{\partial b} = \mathbf{u}_k^{\mathrm{T}}\left(\frac{\partial \mathbf{K}}{\partial b} - \lambda_k\frac{\partial \mathbf{M}}{\partial b}\right)\mathbf{u}_k\,. \tag{7.29}$$

There's a delicate thing here: when deriving the formula (7.29), we assumed implicitly that the eigenvalues could be differentiated with respect to a design parameter. This assumption is true if λ_k is a simple (not multiple) eigenvalue [12]. If there were multiple eigenvalues, the sensitivity analysis would be much more complicated. Suppose that the dependencies of two eigenvalues λ_1 and λ_2 on a varied parameter b look like plots in Fig. 7.5.

At $b = b_0$ the curves $\lambda_1(b)$ and $\lambda_2(b)$ cross each other, and we have a double eigenvalue $\lambda_1(b_0) = \lambda_2(b_0)$. As the eigenvalues are numbered in the ascending order ($\lambda_1 \leq \lambda_2$), crossing the point $b = b_0$ will make the curves exchange their indexes (places). Actually, the dependencies $\lambda_1(b)$ and $\lambda_2(b)$ will have the form shown as a dash line in Fig. 7.5, taking into account their arrangement in the ascending order of magnitudes. It is evident that the multiplicity point b_0 is a singular one where the analyticity of the eigenvalues is violated. Though, unilateral (directional) derivatives exist all the same, which enables us to perform the analysis even in this case. We do not address this, analytically delicate and theoretically interesting, issue here, referring an inquisitive reader to papers [12, 19, 25].

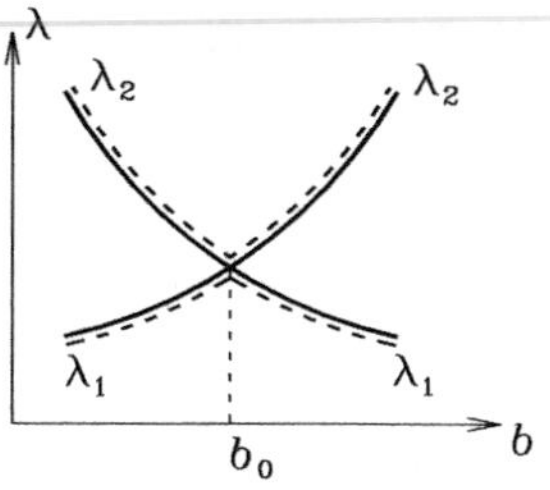

Fig. 7.5. Multiple eigenvalues

As the main point of our interest is a numerical analysis technique, we can recommend the following practical approach to avoid complications with singularities when dealing with multiple eigenvalues. This approach is based on shifting from a singular point by a small parametric magnitude ε and solving the eigenvalue problem (7.25) twice: both at $b = b_0 - \varepsilon$ and $b = b_0 + \varepsilon$.

Now let's begin finding the sensitivity coefficients of eigenvectors corresponding to multiple eigenvalues. It is not possible to find these coefficients directly by using (7.26) because the matrix factor $(\mathbf{K} - \lambda_k \mathbf{M})$ that stands on the left before the desirable vector $\partial \mathbf{u}_k / \partial b$, is an assuredly degenerate matrix. In this connection let us try to find our vector of interest as an expansion over a system of eigenvectors of the original problem. In other words, let us consider a representation

$$\frac{\partial \mathbf{u}_k}{\partial b} = \sum_{i=1}^{n} \alpha_i \mathbf{u}_i \tag{7.30}$$

with coefficients α_i unknown so far.

Multiplying (7.26) by $\mathbf{u}_j^{\mathrm{T}}$ on the left, we have this, with (7.30) taken into account,

$$\mathbf{u}_j^{\mathrm{T}}\left(\frac{\partial \mathbf{K}}{\partial b} - \lambda_k \frac{\partial \mathbf{M}}{\partial b}\right)\mathbf{u}_k - \frac{\partial \lambda_k}{\partial b}\mathbf{u}_j^{\mathrm{T}}\mathbf{M}\mathbf{u}_k + \mathbf{u}_j^{\mathrm{T}}(\mathbf{K} - \lambda_k \mathbf{M})\sum_{i=1}^{n} \alpha_i u_i = 0\,. \tag{7.31}$$

Now using (7.28), we find the following at $j \neq k$

$$\alpha_j = \frac{1}{\lambda_k - \lambda_j}\mathbf{u}_j^{\mathrm{T}}\left(\frac{\partial \mathbf{K}}{\partial b} - \lambda_k \frac{\partial \mathbf{M}}{\partial b}\right)\mathbf{u}_k \quad (j = 1,\ldots,n;\ j \neq k)\,. \tag{7.32}$$

As λ_k is a simple eigenvalue ($\lambda_j \neq \lambda_k$ at $j \neq k$), the formula (7.32) gives us all coefficients of the expansion (7.30), except for the coefficient α_k. In order to find α_k, we take the first condition of orthogonality (7.28) written for the vector $\mathbf{u}_k$ and differentiate it with respect to the parameter b. The result is

$$2\mathbf{u}_k^{\mathrm{T}}\mathbf{K}\frac{\partial \mathbf{u}_k}{\partial b} + \mathbf{u}_k^{\mathrm{T}}\frac{\partial \mathbf{K}}{\partial b}\mathbf{u}_k = \frac{\partial \lambda_k}{\partial b}\,. \tag{7.33}$$

After substituting the representations (7.29) and (7.30) to (7.33) and taking into account the orthogonality conditions (7.28), we have

$$\alpha_k = -\frac{1}{2}\mathbf{u}_k^{\mathrm{T}} \frac{\partial \mathbf{M}}{\partial b} \mathbf{u}_k \, . \tag{7.34}$$

It is interesting to note that if the mass matrix does not depend on the varied design parameter b (that's a pretty typical situation) then the coefficient α_k is equal to zero as follows from the formula (7.34). It means geometrically that if the mass matrix is independent from the b parameter then the vectors $\mathbf{u}_k$ and $\partial \mathbf{u}_k/\partial b$ are orthogonal both by the metric generated by the stiffness matrix $\mathbf{K}$ and the mass, i.e.

$$\mathbf{u}_k^{\mathrm{T}} \mathbf{K} \frac{\partial \mathbf{u}_k}{\partial b} = 0, \qquad \mathbf{u}_k^{\mathrm{T}} \mathbf{M} \frac{\partial \mathbf{u}_k}{\partial b} = 0 \, . \tag{7.35}$$

As can be seen, the first relation in (7.35) follows immediately from (7.33) and (7.29) while the second one follows from the first one seeing that $\mathbf{u}_k^{\mathrm{T}} \mathbf{M} = \mathbf{u}_k^{\mathrm{T}} \mathbf{K}/\lambda_k$. In a general case we have this final result:

$$\frac{\partial \mathbf{u}_k}{\partial b} = \sum_{\substack{j=1 \\ j\neq k}}^{n} \frac{1}{\lambda_k - \lambda_j} \mathbf{u}_j \mathbf{u}_j^{\mathrm{T}} \left(\frac{\partial \mathbf{K}}{\partial b} - \lambda_k \frac{\partial \mathbf{M}}{\partial b} \right) \mathbf{u}_k - \frac{1}{2} \mathbf{u}_k \mathbf{u}_k^{\mathrm{T}} \frac{\partial \mathbf{M}}{\partial b} \mathbf{u}_k \, . \tag{7.36}$$

The formula (7.36) defines the coefficients of sensitivity of kth natural oscillation mode via all modes of the unmodified system. It requires the full eigenvalue problem to be solved which is a really toilsome computational operation.

Let us show that we can derive the expressions of the sensitivity coefficients of the eigenvector $\mathbf{u}_k$ without determining all the other eigenvalues and eigenvectors.

We introduce a vector $\mathbf{w}_k$ equal to the desirable vector $\partial \mathbf{u}_k/\partial b$ minus its component $\alpha_k \mathbf{u}_k$ in the expansion (7.30). In other words, we define

$$\mathbf{w}_k = \frac{\partial \mathbf{u}_k}{\partial b} - \alpha_k \mathbf{u}_k \, , \tag{7.37}$$

and (7.30) makes the vector $\mathbf{w}_k$ a linear combination of the other eigenvectors of the original problem, i.e.

$$\mathbf{w}_k = \sum_{\substack{j=1 \\ j\neq k}}^{n} \alpha_j \mathbf{u}_j \, . \tag{7.38}$$

The new desirable vector $\mathbf{w}_k$ belongs to a linear manifold L built on a set of $n-1$ vectors $\{\mathbf{u}_i\}$ where the index i passes all values from 1 to n except for $i = k$. According to (7.28), this manifold is a hyper-plane orthogonal to the vector $\mathbf{Mu}_k$. It means that for any $\mathbf{v}$ from L the following holds:

$$\mathbf{v}^{\mathrm{T}}\mathbf{K}\mathbf{u}_k = 0, \qquad \mathbf{v}^{\mathrm{T}}\mathbf{M}\mathbf{u}_k = 0 \, . \tag{7.39}$$

In the hyper-plane L we choose an (arbitrary) set of linearly independent vectors $\mathbf{v}_1,\ldots,\mathbf{v}_{n-1}$ and expand the desirable vector $\mathbf{w}_k$ over these vectors as a basis of L, i.e.

$$\mathbf{w}_k = \sum_{i=1}^{n-1} \beta_i \mathbf{v}_i \ , \tag{7.40}$$

where β_i ($i = 1$,..., $n-1$) are scalar values to be found.

Let's return to the equality (7.26) and multiply it on the left by the vector $\mathbf{v}_j^{\mathrm{T}}$

$$\mathbf{v}_j^{\mathrm{T}}\left(\frac{\partial \mathbf{K}}{\partial b} - \lambda_k \frac{\partial \mathbf{M}}{\partial b}\right)\mathbf{u}_k - \frac{\partial \lambda_k}{\partial b}\mathbf{v}_j^{\mathrm{T}}\mathbf{M}\mathbf{u}_k + \mathbf{v}_j^{\mathrm{T}}(\mathbf{K} - \lambda_k \mathbf{M})\left(\alpha_k \mathbf{u}_k + \sum_{i=1}^{n-1}\beta_i \mathbf{v}_i\right) = 0\,, \tag{7.41}$$

which yields a system of $n-1$ equations with respect to coefficients β_i of the expansion (7.40) after taking into account the orthogonality conditions (7.39):

$$\sum_{i=1}^{n-1} \beta_i \mathbf{v}_j^{\mathrm{T}} (\mathbf{K} - \lambda_k \mathbf{M})\mathbf{v}_i = -\ \mathbf{v}_j^{\mathrm{T}} (\frac{\partial \mathbf{K}}{\partial b} - \lambda_k \frac{\partial \mathbf{M}}{\partial b})\mathbf{u}_k \qquad (j = 1,..., n-1)\,. \tag{7.42}$$

Making a vector $\boldsymbol{\beta} = \|[\beta_1, \beta_2, ..., \beta_{n-1}]\|^{\mathrm{T}}$ out of the coefficients β_i, we can rewrite the equation system (7.42) in a matrix form

$$\mathbf{V}^{\mathrm{T}}(\mathbf{K} - \lambda_k \mathbf{M})\mathbf{V}\boldsymbol{\beta} = -\ \mathbf{V}^{\mathrm{T}}\left(\frac{\partial \mathbf{K}}{\partial b} - \lambda_k \frac{\partial \mathbf{M}}{\partial b}\right)\mathbf{u}_k\,. \tag{7.43}$$

Here the matrix $\mathbf{V}$ of the size $n \times (n-1)$ consists of vectors $\mathbf{v}_i$ so that i^{th} column of the matrix $\mathbf{V}$ is the vector $\mathbf{v}_i$.

Making use of the fact that the eigenvalue λ_k corresponds to only one (up to a multiplier) eigenvector $\mathbf{u}_k$, one can prove that the matrix $\mathbf{W} = \mathbf{V}^{\mathrm{T}}(\mathbf{K} - \lambda_k \mathbf{M})\mathbf{V}$ is not degenerate, thus a solution of the equation system (7.43) exists. Having found the vector $\boldsymbol{\beta}$, it is easy also to find the desirable vector $\partial \mathbf{u}_k/\partial b$, so that's what we have after making all necessary substitutions

$$\frac{\partial \mathbf{u}_k}{\partial b} = -\frac{1}{2}\mathbf{u}_k \mathbf{u}_k^{\mathrm{T}} \frac{\partial \mathbf{M}}{\partial b}\mathbf{u}_k - \mathbf{V}\mathbf{W}^{-1}\mathbf{V}^{\mathrm{T}}\left(\frac{\partial \mathbf{K}}{\partial b} - \lambda_k \frac{\partial \mathbf{M}}{\partial b}\right)\mathbf{u}_k\,. \tag{7.44}$$

The formula (7.44) gives a final solution to the problem, and all that we have to do is to choose a method for building some basis in the L hyper-plane. Note that the L basis in question does not have to be orthogonal, nor a unit one.

Let's consider a simplest basis in the full space of the dimension n, formed by the vectors $\mathbf{e}_1,...,\mathbf{e}_n$, where each of the vectors $\mathbf{e}_i$ contains zeros except for its ith component which we assume to be equal to one. After complementing this system of vectors with a vector of the unit length $\mathbf{g} = \mathbf{M}\mathbf{u}_k / \|\mathbf{M}\mathbf{u}_k\|$, we can choose a vector $\mathbf{e}_j$ from this system of $n+1$ vectors so that it be a linear combination of the other vectors, that is,

$$\mathbf{e}_j = g_1/g_j\mathbf{e}_1 + g_2/g_j\mathbf{e}_2 + ... + g_{j-1}/g_j\mathbf{e}_{j-1} + g_{j+1}/g_j\mathbf{e}_{j+1} + ... + g_n/g_j\mathbf{e}_n - 1/g_j\mathbf{g}\ , \tag{7.45}$$

where the quantities g_i are respective components of the vector $\mathbf{g}$, and the coefficient g_j must be nonzero. Evidently, the set of vectors $\{\mathbf{g}, \mathbf{e}_1, ..., \mathbf{e}_{j-1}, \mathbf{e}_{j+1}, ..., \mathbf{e}_n\}$ is also a basis in the original n^{th} dimensional space.

To build a basis in the hyper-plane L orthogonal to the vector $\mathbf{g}$, a known Gram-Schmidt orthogonalization technique suggests itself immediately. It is to be applied to the system of linearly independent vectors $\{\mathbf{g}, \mathbf{e}_1, ..., \mathbf{e}_{j-1}, \mathbf{e}_{j+1}, ..., \mathbf{e}_n\}$ beginning from the vector $\mathbf{g}$. Though, in our case we can make things simpler. Indeed, it suffices to project the vectors $\mathbf{e}_1, ..., \mathbf{e}_{j-1}, \mathbf{e}_{j+1}, ..., \mathbf{e}_n$ onto the hyper-plane L to get the required basis $\mathbf{v}_1, ..., \mathbf{v}_{n-1}$. So, the vectors $\mathbf{v}_i$ can be the following:

$$\mathbf{v}_1 = \mathbf{e}_1 - g_1\mathbf{g}\,, \ldots, \mathbf{v}_{j-1} = \mathbf{e}_{j-1} - g_{j-1}\mathbf{g}\,,\ \mathbf{v}_j = \mathbf{e}_{j+1} - g_{j+1}\mathbf{g}\,, \ldots, \mathbf{v}_{n-1} = \mathbf{e}_n - g_n\mathbf{g}. \quad (7.46)$$

The index j for which the expansion (7.45) is performed can be found from the condition $|g_j| = \max |g_i|$ where the maximum is to be taken by absolute values of all components of the vector $\mathbf{g}$.

Note that there is another way to avoid the solution of the full eigenvalue/eigenvector problem in the general case of an asymmetric matrix, one suggested in the paper [19]. There they give a method of calculation based on knowing only the eigenvector in question, just as described here. Though, an auxiliary system of n+1 equations must be solved to obtain the solution by that method.

7.4 Estimating Extra Stresses Caused by Varied Stiffness

A variation of stiffness of elements in a statically indeterminate system results in changed internal forces/stresses, and designers want to be able to estimate this effect quantitatively. Let us consider the stated problem by the example of a hinged bar structure and derive all appropriate design formulae.

We take general expressions (7.9) of the sensitivity coefficients of the internal stresses and confine ourselves to the situation when the vector of external loads referred to nodes does not depend on the design variables which are cross-section areas of the bar members in the case of interest.

Assuming $b_k = A_k$ in (7.9) where A_k is the cross-section area of kth member and taking into account all that was said, we have the following

$$\frac{\partial \mathbf{s}}{\partial A_k} = (\mathbf{I} - \mathbf{F}\mathbf{Q}^{\mathrm{T}}\mathbf{K}^{-1}\mathbf{Q})\,\frac{\partial \mathbf{F}}{\partial A_k}\,\boldsymbol{\Delta}\,. \quad (7.47)$$

The intrinsic stiffness matrix $\mathbf{F}$ of a truss-type structure is a diagonal matrix with the components

$$\mathbf{F} = \|[\delta_{ij} f_j]\|, \qquad f_j = E_j A_j / l_j\,,\ \ (i, j = 1, \ldots, m)\,, \quad (7.48)$$

where E_j and l_j are the elasticity modulus and the length of jth bar, respectively. Here and below, m denotes the number of the bars in the truss. From (7.48) it follows immediately that

$$\frac{\partial \mathbf{F}}{\partial A_k} = \|[\delta_{ik} f_k / A_k]\|\,, \quad (7.49)$$

in other words, this matrix consists of zeros solely, except for an element at the intersection of kth column and k^{th} row, and this element is equal to f_k/A_k . Taking into account the expression (3.9) $\mathbf{s} = \mathbf{F}(\mathbf{Q}^{\mathrm{T}}\mathbf{u} + \mathbf{d})$ for calculating the stresses in a truss members, from (7.49) follows

$$\frac{\partial \mathbf{F}}{\partial A_k}\,\boldsymbol{\Delta} = \frac{s_k}{A_k}\,\mathbf{e}_k \tag{7.50}$$

where $\mathbf{e}_k$ is a vector that consists of zeros except for k^{th} components equal to one.

The substitution of (7.50) to (7.47) gives

$$\frac{\partial \mathbf{F}}{\partial A_k} = \frac{s_k}{A_k}\,(\mathbf{I} - \mathbf{F}\mathbf{Q}^{\mathrm{T}}\mathbf{K}^{-1}\mathbf{Q})\,\mathbf{e}_k\,. \tag{7.51}$$

Comparing (7.51) with (3.12), we notice that the matrix standing here in parentheses can be expressed via the pseudo-compliance matrix, namely

$$\frac{\partial \mathbf{s}}{\partial A_k} = \frac{s_k}{A_k}\,(\mathbf{I} - \mathbf{F}\mathbf{H})\,\mathbf{e}_k\,. \tag{7.52}$$

Further, denoting the components of the matrix $(\mathbf{I} - \mathbf{FH})$ as μ_{ij} and treating the stress increments Δs_r as differentials, which is quite acceptable if the cross-section areas ΔA_k do not change noticeably, we have

$$\Delta s_r = \sum_{k=1}^{m} \mu_{rk}\,\frac{s_k}{A_k}\,\Delta A_k\,. \tag{7.53}$$

If in a hinged structure the cross-section areas of all bars increase proportionally to one coefficient, which it is convenient for us to denote as $1+\beta$, then all components of the intrinsic stiffness matrix $\mathbf{F}$ will be multiplied by the same coefficient. So, the displacements $\mathbf{u}+\Delta\mathbf{u}$ and stresses $\mathbf{s}+\Delta\mathbf{s}$ in this modified system will be

$$\begin{aligned}\mathbf{u} + \Delta\mathbf{u} &= (1+\beta)^{-1}\mathbf{K}^{-1}\mathbf{p} - \mathbf{K}^{-1}\mathbf{Q}\mathbf{F}\mathbf{d},\\ \mathbf{s} + \Delta\mathbf{s} &= \mathbf{F}\mathbf{Q}^{\mathrm{T}}\mathbf{K}^{-1}\mathbf{p} + (1+\beta)(\mathbf{I} - \mathbf{F}\mathbf{Q}^{\mathrm{T}}\mathbf{K}^{-1}\mathbf{Q})\mathbf{F}\mathbf{d}\,,\end{aligned} \tag{7.54}$$

which follows directly from the formulae (3.8) and (3.9).

Confining ourselves to the case of force influences only, we should assume the dislocations $\mathbf{d}$ to be equal to zero, and as (7.54) shows, it won't change the stresses in the members of the system, i.e. $\Delta\mathbf{s} = 0$. The current modification of the system corresponds to incrementing the cross-section areas of all bars proportionally to the same parameter β, i.e.

$$\Delta A_k \,/\, A_k = \beta \qquad (k = 1,\dots, m)\,. \tag{7.55}$$

So, when analyzing the effect of arbitrary changes ΔA_k of the bar cross-section areas on the stress increments Δs_r, we can distract an arbitrary constant β from the ratio $\Delta A_k \,/\, A_k$, and the stress increments will remain the same. In other words, the formula (7.53) can be written in a little bit more general way (with an arbitrary constant β included)

$$\Delta s_r = \sum_{k=1}^{m} \mu_{rk}\,s_k\left(\frac{\Delta A_k}{A_k} - \beta\right). \tag{7.56}$$

The correctness of the formula (7.56) at $\mathbf{d} = \mathbf{0}$ can be proved by formal matrix transformations, too. As one can see, the equivalency of the formulae (7.53) and (7.56) can be reduced to the matrix equality

$$(\mathbf{I} - \mathbf{FH})\mathbf{s} = \mathbf{0}\,. \tag{7.57}$$

Substituting (3.12) and (3.9) here, and assuming $\mathbf{d} = \mathbf{0}$, we make sure that (7.57) holds.

Let us introduce a vector $\mathbf{h}(\beta)$ with components $h_k = s_k(\Delta A_k/A_k - \beta)$. The matrix form of representation of the stress increment vector $\Delta\mathbf{s}$ is now

$$\Delta\mathbf{s} = (\mathbf{I} - \mathbf{FH})\mathbf{h}(\beta)\,. \tag{7.58}$$

Making use of the arbitrary choice of the factor β to minimize the norm of the vector $\mathbf{h}$

$$\|\mathbf{h}(\beta)\| = \sqrt{\sum_{k=1}^{m} s_k^2 \left(\frac{\Delta A_k}{A_k} - \beta\right)^2}\,, \tag{7.59}$$

we have

$$\beta = \frac{\sum_{k=1}^{m} s_k^2 \dfrac{\Delta A_k}{A_k}}{\sum_{r=1}^{m} s_r^2}\,. \tag{7.60}$$

To estimate the maximum absolute change of the stress, we use a known inequality

$$\max_k |\Delta s_k| \le \sqrt{\sum_{k=1}^{m} \Delta s_k^2} = \|\Delta\mathbf{s}\|\,. \tag{7.61}$$

Then, from (7.58) it follows that

$$\|\Delta\mathbf{s}\| \le \|(\mathbf{I} - \mathbf{FH})\|\;\|\mathbf{h}(\beta)\|\,. \tag{7.62}$$

The matrix $\mathbf{FH}$ is idempotent which can be verified directly by substituting the formulae (3.12) and (3.7) to the definition of the idempotency criterion $(\mathbf{FH})^2 = \mathbf{FH}$. The matrix $(\mathbf{I} - \mathbf{FH})$ is idempotent together with $\mathbf{FH}$. It is known that the norm of any nonzero idempotent matrix is equal to one. Hence an estimate

$$\max{}_k |\Delta s_k| \le \|\mathbf{h}(\beta)\|\,, \tag{7.63}$$

that holds at any β including such β that minimizes the norm of the vector $\mathbf{h}$. Now substituting (7.59) and (7.60) to the estimate (7.63), we have the final expression

$$\max_k |\Delta s_k| \le \sqrt{\sum_{k=1}^{m} s_k^2 \left(\frac{\Delta A_k}{A_k} - \frac{\sum_{p=1}^{m} s_p^2 \Delta A_p / A_p}{\sum_{r=1}^{m} s_r^2} \right)^2} \,. \tag{7.64}$$

The inequality (7.64) makes a pretty simple estimation of the maximal possible variation of the stresses as the cross-section areas of the truss members vary.

7.5 Theoretical Estimates in the Case of Uncertain Stiffness Properties

Often, stiffness properties of a system (all or some) are uncertain quantities for which there are no sufficient data to estimate the probability distribution but there is other information to define the uncertainty. In [14] it has been stated that a maximum spread of a centered random variable ε corresponds to a certain distribution density $f(\varepsilon)$ established depending on an available information:

- when a random variable's variance is known not to exceed a value υ, a normal distribution is recommended with the density

$$f(\varepsilon) = \frac{1}{\sqrt{2\pi\upsilon}} \exp\left(-\frac{\varepsilon^2}{2\upsilon} \right); \tag{7.65}$$

- if the variance ε is known to have an average value of σ_0, then a Laplacian distribution should be used, having the density

$$f(\varepsilon) = \frac{1}{2\lambda} \exp\left(-\frac{|\varepsilon|}{\lambda} \right), \qquad \text{where } \lambda = \sigma_0 \sqrt{\frac{2}{\pi}}\,; \tag{7.66}$$

- in the case when the value ε ranges within $-b \le \varepsilon \le b$, a Huber distribution is applied:

$$f(\varepsilon) = \frac{1}{b} \cos^2 \frac{\pi\varepsilon}{2b} \,. \tag{7.67}$$

Adopting an approach used in the paper [22], we are going to show how the solution of a problem with uncertain stiffness properties can be estimated. Let the stiffness properties of a system are such that only their average values $\overline{\mathbf{F}}$ are known along with some limitations so that the uncertain value of $\widetilde{\mathbf{F}}$ can be classified into one of the above listed classes characterized by the "extreme" distribution densities (7.65) — (7.67).

After the system of equations of the slope deflection method with uncertain coefficients is linearized in the vicinity of their average values, we have

$$\overline{\mathbf{K}}\mathbf{u} + \mathbf{B}(\mathbf{u})\,\varepsilon = \mathbf{p}, \tag{7.68}$$

where the partial derivative matrix

$$\mathbf{B}(\mathbf{u}) = \left\| \left[\frac{\partial \mathbf{K}(\mathbf{u})}{\partial \mathbf{F}} \right] \right\|, \tag{7.69}$$

as well as an averaged stiffness matrix $\overline{\mathbf{K}}$, is calculated at $\mathbf{F} = \overline{\mathbf{F}}$. Assuming the second term in the left part of (7.68) to be a random disturbance (noise) ε_0, we state that the solution of (7.68) yields only some estimates $\tilde{\mathbf{u}}_0$ of the displacements that depend on ε_0.

Wishing to obtain an unbiased estimate $\tilde{\mathbf{u}}_0$ for which

$$\int_{-\infty}^{\infty} (\tilde{\mathbf{u}}_0 - \mathbf{u})^{\mathrm{T}} f(\varepsilon_0) d\varepsilon_0 = 0\,, \tag{7.70}$$

and seeing that (7.68) gives

$$\mathbf{u} = \overline{\mathbf{K}}^{-1} (\mathbf{p} - \varepsilon_0)\,, \tag{7.71}$$

we have

$$\begin{aligned} &\int_{-\infty}^{\infty} \left(\tilde{\mathbf{u}}_0 - \overline{\mathbf{K}}^{-1}\mathbf{p} + \overline{\mathbf{K}}^{-1}\varepsilon_0\right)^{\mathrm{T}} f(\varepsilon_0) d\varepsilon_0 = \\ &= \int_{-\infty}^{\infty} \left(\tilde{\mathbf{u}}_0 - \overline{\mathbf{K}}^{-1}\mathbf{p}\right)^{\mathrm{T}} f(\varepsilon_0) d\varepsilon_0 + \int_{-\infty}^{\infty} \varepsilon_0^{\mathrm{T}} \overline{\mathbf{K}}^{-1} f(\varepsilon_0) d\varepsilon_0 = 0\,. \end{aligned} \tag{7.72}$$

The second term in (7.72) is zero due to our assumption of the error being centered and thus having the zero average of distribution. Then (7.72) implies that the unbiased estimate

$$\tilde{\mathbf{u}}_0 = \overline{\mathbf{K}}^{-1}\mathbf{p} \tag{7.73}$$

depends neither on values of the matrix $\mathbf{B}(\mathbf{u})$ nor on the class of distribution of the quantity ε_0.

In order to find the variance matrix $\mathbf{D}_u$ of the estimates $\tilde{\mathbf{u}}_0$, we use the Rao – Kramer inequality

$$(\tilde{\mathbf{u}}_0 - \mathbf{u})^{\mathrm{T}} \mathbf{D}_u (\tilde{\mathbf{u}}_0 - \mathbf{u}) \ge (\tilde{\mathbf{u}}_0 - \mathbf{u})^{\mathrm{T}} \mathbf{R}^{-1} (\tilde{\mathbf{u}}_0 - \mathbf{u}), \tag{7.74}$$

where $\mathbf{R}$ is a Fisher information matrix which is defined as below due to (7.71) and the symmetry of the matrix $\overline{\mathbf{K}}$:

$$\mathbf{R} = \overline{\mathbf{K}}\mathbf{E}\overline{\mathbf{K}}\,. \tag{7.75}$$

Here $\mathbf{E}$ is a diagonal matrix having the components

$$e_{ii} = \int_{-\infty}^{\infty} \frac{\left(f'(\varepsilon_i)\right)^2}{f(\varepsilon_i)} d\varepsilon_i\,, \tag{7.76}$$

and the use of the distributions (7.65) — (7.67) minimizes ε_i and the left part of (7.74) at the same time.

The variances σ_u^2 of the estimates $\tilde{\mathbf{u}}_0$ that the diagonal matrix $\mathbf{D}_u$ consists of can be obtained from the expression

$$\mathbf{D}_u = \int_{-\infty}^{\infty} \left(\bar{\mathbf{K}}^{-1}\varepsilon_0\right) f(\varepsilon_0) \left(\bar{\mathbf{K}}^{-1}\varepsilon_0\right)^{\mathrm{T}} d\varepsilon_0 = \bar{\mathbf{K}}^{-1}\left(\int_{-\infty}^{\infty} \varepsilon_0 f(\varepsilon_0)\varepsilon_0^{\mathrm{T}} d\varepsilon_0\right)\bar{\mathbf{K}}^{-1}. \qquad (7.77)$$

Substituting (7.75) and (7.77) to (7.74) yields

$$\left(\tilde{\mathbf{u}}_0 - \mathbf{u}\right)^{\mathrm{T}} \bar{\mathbf{K}}^{-1}\mathbf{D}_0\bar{\mathbf{K}}^{-1}\left(\tilde{\mathbf{u}}_0 - \mathbf{u}\right) \geq \left(\tilde{\mathbf{u}}_0 - \mathbf{u}\right)^{\mathrm{T}} \bar{\mathbf{K}}^{-1}\mathbf{E}^{-1}\bar{\mathbf{K}}^{-1}\left(\tilde{\mathbf{u}}_0 - \mathbf{u}\right), \qquad (7.78)$$

from where it follows that

$$\sigma_u^2 \geq \frac{1}{e_{ii}} = \int_{-\infty}^{\infty} \frac{f(\varepsilon_i)}{\left(f'(\varepsilon_i)\right)^2} d\varepsilon_i \,. \qquad (7.79)$$

Minimally possible variances which correspond to the equality in (7.79) for the above said “extreme” densities follow:

- for the density (7.65): $\sigma_u^2 = \sigma_0^2$;
- for the density (7.66): $\sigma_u^2 = \sigma_0^2$;
- for the density (7.67): $\sigma_u^2 = b^2/\pi^2$.

The rated values of the displacements make a vector $\mathbf{u}^*$ with its components defined by the formula

$$u_i^* = \tilde{u}_{0,i} + \gamma_p \sigma_u \operatorname{sign}(\tilde{u}_{0,i}), \qquad (7.80)$$

where γ_p is an assuredness factor of the estimate (for the probability $P = 0.997$ one should take $\gamma_p = 3$).

Estimates of values and variances of the stresses are described by the formulae

$$\tilde{\mathbf{s}}_0 = \bar{\mathbf{F}}\mathbf{Q}^{\mathrm{T}}\tilde{\mathbf{u}}_0, \qquad (7.81)$$

$$\mathbf{D}_s = \bar{\mathbf{F}}\mathbf{Q}^{\mathrm{T}}\mathbf{D}_u\mathbf{Q}\bar{\mathbf{F}}^{\mathrm{T}}, \qquad (7.82)$$

and their rated values are, similarly to (7.80),

$$s_i^* = \tilde{s}_{0,i} + \gamma_p \sigma_s \operatorname{sign}(\tilde{s}_{0,i}). \qquad (7.83)$$

7.6 Making Use of Experiment Planning Methods

A classical statement of a structural mechanics problem assumes that all source data (dimensions of a structure, properties of its stiffness, loads et al.) are accurately defined. It implies that the results of the solution must be accurate as well, and it is only an incorrect design model or an inaccurate numerical technique that may cause an error. But actually, any kind of modeling of a real mechanical system using any methods of solving its governing equations causes a noticeable deviation of the analysis results from data measured experimentally (sometimes it can be treated as a success that the data coincide at least by their order of magnitude). Not the least reason for this is an uncertainty of all source data. In most cases this uncertainty can be estimated only by indicating an interval to which the true value of a parameter belongs.

When involving an interval estimation of parameters that describe a design finite-element model of a system, we use an interval of values of the stiffness matrix coefficients and the right part vector. Thus we encounter a problem how to find an interval solution of the system of governing equations. For a linear equation system $\mathbf{Ax} = \mathbf{b}$ with its coefficients and its right parts defined as interval quantities ($\mathbf{A} \in \mathbf{I}\, R^{n\times n}$, $\mathbf{b} \in \mathbf{I}\, R^n$), one should agree upon what is a solution, i.e. a solution set.

According to an analysis presented by Z. Kulpa and K. Roslaniec in [16], we distinguish:

- *a united solution set* defined as

  $$\Sigma(\mathbf{A},\mathbf{b}) = \{\mathbf{x} \in R^n \mid (\exists \mathbf{A}^* \in \mathbf{A})\,(\exists \mathbf{b}^* \in \partial\mathbf{b})\ \mathbf{A}^*\mathbf{x} = \mathbf{b}^*\};$$

  each point of the set being correspondent to a certain combination of values acquired by the interval variables;
- *a boundary solution set* defined as

  $$\Sigma(\partial\mathbf{A},\partial\mathbf{b}) = \{\mathbf{x} \in R^n \mid (\exists \mathbf{E}^* \in \partial\mathbf{A})\,(\exists \mathbf{e}^* \in \partial\mathbf{b})\ \mathbf{E}^*\mathbf{x} = \mathbf{e}^*\};$$

  here $\partial\mathbf{A}$ and $\partial\mathbf{b}$ designate matrices and vectors the elements of which acquire their values only from the boundaries not from the interior parts of the intervals. Points of this set correspond to combinations of such boundary values of the interval variables;
- *a parametric solution set* defined as

  $$\Sigma(\mathbf{A}(\mathbf{p}),\mathbf{b}(\mathbf{p})) = \{\mathbf{x} \in R^n \mid (\exists \mathbf{p}^* \in \mathbf{p})\ \mathbf{A}(\mathbf{p}^*)\mathbf{x} = \mathbf{b}(\mathbf{p}^*)\};$$

  here a vector of parameters $\mathbf{p}$ is introduced. Each of those passes through the interval [0,1] and realizes variations of its dependent variables from their lower to their upper boundaries. Points of the set correspond to possible values of components of the parametric vector.

Fig. 7.6 shows all three sets for interval equations with their matrix and vector

$$\left[\!\!\left[\begin{matrix} [0.1] & [1.2] \\ -2 & [-1.2] \end{matrix}\right]\!\!\right] \qquad \left[\!\!\left[\begin{matrix} [0.2] \\ 0 \end{matrix}\right]\!\!\right]$$

The matrix $\mathbf{A}(\mathbf{p})$ and the right part vector $\mathbf{b}(\mathbf{p})$ are parametrized using $p_1, p_2 \in [0,1]$ as

$$\left[\!\!\left[\begin{matrix} p_1 & 1+p_1 \\ -2 & 3p_1-1 \end{matrix}\right]\!\!\right] \qquad \left[\!\!\left[\begin{matrix} 2p_2 \\ 0 \end{matrix}\right]\!\!\right]$$

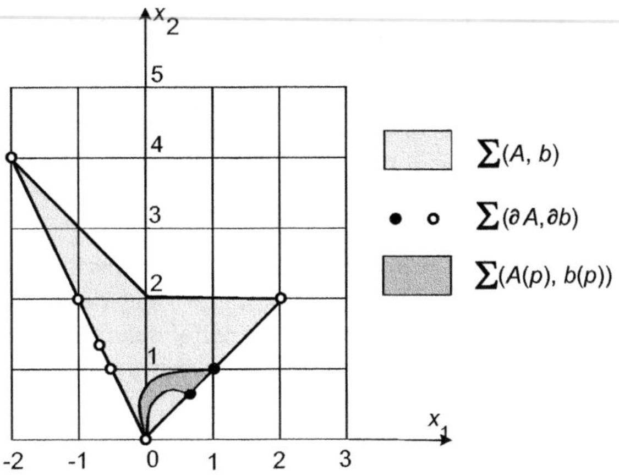

Fig. 7.6. Solution sets of an interval equation system

The **SCAD** system includes an option of searching for a *boundary solution set* implemented by a special mode which helps process results of an analysis of a few close versions of a model.

The closeness of the versions should be understood in the sense that those are topologically similar, include the same number of nodes and elements, and allow only certain differences between the involved models being compared:

- different types of elements can be used in the different models, including an "empty" type to imitate the absence of elements without changing their total number;
- stiffness properties of finite elements can be altered, including making some rigidities equal to zero;
- different ways of constraint imposition and/or specification of junctions between elements and nodes are allowed (such as embedded hinges or perfectly rigid inserts).

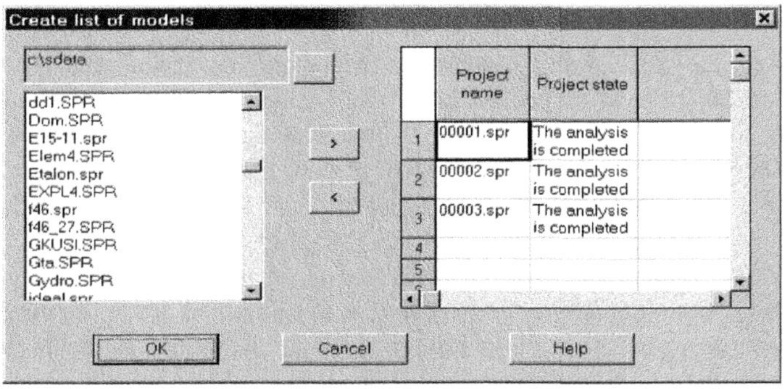

Fig.7.7. A dialog of the mode

The source data of this mode include only a list of files keeping solutions of particular problems (Fig.7.7).

The system performs a logical check in order to make sure the series of problems possesses the feature of closeness as declared. If this check is positive, the results of all analyses of the different versions of the model are merged into one array as though a single model were analyzed under each loading specified in every previous analysis.

With a super-problem thus obtained, one can then perform the modes of determining design stress combinations (DSC), assign design load combinations (DLC), select a concrete reinforcement, check or select steelwork cross-sections. One should establish a logical connection between the versions of loading specified for each particular subtask. For example, a common rule is that the dead weight loads included in all subtasks should be marked as mutually exclusive to avoid the summation of the same weights.

A demonstrative example is an analysis of a beam on an elastic foundation (Fig.7.8, *a*) with its average bedding factor 350 t/m^2 that can vary between 300 t/m^2 and 400 t/m^2. Results of the analysis are presented in Fig. 7.8, *b* and 7.8, *c*.

Searching for a solution set is a pretty toilsome operation. As a rule, it is based on a re-calculation of a system at different values of a parameter specified on an interval. Considering an immense performance and a convenient interface of the contemporary structural analysis software, such estimation is not hard to make, though the variation of a few parameters instead of only one increases the amount of the exhaustive search dangerously.

The latter circumstance fosters us to turn to experiment planning methods such as a factorial design with which the whole experiment planning theory began [9]. In the course of a research, an approximate dependence of a certain result the researcher is interested with on a variation of determinative parameters is to be constructed. It is assumed that for every parameter x_i an interval of variation is to be defined, with its upper b_i and lower a_i levels

$$a_i \le x_i \le b_i \quad (i = 1, \dots , n) \tag{7.84}$$

and for any value of x_i from this set the value of the result of interest $Y(x_1, \dots , x_n)$ can be obtained.

Unlike classical planning techniques where only one parameter is varied while the others are kept at a specific level, the factorial design supposes that all parameters should be varied at the same time. This approach enables one to investigate complex systems where the effects of particular parameters are hard, or just impossible, to separate from one another. Also, a correctly organized multifactorial experiment is better by accuracy than the classical ones.

The improvement is due to the fact that in an n-dimensional space of the parameters x_i the points where the experiment is conducted are at the maximum possible distance from one another [18] (p. 144), and this can be achieved by using the properties of a multi-dimensional space without increasing the variation interval (7.84) of each particular variable. As can be seen, the variation of each particular variable defines the said distances as $(b_i - a_i)$, i.e. as the edge lengths of an n-dimensional parallelepiped, and in a multi-dimensional factorial experiment these points are located in vertices of the same parallelepiped and the distances between them correspond to its diagonals.

(*a*)

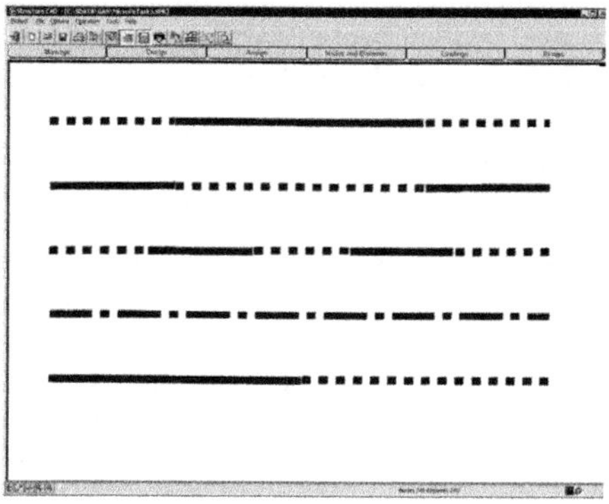

(*b*)

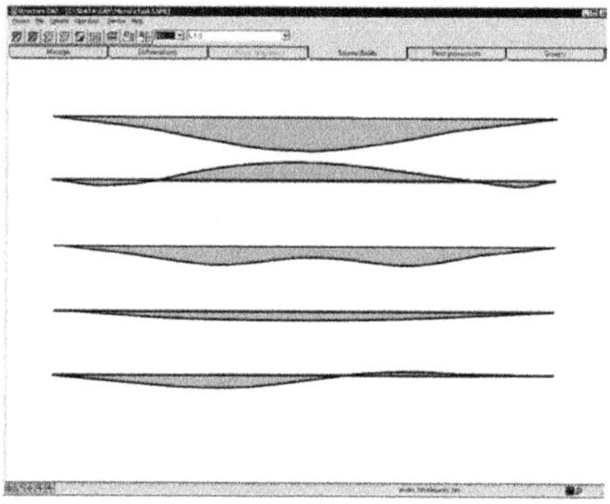

(*c*)

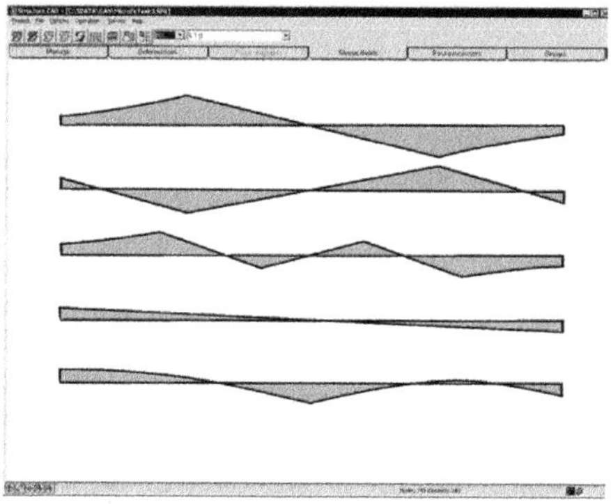

Fig.7.8. An effect of a variable stiffness of the elastic foundation: (*a*) a distribution of stiffness properties (the dotted line is $C = 300$ t/m^2, the solid line is 400 t/m^2, the dash line is 350 t/m^2); (*b*) bending moment curves; (*c*) lateral force curves (the order of the curves corresponds to that of the scheme (*a*) designs).

The calculation is based on a so-called planning matrix with its elements equal to +1 or –1 depending on the level the parameter in question is at. Such matrices have been well studied and are presented in the literature on the subject. A *full factorial experiment* realizes all possible combinations of levels of all variable parameters. Thus, in a two-level scheme (only upper and lower levels are involved) 2^n portions of calculations should be performed.

This number can be too great even at a medium n. Therefore one often uses fractional factorial designs of the type 2^{n-1}. These designs are built in the following way. For $(n-1)$ key parameters one builds a full factorial design to investigate all possible values of theirs. For each of the other parameters the planning matrix specifies values equal to products of any number (at least two) of indexes assigned to the key factors. It shows that we deem the respective interconnections (double, triple etc.) taken into account by the product to be essential. It is supposed that the key parameters are to be selected and their interrelated combinations are to be detected using some a priori qualitative considerations. After that it is just a routine to construct the planning matrix. For example, a 2^{5-2} fractional factorial design can be constructed in the way presented in Table 7.3.

Table 7.3.

Analysis	Key factors			(x_4)	(x_5)
	x_1	x_2	x_3	x_1x_2	$x_1x_2x_3$
1	+1	+1	+1	+1	+1
2	–1	+1	+1	–1	–1
3	+1	–1	+1	–1	–1
4	–1	–1	+1	+1	+1
5	+1	+1	–1	+1	–1
6	–1	+1	–1	–1	+1
7	+1	–1	–1	–1	+1
8	–1	–1	–1	+1	–1

The results of the analyses enable us to obtain approximant expressions for describing dependencies of the stress/strain distribution in the system vs. values of variable parameters. Of course, the governing equations themselves provide such description, but here we talk about an explicit approximation. To build such, most often interpolation techniques are used, involving a uniform (Chebyshev's) or mean-square approximation of a function in points of a specified region of the design variables space. To be definite, further we assume that the variable parameters are stiffness properties of the system.

Consider a sequence $\mathbf{X}^{(1)}, \dots, \mathbf{X}^{(n)}$ of points from the search region in which the structure has been analyzed and values of the properties $P_e(\mathbf{X}^{(k)})$ to be approximated have been found. Those may include stresses, displacements, design combinations thereof etc. Let's build the approximant expressions in the form

$$P = \sum_{i=0}^{m} a_i \, \varphi_i(\mathbf{X}), \tag{7.85}$$

where $\varphi_i(\mathbf{X})$ are given coordinate functions and a_i are unknown coefficients.

The mean-square approximation involves the search for such values of $a_i\,(i = 0,1,\ldots,m)$ that minimize the mean-square error

$$E = \sum_{k=1}^{n}\left\{P_e(\mathbf{X}^{(k)}) - \sum_{i=0}^{m} a_i\varphi_i\left(\mathbf{X}^{(k)}\right)\right\}^2. \tag{7.86}$$

The necessary conditions of the minimum $\partial E/\partial a_j = 0$ yield a system of equations with respect to the desirable coefficients a_i as follows:

$$\sum_{k=1}^{n}\left\{P_e(\mathbf{X}^{(k)}) - \sum_{i=0}^{m} a_i\varphi_i\left(\mathbf{X}^{(k)}\right)\right\}\varphi_j\left(X^{(k)}\right) = 0 \quad (j = 1,\ldots,\ m). \tag{7.87}$$

The quality of approximation is much dependent on a successful selection of the coordinate functions $\varphi_i(\mathbf{X})$. Investigation of explicit dependencies of key parameters of stress/strain distributions shows that rational expressions can be recommended for this purpose:

$$P_j = [Q^p(a_i,\mathbf{X})] \,/\, [Q^q(a_i,\mathbf{X})] \tag{7.88}$$

in cases when the design variables are stiffness properties [12]. In (7.88) the designation Q^p is used to denote an algebraic polynomial of the order p.

Most often, one uses rational expressions (Pade approximations [1]) of the form Q^0/Q^1, Q^1/Q^1 or Q^1/Q^2 where the number of the unknown coefficients a_i is relatively small. In some cases it is reasonable to employ inverse variables $z_i = 1/x_i$ in order to improve the approximation accuracy.

The coefficients a_i can be searched for by solving a quadratic error function minimization problem:

$$E_1 = \sum_{k=1}^{N}\left\{P_e\left(Z^{(k)}\right) - \left[\frac{Q^p\left(a_i, Z^{(k)}\right)}{Q^q\left(a_i, Z^{(k)}\right)}\right]\right\}^2. \tag{7.89}$$

Fig. 7.9 presents one of examples for which different approximation options were tried. The variable parameter (in a direct form) was the moment of inertia of the column with a pretty wide range of variation. The results of the analysis are presented in Table 7.4.

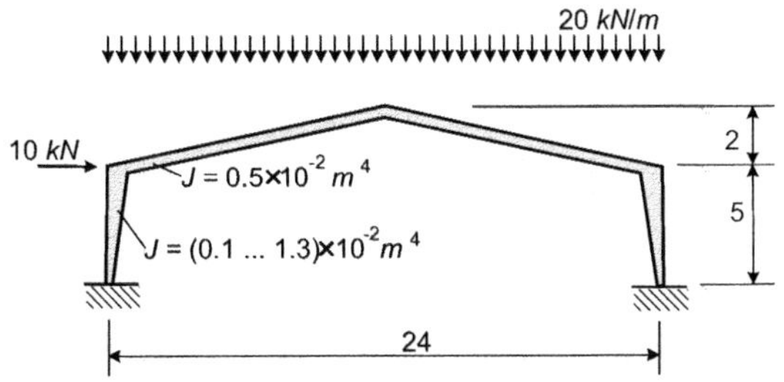

Fig. 7.9. An example of a stress and displacement approximation

Table 7.4.

Type of functions used	Variables	Approximation error, %	
		Stresses	Displacements
Linear	J_{col}	9.9 — 68.3	36.4 — 101.0
Linear	$1/J_{col}$	4.3 — 52.1	17.0 — 62.0
Quadratic	J_{col}	5.9 — 19.6	24.8 — 103.0
Quadratic	$1/J_{col}$	1.7 — 11.3	1.7 — 8.2
Rational Q^1/Q^1	J_{col}	0.03 — 3.27	3.8 — 57.4
Rational Q^1/Q^1	$1/J_{col}$	0.02 — 0.44	0.3 — 10.1
Pade approximation Q^1/Q^1	J_{col}	3.4 — 7.50	48.2 — 62.5
Pade approximation Q^0/Q^1	J_{col}	27.7 — 36.1	2.8 — 8.4

Designation: J_{col} is a moment of inertia of the frame's column cross-section.

It is useful to notice that an approximation like (7.88) can be used not just for components of a displacement or stress vector but for other parameters too. The paper [15] used a linear approximation to represent design combinations (envelopes) of internal stresses in a system subject to multiple loadings.

Another example is given in the paper [6] dealing with a system that includes a strutting of flexible cables and beams. There approximant expressions for a maximum displacement, maximum bending moment in the beam members, and a maximum tension in the flexible cables have been obtained.

7.7 Limit Equilibrium under an Uncertain Load

The problem of determining a limit load for an ideal elastic-plastic system can be formulated as a problem of linear programming, as it is known (see [7, 23], for example). The static formulation is

$$\max_{x_0, x_i} \left\{ x_0 \,\middle|\, S_i^+ \geq n_i^p x_0 + \sum_{j=1}^{m} n_{ij} x_j \geq S_i^-,\ i = 1,\ldots,k \right\}, \tag{7.90}$$

where x_0 denotes a load intensity parameter; S_i^+, S_i^- are respective upper and lower values of the ultimate stress in i^{th} section; n_i^p and n_{ij} are stresses in i^{th} section of the principal system of the force method caused by the loads and by the unit influence of jth unknown; m is the redundancy; k is the number of dangerous sections.

In cases when relationships between the external forces cannot be specified accurately, i.e. there is an uncertainty of the loading, the known information may include the intervals $n_i^{\min} \le n_i^p \le n_i^{\max}$ $(i = 1,...,k)$. This information can be easily obtained or estimated a priori on the basis of simplest physical considerations regarding the nature of the loads and conditions of the structure's existence. Then it becomes natural to formulate the problem as a mini-max one

$$\min_{n_i^p} \max_{x_0, x_j} \left\{ x_0 \left| S_i^+ \ge n_i^p x_0 + \sum_{j=1}^{m} n_{ij} x_j \ge S_i^-; n_i^{\min} \le n_i^p \le n_i^{\max}; i = 1,...,k \right. \right\}, \qquad (7.91)$$

which corresponds to something like a "game against the nature" [27]. In other words, we have to find a maximal possible value of the limit load under worst assumptions of relationships between external forces applied to the system.

It has been proved [21] that a linear programming problem equivalent to the game (7.91) can be constructed, the problem being how to maximize the parameter x_0 under the following limitations:

$$\left. \begin{array}{c} S_i^+ \ge n_i^{\max} \alpha_i^+ - n_i^{\min} \alpha_i^- + \sum_{j=1}^{m} n_{ij} x_j \ge S_i^-; \\ x_0 - \alpha_i^+ + \alpha_i^- = 0; \\ \alpha_i^+ \ge 0; \ \alpha_i^- \ge 0; \end{array} \right\} \ i = 1,...,k \ . \qquad (7.92)$$

It means that the solution of the problem occurs at boundary values of loads applied to the system.

Let's consider an example of a two-span beam schematically shown in Fig. 7.10, *a*.

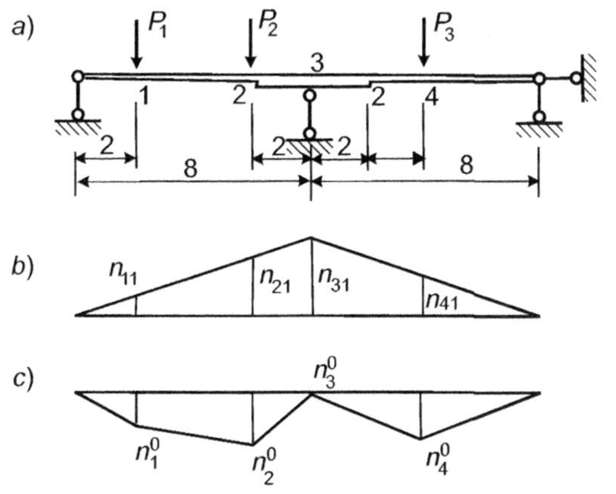

Fig. 7.10. A schematic of an example analysis

The beam is loaded by three concentrated forces which vary in the ranges $0.75x_0 \le x_0P_1 \le 1.25x_0$; $2.00x_0 \le x_0P_2 \le 4.00x_0$ and $2.50x_0 \le x_0P_3 \le 3.50x_0$ up to a load intensity factor x_0. The limit stresses in four cross-sections are:

$$S_1^+ = S_2^+ = S_4^+ = -S_1^+ = -S_2^+ = -S_4^+ = 0.8;\; S_3^+ = -S_3^+ = 1.0\,.$$

Using moment diagrams of the principal system obtained by adding a hinge to the intermediate support, with their values $n_{11} = 0.25$; $n_{21} = 0.75$; $n_{31} = 1.0$ and $n_{41} = 0.5$, we can obtain the interval values

$$2.125 \le n_1^p \le 3.875;\; 3.375 \le n_2^p \le 6.625;\; n_1^p = 0;\; 5.000 \le n_4^p \le 7.000,$$

where the designations are

$$n_1^p = 1.5P_1 + 0.5P_2;\; n_2^p = 0.5P_1 + 1.5P_2;\; n_3^p = 0;\; n_4^p = 2.0P_3\,.$$

The conditions of the linear programming problem (7.92) are formulated as

$$\begin{aligned}
0.8 \ge\; & 3.875\alpha_1^+ \; -2.125\alpha_1^- \; -0.250x_1 \ge -0.8;\\
0.8 \ge\; & 6.625\alpha_2^+ \; -3.375\alpha_2^- \; -0.750x_1 \ge -0.8;\\
1.0 \ge\; & \qquad\qquad\qquad\qquad\quad -1.000x_1 \ge -1.0;\\
0.8 \ge\; & 7.000\alpha_4^+ \; -5.000\alpha_4^- \; -0.500 \ge -0.8;\\
& x_0 + \alpha_1^- - \alpha_1^+ = 0;\; x_0 + \alpha_2^- - \alpha_2^+ = 0;\; x_0 + \alpha_4^- - \alpha_4^+ = 0;\\
& \alpha_1^+ \ge 0;\; \alpha_2^+ \ge 0;\; \alpha_4^+ \ge 0;\; \alpha_1^- \ge 0;\; \alpha_2^- \ge 0;\; \alpha_4^- \ge 0,
\end{aligned}$$

and its solution yields $x_0 = 0.1857$. It is interesting to note that if the ratio between the loads is $P_1{:}P_2{:}P_3 = 1{:}3{:}3$, which corresponds to the middles of the uncertainty intervals, then the solution of the problem (7.92) will yield $x_0 = 0.2195$. The difference $\Delta x_0 = 0.031$ is a "price of uncertainty" [27]. The closer are estimates of the load variation ranges, the less is this one.

The key point is the very fact that there is a "price to be paid" for the uncertainty. It evidences that an additional information has a certain value, and the effort/resources spent to obtain this information should be comparable with its value. Generally speaking, the necessity to compare the amount of additional effort, including information resources, with the possible value of the result has been long recognized informally by people in such sayings as "the game is not worth the candle" (an extra effort may give a trifle gain) or "having a potluck" (confining oneself to what one already possesses).

References

1. Baker GA Jr, Graves-Morris PR (1985). Pade Approximants. Cambridge Univ. Press,. Cambridge
2. Bayer J (1966) Estimation of Uncertainties of Theoretical Natural Frequencies and Mode Shapes of Structures. Engineering Mechanics, vol. 3, 5:323–330

3. Birbraer AN (1998) Seismic analysis of structures (in Russian). Nauka Publishing House, Sankt-Petersburg
4. Blekhman II, Myshkis AD, Panovko YG (1983) Mechanics and applied mathematics: Logic and peculiarities of mathematical applications (in Russian). Nauka Publishing House, Moskow
5. Bondy H (1967) Assumption and myth in physical theory. Cambridge At the University Press, Cambridge
6. Chyornaya EG (1973) Empirical type dependencies of maximum deflections and stresses in nets made of cables and beams (in Russian). In: Basics of theory of instantaneously stiff systems. LenZNIIEP, Leningrad, pp. 48–53
7. Čyras AA (1982) Mathematical models of analysis and optimization of elastic-plastic systems (in Russian). Mokslas publishing house , Vilnius
8. Drozdov YP (1971) Reliability of flexible ferroconcrete columns (in Russian). Concrete and Reinforced Concrete, 4:16–20
9. Ermakov SM (ed) (1983) Mathematical theory of experiment planning (in Russian). Nauka, Moscow
10. Fox RL, Kapoor MP (1968) Rates of Change of Eigenvalues and Eigenvectors. AIAA Journal, 6:2426–2429
11. Gordeyev VN, Artemenko VV, Minkovich EI (1989) Choosing a worst possible load combination as a solution of a multi-criteria optimization problem (in Russian). In: Numerical methods of analysis and optimization of structural constructions, Proc. of Kucherenko Central Research Institute of Structural Constructions, Moscow, pp. 26–32
12. Grebenyuk GI, Yankov EV (1989) Approximation of state parameters of bar structures by rational functions (in Russian). In: Proc. of high schools, Structural engineering and architecture, 4:16-19
13. Haug EJ, Choi KK, Komkov V (1986) Design sensitiviti analysis of structural systems. Academic Press, Orlando San Diego New York Austin London Montreal Sydney Tokio Toronto
14. Huber PJ (1964) Robust estimation of a location parameter. Ann. Mathematical Statistics, 35:. 73–101
15. Kalinina LG, Perelmuter AV (1985) On the about optimal desidn of structures (in Russian). In:Abovsky NP (ed) Space structures in Krasnoyarsk region, Krasnoyarsk politechnical institute press, Krasnoyarsk
16. Kulpa Z, Roslaniec K (1999) Solutionsets for systems of linear interval equations In: Proc. of the XIV Polish Conference on Computer Methods in Mechanocs, 26–28 May 1999, Rzeszów, Poland. Rzeszów University of Technology, Rzeszów: pp. 195–196
17. Malov VY, Pochtman YM (1987) Fuzzy estimation of nodal joint properties in models of optimal design (in Russian). In: Proc. of high schools. Structural engineering and architecture, 10:13-16
18. Nalimov VV (1971) Theory of experimenting (in Russian).Nauka, Moscow
19. Nelson RB (1976) Simplified calculation of eigenvector derivatives. AIAA Journal, 14:1201-1205
20. Perelmuter AV (1969) Principles of analysis of cable-bar systems (in Russian) Stroyzdat, Moskow
21. Perelmuter AV (1972) A limit equilibrium of an ideal elastic-plastic system in conditions of uncertainty (in Russian). In: Structural mechanics and analysis of structures, 5:23–25

22. Podolsky DM (1984) Analysis of structural systems with uncertain stiffness properties (in Russian). In: Reliability and durability of machines and structures, 6:78-86
23. Protsenko AM (1982) Theory of elastic-perfectly-plastic systems (in Russian). Nauka, Moscow
24. Rabinovich IM (1950) A course in structural mechanics. Part 1 (in Russian). Stroyizdat, Moscow-Leningrad
25. Rodgers LC (1970) Derivaties of Eigenvalues and Eigenvectors. AIAA Journal, vol.8, 5: 943–944
26. Seismic Analysis of Safety-Related Nuclear Structures and Commentary of Standard for Analysis of Safety-Related Nuclear Structures (1986). ASCE Standard, Sept., 1986.
27. Vorobiov NN (1967) Applications of the game theory to engineering. In: Proc. IV Internationaler Kongress über Anwendungen der Mathematik in den Ingenierwissenschaften.Weimar, 1:411–422.
28. Yusupov AK, Davletkhanova AD (1983) Analysis of plane frames on a statistically heterogeneous elastic foundation (in Russian). In: Structural mechanics and analysis of structures, 6:14–18

8 A Review of some Problem Classes

... a computer program does not make a good engineer, only a good engineer should use a computer program!

Dr. Leroy Z. Emkin. Computer Aided Engineering: Dangers / Ethics / Quality and a return to engineering common sense.— 1997. — January 15.

8.1 Hinged Bar Systems

A hinged bar system (a truss) is one of most widely usable design models. At the same time, it never happens that the nodal joints in respective structures are true hinges (Fig. 8.1,*a*), moreover, their axes might not even intersect at the same point (Fig. 8.1,*b*). One of the following circumstances (or the whole set of them) is conventionally used as an excuse of employing a hinged model:

- bars are so much flexible that resist very little to bending moments;
- all external loads are applied to nodes, therefore the longitudinal stresses are prevailing.

The following consideration is more careful [30]. If there is a structure having rigid nodes, its *hinged model* can be taken into consideration, too. The latter's difference from the former is that its rigid nodes are replaced by perfect hinges. The question is whether the hinged model can be used to simulate the structure as its design model. To answer this question, note that bars of the original system may experience both tension and bending while those of the hinged model are subject to tension only.

a) b)

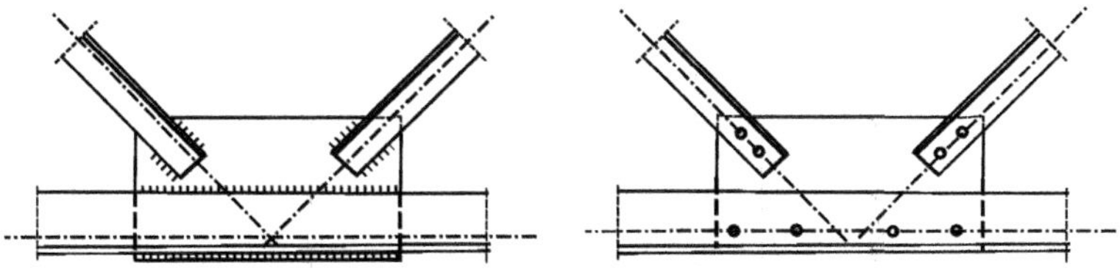

Fig. 8.1. Nodes of a truss

The order of magnitude of the flexural rigidity of a bar is EI/l^3, while that of the tension/compression is EA/l (I is the moment of inertia; A is the area). For a thin bar, the following holds

$$\frac{EI/l^3}{EA/l} = \frac{r^2}{l^2} = \frac{1}{\lambda^2} << 1 \,, \tag{8.1}$$

where $r^2 = I/A$ is the square of the inertia radius, $\lambda = r/l$ is the slenderness ratio of the bar.

If the hinged model is unchangeable, then any load on it can be resisted (and balanced), and elongations of the first order of smallness will correspond to such load. As can be seen, (8.1) implies that the bending deformation not taken into account by the hinged model contributes little to the stiffness of the structure, so the external forces are resisted chiefly by the tensioned bars. It is a different matter if the hinged model is changeable. In this case there are loads (unbalanced by the hinged model) which can be resisted only by flexural bars.

Of course, the above consideration is purely qualitative and should not be taken for granted. Thus, if the lower chord of a truss having an unchangeable hinged structure (see Fig. 8.2, *a*) has been designed much stiffer than its upper chord and its web, then one should take into account its continuity and assume its model to be composite (Fig. 8.2, *b*).

Sometimes one adopts the following recommendations to be the background for the modeling: limit the ratio of the cross-section's lateral size to the bar length by the value 1/10...1/12 obtained for bridge trusses in the course of a research conducted by B.E. Paton [25]. Often one fails to remember that these recommendations relate to systems with webs of certain types loaded in a certain way and made of soft steel, i.e. the recommendations are not absolutely acceptable in every particular case.

A hinged model facilitates a manual calculation so much that one uses it at the earliest opportunity. Though, it is nothing difficult at all to allow for rigid nodes in the computer analysis, so the very problem seems to vanish. But it encumbers the data set of the analysis results with an inessential information (or, exactly, it may do so). Therefore it is hardly possible that the scope of application of hinged bar models will narrow, and a serious attention should be paid to the background of their usage.

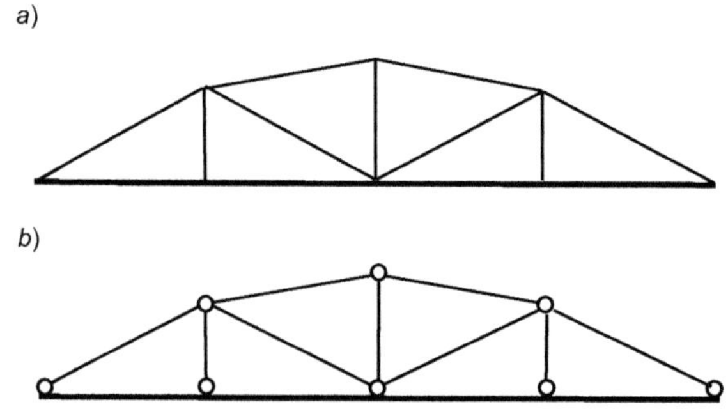

Fig. 8.2. Design models of a truss

One should be more attentive in cases when his software both performs a static/dynamic analysis and selects cross-sections of members. It may prove that the assumption of a sufficient slenderness of the members made by the designer does not hold for cross-sections selected automatically.

Another unpleasant peculiarity of design models where truss structures are formally treated as framed ones is a worsened conditionality of the linear algebraic equation system caused by a drastic difference between the orders of magnitude of coefficients that describe the flexural and tensional deformation components. In cases when one suspects a possibility for a substantial accuracy loss due to the use of a rigid-node truss model while wanting to know bending stresses in the truss members, one may follow this procedure. First, analyze the truss as a hinged system and thus find linear displacements of all its nodes. Second, consider the same system with rigid nodes restrained from all linear displacements and impose the external influences in the form of the given nodal displacements obtained at the first stage, in order to find flexural components of the stresses in the truss members.

Besides neglecting the flexural components in the analysis of hinged bar systems, one does not take into account the compliance of members due to shear in most bar system. It is a long-lasting tradition, too, explained in standard courses of the structural mechanics by a negligible contribution of the shear component to the Mohr's integral. Also, it should be stipulated that the shear deformation must not be neglected when considering a system with anisotropic members of a small shear rigidity such as relatively short wooden beams or double-tee bars where the tangential stresses in thin walls may become noticeable [28].

The presence of eccentricities caused by an inaccurate centering of bars in nodes and random bowing of the members results in the truss members experiencing a compression or tension together with bending even though there is no off-node loading. To allow for this circumstance, sometimes one complements the tension/compression stresses with bending stresses calculated with a frame model, trying to improve the accuracy in this way. As it has been shown in [9], such a complementary analysis improves nothing if the linear theory is used (a non-deformed shape analysis). If there are substantial longitudinal forces in the members due to effects of combined bending and compression (this is also called a "second-order analysis"), the flexural rigidity of tensioned bars increases noticeably while that of compressed bars decreases, and this causes the moments to redistribute and possibly change their values to a great extent.

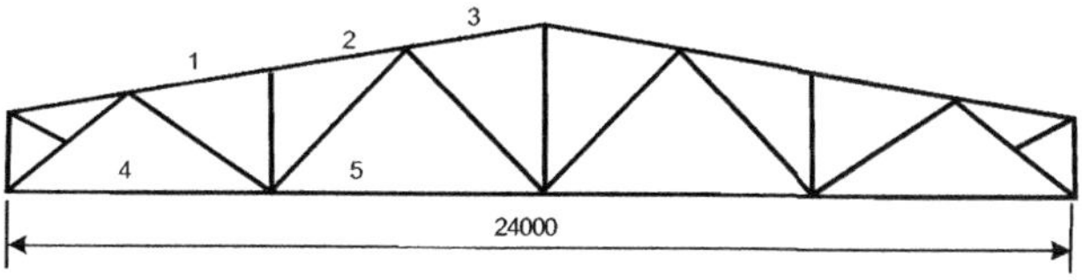

Fig. 8.3. A truss design

For a roof truss designed in compliance with the standard PK-01-125 (Fig. 8.3), the analysis yielded results presented in Table 8.1. This table uses the designation of M_1

for the moments calculated without the longitudinal forces taken into account, i.e. according to a typical analysis procedure of framed bar systems, and the designation of M_2 for those taking into account the transverse longitudinal bending when the stiffness matrix components are functions of a parameter ν where $\nu^2 = Nl^2/EI$.

Table 8.1.

Element	Bending moment, [kg×cm]		
	M_1	M_2	M_2 / M_1
1	8710	14930	1.715
2	10350	16000	1.545
3	17220	22180	1.285
4	12850	7800	0.607
5	18570	6220	0.335

8.2 Erection

The process of creating a complex structure is, generally, a multi-phase procedure involving a certain sequence of structural assembling operations. The latter may include successive works on installation or removal of some members of the structure, installation/removal of ballast weights, regulation of members' lengths, modification of constraints etc. Also, at some points of this process one may perform validation measurements of the internal stresses in the members and/or displacements in particular locations. Each operation of this kind defines a *step* of the erection. Steps that relate to the same design model make a *phase* of the erection. So, in our terms a single erection phase may include one or more erection steps.

Usually, the developer of a structure can specify his desirable stress/strain distribution to be obtained in the final assembled and pre-stressed structure under a certain load (such as its dead weight), or at least he can do it with some degree of accuracy. Also, a set of *regulated* parameters of the system is often involved: a list of members the length of which can be varied, a list of nodes wherein a ballast load can be placed etc. Some of those regulated parameters may be specified beforehand while others should be determined by calculation on the basis of the desired pre-stress distribution. The latter may be restrained by limitations that both bound their values and define their signs. These limitations are specified by the structure's designer on the basis of considerations related to the physical nature of the regulating effects. For example, the regulating effect of a surcharge ballast has a certain direction and sign, and values of dislocations of cable stays must be strictly positive because the cables do not resist to dislocations of the opposite sign[1].

[1] Talking about a flexible cable, we mean a hypothetic case of no lateral load on the one, or a load of such kind that can be safely neglected in the model.

Most actions performed during the erection modify the design model and/or the stress-strain distribution in the system.

Influences on the system that define its stressed and deformed state can be divided into two classes:

- *known influences,* for which the magnitudes, application points and principles of action are known beforehand;
- *unknown controlling parameters,* for which only the principles of action and their application points are known while the magnitudes of the influences are yet to be determined.

In both cases the influences may include concentrated or distributed loads, temperature, dislocations etc. The influences may be grouped (generalized) ones, then the unknown controlling parameter may be their common scale factor that defines the intensity of such grouped influence.

Stresses in elements or displacements of particular locations of the system which the designer would like to keep within certain, previously known limits during various phases of the erection (particularly, after its completion too) under a particular load acting upon the system, will be further called *controllable parameters*. Values of the controlling parameters are to be determined on the basis of limitations imposed on magnitudes of the controllable parameters.

Here we can present a set of elementary actions the sequence of which constitutes all needed transformations of a design model to determine the stress/strain distribution in the course of a structure's erection and pre-stressing. Generally, such elementary actions can be selected in a variety of ways. The list below [22] is peculiar because each item of it involves a simplest possible action of transforming the slope deflection method's governing equations or of determining the stress/strain fields.

We should note at once that actually one must distinguish clearly between the ultimate stress/strain distribution in the system that arises after each erection phase together with all its previous phases, and an incremental stress/strain distribution caused by additional influences upon the structure belonging to a particular erection phase only. The controllable parameters are surely those of the ultimate stress/strain state, while each phase of calculation is to determine the incremental stresses and strains. It should be emphasized that we assume the system to possess an ability of remembering the ultimate stress/strain distribution which enables it to store and inherit all accumulated stresses and strains.

So, our set of elementary actions includes the following items.

(A) ***Loading the system by a known influence*** which is a set of given loads, dislocations and thermal effects (Fig. 8.4).

This action regards a certain step of the erection during which the loading in question takes place. The change in the stress/strain state caused by it arises during the current (say, k^{th}) step and is inherited by all subsequent steps from $(k+1)^{th}$ to n^{th}. Here and further in this section, n refers to No. of the latest erection step.

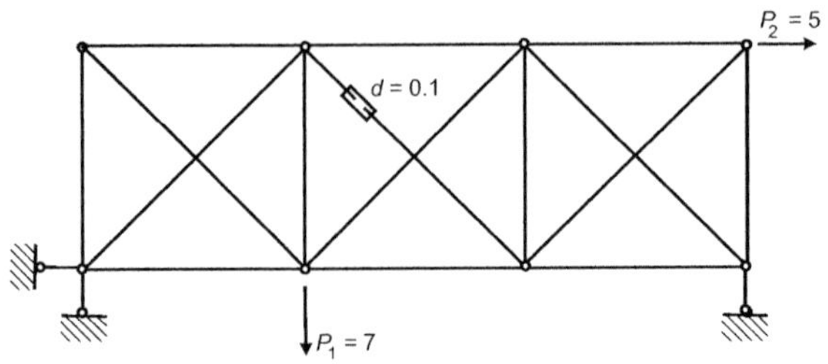

Fig. 8.4. Loading a system by known influence factors

Sometimes at a later step of the erection, say, at the step l where $l > k$, the system is subjected to a "canceling" load having the same nature as that of the kth step above but opposite by its sign. Then it is convenient to treat this canceling load as an independent one because otherwise the kth erection step's load would be inherited by steps up to l^{th} inclusive and lost by steps l+1 to n.

(B) ***Loading the system by an unknown load (controlling influence)***: this action is not very different from that described above, except for the fact that the value of the controlling parameter is so far unknown therefore we should take its unit value to perform the analysis instead of an actual one. The result is that we will have an auxiliary result (influence factors) rather than a particular stress/strain change. Further this result can be used to determine magnitudes of the controlling parameters and the stresses/strains themselves.

The inheritability of the stress/strain state and its localization within an isolated super-element (or an isolated fragment of other type) are the same as before.

A more general case is when the modified structure is loaded by a controlling influence equal to a controlling parameter used before (in a different model) and taken with a factor (Fig. 8.5).

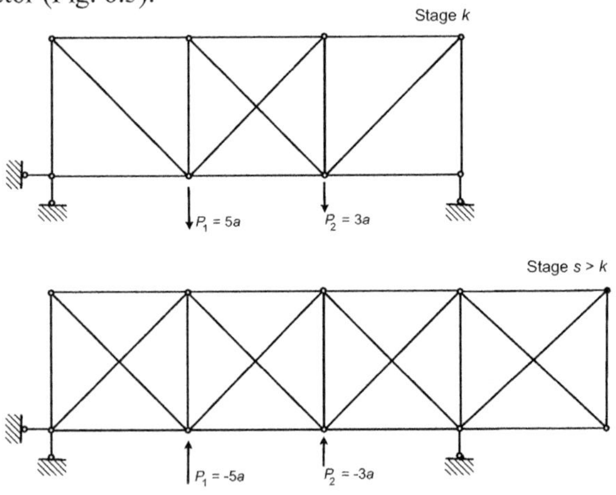

Fig. 8.5. Loading a structure by a controlling influence

The most often case is to assume this coefficient to be minus 1 which corresponds to an unloading (a removal of ballast weights). Though, generally one can adopt other values of it (such as +1.0 to double the ballast).

In the case in question, the stress/strain distribution regards the steps s to n where $s > k$ is the number of the step at which the load is introduced. The modified stress/strain field is located within the modified system rather than its previous state where the previous controlling parameter was used.

(C) ***Installing an unstressed external constraint*** in a node of the system to restrain it from a particular incremental displacement or rotation. It should be noted that here we require the increment of a displacement (Fig. 8.6) rather than the displacement/rotation itself (i.e. making it zero) to be prohibited.

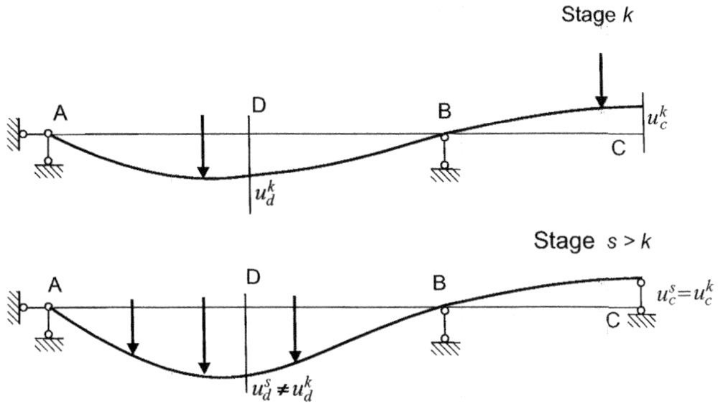

Fig. 8.6. Installing an unstressed constraint

The latter action would require a constraint with some pre-stress (a stressed constraint) to be installed.

The unstressed constraint installed modifies the model but not its stress/strain distribution which remains the same as that at the previous step.

(D) ***Removing an external constraint*** (Fig. 8.7). In this case the stress/strain distribution changes together with the model. The reason for this is that generally the constraint can be stressed. So, one has to zero the stress in the constraint before removing it.

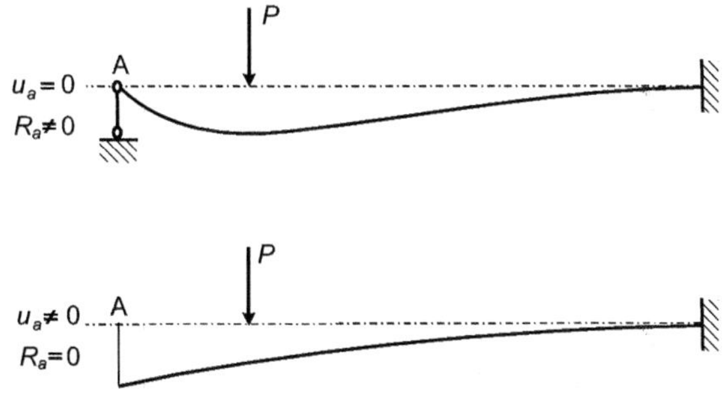

Fig. 8.7. Removing an external constraint

Here the stress in the constraint being removed is a controllable parameter, and the appropriate control conditions should be satisfied accurately.

Another approach to the removal of a stressed constraint is also feasible. It follows from the release principle: the equilibrium holds if the reaction of a removed constraint is applied to the system released of that constraint. Then the reaction of the removed constraint should be canceled in the new system without the constraint. It can be done by applying a force equal to the reaction by its magnitude but opposite by its sign, to the node whereto the constraint used to adjoin.

(E) ***Mounting an element*** is possible both without and with a pre-stress (the latter can be implemented in a number of ways). This action always changes the design model and can change the stress/strain distribution in the system differently depending on the selected method of pre-stressing. A variation of this action is to specify an unknown value of the pre-stress thus making it a controlling parameter. Then, just as in the action of loading the system with an unknown force, the analysis of a unit influence is to be performed. The result of it will be auxiliary data usable for calculating the stress/strain state and controlling parameters.

(F) ***Removing an element or a super-element from the system*** both modifies the design model and makes it necessary to re-calculate the stress/strain distribution. The transformed system must not experience any influence of its removed part, therefore this elementary action consists of two essential steps:

- making the interaction forces between the removed and remaining parts of the system equal to zero;
- modifying the design model.

In order for one to be able to perform the first step of the action, the structure must include devices to regulate the stress in the part to be removed, at least in the vicinity of its junction with the remaining part of the system (screw coupling, wedges, systems of ballast weights etc.). Selecting appropriate values of the controlling parameters, one can achieve zero values of the interaction forces if those have not been zero before. Note that the interaction forces have a peculiar feature as controlling parameters: the control requirements (zero values) must be satisfied exactly due to laws of nature, while in other cases it would be quite acceptable that the control requirements were satisfied approximately.

The elementary action that we are discussing is a kind of generalization of the "removing an external constraint" action. Another approach can be used to implement it: imagine that the elements are removed from the system, and their influence on the remaining part of the structure is replaced by forces N_j which existed in the removed elements; finally, suppress these forces by applying the loads $P_j = -N_j$.

(G) ***Adding an internal constraint*** prohibits the change of a certain mutual displacement between a node of the system and an end section of a member adjoining the nodes (for example, if a hinged joint should be changed into a rigid one, then the mutual rotation is to be prohibited). Such a constraint is allowed when the finite element possesses the respective degree of freedom and allows it to be specified as "displacement possible/impossible".

(H) ***Removing an internal constraint*** is a particular case of the "removing an element from the system" action. As a rule, the internal constraint is stressed, therefore the respective interaction force should be zeroed similarly to the cases discussed above.

(I) ***Checking the stress/strain parameters*** is needed when unknown controlling parameters participate in creating the pre-stressed state. The essence of the action is that it imposes limitations in the form of equalities with tolerances of the accuracy, or in the form of bilateral inequalities that relate to displacements of a certain group of nodes or stresses in some elements of the system. Accordingly, values of controlling parameters are to be calculated in order to fulfill the limitations introduced. This constitutes the check mentioned above. The check may relate to particular phases of the erection or to the completely erected structure.

8.2.1 Genetic Nonlinearity

All analyses related to the erection of a system and creation of a pre-stressed state in it are performed, as a rule, with conventional assumptions of the linear structural mechanics at each step of the erection. But in general the mechanical problem is nonlinear due to changes made to the model during transitions from one step to another. This nonlinearity caused by the history of the system's creation can be called *genetic*.

At each step of the erection, any classical structural analysis method can be involved. Seeing the specific nature of the multi-step analysis, we think it is useful to represent these methods in such form as to reflect the variability of the system. For example, the governing equations of the slope deflection method will be

$$\mathbf{K}_r \Delta\mathbf{u}_r = \Delta\mathbf{q}_r, \tag{8.2}$$

where $\mathbf{K}_r$ is the system's stiffness matrix at r^{th} step, $\Delta\mathbf{u}_r$ and $\Delta\mathbf{q}_r$ are respective vectors of incremental displacements and incremental nodal loads related to the rth step. Knowing $\Delta\mathbf{u}_r$, we can determine the stress increments $\Delta\mathbf{s}_r$ and obtain values of the displacements $\mathbf{u}_r$ and stresses $\mathbf{s}_r$ accumulated at all r steps

$$\mathbf{u}_r = \mathbf{u}_{r-1} + \Delta\mathbf{u}_r, \tag{8.3}$$

$$\mathbf{s}_r = \mathbf{s}_{r-1} + \Delta\mathbf{s}_r. \tag{8.4}$$

This step-by-step summation of the stress/strain components should be performed because of the system's memory mentioned above. The relevant entitlement for the relations (8.3) and (8.4) could be *laws of inheritance of erection states* of a structure. The genetic nonlinearity of the problem is caused by the requirement that the linear relations (8.2) and the laws of inheritance (8.3) — (8.4) must hold simultaneously.

It is obvious that influences related to different steps of the same erection phase are described by conventional linear mechanical laws: according to the terms we have agreed before, the design model of a structure changes only from one erection phase to another. In this connection, we imply that the displacements $\Delta\mathbf{u}_r$ and stresses $\Delta\mathbf{s}_r$ are those accumulated since the completion of the last step of the previous erection phase. In order to make things simpler and not to introduce a double numbering of steps and phases, we will assume that every phase includes only one erection step. This assumption is actually not a limitation because we can always increase the number of phases up to the total number of steps.

Again, we emphasize that in certain cases some loads work only at r^{th} step of the erection and are removed at the steps that follow. For example, this situation is typical for a cantilever erection of a structure where the weight of the crane equipment is taken into account when forming the vector $\Delta\mathbf{q}_r$, the cranes being arranged exactly as this step of erection requires. As the new $(r+1)$th erection step begins, the nodal load vector is formed for a new arrangement of the crane equipment, and one should remember to apply also negative loads at the step $(r+1)$ to cancel the crane load upon the system applied at the previous step. If one does not do it, the inheritance laws (8.3) and (8.4) will fail to hold.

In cases when the influences upon the structure are loads/dislocations specified up to an unknown factor α_k (controlling parameters), the analysis should be performed at the unit value $\alpha_k = 1$. The result of such will be a set of influence numbers $\mathbf{U}_k$ and $\mathbf{S}_k$ of the displacements and stresses using which all needed expressions can be derived.

Generally, every particular erection step may involve a set of controlling parameters rather than a single one. Let us introduce the following designations: m_k is the total number of controlling parameters which the designer can handle at k^{th} erection step, α_j^k is j^{th} controlling parameter referred to k^{th} step. Due to the linear character of the problem at each erection step with respect to incremental stress/strain components, and due to the inheritance laws (8.3) and (8.4), this relation holds:

$$T_i^r = \sum_{k=1}^{r} \sum_{j=1}^{m_k} W_{ij}^{rk} \alpha_j^k + \overline{W}_i^r \,, \tag{8.5}$$

that establishes a connection between controllable parameters T_i^r and controlling parameters α_j^k using an influence number matrix $\mathbf{W}$.

Components of the vector $\mathbf{T}$ are the stresses $\mathbf{s}_i^r$ and displacements $\mathbf{u}_i^r$ which must be controlled at r^{th} step of erection. The matrix $\mathbf{W}$ consists of the coefficients α_j^k (j^{th} controlling parameter at k^{th} step) of the controlling parameter influence on T_i^r (i^{th} control parameter at r^{th} step). The vector $\overline{\mathbf{W}}$ contains, as its components, those values of the controllable parameters which have been accumulated by the step r from the action of known loads.

Evidently, the controlling parameters introduced at the step k, cannot influence the controllable parameters (as well as other stress/strain components) at the steps preceding the emergence of the controlling influence, thus the elements of the influence matrix $W_{ij}^{rk} = 0$ at $r < k$ for all i and j.

Accordingly, the matrix **W** is a blocked triangular one such as shown in Table 8.2.

Table 8.2.

	$k=1$	$k=2$	...	$k=s$	...	$k=n$
$r=1$	$\mathbf{W}^{11}$	**0**	...	**0**	...	**0**
$r=2$	$\mathbf{W}^{21}$	$\mathbf{W}^{22}$	...	**0**	...	**0**
...	...	...	...	...	...	...
$r=l$	$\mathbf{W}^{l1}$	$\mathbf{W}^{l2}$	...	$\mathbf{W}^{ls}$	...	**0**
...	...	...	...	...	...	...
$r=n$	$\mathbf{W}^{n1}$	$\mathbf{W}^{n2}$	...	$\mathbf{W}^{ns}$	...	$\mathbf{W}^{nn}$

Here a block of the matrix with superscripts r and k refers to a check step r and a regulation step k. The dimensions of the blocks are determined by the number of controlling and controllable parameters introduced at each step.

The influence matrix W has the dimensions of $M \times N$ where M is the total number of controlling parameters. We can write the system of equations (8.5) in a matrix form after forming the full vector of controllable parameters **T** and the full vector of controlling parameters $\boldsymbol{\alpha}$

$$\mathbf{T} = \mathbf{W}\boldsymbol{\alpha} + \bar{\mathbf{W}} \,. \tag{8.6}$$

Let the matrix **W** have the rank N, that is, we can extract N linearly independent rows from this matrix. Without losing the generality, we can assume that the first N rows of the matrix **W** are linearly independent which is achievable by a simple transposition of the rows in the system (8.6) provided that rank $\mathbf{W} = N$.

Now let's represent (8.6) in a block form

$$\left\| \begin{bmatrix} \mathbf{T}_1 \\ \mathbf{T}_2 \end{bmatrix} \right\| = \left\| \begin{bmatrix} \mathbf{W}_1 \\ \mathbf{W}_2 \end{bmatrix} \right\| \boldsymbol{\alpha} + \left\| \begin{bmatrix} \bar{\mathbf{W}}_1 \\ \bar{\mathbf{W}}_2 \end{bmatrix} \right\| , \tag{8.7}$$

where $\mathbf{W}_1$ is a non-degenerate square matrix of the order N. The solution of the system (8.7) yields

$$\boldsymbol{\alpha} = \mathbf{W}_1^{-1}(\mathbf{T}_1 - \bar{\mathbf{W}}_1) , \tag{8.8}$$

the vectors $\mathbf{T}_1$ and $\mathbf{T}_2$ being related via the linear dependencies

$$\mathbf{T}_2 = \mathbf{W}_2\mathbf{W}_1^{-1}(\mathbf{T}_1 - \bar{\mathbf{W}}_1) + \bar{\mathbf{W}}_2 \,. \tag{8.9}$$

For each of the controllable parameters a corridor of allowable values should be specified in the form of bounding inequalities

$$T_i^{\min} \le T_i \le T_i^{\max} \quad (i = 1,\dots, M). \tag{8.10}$$

Note that the limitation of the controllable parameters in the form of equalities is a particular case of the allowable corridor with its width equal to zero, i.e. at $T_i^{\min} = T_i^{\max}$ for some index i.

The set of first N limitations in (8.10) referring to components of the vector $\mathbf{T}_1$ defines an N-dimensional parallelepiped P_1, while the last $(M–N)$ limitations in (8.10) refer to components of the vector $\mathbf{T}_2$ and form a rectangular parallelepiped P_2 in a space of the dimensionality. If there is at least one point from P_1 which can be mapped to a point belonging to P_2 by the linear transformation (8.9), then the problem of finding the controlling parameters $\boldsymbol{\alpha}$ is solvable. Otherwise there can be no such set of the controlling parameters $\boldsymbol{\alpha}$ at which all controllable parameters fit in the corridor of allowable values (8.10), thus the design decisions need changing.

The authors should admit they do not know of any software capable of solving design problems related to erection states of structures in this general formulation. Therefore practical engineers have to adopt the set of controlling parameters $\boldsymbol{\alpha}$ on the basis of their intuition, design experience etc., and then check the fulfillment of controllable limitations such as (8.10). In particular, the first iteration of the solution is often the central point of the parallelepiped P_1, that is,

$$T_i = \tfrac{1}{2}\,(T_i^{\min} + T_i^{\max}) \qquad (i = 1,\ldots, N). \tag{8.11}$$

If necessary, the controlling parameters are then corrected, and the calculation repeats again.

It is quite possible (and even highly probable) that the image of the rectangle P_1 when transformed according to (8.9) crosses the rectangle P_2 in more than one point. This fact corresponds to a non-uniqueness of the solution, that is, there can be multiple allowable vectors of controlling parameters $\boldsymbol{\alpha}$ that meet the requirements. In such case the engineer can use this multiplicity to optimize the structure. The optimization depends strongly on a particular design task, and its goal function can by no means be formulated in any general way. Therefore we do not address this matter here. To perform the erection task itself, it suffices to find at least one allowable solution that satisfies the limitations (8.10).

Finally, let us note the case when the matrix $\mathbf{W}$ rank is less than the number of the controlling parameters N. It introduces nothing new to the formulation of the problem because the matrix can always be complemented with additional linearly independent rows so that the rank of this extended matrix be equal to N. The introduction of these new rows into the matrix $\mathbf{W}$ is equivalent to extending the list of the controllable parameters up to the length $M + N - \mathrm{rank}\mathbf{W}$ where $\mathrm{rank}\mathbf{W}$ is the rank of the original matrix (before its complementing with linearly independent rows). In order for the new controllable parameters formally introduced not to affect the solution, one should specify the infinite width of their corridor in both directions.

Most often, in the course of the erection one confines oneself to a set of controlling parameters that define an internal self-stressed state of a structure. As was shown in Chapter 3, the number of possible linearly independent self-stressed states of a discrete structure is equal to its degree of redundancy. Therefore the number of controlling parameters is not to exceed the degree of redundancy. On the other hand, the number of controllable parameters N can be much greater: the designer wants to ensure the sufficient strength of all members of the structure. If the formulation of the problem uses point definitions of allowable values of the controllable parameters (the respective corridor is zero wide) rather than corridors, then the system of equations (8.6) may have no solution.

The solution of an over-determined system of equations (8.6) should not be repulsed unquestionably. It is possible that the engineer has his own particular considerations regarding a set of controllable parameters to choose. His interests can be accommodated by solving the over-determined system (8.6) in the sense of a mean-square approximation. This is equivalent to multiplying (8.6) on the left by the transposed matrix $\mathbf{W}^T$ which gives a system of equations of the order N with a positive definite matrix $\mathbf{W}^T\mathbf{W}$

$$\mathbf{W}^T \mathbf{T}_0 = \mathbf{W}^T \mathbf{W}\boldsymbol{\alpha} + \mathbf{W}^T \bar{\mathbf{W}} \,. \tag{8.12}$$

Here $\mathbf{T}_0$ denotes a vector of the controllable parameters with predefined values. Of course, the solution of the system (8.12) won't satisfy the original system (8.6) exactly. But here the designer can return to the corridor concept and use (8.6) to restore values of the controllable parameters by the values of the controlling parameters taken from the solution of (8.12)

$$\mathbf{T} = \mathbf{W}\,(\mathbf{W}^T \mathbf{W})^{-1}\, \mathbf{W}^T (\mathbf{T}_0 - \bar{\mathbf{W}}) \,. \tag{8.13}$$

If some components of the vector $\mathbf{T}$ obtained from (8.13) deviate beyond the allowable spread around the components of the vector $\mathbf{T}_0$, then the task of finding the proper vector $\boldsymbol{\alpha}$ can be accomplished by means of a weighted method of least squares. The more the components of the vector $\mathbf{T}$ in (8.13) deviate from the boundary of its corridor, the greater weight should be assigned to the respective equation of the original system (8.6). Other approaches to the formation of the weights are feasible too. For example, such may involve an importance rating of parameters. Sometimes the importance can be judged by the width of the allowable corridor of values. The narrower the corridor, the more important it is for the user to keep the respective value within its boundaries.

The simplest erection analysis takes place when the whole sequence of actions is predefined and no controlling parameters are to be found. The analyst just follows a "natural order of things" and takes the obtained results for granted. These results may be substantially different from those obtained under the conventional assumption that the system was created at once in its full configuration and only then loads were applied.

The necessity to take the erection sequence into account can be illustrated by a simple example of a two-span three-story frame analysis. As each story is being erected, the respective crossbar is attached to the columns via hinges and carries the load of 2.0 t/m. After that the crossbar joints are made monolithic, and the crossbars are loaded by the floor slabs which create the extra weight of 2.0 t/m. In this way all stories are erected (Fig. 8.8).

A successive analysis of the structure at the erection steps (*a*) ... (*f*) and the summation of the obtained results yields a bending moment diagram shown in the left part of Fig. 8.9. For comparison, there is a bending moment distribution on the right which would arise in the ready-made system were its crossbars subjected to the load of 4.0 t/m.

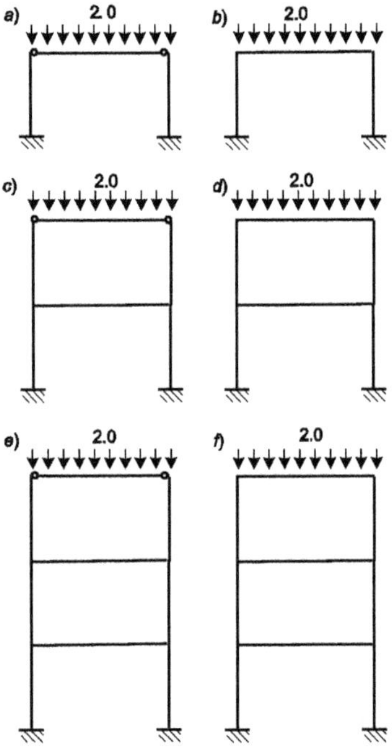

Fig. 8.8. Erection states of a plane frame

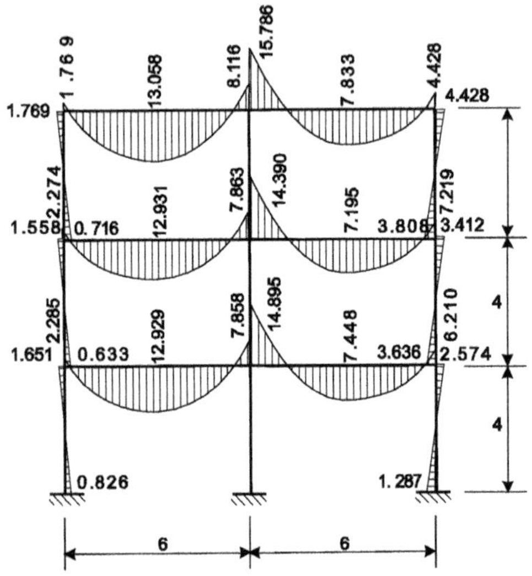

Fig. 8.9. Comparative diagrams of bending moments in a frame

Another characteristic example that shows how different these two approaches can be is given in [14]. The results relate to an analysis of a ferroconcrete spherical shell, square in its plane projection, assembled of elements 3×3 m big without scaffolds. The order of its erection is indicated by numbers on the schematic of the shell's quarter (see Fig. 8.10, *a*). The distribution of deflections at different steps of erection shown in Fig. 8.10, *b* demonstrates a substantial discrepancy between this analysis and the one that ignores peculiarities of the segmental cantilevering of the shell.

(*a*)

(*b*)

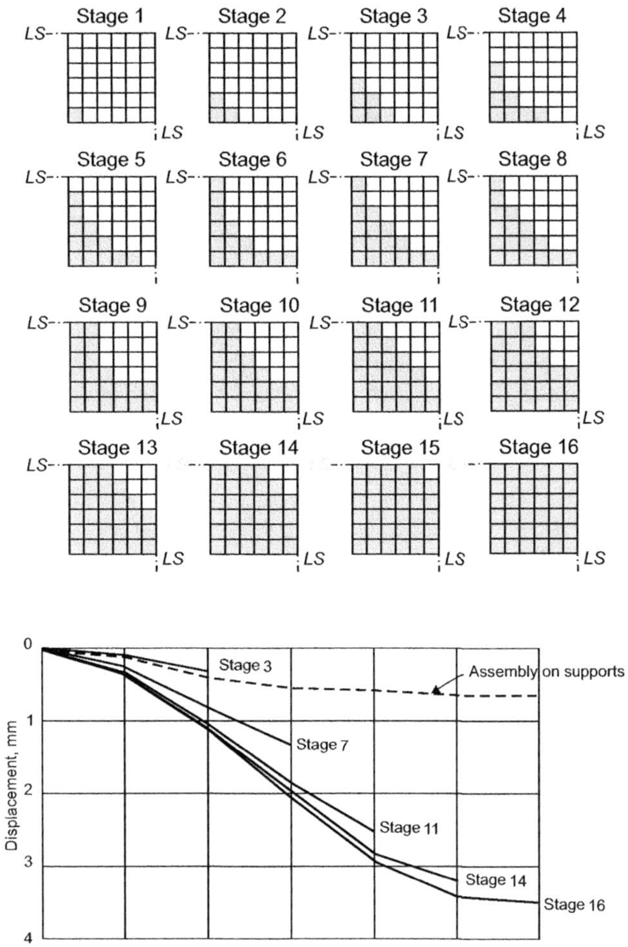

Fig. 8.10. Results of analysis of a shell erected by cantilevering

The erection using scaffolds could imitate to some extent the conventional approach to the analysis of structures as ones having no history and created "at once". It is easy for one to see the result of such an approach. This one has become quite common in the design practice and corresponds to a standard hypothesis of the elasticity theory (sometimes not even stated explicitly in educational courses): a "natural state hypothesis", or an assumption of no dislocation-caused stresses. The apparent reason for this is that the educational courses deal with two-dimensional and three-dimensional continuous problems for single connected areas where no dislocation stress may arise as shown in [23]. That's what V.V. Novozhilov says in this regard [23]:

"*The emergence of dislocation stresses is always related to a transformation of a stress-free body into another body with its degree of connectivity greater at least by one than that of the original body.*"

By the way, this is one of delicate and key differences between discrete and one-dimensional continuous systems, on one hand, and two/three-dimensional continuous systems, on the other hand, which is somehow missed by standard course of the mechanics of solids and structures. Really, the very concept of static determinacy in discrete and continuous bar systems implies the possibility of existence of a self-stressed state. At the same time, continuous two- and three-dimensional bodies are statically indeterminate by definition: their equilibrium equations are not sufficient to calculate the stress tensor components, at the same time no self-stressed state is feasible in single connected areas.

The majority of real structures are not single connected or could be such only when being created. The available experience of erection state analysis sequences shows that often one cannot help but consider a stress/strain state determined by the history of the structure's emergence and "embedded" in the system.

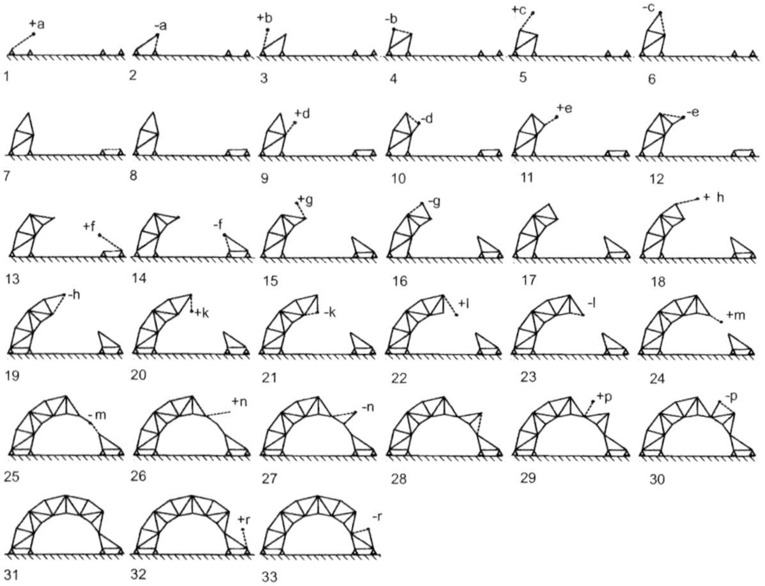

Fig. 8.11. A sequence of erection

Note that the investigation of the erection process can be based on transformations performed with pseudo-inverse Moore – Penrose matrices. Here we refer to the paper [26] dealing with a problem of determining a sequence of erection to minimize the number of additional braces that prevent a partially erected structure from being moveable.

Fig. 8.11 shows a result of such solution for an example of a simple arch system. This solution allows for an additional requirement that the temporary braces should both ensure the fixation of the structure and warrant that a certain intensity of stress in elements caused by their dead weight be not exceeded.

8.3 Pre-stressing

Sometimes a certain preliminarily planned pre-stressed state should be artificially created in a structure. The most popular technique of pre-stressing is an installation of special stressed elements [26].

It should be noted that when a stressed element is introduced into a structure, one should estimate the employed pre-stressing technique with all possible care. An underestimation of particularities of actions actually undertaken in the course of the erection can result in egregious blunders.

Let us demonstrate the statement by the example of a simple system shown in Fig. 8.12 where the mounted element having the tension/compression rigidity $EA = 20000$ is located between the nodes A and B, and the "system" is represented by a bar with its tension/compression rigidity $EA = 2000$, its flexural rigidity $EI = 2000$, and with infinitely rigid cantilevers. One can imagine a few choices for mounting this pre-stressed element. The choices are explicated below. The magnitude of the pre-stress will be assumed $P_0 = 2$ for the sake of determinacy.

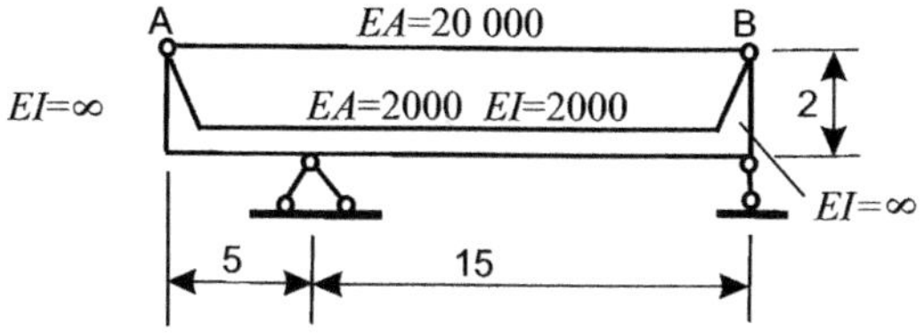

Fig. 8.12. An illustration of a pre-stressed element analysis

Choice **A**: Pulling the element "to the system".

In this case the unlinked element is pulled by the force $P_0 = 2$ according to a scheme shown in Fig. 8.13, *a*. After that the cut's ends are linked (Fig. 8.13, *b*) and the load $P_0 = 2$ is eliminated by applying the forces P_0 of the opposite sign to the common point of the linked structure (Fig. 8.13, *c*). The latter operation does not change the stress distribution (in the case considered) because two equal and oppositely directed forces applied to the same point of the structure are equivalent to no force at all. The stress distribution in the system is shown as diagrams of internal stresses in Fig. 8.14.

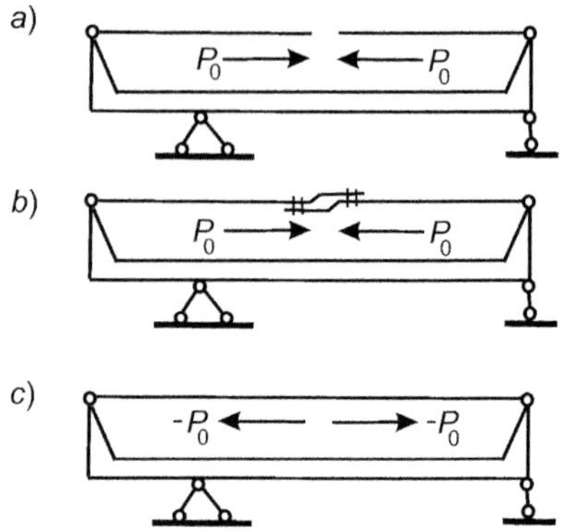

Fig. 8.13. A sequence of pulling "to the system"

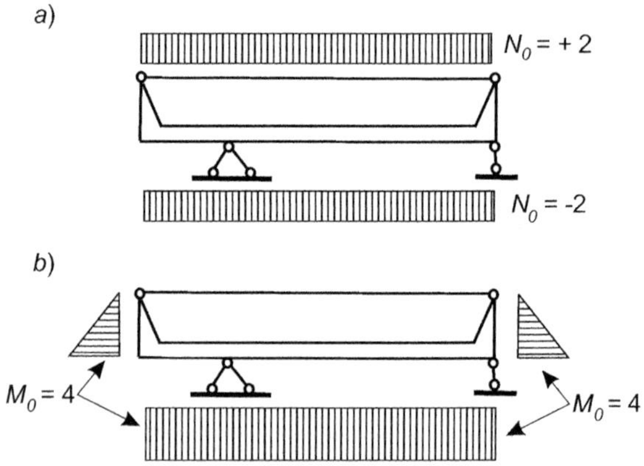

Fig. 8.14. Results of pulling "to the system"

Choice **B**: Pulling the element "to the ground" with linking to the system.

In this case the element is pulled by the forces $P_0 = 2$ which are balanced by the "ground" reactions in additional nodes A_1 and B_1 (Fig. 8.15, *a*). Then, this element already stressed is linked to the system (Fig. 8.15, *b*) which acquires no extra stress thereby.

Detaching the element AB from the nodes A_1 and B_1 can be represented as replacing the constraints AA_1 and BB_1 by their reactions (Fig. 8.15, *c*) and further eliminating these reactions by applying the opposite-sign forces. The components of these forces acting on the nodes A_1 and B_1 (the "ground") do not cause any stress in the system, while the components acting on the nodes A and B cause internal stresses to be taken into account.

The overall stress state is shown in Fig. 8.16 as diagrams of longitudinal forces.

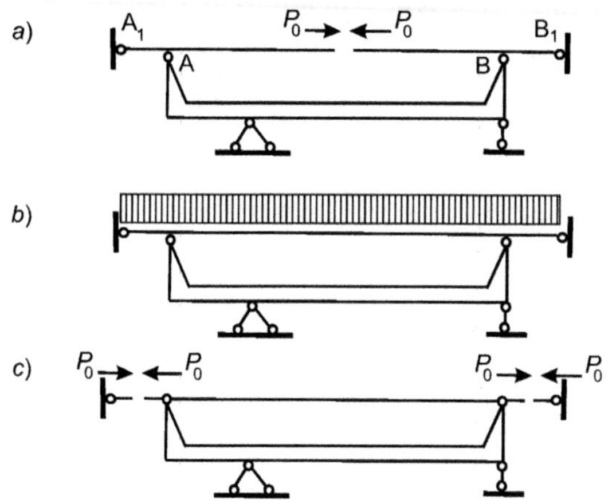

Fig. 8.15. Pulling "to the ground"

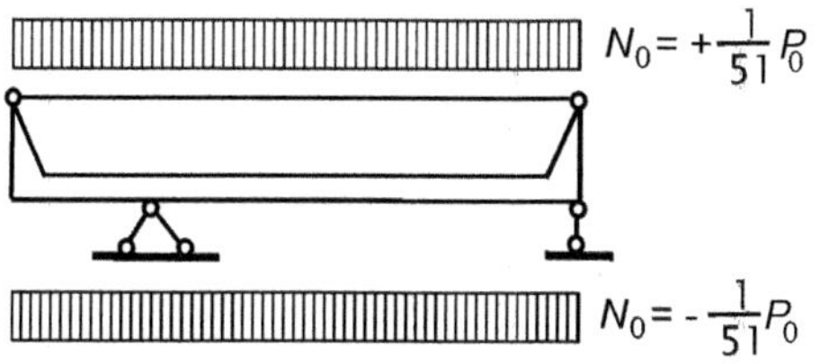

Fig. 8.16. Results of the analysis with the element pulled "to the ground"

<u>Choice **C**: Pulling the element "to the ground" partially.</u>

This option can be implemented according to a scheme shown in Fig. 8.17 or that in Fig. 8.18. Diagrams of the longitudinal forces in the pre-stressed state are shown in Fig. 8.17, *b* and Fig. 8.18, *b*, respectively.

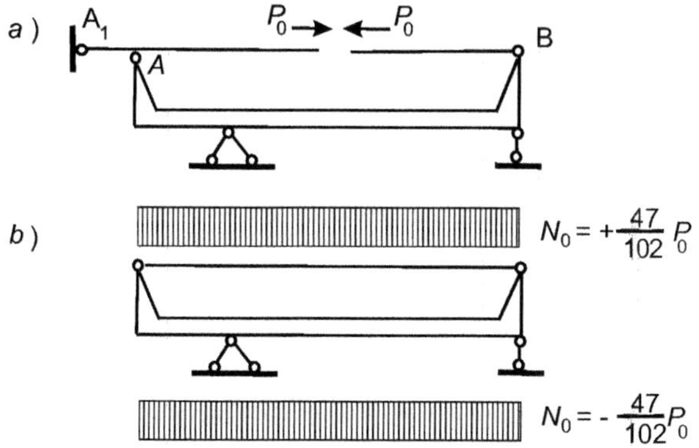

Fig. 8.17. Pulling "to the ground" partially (**C**-1 option)

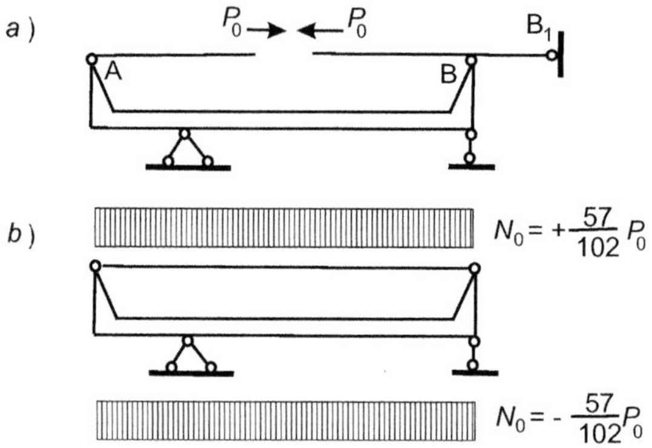

Fig. 8.18. Pulling "to the ground" partially (C-2 option)

Comparing the results of the analyses performed, one can see how at the same magnitude of the active load $P_0 = 2$ the stressed state of the system may vary substantially depending upon a pre-stressing scheme, ranging between 1/51 and 2.0, and $N_0 = P_0$ only when "pulling to the system". This is no occasional result because pulling to the system can be treated as a physical implementation of a known mechanical principle of constraint releasing (any constraint can be released without breaking the equilibrium if replaced by the reaction thereof), though applied inversely.

8.4 Structures with Hydraulic Jacks

The erection of heavy structures often involves a system of hydraulic jacks because the jacks are perfect mechanical devices to equalize forces transferred to supporting nodes. Therefore it makes sense to discuss peculiarities of the structural analysis of hydraulic jack systems in the section that relates directly to other design problems of structural erection states. Though, we should say the analysis of structures supported by the hydraulic jacks has a much wider scope of application than just the erection analysis. Note that similar problems arise when designing canal-lifts and some types of crane/hoist mechanisms.

To get off direct needs of structural engineering for a while, we'd like to note also some problems of geomechanics and biomechanics dealing with bodies which include cavities filled with a compressible or an ideal incompressible fluid (such as a problem of stress/strain analysis of an eyeball). By meshing a continuous problem of the stress/strain calculation in a two- or three-dimensional body having a cavity filled with liquid, we come to the same equations as those governing the models of structures supported by systems of hydraulic jacks.

So, let an elastic structure be supported by hydraulic jacks combined into a few independent groups (Fig. 8.19).

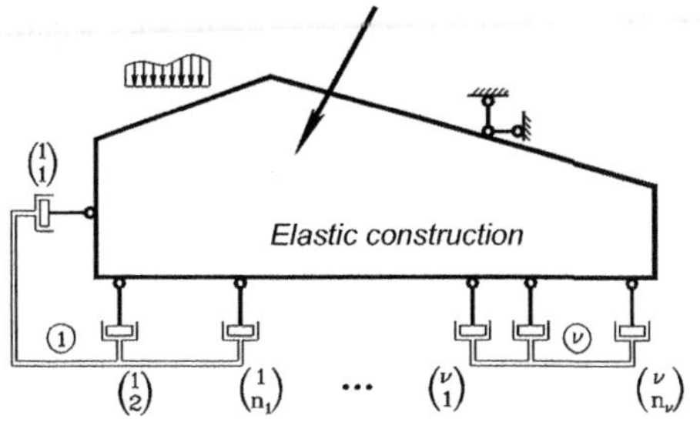

Fig. 8.19. An elastic structure supported by hydraulic jacks

Let the number of groups of jacks be ν, and the number of the jacks in a certain group No. α (α = 1,...,ν) be n_α. We number the jacks within each group by assigning an index pair (α, *i*) to every jack so that the first index indicate the number of a group and the second one the number of a jack within the group. For the purpose of visual convenience, in Fig. 8.19 the index pairs are presented vertically, the first index (No. of group) at the top and the second one (No. of a jack within the group) at the bottom. To simplify the system of indexes, we also assume to use Greek letters to indicate the jack groups and Latin letters for indicating jacks within groups.

The following notation will be used:

- $x_{\alpha,i}$ is a stress in the jack (α,*i*) which we assume positive when the piston is compressed;
- $z_{\alpha,i}$ is a displacement of the structure's point rigidly connected to the jack (α,*i*) in the direction of the piston's movement (it is positive when the volume of the liquid in the hydraulic cylinder increases);
- $F_{\alpha,i}$ is an area of the hydraulic jack's piston (α,*i*);
- p_α is a pressure of liquid in the jack group α.

The stress $x_{\alpha,i}$ in each jack of the group α is proportional to the piston area $F_{\alpha,i}$

$$x_{\alpha,i} = p_\alpha F_{\alpha,i} \,. \tag{8.14}$$

The condition of equal liquid pressures p_α in every jack of the same group gives

$$c_{\alpha,1}\, x_{\alpha,1} = c_{\alpha,2}\, x_{\alpha,2} = \ldots = c_{\alpha,n_\alpha} x_{\alpha,n_\alpha}\,, \qquad c_{\alpha,i} = F_{\alpha,0} \,/\, F_{\alpha,i}\,, \tag{8.15}$$

where $F_{\alpha,0}$ is a constant with the dimension of area.

The condition of a constant total volume of the liquid in each jack group implies the following relationship between displacements of jacks of the same group

$$f_{\alpha,1} z_{\alpha,1} + f_{\alpha,2} z_{\alpha,2} + \ldots + f_{\alpha,n_\alpha} z_{\alpha,n_\alpha} = \Delta_\alpha, \qquad f_{\alpha,i} = F_{\alpha,i} \,/\, F_{\alpha,0}\,, \tag{8.16}$$

where Δ_α is a known value determined by the volume of liquid additionally pumped to the hydraulic system of αth jack group.

Systems of the type being considered can be sometimes analyzed using a lever mechanism based model. For example, a group of eight jacks can be represented by a lever model in the way shown in Fig. 8.20. The relations between the lever arms should be selected so that the conditions (8.15) hold. For a big number of jacks, the lever model will contain a lot of extra elements and nodes.

Note that the lever mechanism consists of perfectly rigid elements, and if we use null-compliant elements to describe those, then the number of elements and nodes will be greater again. Also, in some situations the directions of the pistons of jacks in the same group are not parallel to one another (see the first group of jacks in Fig. 8.19). In such situation the lever model can hardly be constructed at all.

The said circumstances make it reasonable to develop special techniques to analyze structures which belong to the category in question. Let us first consider the problem with the following assumptions:

- (I) the principal system of the force method formed from the original structure by removing all jacks is kinematically unchangeable;
- (II) the fluid in the hydraulic system is incompressible.

As the stresses in all jacks of the same group are related as stated by (8.15), we can treat the set of n_α forces $x_{\alpha,i}$ as a single generalized force. Its definition is

$$X_\alpha = c_{\alpha,i}\, x_{\alpha,i} \qquad (i = 1,\ldots, n_\alpha)\,. \tag{8.17}$$

Let $g_{\alpha\beta,ij}$ be a displacement in the principal system in the direction of the (α,i) jack's piston caused by the force $x_{\beta,j} = 1$.

The compliance matrix of the principal system that corresponds to the generalized forces X_α will be denoted as $\|[g_{\alpha\beta}]\|$. The quantity $g_{\alpha\beta}$ is a virtual work done by the set of forces $c_{\alpha,1}x_{\alpha,1} = 1,\ldots, c_{\alpha,n_\alpha} x_{\alpha,n_\alpha} = 1$ at their respective displacements caused by the forces $c_{\beta,1}x_{\beta,1} = 1,\ldots,\ c_{\beta,n_\beta} x_{\beta,n_\beta} = 1$ applied to the principal system all at the same time, therefore

$$g_{\alpha\beta} = \sum_{i=1}^{n_\alpha}\sum_{j=1}^{n_\beta} g_{\alpha\beta,ij}\, f_{\alpha,i}\, f_{\beta,j}\ . \tag{8.18}$$

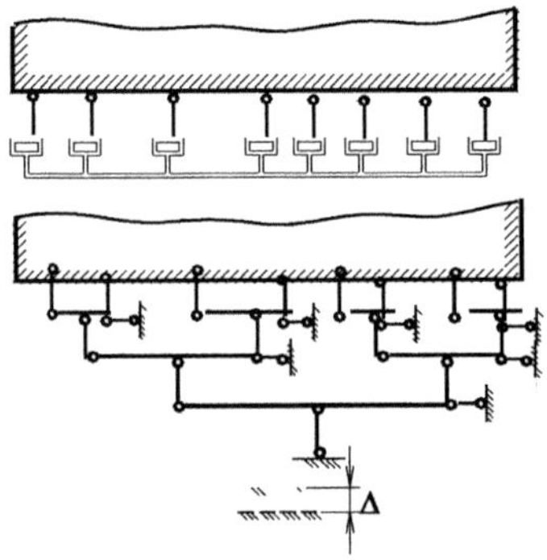

Fig. 8.20. A lever model of a hydraulic system consisting of eight jacks

Let $\Delta_{\alpha,ip}$ be the displacement in the principal system caused by the external load. Then the displacement $\Delta_{\alpha p}$ that corresponds to the generalized force X_α will be determined by the formula

$$\Delta_{\alpha p} = \sum_{i=1}^{n_\alpha} \Delta_{\alpha,ip} f_{\alpha,i} \,. \tag{8.19}$$

The complementary work in the system (a Castigliano functional K) can be represented as a quadratic form of the generalized forces X_α. Thus

$$\mathsf{K} = \frac{1}{2}\sum_{\alpha=1}^{\nu}\sum_{\beta=1}^{\nu} g_{\alpha\beta}\, X_\alpha X_\beta + \sum_{\alpha=1}^{\nu} \Delta_{\alpha p} X_\alpha - \sum_{\alpha=1}^{\nu} \Delta_\alpha X_\alpha \,. \tag{8.20}$$

The last sum in (8.20) is a work done by the forces $x_{\alpha,i}$ at the displacements of the jack pistons. One can see that

$$\sum_{\alpha=1}^{\nu}\sum_{i=1}^{n_\alpha} x_{\alpha,i} z_{\alpha,i} = \sum_{\alpha=1}^{\nu}\sum_{i=1}^{n_\alpha} c_{\alpha,i} x_{\alpha,i} f_{\alpha,i} z_{\alpha,i} = \sum_{\alpha=1}^{\nu} X_\alpha \sum_{i=1}^{n_\alpha} f_{\alpha,i} z_{\alpha,i} = \sum_{\alpha=1}^{\nu} \Delta_\alpha X_\alpha \,. \tag{8.21}$$

The stationarity condition of the functional K yields canonic equations of the force method

$$\sum_{\beta=1}^{\nu} g_{\alpha\beta} X_\beta + \Delta_{\alpha p} - \Delta_\alpha = 0 \quad (\alpha = 1,\ldots,\nu)\,. \tag{8.22}$$

Now let's demonstrate that the fluid balance (8.16) holds for the displacements $z_{\alpha,i}$ calculated by the forces X_α that satisfy the system of equations (8.22). We have

$$z_{\alpha,i} = \sum_{\beta=1}^{\nu}\sum_{j=1}^{n_\beta} g_{\alpha\beta,ij} x_{\beta,j} + \Delta_{\alpha,ip} \tag{8.23}$$

and thus

$$\sum_{i=1}^{n_\alpha} f_{\alpha,i} z_{\alpha,i} =$$

$$= \sum_{i=1}^{n_\alpha}\sum_{\beta=1}^{\nu}\sum_{j=1}^{n_\beta} g_{\alpha\beta,ij} f_{\alpha,i} f_{\beta,j} X_\beta + \sum_{i=1}^{n_\alpha} \Delta_{\alpha,ip} f_{\alpha,i} = \sum_{\beta=1}^{\nu} g_{\alpha\beta} X_\beta + \Delta_{\alpha p} = \Delta_\alpha \,. \tag{8.24}$$

Now let's discard the limiting assumption (I). If removing all jacks results in a changeable system having, say, μ rigid-body degrees of freedom, then the principal system can be made of that by introducing extra μ constraints to restore the kinematical stability. This gives us a conditional extremum problem for the functional K at additional restrictions in the form

$$\sum_{\beta=1}^{\nu} a_{\gamma\beta} X_\beta + a_{\gamma p} = 0 \qquad (\gamma = 1,\ldots,\mu)\,, \tag{8.25}$$

which require that zero stresses appear in the additionally introduced constraints that eliminate the changeability of the principal system. Here $a_{\gamma\beta}$ and $a_{\gamma p}$ are stresses in γ^{th} additional constraint caused by the unit generalized force $X_\beta = 1$ and the external load, respectively.

The Lagrangian multiplier method gives the following equations, instead of the system (8.22),

$$\sum_{\beta=1}^{\nu} g_{\alpha\beta} X_\beta + \sum_{\gamma=1}^{\mu} a_{\gamma\alpha}\lambda_\gamma + \Delta_{\alpha p} - \Delta_\alpha = 0 \qquad (\alpha = 1,\ldots,\nu), \tag{8.26}$$

which should be considered together with (8.25) as a single system of linear algebraic equations of the dimension ($\nu+\mu$).

The Lagrangian multipliers λ_γ are displacements in the original system in the direction of the additional constraints.

The quantity Δ_α that occurs above can be determined by the condition of the constant volume of the fluid

$$\sum_{i=1}^{n_\alpha} F_{\alpha,i} z_{\alpha,i} = V_\alpha = F_{\alpha,0}\Delta_\alpha, \tag{8.27}$$

where V_α is the volume of the fluid additionally pumped to the hydraulic system of αth group of the jacks. This implies that

$$\Delta_\alpha = V_\alpha / F_{\alpha,0}. \tag{8.28}$$

If we need to allow for the compressibility of the liquid thus canceling the limitation (II), we should consider the relation

$$x_{\alpha,i} = EF_{\alpha,i} \frac{V_\alpha - \sum_{j=1}^{n_\alpha} F_{\alpha,j} z_{\alpha,j}}{V_{\alpha 0}}, \tag{8.29}$$

where $V_{\alpha 0}$ is the original volume of the liquid in the hydraulic system of αth group of jacks, E is the elastic modulus of dilatation of the fluid. Using (8.16) and (8.17) with the last relation, we have

$$\Delta_\alpha = \frac{V_\alpha}{F_{\alpha,0}} - \frac{V_{\alpha 0}}{F_{\alpha,0}} \frac{X_\alpha}{EF_{\alpha,0}}. \tag{8.30}$$

Substituting (8.30) to (8.26) gives the following system of equations that governs the problem:

$$\sum_{\beta=1}^{\nu} (g_{\alpha\beta} + h_{\alpha\beta}) X_\beta + \sum_{\gamma=1}^{\mu} a_{\gamma\alpha}\lambda_\gamma + \Delta_{\alpha p} - \Delta_\alpha = 0 \qquad (\alpha = 1,\ldots,\nu),$$

$$\sum_{\beta=1}^{\nu} a_{\gamma\beta} X_\beta + a_{\gamma p} = 0 \qquad (\gamma = 1,\ldots,\mu), \tag{8.31}$$

where

$$h_{\alpha\beta} = \delta_{\alpha\beta} \frac{V_{\alpha 0}}{F_{\alpha,0} EF_{\alpha,0}}, \tag{8.32}$$

$\delta_{\alpha\beta}$ being the Kronecher's delta, and the value Δ_α is determined by the formula (8.28).

We do not consider here an explanatory example of such analysis, referring the readers to our paper [27] which contains a respective example.

8.4.1 Liquid Finite Element

The method of analysis of an elastic system supported by hydraulic jacks that we have discussed above is based on the method of forces. Sometimes this technique complies poorly with computer programs for structural analysis because most of those are based on the slope deflection method.

Let's demonstrate how we can build governing equations of the slope deflection method and allow for the hydraulic jacks at the same time. To do it, let us treat a group of jacks as a peculiar type of finite element and refer to such as a *liquid finite element.*

The full potential energy L accumulated by a deformation of a compressible fluid in α^{th} group of hydraulic jacks with its pressure p_α, can be represented as

$$\mathsf{L} = \tfrac{1}{2}\int p_\alpha \varepsilon_\alpha dV . \tag{8.33}$$

Here ε_α is a relative dilatation, and the integration is to be performed over the whole volume V of the fluid in α^{th} hydraulic system.

Seeing that

$$\varepsilon_\alpha = \frac{\sum_{j=1}^{n_\alpha} F_{\alpha,j} z_{\alpha,j} - F_{\alpha,0}\Delta_\alpha}{\mathrm{V}_{\alpha 0}}, \tag{8.34}$$

and the pressure relates to ε_α according to the elasticity law

$$p_\alpha = E\,\varepsilon_\alpha , \tag{8.35}$$

the substitution of (8.34) and (8.35) to (8.33) will make L a quadratic function of the jack piston displacements (up to a constant independent of $z_{\alpha,i}$), that is, a Lagrangian functional

$$\mathsf{L} = \frac{1}{2}E\frac{\sum_{i=1}^{n_\alpha}\sum_{j=1}^{n_\alpha} F_{\alpha,i}F_{\alpha,j} z_{\alpha,i} z_{\alpha,j} - 2F_{\alpha,0}\Delta_\alpha \sum_{i-1}^{n_\alpha} F_{\alpha,i} z_{\alpha,i}}{\mathrm{V}_{\alpha 0}}, \tag{8.36}$$

and the desirable components of the stiffness matrix of the liquid finite element can be found as second partial derivatives

$$k_{\alpha,ij} = \frac{\partial^2 \mathsf{L}}{\partial z_{\alpha,i}\partial z_{\alpha,j}} = \frac{E}{\mathrm{V}_{\alpha 0}} F_{\alpha,i}F_{\alpha,j} \quad (i = 1,\ldots, n_\alpha) . \tag{8.37}$$

The elements of the load vector related to the effect of the given displacement Δ_α are also not hard to find

$$k_{\alpha,ip} = -\frac{\partial^2 L}{\partial \Delta_\alpha \partial z_{\alpha,i}} = \frac{E}{V_{\alpha 0}} F_{\alpha,i} F_{\alpha,o} \qquad (i = 1,\ldots, n_\alpha)\,. \tag{8.38}$$

The formula (8.37) and (8.38) determine all needed stiffness properties of a liquid finite element that possesses n_α degrees of freedom.

We emphasize that the formula given above determine the components of the stiffness matrix $\mathbf{K}_\alpha = \|[k_{\alpha,ij}]\|$ $(i,j = 1,\ldots, n_\alpha)$ and the load vector $\mathbf{p}_\alpha = \|[k_{\alpha,ip}]\|$ for nodal displacements of a liquid finite element that corresponds to α^{th} group of hydraulic jacks co-directed with the jack pistons. The liquid finite element does not resist in the transverse direction to the pistons. This circumstance should be taken into account when forming the global stiffness matrix **K** and the nodal load vector **p** of the structure which will possibly require a preliminary coordinate transformation of $\mathbf{K}_\alpha$ and $\mathbf{p}_\alpha$ if the piston directions are different from those of the local (nodal) coordinate system axes. We do not address this procedure here in any more detail because it differs little from standard techniques of coordinate transformations applied to finite element stiffness matrices.

Finally, we would like to note that the slope deflection method degenerates at $E\to\infty$ (an incompressible fluid) while the equations of the force method fail to work at $E\to 0$ (a gas cushion used in pneumatic constructions).

8.5 A "Structure – Foundation" Model

8.5.1 Allowing for a Distributive Ability of Soil

The analysis of structures on elastic foundations includes an issue how to allow for distributive properties of a foundation ignored by a simplest Winkler simulation (a piano-key model). Most real soils possess a distributive ability which enables them to resist to a load with their adjacent unloaded massifs as well as parts directly subjected to the load. So, in order to allow for the distributive ability of the foundation, one has to, first, use foundation models other than Winkler's, and second, introduce parts of the soil bed beyond the foundation area into the respective design model. The latter is necessary because the work of interaction forces between the structure and its elastic foundation is spent to accumulate the potential energy of the bed both within the structure's site and beyond it.

Drawbacks of the Winkler's mechanical model of foundation are well known to designers. It is principally incapable of describing a real interaction between a structure and its bed soil properly. Its main advantage is its ultimate mathematical simplicity. That's why, though the Winkler's model possesses obvious and incurable vices, it continues being widely popular and applicable. Just ask any engineer (who is not utterly left-handed to his profession) why he confines himself to an analysis with such an imperfect and by no means true-to-life model, and you will have a plethora of answers ranging from discursive explanations to bewildered excuses all threaded by the same motive: it is extremely easy to calculate[2].

Generally, the force of interaction r between a structure and its elastic bed is related to the displacement w of the bed's day surface via an integral dependency (a two-dimensional analysis is under consideration)

$$w(x,y) = \iint_{\Omega} G(x,y,\xi,\eta) r(\xi,\eta) d\Omega \ , \tag{8.39}$$

where $G(x,y,\xi,\eta)$ is a so-called *bed kernel* being actually an influence function, a displacement of the point $P(x,y)$ of the bed's day surface caused by a unit force applied at the point $Q(\xi,\eta)$ in the direction of the normal to the surface. In (8.39) the integration is performed over a two-dimensional region Ω where the function r is nonzero[3].

The best developed models of elastic bedding are those where the sag of the soil's *day surface* caused by the unit force is a surface of revolution. That is, the kernel is actually a function of the distance ρ from the force application point $Q(\xi,\eta)$ to the displacement measurement point $P(x,y)$

$$G(x,y,\xi,\eta) = G(\rho), \qquad \rho^2 = (x-\xi)^2 + (y-\eta)^2 \,. \tag{8.40}$$

Field tests show [15] that the Winkler's model describes the reality better in the cases of oozy, peaty soil or fine-grained water-saturated sand. The kernel G of a Winkler's bed is a Dirac delta function

$$G(\rho) = \delta(\rho)/k \ , \tag{8.41}$$

where k is a so-called *bedding value.*

For a cohesive soil the application of the Winkler's model is utterly baseless and only distorts the real behavior of a structure on an elastic bed. An elastic semi-space model (G.E. Proctor, N.N. Ghersevanov, B.N. Zhemochkin, M.I. Gorbunov-Posadov et al.) treats the bed as an isotropic elastic body of infinite dimensions both along its plane and inwards. The solution of the Boussinesq problem known from the elasticity theory gives us the following elastic semi-space kernel

$$G(\rho) = \frac{1-\nu^2}{\pi E \rho} \ , \tag{8.42}$$

where ν and E are the soil's coefficient of lateral strain and the elasticity modulus respectively. Compared to the Winkler's model, the elastic semi-infinite space overestimates the distributive ability of the soil, which fact was confirmed by numerous experiments [15].

The distributive ability of an elastic bed model can be reduced by assuming the compressible soil body to have a finite thickness H rather than to be infinitely

---

[2] It sounds like an old joke telling of a tipsy man on his knees under a street-lamp by night looking for his lost purse. When he is asked how he can be sure the purse is near the lamp-post, the guy answers he has lost his purse somewhere else but he is looking for it here where light is better.

[3] The r function is sometimes called a *back pressure* of an elastic bed.

deep. Thus the kernel will be a function of both the distance ρ and H as a parameter. This gives us a finite-thickness elastic layer model of the bed. The kernel of the finite-thickness layer is a complex function that depends also on boundary conditions specified at the *night surface* of the elastic bed[4]. One can use the finite element method to analyze a finite-thickness layer and a structure upon it as a single integrated design model. Though, the computational effort needed for this will be greater by an order of magnitude due to an increased dimensionality of the problem (the two-dimensional flexural plate analysis becomes three-dimensional when involving the elastic layer underneath). Again, a notorious "malediction of dimensionality" comes to mind.

8.5.2 A Two-parametric Bed Model

For today, a few tens of suggestions how to improve the mechanical models of the soil bed have been advanced. Apparently, the next simplest model in its mathematical formulation after the Winkler's model is a two-parametric elastic bed model suggested by Pasternak [24], Filonenko-Borodich [12], Vlasov and Leontiev [36] independently and on the basis of totally different assumptions. On one hand, the two-parametric bed model eliminates the key drawback of the Winkler's model (it allows for a distributive ability of the soil), on the other hand, it complicates the mathematical formulation of the problem very little comparing to the Winkler's model.

To describe a two-factor bed, the kernel is assumed to be a modified second-kind zero-order Bessel function (sometimes called a McDonald function) K_0 [24], namely

$$G(\rho) = \frac{1}{2\pi C_2} K_0(\rho\sqrt{C_1 / C_2}) \ , \tag{8.43}$$

where C_1 and C_2 are its first and second bedding values. The first one describes the bed's compression stiffness while the second one its shear rigidity.

For a two-parametric elastic bed, the integral relationship (8.39) can be inversed, and the relation between w and r becomes a simple differential dependency

$$r = -C_2\nabla^2 w + C_1 w \ , \tag{8.44}$$

where ∇^2 is the Laplacian operator

$$\nabla^2 = \frac{\partial^2}{\partial x^2} + \frac{\partial^2}{\partial y^2} \ ,$$

[4] Soil experts refer to the upper (exterior) surface of soil on which a superstructure stands as a *day surface*. So, a symmetric term of a *night surface* suggests itself for the lower surface of the soil body that does not experience any displacements. This name is inspired by an everlasting darkness in depths of the Earth.

and it is the simplicity of the differential dependence (8.44) that makes this model so attractive mathematically, thus computationally too.

The two-parametric model, unlike the Winkler's model, predicts a plate's bending under a uniform loading and a sag of the elastic bed beyond the structure's plane projection.

Suppose that a flexural plate with unrestrained edges lies on a two-parametric bed. We can prove that the two-parametric model causes so-called fictitious shear forces to appear on the unrestrained edges of the plate. It is exactly this circumstance that has been the reason for numerous attacks at the two-parametric model by adherents of other models (such as an elastic semi-space or a finite-thickness layer). The latter treat this fact as a mathematical paradox that distorts the physical essence of the phenomenon for the sake of pure mathematics.

Actually, it is not hard to prove that there is no mathematical paradox involved here. For example, let's take an elastic semi-space or a finite-thickness layer. An accurate solution for a plate lying on such bed contains a singularity in the contact stress (the bed's back pressure r) under the edges of the plate: the point value of the back pressure is infinite. To illustrate the point, let us present an expression of the back pressure r arising under the foot of a round stamp of a radius R pressed by a force P into an elastic semi-space [6]

$$r(\rho) = \frac{P}{2\pi R\sqrt{R^2 - \rho^2}}, \tag{8.45}$$

where $\rho \le R$ is the current radial coordinate of the stamp's point under which the back pressure r is calculated.

The order of the singularity depends on the model used, and all the two-parametric model is peculiar of is that the singularity is a Dirac delta function that represents a concentrated force mathematically. But if one is not confounded by an infinite back pressure function at the edge of the plate, one need not be so anxious about the concentrated force encountered in the same function and at the same point!

A symptomatic fact is a renewal of the interest to the two-parametric elastic bed model by soil mechanics experts [21]. The model itself has been further developed, and a discrete counterpart of the two-parametric elastic bed model has been addressed in publications [13]. Such a discrete two-parametric model is shown in Fig. 8.21 where a one-dimensional problem is shown for the sake of simplicity.

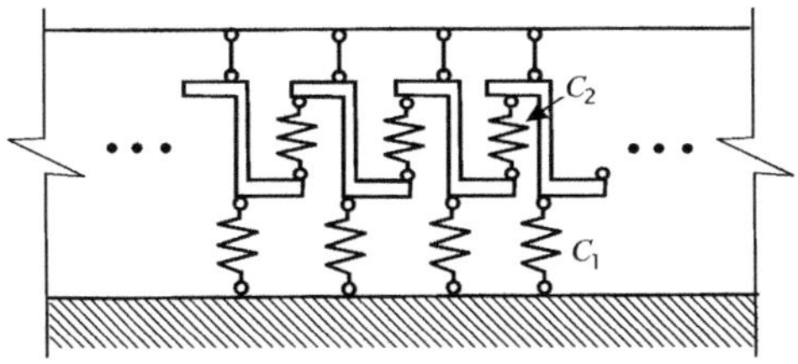

Fig. 8.21. A discrete analogue of the two-parametric model

In this model Z-shaped perfectly rigid elements are connected to the ground by a set of springs being a discrete analogue of the C_1 coefficient (one describing the bed's compression rigidity), while springs installed between adjacent Z-shaped elements simulate the coefficient C_2 (one characterizing the shear rigidity of the bed).

An unquestionable computational convenience of the two-parametric model is the fact that the model is seamlessly integrated with the standard finite element analysis procedure without adding any new degrees of freedom to the design model of a structure together with its bed. All that one has to do is to make corrections to the expressions of the stiffness matrices of finite elements of a flexural plate lying on an elastic bed. These corrections are easy for both the thin plate and the medium-thickness plate theories [10].

In order to outline the general approach of the finite element analysis procedure as related to structures lying on an elastic two-parametric bed, we employ an expression of the full potential energy of the system "structure + bed".

$$\mathsf{L} = \mathsf{E}_k + \mathsf{E}_0 - \Pi \ , \tag{8.46}$$

where E_k is an energy accumulated by the structure's deformation itself, E_0 is that accumulated by the elastic bed, Π is the work of external forces.

It can be seen in (8.46) that the elastic bed included in the combined system is allowed for by the expression of E_0.

The expression of the energy E_0 of a two-parametric bed model can be easily derived in a formal way by restoring a functional for which the differential equation (8.44) is its Euler equation. Though, we'd like to see to interests of the potential reader of this book who seems to be rather a mechanics guy than a mathematician. Therefore we will derive the E_0 energy expression by considerations of a mechanical nature. With this purpose in mind, we turn to an interpretation of the two-parametric model suggested by Filonenko-Borodich [12].

According to Filonenko-Borodich, the two-parametric model of an elastic bed in the one-dimensional case is an inextensible thread, infinite in both directions, tensioned by the force C_2 and connecting top ends of continuously arranged springs with the distributed stiffness C_1 — a so-called *laminar* model. In the two-dimensional case the laminar model is replaced by a *membrane* model where an inextensible membrane with the tension C_2 is used instead of the thread. It can be imagined how the thread tension C_2 is created by a mechanical device where one of its ends is restrained from displacements and a weight C_2 is suspended from the other end of the thread roved through a pulley to create a proper tension — see Fig. 8.22.

As the surface of the elastic bed sags which is measured by $w(x)$, the energy E_0 consists of two components:

$$\mathsf{E}_0 = \mathsf{E}_{01} + \mathsf{E}_{02} \ , \tag{8.47}$$

where E_{01} is an energy of the sagged springs, E_{02} is a potential energy accumulated by the suspended weight lifted by a certain height Δ. It is easy to calculate each component.

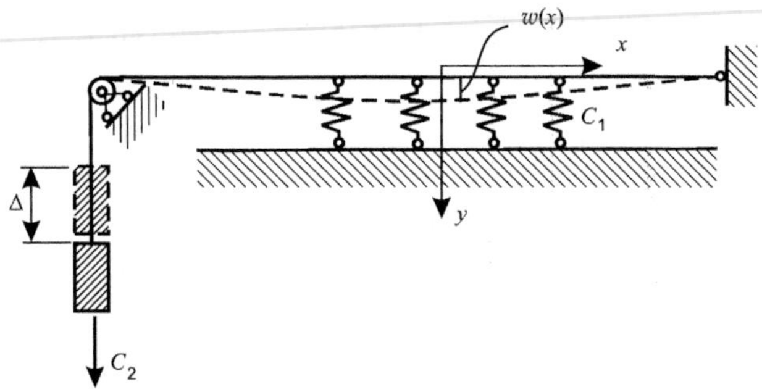

Fig. 8.22. To the determination of the elastic bed energy

The result is

$$\mathsf{E}_{01} = \frac{1}{2} C_1 \int_{-\infty}^{\infty} w^2 dx\,, \qquad \mathsf{E}_{02} = C_2 \Delta = \frac{1}{2} C_2 \int_{-\infty}^{\infty} \left(\frac{dw}{dx} \right)^2 dx\,. \qquad (8.48)$$

Generalizing the obtained expressions of the energy E_0 to the two-dimensional case and allowing for a variation of the coefficients C_1 and C_2 over the plane, we have

$$\mathsf{E}_0 = \frac{1}{2} \iint_{\Omega} [C_1 w^2 + C_2 (w_{,x}^2 + w_{,y}^2)] d\Omega + \frac{1}{2} \iint_{\Omega_e} [C_1 w^2 + C_2 (w_{,x}^2 + w_{,y}^2)] d\Omega\,. \qquad (8.49)$$

Here Ω is a two-dimensional area in the plane (X,Y) in which the structure touches its elastic bed (we assume that the structure contacts with the bed via a flexural foundation slab); Ω_e is an area of the (X,Y) plane external to the area Ω, that is, a complement of the Ω area to the whole plane. The character w designates a displacement of the structure's foundation slab normal to the surface of the bed, and subscripts after commas denote partial derivatives with respect to coordinates. The first integral in (8.49) is a basis to correct the stiffness matrix of the foundation slab's finite elements, while the integral over Ω_e in (8.49) allows for the bed's response beyond the slab's foot.

The part of the bed located in the exterior area Ω_e can be analyzed using "*semi-infinite finite elements*", *wedge*-shaped or *strip*-shaped, such as suggested in [10]. For example, these peculiar semi-infinite elements are implemented in the **SCAD** software. Such elements enable one to simulate the whole surrounding of the Ω area if its is convex and polygonal (Fig. 8.23).

One should understand clearly that these finite elements are able to provide only an approximate description of the behavior of the bed beyond the slab's foot. The vertical displacements of the bed's day surface vary in the direction of the slab's boundary normal according to an exponential law which describes their asymptotic behavior in the infinity quite well, but that's the only thing it does. When gradually decreasing the dimensions of the mesh only within the slab's area and keeping all other parameters intact, we thereby change only one of the sizes (namely the width) of the strip-shaped semi-infinite elements.

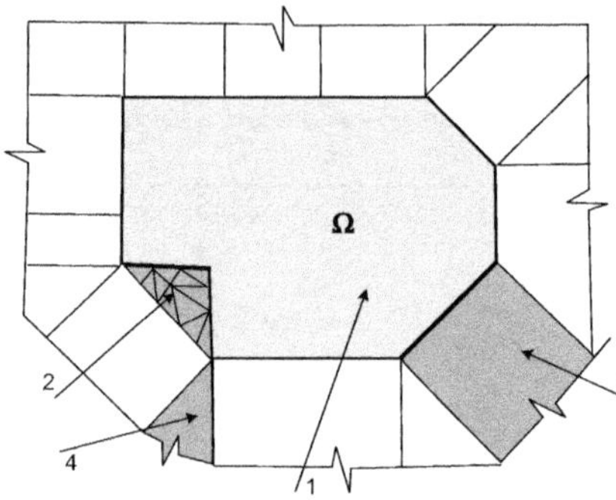

Fig. 8.23. Arrangement of the off-contour wedge-like and strip-like finite elements: (1) the slab; (2) a complement of the Ω area to a convex one; (3) a strip element; (4) a wedge element

This strategy of using a sequence of nested meshes does not provide a convergence to an accurate solution of the problem. Rather, it provides a convergence to a solution of a close problem that differs from the original one due to a certain distortion of the bed's response in the exterior of Ω. Of course, the engineer is always able to reduce the effect of such distortion by surrounding the area Ω from outside by two or three (or more) rows of conventional finite elements with zero flexural stiffness, but practical analysts use this technique rarely. It is important for an analyst to understand principles of origination and existence of the distorting effect of the semi-infinite elements in order to solve his problems consciously rather than formally. As for the software tools, they can be of help for an engineer by providing needed estimates at least in average (energetically) if the software used is capable of calculating the deformation energy of both the whole structure and its arbitrary selected part[5]. It is worth noticing also that the employment of semi-infinite elements is not the only way to allow for the contribution of the exterior elastic bed to the whole structure's behavior.

For example, to solve this problem G.S. Kolosova [17] used a technique of complementing the Ω area to a circular area Ω_1 that surrounded the former, $\Omega \subseteq \Omega_1$. Inside Ω_1 the solution was built by the conventional finite element analysis, while in the exterior of Ω_1 it is done analytically by expanding into a trigonometric series by the contour coordinate, with a subsequent matching of these two solutions. In spite of this technique being theoretically attractive, it is not implemented in any industry-oriented software, as far as the authors are aware.

Returning to semi-infinite elements, we should note that an area can be polygonal only to a certain extent. If the Ω area is non-convex or not single connected, it should be complemented to a convex single connected area by limited-size finite elements. The thickness of the slab in the added parts should be assumed zero.

[5] We draw the attention of the software developers to this circumstance as one more argument "pro" equipping commercial software tools with a special function for the calculation of the deformation energy of a structure and its parts.

The use of available finite elements on an elastic bed (bars, plates, shells) and special off-contour elements only does not enable creating an arbitrary design model of an elastic-bed-based structure. In particular, there can be difficulties when trying to build a model of a dam subject to plane strain if the software used (such as **SCAD**) cannot use elements like a deep beam on an elastic bed.

The issue can be resolved in a simple way by including bar elements on an elastic bed between the dam's boundary and the soil. The rigidity of such bars can be zero. In the same way a zero-rigidity slab can be put under a spatial body part of the design model. The same technique of introducing zero-rigidity plate elements into a model is employed when it is necessary to allow for a mutual influence of two separate (but adjacent) structures on an elastic bed. In this case the gap between the two spatially separated structures is filled with fictitious plate elements of zero rigidity, and the joint response of the structures is provided by their elastic bed only.

8.6 Assigning Properties of a Two-parametric Elastic Bed

The most reasonable way to assign the coefficients C_1 и C_2 of a two-parametric elastic bed seems to be that based on data of field tests.

It is known [24] that the problem of a round stamp of a radius R lying on an elastic bed with two bedding factors under an eccentric force P (Fig. 8.24) has the following solution that describes the sag w_o at the center of the stamp and the stamp's slope φ

$$w_0 = \frac{P}{\pi R^2 C_1} \frac{\xi K_0(\xi)}{\xi K_0(\xi) + 2K_1(\xi)}, \qquad \varphi = \frac{4Pe}{\pi R^4 C_1} \frac{\xi K_1(\xi)}{\xi K_1(\xi) + 4K_2(\xi)}, \tag{8.50}$$

where these designations are introduced in addition to those already known: e is the eccentricity of the P force application point; K_1 and K_2 are first and second order McDonald functions of a dimensionless argument ξ

$$\xi = R\sqrt{C_1 / C_2}\,. \tag{8.51}$$

The two equations of the system (8.50) yield

$$\frac{w_0 e}{R^2} = f(\xi), \tag{8.52}$$

where the transcendental functions $f(\xi)$ is

$$f(\xi) = \frac{K_0(\xi)}{4K_1(\xi)} \frac{\xi K_1(\xi) + 4K_2(\xi)}{\xi K_0(\xi) + 2K_1(\xi)}\,. \tag{8.53}$$

If an engineer has results of stamp tests at his disposal, then he can determine the dimensionless parameter ξ by solving the transcendental equation (8.52) to substitute it to the first equation of (8.50) and calculate the value of the C_1 coefficient, then use (8.51) to calculate the second coefficient C_2.

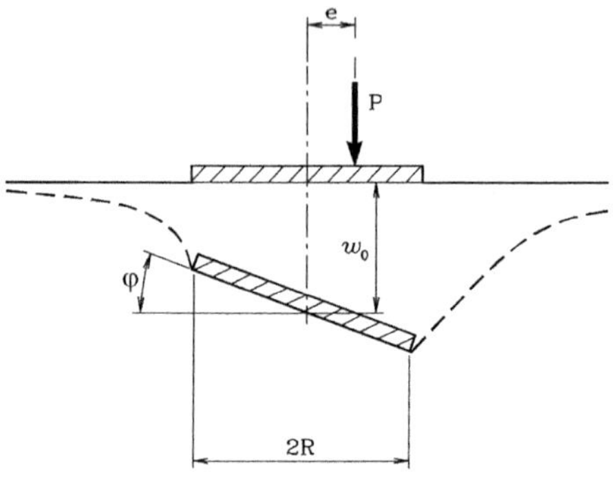

Fig. 8.24. A round stamp under a force and a moment

This is the way recommended by P.L. Pasternak [24] who has built a plot of the function $f(\xi)$ for convenience of the calculation, which we reproduce here in Fig. 8.25.

Evidently, most reliable results are obtained from a series of tests with stamps of various diameters. As Pasternak states justly, if a statistical processing of the coefficients C_1 and C_2 calculated by results of field tests shows a sufficient stability of the coefficients within a construction site, then there will be a reliable basis for numerical values of bedding factors to be included in a design model of a structure and its elastic bed.

Alas, but this is exactly the kind of data (field test results) the engineer seldom has access to, because such tests are performed, at best, only when very responsible structures are to be designed.

Acting in compliance with design codes, one can perform an analysis of a compressible soil body and determine conditional (averaged across the soil thickness) properties of the elastic layer: E, H and ν. Having a computational tool capable of processing a two-parametric elastic bed model, an analyst will necessarily aim to switch from an elastic finite-thickness layer to a two-parametric bed.

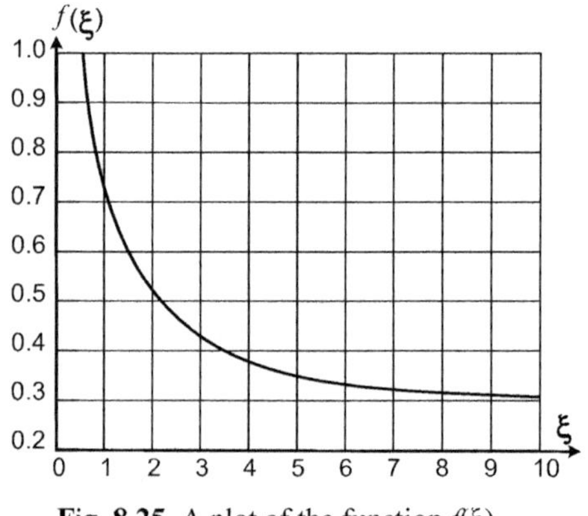

Fig. 8.25. A plot of the function $f(\xi)$

In other words, a problem arises which can be called "*modeling of a model*". The problem is: given the layer parameters E, H and ν, choose two bedding values C_1 and C_2 so that the two-parametric model be a best approximation of the elastic layer model in a certain sense.

The practical need for this kind of modeling has been recognized for a long time, therefore a variety of approaches have been suggested to solve the said problem. Those could be a basis for practical recommendations. For example, two following requirements are used as conditions of an equivalency between the original and replacing models in [8]:

- equal sags of a symmetrically loaded rigid stamp in the original and the replacing models;
- equal back pressures at the central point of the same stamp.

These requirements that make use of an engineering intuition enable an unambiguous determination of a two-parametric bed. At the same time, these requirements are arbitrary to a certain extent. For example, one of them could be replaced by a condition, say, of equal stamp slopes in the original and replacing models, or another condition.

A different approach to the issue of the approximate equivalency of models was adopted by Barvashov [3] who treats the proximity of two models as a proximity of their respective kernels. Due to the fact that the kernels are hard to compare directly, in [3] the kernel proximity condition is replaced by a requirement of proximity of their Hankel integral transforms. This approach is attractive because in a limiting case (totally equivalent transforms) the kernels of the original and the replacing models are equivalent, too, thus the solutions of the respective contact problems coincide. This approach has the following drawback: there is no strict mathematical definition of the measure of proximity between the transforms and kernels.

The paper [31] discusses an attempt to compare two elastic bed models with the energy as a measure of their proximity. It seems already a traditional approach to the construction of practical mechanical models. Without going into mathematical details, we note that the research in [31] builds a general algorithm to establish the values of the parameters C_1 and C_2 by iterations, which warrants the best energetic proximity of two elastic bed models.

It has turned out that the best coefficients C_1 and C_2 in the said energetic sense depend not only on the parameters of the original models but also on the plane dimensions and shape of a structure, and what is most important, on its rigidity. The effect of the rigidity of a flexural foundation slab on the coefficients of the two-parametric elastic bed model can be estimated by plots shown in Fig. 8.26. This figure shows the solution of the problem for a round stamp of the radius R by solid lines, and the dash lines demonstrate a solution regarding the effect of a uniformly distributed load directly on a round spot of the bed of the radius R (the zero rigidity of the slab).

The plots in Fig. 8.26 relate to two functions f_1 and f_2 depending on the relative size of the stamp or a loaded area, that is, on the value R / H.

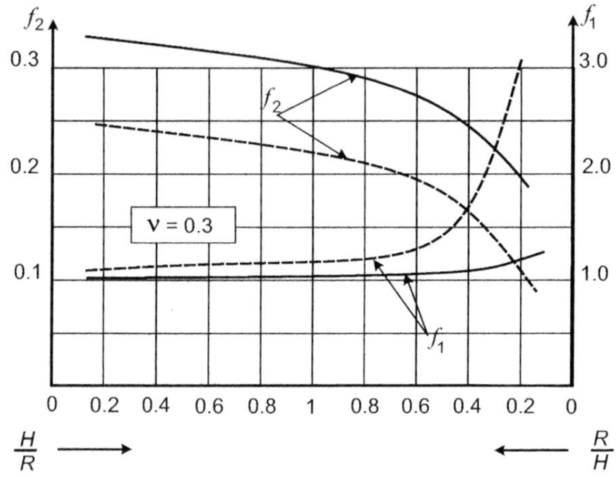

Fig. 8.26. Plots of functions f_1 and f_2

The coefficients C_1 and C_2 are related to values of these functions as follows:

$$C_1 = \frac{E_0}{(1-\nu_0^2)H} f_1 , \qquad C_2 = \frac{E_0 H}{2(1+\nu_0)} f_2 , \tag{8.54}$$

where

$$E_0 = E / (1-\nu^2) , \qquad \nu_0 = \nu / (1-\nu).$$

These diagrams show that at a relatively small depth H of the compressible body the function f_1 (thus the coefficient C_1 too) is little dependent on the slab's rigidity. The function f_2 (hence the coefficient C_2) is more sensitive to variations of the rigidity of the foundation slab.

8.6.1 A "*CCC*" Bed Model

This is a name given by V.A. Barvashov and V.G. Fedorovsky [4] to their elastic bed model. They proposed to improve the two-parametric elastic bed by introducing an additional layer of Winkler's springs with a distributed stiffness C_3 which covered the membrane model by Filonenko-Borodich from above. It is clear a priori that an additional numerical parameter should allow a more accurate description of properties of a real soil bed by a formal mathematical model. Also, this model deprives the opponents of the two-parametric model of their jokers because no fictitious shear forces appear at the edges of foundation slabs lying on a *CCC* bed. Authors of the paper [4] paid attention also to the fact that the application of this model was especially relevant in the case of a piling under the structure's foundation.

There is another important reason why this model can be of interest for designers. The matter is that well-developed CAD programs capable of operating the two-parametric model (such as **SCAD**) do not require any reworking at all to perform structural analyses that involve the three-parametric bedding.

To allow for that, it suffices to change the source data for the analysis: connect a conventionally defined two-parametric part of the elastic bed to the above foundation slab by discrete springs located in nodes of the slab's finite element mesh. As for the mesh of the two-parametric part of the bed, it is created by simply translating the foundation slab's finite element mesh down the vertical. All that was said is enough to understand why the *CCC* model is worth discussing here in more detail.

Let's introduce a new term for our further consideration. Together with the day and night surfaces of the elastic bed, let us define its *evening surface* as a plane where the membrane with the tension C_2 is located as shown in Fig. 8.27 [6].

Let w be the displacements of the day surface of the elastic bed, and v be the displacements of its evening surface. It is easy to notice the following relationships between the back pressure r and displacements of the day/evening surfaces:

$$r = C_3(w - v), \qquad r = -C_2\nabla^2 v + C_1 v \; . \tag{8.55}$$

Let's consider a problem of a round stamp of a radius R pressed into a *CCC* elastic bed by a centrally applied force P. The sag of the stamp will be denoted as w_0. As it is known, in the polar coordinates (ρ,θ) the differential Laplacian operator ∇^2 looks like

$$\nabla^2(\,) = \frac{1}{\rho}\frac{\partial}{\partial\rho}(\rho\frac{\partial()}{\partial\rho}) + \frac{1}{\rho^2}\frac{\partial^2()}{\partial\theta^2} \; , \tag{8.56}$$

and there will be no derivative with respect to the angular coordinate in (8.56) because the load is axially symmetric in the present case. Further, beyond the stamp the back pressure r is zero (the displacements of the day surface follow those of the evening surface $w = v$), and the first formula of (8.55) can be used to determine the back pressure under the stamp by assuming $w = w_0$.

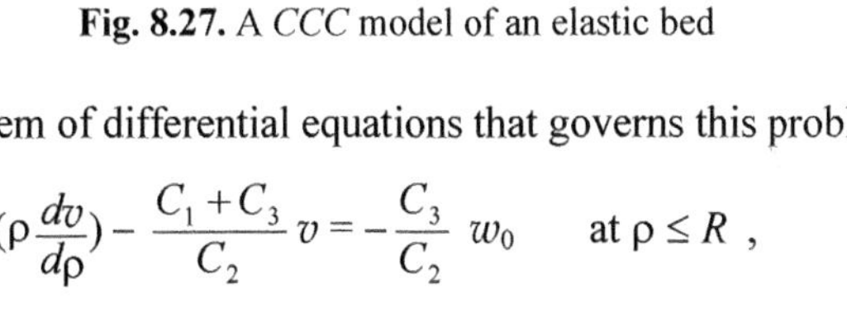

Fig. 8.27. A *CCC* model of an elastic bed

The resulting system of differential equations that governs this problem follows

$$\frac{1}{\rho}\frac{d}{d\rho}(\rho\frac{dv}{d\rho}) - \frac{C_1 + C_3}{C_2}v = -\frac{C_3}{C_2}w_0 \qquad \text{at } \rho \le R \; , \tag{8.57}$$

[6] As everyone knows, the evening twilight is an intermediate state of the nature between the daylight and the night darkness.

$$\frac{1}{\rho}\frac{d}{d\rho}(\rho\frac{dv}{d\rho}) - \frac{C_1}{C_2}v = 0 \qquad \text{at } \rho \geq R\ . \qquad (8.58)$$

Each of the equations (8.57) and (8.58) is a Bessel's differential equation, so the solution v can be represented as linear combinations of modified first-kind and second-kind zero-order Bessel's functions I_0 and K_0 plus a partial solution of an inhomogeneous equation corresponding to (8.57)

$$v(\rho) = \begin{cases} A_1 I_0(\beta\rho) + B_1 K_0(\beta\rho) + \gamma w_0 & \rho \leq R \\ A_2 I_0(\alpha\rho) + B_2 K_0(\alpha\rho) & \rho \geq R \end{cases}, \qquad (8.59)$$

where

$$\alpha^2 = C_1 / C_2, \quad \beta^2 = (C_1 + C_3) / C_2, \quad \gamma = C_3 / (C_1 + C_3). \qquad (8.60)$$

As I_0 is singular at infinity and K_0 is that at zero, two integration constants out of four in (8.59) must be excluded, i.e. $B_1 = 0,\ A_2 = 0$. The other two constants A_1 and B_2 and the stamp's sag w_0, so far unknown, will be determined from the continuity of v at $\rho = R$ and from an integral equilibrium condition. The continuity of v gives

$$\gamma w_0 = B_2 K_0(\xi) - A_1 I_0(\eta), \qquad (8.61)$$

where

$$\xi = \alpha R, \qquad \eta = \beta R. \qquad (8.62)$$

The summed stress in the top layer springs must be equal to the external force P according to the equilibrium condition, that is,

$$P = C_3 \int_0^{2\pi}\int_0^R (w_0 - v)\rho d\rho d\theta\ . \qquad (8.63)$$

Known relationships of the Bessel's function theory state [18]

$$\frac{d}{d\rho}\left[\rho^n I_n(\beta\rho)\right] = \beta\rho^n I_{n-1}(\beta\rho), \quad \frac{d}{d\rho}\left[\rho^n K_n(\alpha\rho)\right] = -\alpha\rho^n K_{n-1}(\alpha\rho)\ . \qquad (8.64)$$

Now (8.63) yields the following after substitutions and integrations

$$(1-\gamma)w_0 = \frac{P}{\pi R^2 C_3} + A_1\frac{2}{\eta}I_1(\eta)\ . \qquad (8.65)$$

Another equilibrium equation will be derived by dissecting the lower spring layer between the night and the evening surfaces and equating the overall stress in these springs to the external load P, which gives

$$P = C_1 \int_0^{2\pi}\left[\int_0^R v(\rho)\rho d\rho + \int_R^\infty v(\rho)\rho d\rho\right]d\theta\ . \qquad (8.66)$$

Integrating (8.66) and taking into account (8.59) and (8.64) gives us an additional relation

$$\gamma w_0 = \frac{P}{\pi R^2 C_1} - A_1 \frac{2}{\eta} I_1(\eta) - B_2 \frac{2}{\xi} K_1(\xi) . \tag{8.67}$$

Equation (8.65) helps determine A_1, and summing (8.65) and (8.67) yields B_2

$$A_1 = \left[(1-\gamma) w_0 - \frac{P}{\pi R^2 C_3} \right] \frac{\eta}{2 I_1(\eta)} ,$$

$$B_2 = \left[- w_0 + \frac{P}{\pi R^2 \gamma C_1} \right] \frac{\xi}{2 K_1(\xi)} . \tag{8.68}$$

Substituting the derived expressions of A_1 and B_2 to the equation (8.61), we have a final expression of the stamp's sag

$$w_0 = \frac{P}{\pi R^2 C_1} \frac{F_1}{\gamma(2\gamma + F_1)} , \tag{8.69}$$

where

$$F_1 = \frac{\xi K_0(\xi)}{K_1(\xi)} + (1-\gamma) \frac{\eta I_0(\eta)}{I_1(\eta)} . \tag{8.70}$$

Note that actually F_1 is a function of two dimensionless parameters ξ and η, because it follows from (8.60) and (8.62) that

$$1 - \gamma = \xi^2 / \eta^2 . \tag{8.71}$$

Now let's turn to the problem of a stamp under a moment load $M = Pe$ (see Fig. 8.24). In this case, as can be seen, the sag functions of the day and night surfaces can be represented as

$$w = \varphi\rho \sin\theta , \qquad v = u(\rho) \sin\theta , \tag{8.72}$$

where u is a new unknown function depending on the radial coordinate only. Note that the angular coordinate θ is counted from the axis about which the moment M works. The result is that instead of the equations (8.57) and (8.58) we come to equations

$$\frac{1}{\rho} \frac{d}{d\rho} (\rho \frac{du}{d\rho}) - (\frac{C_1 + C_3}{C_2} + \frac{1}{\rho^2}) u = -\frac{C_3}{C_2} \varphi\rho \quad \text{at } \rho \le R , \tag{8.73}$$

$$\frac{1}{\rho} \frac{d}{d\rho} (\rho \frac{du}{d\rho}) - (\frac{C_1 + C_3}{C_2} + \frac{1}{\rho^2}) u = -\frac{C_3}{C_2} \varphi\rho \quad \text{at } \rho \le R . \tag{8.74}$$

The general solution of these can be represented by modified first-order Bessel functions

$$u(\rho) = \begin{cases} A_1 I_1(\beta\rho) + B_1 K_1(\beta\rho) + \dfrac{\beta^2 - \alpha^2}{\beta^2} \rho , & \rho \le R \\ A_2 I_1(\alpha\rho) + B_2 K_1(\alpha\rho) , & \rho \ge R \end{cases} . \tag{8.75}$$

To validate the fact that the partial solution of the inhomogeneous equation (8.73) can be represented as a linear function of ρ, namely as $\varphi\rho(\beta^2-\alpha^2)/\beta^2$, perform a direct substitution of it to (8.73).

The condition of the function's limitedness at zero and at infinity gives $A_2 = 0$ and $B_1 = 0$ again, and the continuity of the solution at $\rho = R$ yields the following (earlier designations are used):

$$\varphi R\,\frac{\eta^2-\xi^2}{\eta^2} = B_2K_1(\xi) - A_1I_1(\eta)\ . \tag{8.76}$$

We equate the response moment that appears in the springs of the top layer to the external moment M:

$$M = 2C_3\int_0^{\pi}\int_0^{R}(\rho-u)\sin^2\theta\,\rho^2 d\rho d\theta = C_3\pi\int_0^{R}\left[\varphi\rho\frac{\alpha^2}{\beta^2} - A_1I_1(\beta\rho)\right]\rho^2 d\rho\,. \tag{8.77}$$

Taking (8.64) into account and integrating yields

$$M = C_3\pi R^3\left[\varphi R\frac{\xi^2}{4\eta^2} - A_1\frac{I_2(\eta)}{\eta}\right]. \tag{8.78}$$

The same moment M is created by stresses in the springs of the bottom layer, that is,

$$M = 2C_1\int_0^{\pi}\left\{\int_0^{R}\left[A_1I_1(\beta\rho) + \varphi\rho\frac{\beta^2-\alpha^2}{\beta^2}\right]\rho^2 d\rho + \int_R^{\infty} B_2K_1(\alpha\rho)\rho^2 d\rho\right\}\sin^2\theta d\theta\,. \tag{8.79}$$

hence

$$M = C_1\pi R^3\left[\varphi R\frac{\eta^2-\xi^2}{4\eta^2} + A_1\frac{I_2(\eta)}{\eta} + B_2\frac{K_2(\xi)}{\xi}\right]. \tag{8.80}$$

Note that the integration in (8.79) takes into account the asymptotic behavior $\rho^2K_2(\rho) \to 0$ at $\rho \to \infty$.

Equations (8.78) and (8.80) yield the integration constants

$$A_1 = \left[\varphi R\frac{\xi^2}{4\eta^2} - \frac{(1-\gamma)M}{\gamma\pi R^3C_1}\right]\frac{\eta}{I_2(\eta)},\qquad B_2 = \left(-\varphi R\frac{1}{4} + \frac{M}{\gamma\pi R^3C_1}\right)\frac{\xi}{K_2(\xi)}\ . \tag{8.81}$$

After substituting (8.81) to (8.76) we have the following expression of the stamp's slope

$$\varphi R = \frac{M}{\gamma\pi R^3C_1}\,\frac{4F_2}{\gamma(4\gamma+F_2)}\ , \tag{8.82}$$

where

$$F_2 = \frac{\xi K_1(\xi)}{K_2(\xi)} + (1-\gamma)\frac{\eta I_1(\eta)}{I_2(\eta)}\ . \tag{8.83}$$

So, having the solution (8.69) that describes the stamp's sag and (8.82) that defines its slope, we find that their ratio is

$$\frac{w_0 e}{\varphi R^2} = \frac{(4\gamma + F_2)F_1}{4(2\gamma + F_1)F_2} \, . \tag{8.84}$$

The right part of (8.84) contains a function of two dimensionless parameters ξ and η, therefore we need one more condition to determine those. This additional condition will be the sag w_q of the *CCC* bed's day surface measured at the center of the circle of the same radius R under a uniformly distributed load with the intensity q applied directly to the surface of the elastic bed. Assuming $q\pi R^2 = P$, we have

$$w_q = \frac{P}{\pi R^2 C_3} + \frac{P}{\pi R^2 C_1}\left[1 - \xi K_1(\xi)\right] = \frac{P}{\pi R^2 C_1}\left[\frac{1}{\gamma} - \xi K_1(\xi)\right] \tag{8.85}$$

hence

$$\frac{w_0}{w_q} = \frac{F_1}{(2\gamma + F_1)[1 - \gamma\xi K_1(\xi)]} \, . \tag{8.86}$$

The joint solution of the system of two transcendental equations (8.84) and (8.86) determines values of the desirable parameters ξ and η that enable us to restore all three coefficients C_1, C_2 and C_3. Knowing ξ and η, and the measured sag of the stamp w_0, we can obtain the coefficient C_1 using the formula (8.69). Now by (8.60), (8.62) and (8.71) we can find the other two unknown bedding values

$$C_2 = C_1 R^2/\xi^2 \, , \qquad C_3 = C_1(\eta^2/\xi^2 - 1) \, . \tag{8.87}$$

Now let's address the solution of the nonlinear equation system (8.84) and (8.86). First we rewrite these equations as

$$a = F_a(\xi,\eta) \, , \qquad b = F_b(\xi,\eta) \, , \tag{8.88}$$

where these designations are introduced

$$a = \frac{w_0 e}{\varphi R^2} \, , \qquad b = \frac{w_0}{w_q} \, , \tag{8.89}$$

and

$$F_a(\xi,\eta) = \frac{(4\gamma + F_2)F_1}{4(2\gamma + F_1)F_2} \, , \qquad F_b(\xi,\eta) = \frac{F_1}{(2\gamma + F_1)[1 - \gamma\xi K_1(\xi)]} \, . \tag{8.90}$$

First of all, let us set possible limits of variation of the input parameters a and b. To do that, we calculate these parameters for two ultimate models of an elastic bed in the regard of their distributive ability: the Winkler's model and the elastic semi-space.

This holds for the Winkler's model with the bedding value k

$$w_0 = \frac{P}{\pi R^2 k} \, , \qquad \varphi = \frac{4Pe}{\pi R^4 k} \, , \qquad w_q = w_0 \, , \tag{8.91}$$

wherefrom $a = 1/4$ and $b = 1$.

For an elastic semi-space with the elasticity modulus E and the Poisson ratio ν, the respective formulae are written as ([34,6])

$$w_0 = \frac{(1-\nu^2)P}{2ER}, \qquad \varphi = \frac{3(1-\nu^2)Pe}{4ER^3}, \qquad w_q = \frac{2(1-\nu^2)P}{\pi ER} \tag{8.92}$$

thus $a = 2/3$ and $b = \pi/4$.

So, we have obtained the following limitations of the input parameters a and b

$$0.25 \le a \le 0.667 \quad \text{and} \quad 0.785 \le b \le 1\,. \tag{8.93}$$

The solution of the system consisting of two transcendental equations (8.88) can be obtained by a widely popular **Mathcad 2000** software intended for mathematical problems of this kind and possessing built-in tools for calculating the modified Bessel's functions I_n and K_n. For the purpose of illustration, here we present a short table of values of the desirable parameters ξ and η calculated at different values of the input parameters a and b (see Table 8.3).

Table 8.3.

	$b = \pi/4$	$b = 0.80$	$b = 0.85$	$b = 0.90$	$b = 0.95$
$a = 0.32$	5.336 19.787 **0.835**	3.974 9.250 **0.940**	2.292 3.874 **1.121**	1.305 2.084 **1.046**	0.483 0.959 **0.479**
$a = 0.35$	2.609 6.119 **0.828**	2.339 5.091 **0.845**	1.569 2.988 **0.838**	0.910 1.758 **0.653**	0.309 0.889 **0.213**
$a = 0.40$	1.708 4.581 **0.634**	1.561 4.002 **0.626**	1.079 2.574 **0.547**	0.614 1.609 **0.356**	0.181 0.892 **0.077**
$a = 0/45$	1.325 4.248 **0.497**	1.191 3.746 **0.470**	0.825 2.487 **0.379**	0.458 1.607 **0.217**	0.119 0.936 **0.034**
$a = 0.50$	1.100 4.218 **0.402**	1.012 3.741 **0.386**	0.701 2.511 **0.297**	0.382 1.646 **0.158**	0.091 0.979 **0.020**
$a = 0.55$	0.950 4.319 **0.336**	0.874 3.833 **0.319**	0.605 2.583 **0.237**	0.324 1.709 **0.117**	0.071 1.029 **0.012**
$a = 0.60$	0.840 4.497 **0.286**	0.774 3.986 **0.270**	0.535 2.683 **0.195**	0.283 1.781 **0.092**	0.058 1.078 **0.008**
$a = 2/3$	0.730 4.817 **0.237**	0.674 4.252 **0.223**	0.466 2.842 **0.156**	0.243 1.887 **0.070**	0.046 1.143 **0.005**

Every cell of Table 8.3 presents three numbers of which the top one reflects the value of ξ, the middle one is η, and the lower boldfaced one gives the value of the function $F_3(\xi,\eta)$ needed to calculate the parameter C_1.

As it follows from the formula (8.69),

$$C_1 = \frac{P}{\pi R^2 w_0} F_3, \quad \text{where} \quad F_3 = \frac{F_1}{\gamma(2\gamma + F_1)} \,. \tag{8.94}$$

To conclude this section, we'd like to express a wish that software developers supplement their existing finite element libraries with plate finite elements (both thin and medium-thickness) lying on *CCC* beds.

As we have noticed before, analysts can confine themselves to the use of flexural plate finite elements on an elastic two-parametric bed (if such are available in a software environment) by supplementing their models with springs concentrated in the plate's nodes to discretely simulate the top Winkler's layer of the stiffness C_3. Though, the direct use of the appropriate finite elements is preferable for two reasons:

- first, it facilitates the user's work by not having him manually calculate and input the stiffness of each discrete spring to the model, the stiffness being dependent on the sizes of finite elements included in the star of the node where the spring is installed. An automatic generation of the mesh can make a big problem of this, especially with irregular meshes;
- second, a neat and consistent finite element discretization of an original continuous problem is more preferable than a medley of physical and mathematical dicretizations, at least in the aspect of the solution's error estimation.

In this connection, we'd like to present an expression of the potential energy E_0 which is accumulated in the elastic bed model being discussed:

$$\mathsf{E}_0 = \frac{1}{2} \iint\limits_{\Omega} \left[C_3 (w - v)^2 + C_1 v^2 + C_2 (v_{,x}^2 + v_{,y}^2) \right] \, d\Omega + $$
$$+ \frac{1}{2} \iint\limits_{\Omega_e} \left[C_1 v^2 + C_2 (v_{,x}^2 + v_{,y}^2) \right] \, d\Omega_e \,. \tag{8.95}$$

Having this expression of the deformation energy, one can derive components of the respective stiffness matrices and load vectors in a conventional FEM manner.

Without going into details, we just note that the energy E_0 stored by the elastic bed depends on two functions v and w that represent the displacements of the evening and day surfaces of the elastic bed. If we confine ourselves to the solid-node finite element analysis, we will have to introduce formally a couple of nodes in every nodal point of the flexural plate. Both nodes of each couple of this kind are coincident by their spatial position, but the displacements of one node in a couple are defined by the function v while those of the other one by the function w. By the way, it should be clear that the energy E_k stored in the flexural plate itself and the potential of the external forces Π — see the formula (8.46) — depend only on the function w (it is reasonable to assume that no load can be applied directly to the evening surface of the elastic bed).

8.7 Employing Finite-Element Foundation Models

An electronic book by A.G. Shashkin and K.G. Shashkin [29] notices justly that:

"... *what a peculiar kind of situation we have: a superstructure designer treats a building as one standing on a perfectly incompliant foundation (on a stone rock) and gives reactions obtained by the calculation with this assumption to a foundation guy; the latter applies these reactions to an elastic, elastic-plastic or viscous medium model as if perfectly flexible loads. The questions are: what kind of foundation do we have – a perfectly rigid or a compliant one? What kind of structure do we have – a perfectly flexible one or that of a finite rigidity?*».

One of feasible options suggested in the book is to arrange the iteration process of computation in such way that at each iteration the sags from the bed deformation solution are applied to the building and the bed is subjected to new reactions of the building. The result is that one may have to solve a coupled problem.

But it makes more sense to allow for the coupled behavior of the structure and its foundation in a single model where the foundation is represented by a certain "influence area" with a finite element mesh inside. The exterior boundary of this soil region should be defined so that edge effects at those affect little the stresses under the structure. The rule of thumb is to set the boundary of the "influence area at 3 to 4 maximum dimensions of the foundation but not farther than 1 to 1.5 thickness of the compressible soil layer under the foundation.

A typical example is a model shown in Fig. 8.28 that presents a conventional technique of rarefying the finite element mesh in the area occupied by the foundation at a distance from the superstructure. To assign the distance to the boundaries of the soil region, one can use considerations given in Section 5.1.

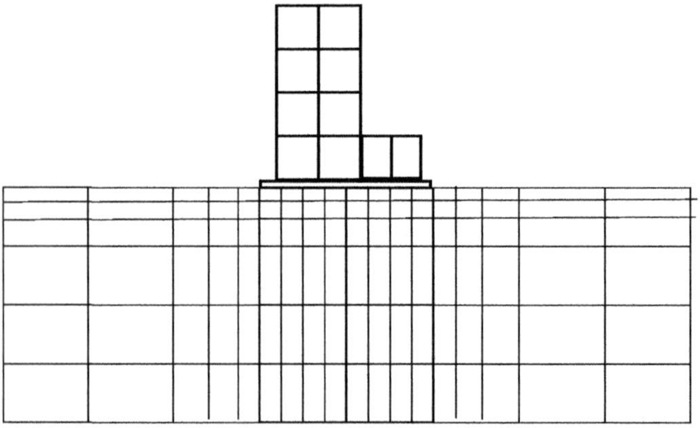

Fig. 8.28. A finite element mesh of an elastic foundation

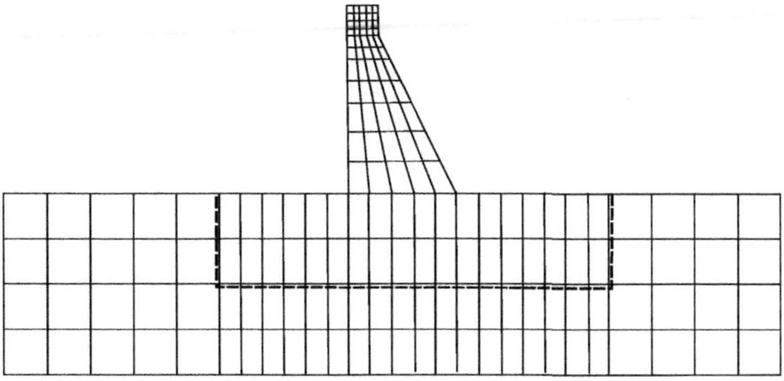

Fig. 8.29. A finite element model with boundaries of a "minor" and "major" areas occupied by the dam's foundation

The lower boundary can be defined at the surface of the bedrock or a layer of soil with its elasticity modulus by an order of magnitude greater than that immediately under the structure's foundation.

This assignment of the soil region's boundaries works quite well in a static analysis, but it may prove incorrect with the dynamics. The matter is that elastic waves that go away freely to the semi-infinite bed, for example, during an earthquake, will reflect from the region's boundaries (a "box effect"). To parry this effect, one has to define special boundary conditions at the boundaries of the region. Particularly, one can use *absorbing boundaries* that dampen waves coming from outside (such as seismic ones) or *passing boundaries* that suppress waves of disturbances caused by the structure [19, 7].

E. Wilson [37] presents data of the solution of a test problem dealing with a dam on a natural bed. Its finite element model is shown in Fig. 8.29. The results are shown in Table 8.4.

Table 8.4.

Parameters	Dam without bed	"Minor" area	"Major" area
Total weight, [t×sec^2/m]	33.41	236.710	1382.04
Periods of natural oscillations, [sec]	0.335 – 0.158	0.404 – 0.210	0.455 – 0.371
Maximum displacement [cm]	1.65	3.25	3.33

Without considering the inertial properties of the foundation, the results become as presented in Table 8.5.

Table 8.5.

Parameters	Dam without bed	"Minor" area	"Major" area
Total weight, [t×sec^2/m]	33.41	33.41	33.41
Periods of natural oscillations, [sec]	0.335 – 0.158	0.400 – 0.210	0.415 – 0.207
Maximum displacement [cm]	1.65	3.22	3.63

It should be noted also that the use of a finite element bed model helps allow for a number of effects such as a complicated terrain. A relevant example taken from a demo version of the **STRAUS7** software is shown in Fig. 8.30.

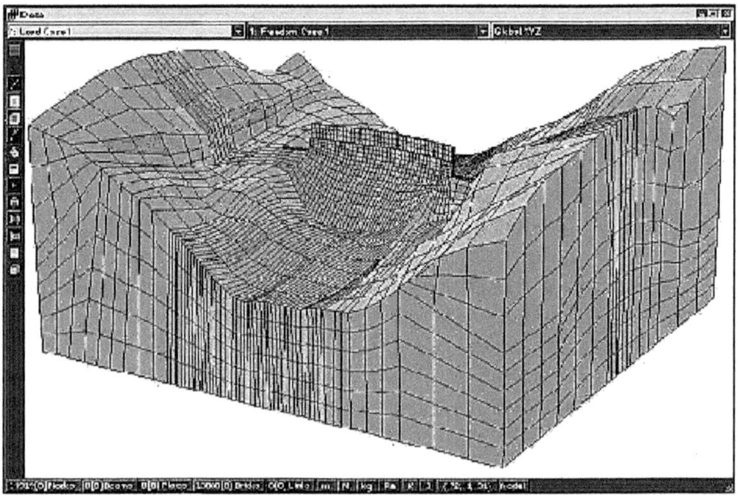

Fig. 8.30. A finite element model of a dam and a soil body

(*a*) (*b*)

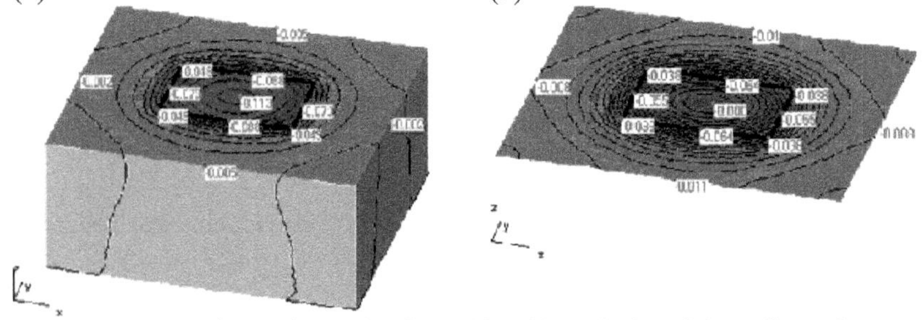

Fig.8.31. Comparative analyses of a plate with different bed models. Isolines of sags (m) are presented: (*a*) a spatial elastic soil model analysis; (*b*) an analysis with a two-parametric soil bed model

Returning to static problems, we'd like to note on a very interesting question how results of analyses performed with different popular models of the elastic bed relate to one another. As an analysis presented in [29] shows, the analysis with the two-parametric (Pasternak's) bed model implemented in a lot of widely used computer programs is often in a good accordance with the analysis that uses a spatial elastic model of soil (Fig. 8.31).

8.8 A Bi-member Model of an Open-section Thin-walled Bar [7]

Thin-walled bars of open section are widely used in engineering structures related to various industries and technologies. A general theory of analysis of thin-walled bars and structures composed of such has been developed thoroughly in works by V.Z. Vlasov and now is well-known. At the same time, structural engineers engaged in computational analysis activities experience certain difficulties with modern computer-aided design tools when they need to analyze mechanical systems including thin-walled bars of open sections. The matter is that the absolute majority of the structural computational software developments based on finite elements take into account up to six degrees of freedom in nodes of discrete systems that correspond to linear displacements and slopes of those nodes as rigid bodies. Meanwhile the thin-walled open-section bar theory requires a seventh degree of freedom to be introduced in nodes adjoining to thin-walled members. This seventh degree of freedom corresponds to a warping component of the nodal displacement. We remind that the warping is an out-of-plane displacement of a cross-section of a thin-walled bar. According to the theory by V.Z. Vlasov, the measure of warping is the change rate of the bar's torsion angle along its longitudinal axis.

Our goal here is to demonstrate a special technique intended to circumvent these difficulties, at the same time confining ourselves to capabilities of a standard finite element analysis software that allows at most 6 degrees of freedom in each node of a design model[8]. As we will demonstrate later, this technique involves the construction of a special model which we call a "*bi-member model of a thin-walled bar*".

8.8.1 Construction of a Bi-member Model

If we consider differential equations of equilibrium in displacements which describe the bending and torsion of a thin-walled open section bar with respect to displacements of the line of the bar's cross-section shear centers, we'll see that the only difference of these equations from their counterparts in the classical theory of

[7] Section 8.8 is written jointly with A.O. Kountsevich.

[8] According to the terminology introduced earlier, we are discussing the feasibility of a solid-node finite element analysis of thin-walled bars.

solid bars is just one equation with respect to the torsion angle θ_x about the longitudinal axis of the bar X. This equation for a thin-walled bar has the following form

$$(EI_\omega\theta_x'')'' - (GI_x\theta_x')' = m_x\,, \tag{8.96}$$

where I_ω is a sectorial inertia moment of the bar's cross-section, I_x is an inertia moment of simple torsion, E and G are the respective elasticity and shear modulus of the bar's material.

The right part of the equation (8.96) contains an expression for the external linear torque taken with respect to the axis crossing the bar section shear centers.

The full torque M_x in a section of a twisted thin-walled bar consists of two summands

$$M_x = M_\omega + H\,, \tag{8.97}$$

where H is a simple torsion moment, an integral characteristic of tangential stresses distributed across the thickness of the section's wall linearly, M_ω is a restrained torsion moment (sometimes called a bending-torsion moment) caused by tangential stresses uniformly distributed across the walls of the bar's cross-section.

It is known that these components of the full torque are related to the torsion angle θ_x via the following differential dependencies

$$M_\omega = -(EI_\omega\theta_x'')'\,, \qquad H = GI_x\theta_x'\,. \tag{8.98}$$

Normal stresses σ of the section that arise in the torsion of the thin-walled bar cause a bi-torque B which is also determined by the torsion angle θ_x as

$$B = -\,EI_\omega\theta_x''\,. \tag{8.99}$$

It is convenient to involve energy considerations in order to build the above mentioned bi-member model of a thin-walled bar that employs six nodal degrees of freedom. With this purpose, let us present an expression of the potential energy of deformation E accumulated by the thin-walled bar in its torsion. We have

$$\mathsf{E} = \frac{1}{2}\int_0^l [E\,I_\omega(\theta_x'')^2 + GI_x(\theta_x')^2]dx\,, \tag{8.100}$$

where l is the length of the bar.

The bi-member model in question will be a mechanical system consisting of two bars of the same length l called the *main* and *fictitious* members, respectively. Both members are classical, that is, subject to assumptions of the Bernoulli – Euler theory of solid bars, therefore they do not require any additional degrees of freedom to be introduced in the design model.

Let (X,Y,Z) be a local coordinate system of the main bar and (X_F, Y_F, Z_F) be that of the fictitious member. The axes X and X_F are longitudinal ones, and the couples (Y, Z) and (Y_F, Z_F) are principal central inertia axes of the main and fictitious bars (Fig. 8.32). Not that it is practically convenient to choose the axis X_F parallel to the axis X.

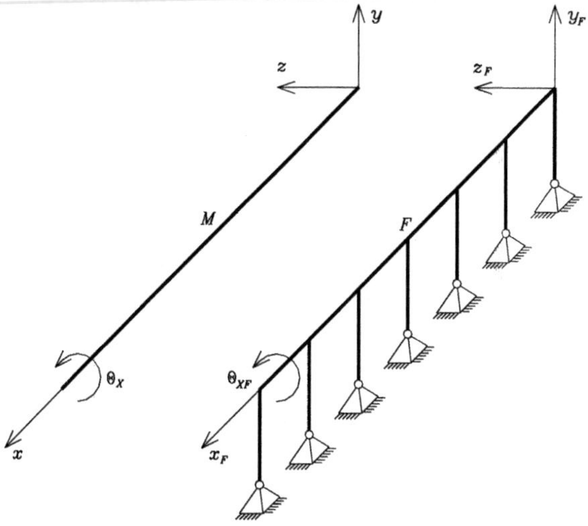

Fig. 8.32. The main and fictitious members

Here and below an additional subscript F denotes quantities referring to the fictitious member while their counterparts relating to the main member are marked by no additional characters.

We establish a dependency between the displacements of the sections of the bars (both the main and fictitious ones) to ensure the equal slopes of the sections with respect to the axes X and X_F, that is,

$$\theta_x = \theta_{xF} . \tag{8.101}$$

Also, the fictitious bar is constrained externally against axial displacements of points of its longitudinal axis — along the axis X_F — and in the direction of one of its principal inertia axes. To be particular, let us assume that the latter constraint prohibits displacements in the direction of Y_F. The linear displacements of the fictitious member along the other principal axis will be related to the torsion angle θ_{xF} as

$$w_F = r\,\theta_{xF} , \tag{8.102}$$

where w_F denotes the displacement of the mass center of the fictitious bar's section along the axis Z_F, and r is a constant that we can handle at our choice.

It is easy to notice that the external constraints imposed on the fictitious bar can be interpreted mechanically as a reinforcement of the bar by a fence of perfectly rigid levers (Fig. 8.33) directed along Y_F and having the length r. The lower ends of the levers are restrained from all linear displacements and rotations about the axis Z_F . Fig. 8.33 does not show the fixation against rotations about Z_F in order not to clutter up the schematic with details.

Geometrical considerations (Fig. 8.33) that follow from the condition of perfect rigidity of the levers imply the equality (8.102).

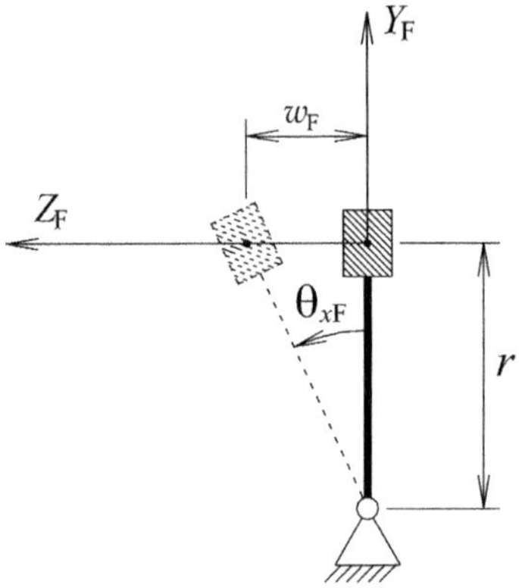

Fig. 8.33. A mechanism of link between the rotation angle and the linear displacement

Thus the torsion of the main bar causes the fictitious bar to move laterally by w_F along the axis Z_F and to bend with respect to the axis Y_F due to the imposed constraints (8.101) and (8.102).

The energy of deformation E in the constructed bi-member model is a sum of energies stored separately by the main and fictitious bars. We assign a torsion stiffness GI_x to the main bar, assume the torsion stiffness of the fictitious bar to be zero and its flexural stiffness with respect to the axis Y_F to be EI_{yF}. Then the energy of deformation E will be

$$\mathsf{E} = \frac{1}{2}\int_0^l [E\, I_{yF}\,(w_F'')^2 + GI_x(\theta_x')^2]dx\,, \tag{8.103}$$

which is totally equivalent to (8.100) when taking into account the relations (8.101) and (8.102), and defining the flexural stiffness of the fictitious bar as

$$EI_{yF} = EI_\omega \,/\, r^2\,. \tag{8.104}$$

Thus the constructed bi-member model is equivalent by energy to the original thin-walled bar.

Now let us demonstrate that the internal stresses in the fictitious bar can be interpreted as generalized stresses of the original thin-walled bar which arise in its restrained torsion.

The bending moment with respect to the axis Y_F of the fictitious member can be written as

$$M_{yF} = -\,EI_{yF}\,w_F''\,, \tag{8.105}$$

wherefrom we obtain the following taking into account the same dependencies (8.101), (8.102), (8.104) and (8.99):

$$B = r\,M_{yF}\,. \tag{8.106}$$

Further, the shear force Q_{zF} is a derivative of the moment M_{yF}, that is,

$$Q_{zF} = M'_{yF} \,, \tag{8.107}$$

and the bending-torsion moment M_ω, in its turn, is found by differentiating the bi-torque B which is implied by (8.98) and (8.99). Hence

$$M_\omega = r\, Q_{zF} \,. \tag{8.108}$$

So, the bi-torque B and the restrained torsion moment M_ω are equivalent, up to the factor r, to the respective bending moment M_{yF} and the shear force M_{yF} that arise in the fictitious bar. For the purpose of practical calculations it is convenient to assume $r = 1$ in an adopted system of linear measurement units.

It is natural that we cannot assure the dependencies (8.101) to hold continuously along the whole X axis from 0 to l in a discrete bi-member model. Though, we can divide the main bar lengthwise into a few (say, n) segments and thus install n+1 nodes on it, including the start and end nodes. Let those be nodes $M_1, M_2,\ldots, M_{n+1}$. Then, dividing the fictitious bar into the same segments forms the respective nodes $F_1, F_2,\ldots, F_{n+1}$ on it. Now the continuous dependencies (8.101) can be replaced approximately by discrete ones specified at a finite set of points with coordinates $x_1, x_2,\ldots, x_{n+1}$, in n+1 nodes of the main and fictitious bars of the bi-member model. The respective rigid levers in the system with the discrete ties will remain also in the nodal points only.

A question may arise: where in this bi-member model is that notorious seventh degree of freedom of thin-walled bar nodes which we mentioned at the beginning of this section? The answer is simple: the seventh degree of freedom related to the warping θ'_x has not utterly disappeared, it just has changed its appearance by having become θ_{yF}. Indeed, the following holds for the bi-member model

$$\theta'_x = \theta'_{xF} = w'_F \,/\, r = -\,\theta_{yF} \,/\, r \,. \tag{8.109}$$

Although θ'_x cannot be referred directly to displacements of nodes as rigid bodies, the quantity θ_{yF} can be naturally interpreted as rotations of nodes of the fictitious bar with respect to the axis Y_F. This circumstance lets us analyze a thin-walled bar under limitations of conventional software developments that operate solid nodes.

Suppose that we use the **SCAD** software. There is a built-in "displacement merging" tool implemented in its environment to specify constraints like

$$\theta_x(x_i) = \theta_{xF}(x_i) \,, \tag{8.110}$$

which ensure equal slopes of respective nodes of the main and fictitious bars. The **GTSTRUDL** software provides the same effect by commands of its input language JOINT TIES and SLAVE RELEASES. The former command provides the equality of all displacements of a given couple of nodes (the nodes M_i and F_i in our case), and the latter releases some degrees of freedom of a tied node couple restoring their partial independence.

By the way, it is convenient to use such capability of CAD software as specification of eccentricities in node-bar ties (infinitely stiff inserts) in order to introduce perfectly rigid levers attached to nodes of the fictitious bar. In this case the nodes of the fictitious bar F_i will be placed along a line that connects the lower ends of the levers rather than their centroid line (one that crosses the mass centers of the cross-sections).

If the employed software provides a capability of specifying a shift of the shear center axis from the centroid axis of a bar (particularly, **GTSTRUDL** provides this one), then one can omit the eccentricities of the levers in the source data and replace those by a quantity $e_{yF} = r$ where e_{yF} is the shift of the shear center axis of the fictitious bar with respect to its centroid axis. The former technique is used in the example 1 given below, and the latter in the example 2.

Example 1

A cantilever beam of the channel section and length $l = 4$ m is loaded by a concentrated force $P = 13.13$ t acting in the vertical plane that goes through the axis of mass centers of the beam's cross-sections (Fig. 8.34).

The following geometrical properties of the section are assumed:

$$I_x = 1.34 \cdot 10^{-6}\ \text{м}^4, \quad I_y = 5.333 \cdot 10^{-5}\ \text{м}^4, \qquad I_z = 37.33 \cdot 10^{-5}\ \text{м}^4,$$

$$A_x = 1.2 \cdot 10^{-2}\ \text{м}^2, \quad I_\omega = 1.524 \cdot 10^{-6}\ \text{м}^6, \qquad e_z = 0.1523\ \text{м}\,.$$

The material constants of the beam are

$$E = 2.1 \cdot 10^7\ \text{т/м}^2, \qquad G = 0.84 \cdot 10^7\ \text{т/м}^2\,.$$

The accurate solution of the problem obtained by integrating the equation (8.96) yields the following results

$$\theta_x(l) = -0.41629\,, \qquad B(0) = -3.3142\ \text{tm}^2\,,$$

and the vertical displacement of the P force application point is equal to

$$v(l) = -\,Pl^3/(3EI_z) + \theta_x(l)e_z = -\,0.03573 - 0.06340 = -\,0.09913\ \text{m}.$$

Assuming $n = 8$ in this problem with equal lengths of the segments, we have the following source data for the **GTSTRUDL** software describing a bi-member discrete model.

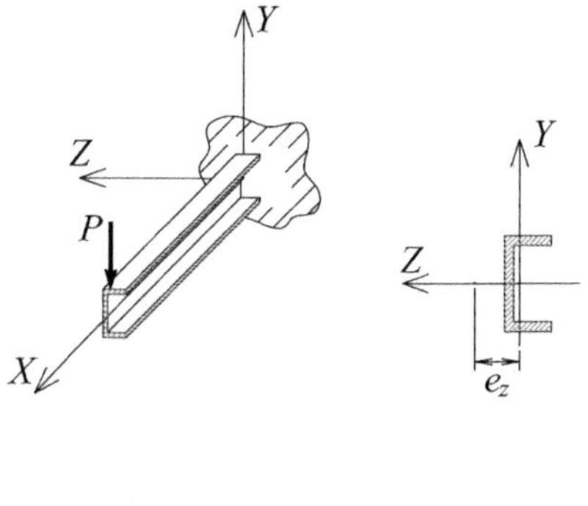

Fig. 8.34. A schematic of the example 1

Source data for the example 1 at $n = 8$

```
GTSTRUDL      'example 1'
UNITS  M  MTON  RAD
$..........................................................
$ THIN WALLED CANTILEVER BEAM OF OPEN CROSS SECTION n = 8
$**********************************************************
GEN 2 JOINTS LIST
    'M1'    0       0       0
    'F1'    0      -1       1
REPEAT 8 TIMES ID 1 X 0.5
STATUS SUPPORT 'M1' 'F1' TO 'F9'
JOINT       RELEASES
    'F2' TO 'F9' MOM X Y
CONSTANTS
     E     2.1E7;           G 0.84E7;      POISSON  0.25
TYPE   SPACE FRAME
   GEN 2 MEMBER LIST
          'BEAM1' 'M1'                'M2'
          'FICT1' 'F1'                'F2'
   REPEAT 7 TIMES ID 1 F 1 T 1
MEMBER ECCENTR
    'FICT1' TO 'FICT8' START 0 1 0 END 0 1 0
MEMBER      PROPERTIES
    'BEAM1' TO 'BEAM8'  AX 120E-4   IX 134E-8   IY 5333E-8 -
                            IZ 37330E-8  EZ 0.1523
    'FICT1' TO 'FICT8'  AX 1E-20    IX 1E-20    IY 1.524E-6 -
                            IZ 1E-20
JOINT  TIES
    'M2' TO 'M9' EQUAL     'F2' TO 'F9'
SLAVE RELEASES
    'M2' TO 'M9'   FOR X Y Z       MOM Y Z
LOAD        1
   JOINT LOADS
    'M9' FOR Y  -13.132
STIFFNESS  ANALYSIS
LIST DISPL;         LIST REACT;     LIST FOR
```

The analysis gives the following values:

$$\theta_x(l) = -0.41728,\quad M_{yF}(0) = B(0) = -3.3016\ \mathrm{tm}^2,\quad v(l) = -0.09929\ \mathrm{m},$$

which is pretty close to accurate results.

For the purpose of comparison, here are results of the calculation with the beam divided into four segments:

$$\theta_x(l) = -0.42016,\quad M_{yF}(0) = B(0) = -3.2707\ \mathrm{tm}^2,\quad v(l) = -0.09973\ \mathrm{m}.$$

8.8.2 A Bi-member Model of a Thin-walled Bar Reinforced by Lateral Slats

Constructions of bridge spans, frames of buildings, frames of bodies of various mobile vehicles and other engineering structures that may include thin-walled bars of open profiles, often involve a reinforcement of thin-walled bars by lateral slats. The purpose of the slats is to increase the rigidity of a bar in torsion.

Below we will demonstrate how the bi-member model should be modified to allow for the influence of the lateral slats on the response of the thin-walled bar.

Let a thin-walled bar of an open section be reinforced by a lateral slat as shown in Fig. 8.35, *a*.

We assume that the lateral slat connects points A and B of the bar's cross-section, and the distance between the points (the slat's length) is denoted as a. Let h be the thickness of the slat, b its width (the size of the slat's section in the direction of the X axis).

It turns out that it suffices to introduce a clock spring into a node of the fictitious bar exactly at the point X_F where the slat is located (Fig. 8.35, *b*) in order to allow for the effect of the slat on the bi-member model. This clock spring is to allow for the rotation of the respective node F of the bar about the axis Y_F. Let us denote the stiffness of the spring as k and choose it so as to comply with the condition of an energetic equivalency between the thin-walled bar with the slat and its bi-member model.

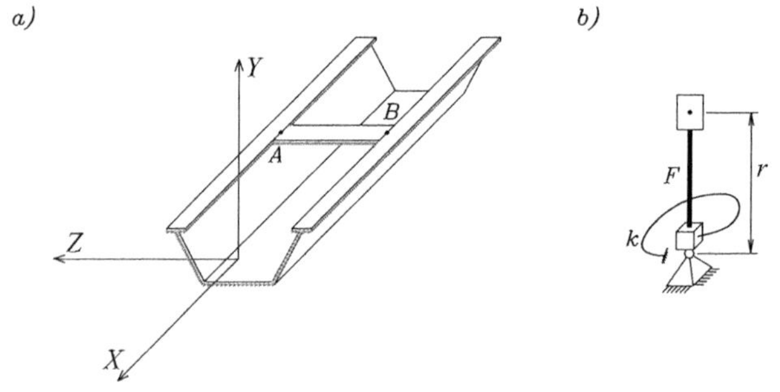

Fig. 8.35. A thin-walled bar reinforced by a lateral slat

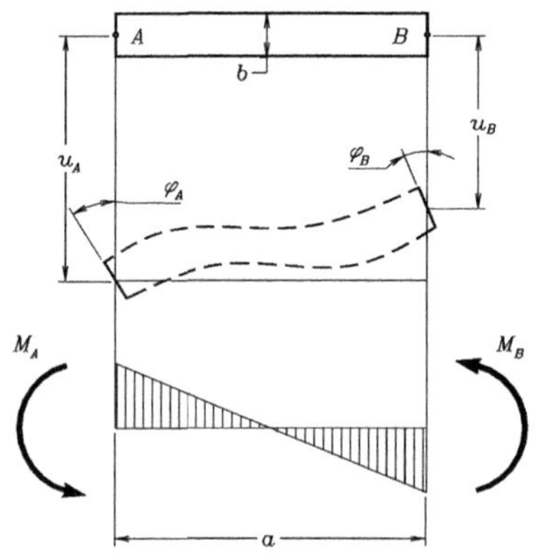

Fig. 8.36. A deformation of the slat

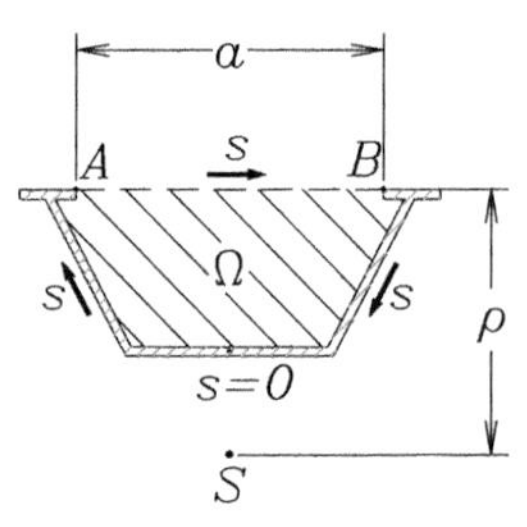

Fig. 8.37. An arc coordinate

In the course of the slat-reinforced thin-walled bar's deformation, the end sections of the slat acquire linear displacements u_A and u_B along the X axis and rotations by certain angles φ_A and φ_B (Fig. 8.36).

Let s be an arc coordinate of an arbitrary point of the bar's cross-section profile counted from a zero reference point $s = 0$ in some adopted positive direction (Fig. 8.37). We denote $u = u(x,s)$ and $v = v(x,s)$ the respective longitudinal and tangential displacements of points of the median surface of the thin-walled bar. The displacement u is assumed positive if its direction is that of the increasing x coordinate along the X axis. The tangential displacement v is positive if in its direction the s coordinate increases.

The theory of thin-walled open-section bars states that the longitudinal displacement u in the torsion of the bar is proportional to the warping θ'_x, or exactly

$$u(x,s) = -\theta'_x\omega(s)\,, \tag{8.111}$$

where $\omega(s)$ is a sectorial coordinate of a point of the section having the arc coordinate s.

Following (8.111), we find that the respective displacements u_A and u_B which we are interested with (Fig. 8.36) can be determined as

$$u_A = -\theta'_x\omega_A\,, \qquad u_B = -\theta'_x\omega_B\,, \tag{8.112}$$

where ω_A and ω_B are sectorial coordinates of the points A and B of the slat's attachment.

Actually, we will be interested with the relative longitudinal displacement of the points A and B, that is, the quantity

$$\Delta = u_A - u_B = -\theta'_x\,(\omega_A - \omega_B)\,. \tag{8.113}$$

It can be proved that the difference of the sectorial areas ω_A and ω_B is expressed by the following simple relation

$$\omega_A - \omega_B = \Omega - \rho a\,, \tag{8.114}$$

where Ω is the doubled area swept by a closed contour that appears after connecting the points A and B by a straight line, and ρ is the distance from the shear center S to the line that crosses the points A and B (Fig. 8.37). In other words,

$$\Delta = -\theta'_x\,(\Omega - \rho a)\,. \tag{8.115}$$

Further, geometrical considerations yield that the rotations of the end sections of the slat φ_A and φ_B in the plane of the slat are determined by the partial derivatives

$$\varphi_A = \frac{\partial v}{\partial x}(x_s, s_A)\,, \quad \varphi_B = \frac{\partial v}{\partial x}(x_s, s_B)\,, \tag{8.116}$$

where x_s is a longitudinal coordinate of the points A and B, s_A and s_B are arc coordinates of the same points.

The theory of thin-walled open-section bars uses an assumption of no shear in the median surface of the bar, that is,

$$\gamma = \frac{\partial u}{\partial s} + \frac{\partial v}{\partial x} = 0\,, \tag{8.117}$$

therefore it follows from both here and (8.111), (8.116) that

$$\varphi_A = \theta'_x(x_s)\,\omega'(s_A), \qquad \varphi_B = \theta'_x(x_s)\,\omega'(s_B)\,. \tag{8.118}$$

Note that in the formulae (8.118) the derivative is taken in the direction of the increasing s coordinate along the axis of the slat. It is easy to notice that

$$\omega'(s_A) = \omega'(s_B) = \rho\,, \tag{8.119}$$

therefore

$$\varphi = \varphi_A = \varphi_B = \theta'_x\rho\,. \tag{8.120}$$

The formulae (8.115) and (8.120) define all end displacements of the slat shown in Fig. 8.36. The result is that the diagram of the moments M arising in the slat will look as shown at the bottom of Fig. 8.36, and

$$M_A = M_B = EI\,(\Delta - a\varphi)\,\delta/a^2\,, \qquad \delta = 6a^2GF/(12EI + a^2GF)\,, \tag{8.121}$$

where I is the moment of inertia of the slat's section, F is the slat's cross-section area in shear (taking into account the cross-section shape factor).

For a rectangular-section slat with the sizes assumed before, we have

$$I = hb^3/12\,, \qquad F = 5hb/6\,. \tag{8.122}$$

Now we use (8.115) and (8.120) to rewrite the expression of the bending moments at the ends of the slat as

$$M_A = M_B = -EI\,\theta_x'\,\Omega\delta/a^2\,. \tag{8.123}$$

Knowing the distribution of the bending moments M in the slat (Fig. 8.36), we can calculate the deformation energy E_s stored by the slat

$$\mathsf{E}_s = \int_0^l \left(\frac{M^2}{2EI} + \frac{Q^2}{2GF}\right) ds = EI\,(\theta'_x)^2\,\Omega^2\delta/a^3\,, \tag{8.124}$$

where Q is the shear force in the slat

$$Q = 2M_A\,/\,a = -2EI\,\theta'_x\,\Omega\delta/a^3\,. \tag{8.125}$$

As it was said before, in the bi-member model the slat is to be replaced by a clock spring of the stiffness k, so the energy accumulated by the spring is determined by the expression

$$\mathsf{E}_s = k(\theta_{y\mathrm{F}})^2/2\,. \tag{8.126}$$

As the following equality holds for a bi-member model

$$\theta_{y\mathrm{F}} = -\,w'_{\mathrm{F}} = -r\,\theta'_x\,, \tag{8.127}$$

we obtain this from here and from (8.126)

$$\mathsf{E}_s = kr^2 (\theta'_x)^2 / 2 \ . \tag{8.128}$$

Equating the expressions of E_s from (8.128) and (8.124) to establish an energy equivalency, we find the final stiffness of the clock spring:

$$k = \frac{2EI\delta}{a^3}\left(\frac{\Omega}{r}\right)^2 . \tag{8.129}$$

The shear force Q in the slat that is of interest for an analyst, can be determined via the moment M_{ys} arising in an elastic spring. We have

$$M_{ys} = k\theta_{yF} = -\frac{2EI\delta}{a^3}\left(\frac{\Omega}{r}\right)^2 r\,\theta'_x \ . \tag{8.130}$$

By comparing (8.125) and (8.130) we obtain

$$Q = M_{ys}\, r / \Omega \ . \tag{8.131}$$

Also note that the product $M_{ys}r$ defines the value of the jump in the bi-torque of the thin-walled bar that appears in the section of the slat's location. No further details will be discussed.

Example 2

The example is an analysis of a side framing girder discussed in the book by V.Z. Vlasov [35]. The cross-section of the girder consisting of two channel beams is shown in Fig. 8.38. The same figure shows the position of the mass center of the section C, principal central inertia axes of the beam's cross-sections Y and Z, and the position of the shear center S. All sizes in Fig. 8.38 are given in centimeters.

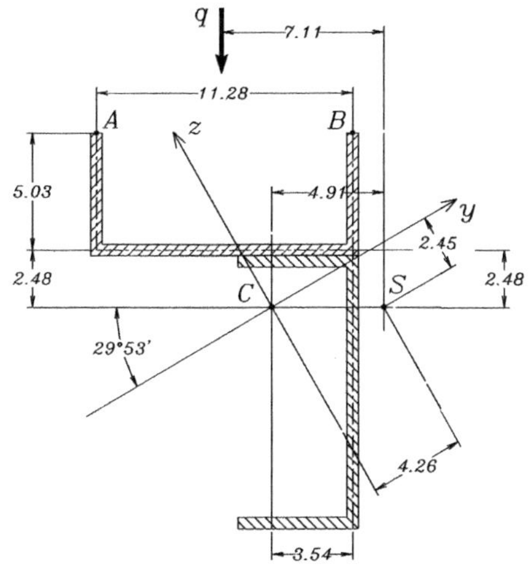

Fig. 8.38. A girder's cross-section for Example 2

The geometrical properties of the cross-section calculated by V.Z. Vlasov are as follows:

$$I_x = 6.17 \cdot 10^{-8}\ \mathrm{m}^4, \quad I_y = 976.8 \cdot 10^{-8}\ \mathrm{m}^4, \quad I_z = 383.8 \cdot 10^{-8}\ \mathrm{m}^4,$$

$$F_x = 30.72 \cdot 10^{-4}\ \mathrm{m}^2,\ I_\omega = 4829 \cdot 10^{-12}\ \mathrm{m}^6,\ e_y = 0.0425\ \mathrm{m}, \quad e_z = -\,0.0248\ \mathrm{m}.$$

The material properties are the same as in Example 1. The length of the span is $l = 3$ m. The girder is freely supported at its ends and loaded by a uniformly distributed vertical load of the intensity $q = 0.9$ t/m applied with the eccentricity $e = 0.0711$ m with respect to the shear center. One slat is welded to each end of the girder, of the following sizes:

$$a = 0.1128\ \mathrm{m}, \qquad h = 0.006\ \mathrm{m}, \qquad b = 0.10\ \mathrm{m}$$

thus having the following properties

$$F = 5 \times 6 \times 1 \times 10^{-4}/6 = 5.0 \cdot 10^{-4}\ \mathrm{m}^2, \qquad I = 50 \cdot 10^{-8}\ \mathrm{m}^4.$$

The doubled area between the contour of the cross-section and the closing slat (Fig.8.38) is

$$\Omega = 2 \times 11.28 \times 5.03 \times 10^{-4} = 113.5 \cdot 10^{-4}\ \mathrm{m}^2.$$

Assuming $r = 1$m and performing calculations, we have

$$\delta = \frac{6 \cdot 0.1128^2 \cdot 0.4E \cdot 5 \cdot 10^{-4}}{12E \cdot 50 \cdot 10^{-8} + 0.1128^2 \cdot 0.4 \cdot E \cdot 5 \cdot 10^{-4}} = 4.487\,,$$

$$k = \frac{2.1 \cdot 10^7 \cdot 50 \cdot 10^{-8}}{0.1128^3} \cdot 2 \cdot 4.487 \cdot \left(\frac{113.5}{1} \cdot 10^{-4} \right)^2 = 8.458\ \mathrm{tm}.$$

The end segments of the bi-member model's bars are assumed to be 5 cm long, and all intermediate segments are equal to one another and 29 cm long. This division makes the axes of the slats align with the nodes of the discrete bi-member model.

The result of the analysis includes, particularly, moments in the model springs $M_{ys} = 0.0304$ tm, thus the shear force in the slat is

$$Q = 0.0304 \cdot \frac{1}{113.5 \cdot 10^{-4}} = 2.68\ \mathrm{t}\,.$$

V.Z. Vlasov obtained for this same problem the value $Q = 2.73$ t.

Here follows a description of source data for this problem, with $n = 12$.

Source data of Example 2 at $n = 12$

```
GTSTRUDL      'example 2'
UNITS  M  MTON  RAD
$...........................................................
$ FREE SUPPORTED BEAM OF OPEN CROSS SECTION WITH 2 SLATS
$                          n = 12
$***********************************************************
GEN 2 JOINTS LIST
     'M1'  0        0        0
     'F1'  0       -1        0
REPEAT 1 TIMES ID 12 X 3
GEN 2 JOINTS LIST
     'M2'  0.05     0        0
     'F2'  0.05    -1        0
REPEAT 10 TIMES ID 1 X 0.29
STATUS SUPPORT 'M1' 'M13' 'F1' TO 'F13'
JOINT      RELEASES
     'M1'              MOM Y Z
     'M13' FOR X       MOM Y Z
     'F1' 'F13'        MOM Y
     'F2' 'F12'       'MOM X KMY 3.3682
     'F3' TO 'F11' MOM X Y
CONSTANTS
     E    2.1E7;         G 0.84E7;        POISSON  0.25
     BETA -1.0492 ALL BUT 0.0 'FICT1' TO 'FICT12'
TYPE  SPACE FRAME
     GEN 2 MEMBER LIST
           'BEAM1'         'M1'              'M2'
           'FICT1'         'F1'              'F2'
     REPEAT 11 TIMES ID 1 F 1 T 1
MEMBER     PROPERTIES
 'BEAM1' TO 'BEAM12' AX 30.72E-4  IX 6.17E-8   IY 976.8E-8 -
                     IZ 383.8E-8  EY 0.0425    EZ -0.0248
 'FICT1' TO 'FICT12' AX 1E-20     IX 1E-20     IY 4829E-12 -
                     IZ 1E-20      EY 1.0
JOINT      TIES
     'M2' TO 'M12' EQUAL 'F2' TO 'F12'
SLAVE RELEASES
     'M2' TO 'M12'         FOR X Y Z        MOM Y Z
LOAD       1
     MEMBER LOADS FOR Y GLOBAL
        'BEAM1' TO 'BEAM12' UNI -0.9
     MEMBER LOADS MOM X GLOBAL
        'BEAM1' TO 'BEAM12' UNI  0.0198
STIFFNESS  ANALYSIS
LIST DISPL;       LIST REACT;     LIST FOR
```

8.8.3 A Thin-walled Bar Reinforced by a Lateral Diaphragm

The reinforcement of a thin-walled bar by a lateral diaphragm can be managed by the bi-member model used to describe the reinforcement by lateral slats. The only difference relates to the determination of the stiffness k of the clock spring that resists to the rotation of the respective node F of the fictitious bar about the axis Y_F.

As one knows (see book by V.Z. Vlasov cited above), a lateral diaphragm of a thickness h causes a jump in the bi-torque distribution with the following value B

$$B = -\frac{Eh^3}{12(1+\nu)}\Omega\,\theta_x' , \tag{8.132}$$

where Ω is the doubled area of the diaphragm plate and ν is the Poisson ratio[9].

On the other hand, we have already found out that the jump in the bi-torque distribution of our model is expressed by the product $M_{ys}r$. This and (8.109) give

$$M_{ys} = \frac{Eh^3}{12(1+\nu)}\frac{\Omega}{r^2}\theta_{y\mathrm{F}} . \tag{8.133}$$

Now it is clear that one should assume the stiffness of the clock spring corresponding to the diaphragm plate that reinforces the bar, to be equal to the factor at the rotation angle $\theta_{y\mathrm{F}}$ in the right part of the formula (8.133), that is,

$$k = \frac{Eh^3}{12(1+\nu)}\frac{\Omega}{r^2} . \tag{8.134}$$

We do not go into further details here.

8.8.4 A Mathematical Interpretation of the Bi-member Model and its Discrete Scheme

It may seem at first sight that the discrete counterpart of the bi-member model is just a version of the conventional finite element analysis. Though, if we return to the results of Example 1 and calculate the work of the external forces equal to the potential energy accumulated by the mechanical system in its deformation according to the Clapeyron theorem, then we will have

- E = 13.13×0.09913×0.5 =0.6508 tm — an accurate value of the energy,
- E_8 = 13.13×0.09929×0.5 =0.6518 tm — the discrete model's energy at $n = 8$,
- E_4 = 13.13×0.09973×0.5 =0.6547 tm — the discrete model's energy at $n = 4$.

Thus the discrete model used converges to the accurate solution of the problem by energy, as the number of the division segments grows (or the maximum segment length falls), from below rather than from above as it usually takes place in a standard finite element analysis, because

$$-\mathsf{E} > -\mathsf{E}_8 > -\mathsf{E}_4 . \tag{8.135}$$

In order to clear up this circumstance, let us turn to a variational formulation of the torsion analysis of a thin-walled open-section bar. With this purpose we involve the expression of the full potential energy of the system, the Lagrangian functional L. We have

[9] V.Z.Vlasov gives this formula as (4.13) without the minus sign due to a different rule of signs adopted by him.

$$\mathsf{L}(\theta_x) = \frac{1}{2}\int_0^l [\,EI_\omega(\theta''_x)^2 + GI_x(\theta'_x)^2\,]\,dx - \int_0^l m_x\theta_x dx\ . \tag{8.136}$$

In the equilibrium position the functional L acquires its minimal value. The minimum is searched for in a set of functions θ_x having the second generalized quadratically summable derivative and satisfying the principal (kinematical) boundary conditions. The Euler equation of the functional L is the equation (8.96).

It is easy to notice that the variational statement of the problem can be easily reformulated as a conditional extremum problem, namely the problem of searching for a point of minimum of the functional L_B that depends on two unknown functions

$$\mathsf{L}_\mathrm{B}(\theta_x, \theta_{x\mathrm{F}}) = \frac{1}{2}\int_0^l [\,EI_\omega(\theta''_{x\mathrm{F}})^2 + GI_x(\theta'_x)^2\,]dx - \int_0^l m_x\theta_x dx \tag{8.137}$$

at additional limitations

$$\theta_x(x) = \theta_{x\mathrm{F}}(x)\ . \tag{8.138}$$

Introducing formally a new variable $w_\mathrm{F} = r\theta_{x\mathrm{F}}$ and assuming $EI_{y\mathrm{F}} = EI_\omega/r^2$, we rewrite the functional L_B as

$$\mathsf{L}_\mathrm{B}(w_\mathrm{F},\theta_x) = \frac{1}{2}\int_0^l EI_{y\mathrm{F}}\,(w_\mathrm{F}'')^2\,dx + \frac{1}{2}\int_0^l GI_x\,(\theta_x')^2 dx - \int_0^l m_x\theta_x dx\ , \tag{8.139}$$

the minimum of which is to be searched for in the set of functions w_F and θ_x satisfying the additional constraint condition

$$w_\mathrm{F}(x) = r\,\theta_x(x)\ . \tag{8.140}$$

The variational formulation (8.139), (8.140) is in exact accordance with the bi-member model. The first integral in (8.139) describes the flexure of the fictitious bar while the second one determines the simple torsion of the main bar.

Suppose that the constraint equations (8.140) hold in predefined points only (nodes of the bi-member model) rather than within a continuous interval:

$$w_\mathrm{F}(x_i) = r\,\theta_x(x_i)\ , \qquad (i = 1,\ldots, n{+}1)\ . \tag{8.141}$$

Evidently, the set of functions w_F, θ_x that satisfy the discrete conditions (8.141) is an expansion of the set of functions that satisfy the continuous conditions (8.140). In other words, the constraints are weaker in the discrete model. When a set of admissible functions is expanded, the minimum of a functional cannot help decreasing.

As the minimum of the full potential energy of the system is equal to the value of its deformation energy E taken with the minus sign, we immediately have the estimate

$$-\,\mathsf{E} > -\,\mathsf{E}_n\ , \tag{8.142}$$

which was noticed by us when discussing the results of Example 1.

Again, we emphasize that it is the constraint equations which are subjected to a discretization rather than the functional itself, unlike the standard procedure of the finite element analysis.

To conclude this section, we'd like to note that actually the discrete constraints (8.141) are nothing but possible modifications of *poly-constraints* discussed earlier. Therefore the bi-member model of a thin-walled bar is one more reminder about the advantages of a software equipped with this powerful tool of building structural design models. Though, strictly speaking, there is no urgent necessity to operate general poly-constraint equations in order to handle bi-member models. It is enough to be able to equate some of the displacements in a chosen couple of nodes. This is exactly the way the **SCAD** and **GTSTRUDL** systems operate.

The approach presented above has enabled the implementation of the model by V.Z. Vlasov that deals with a thin-walled open-section bar, in order to refine a standard approach of engineering mechanics. Though, in the practice of structural design and analysis this level of detailing may prove insufficient too, especially in cases when one needs to allow for a distorted bar cross-section shape or other delicate effects. Then one has to simulate a thin-walled bar by a set of shell finite elements to allow for almost any possible peculiarities of a bar's behavior. A typical example is shown in Fig. 8.39 that represents a fragment of the design model of a block of main beams of a shelter structure erected above the destroyed reactor of the Chernobyl nuclear power plant in 1986. Its structural assessment performed in ten years or so required that all detected faults of the design and damages caused by the erection and operation be allowed for properly. Also, a design model of this level of detailing enabled a correct description of possibilities for a local and global loss of stability.

Generally, when we have to analyze particular members of thin-walled structures, it may prove quite reasonable and efficient to switch from the thin-walled bar theory to a conventional finite element model consisting of plate elements. One of demonstrative results of this kind of analysis is presented in Fig. 8.40.

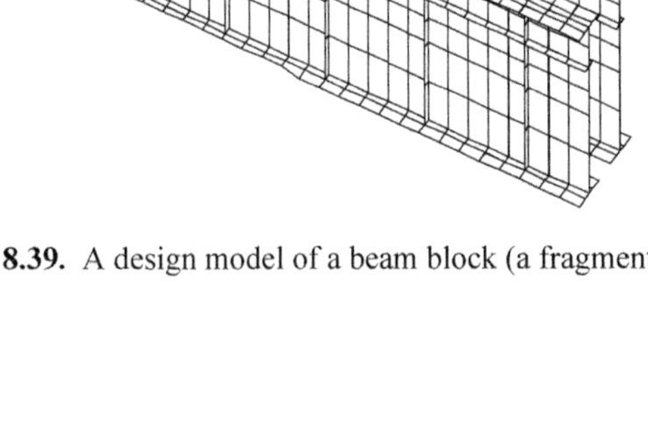

Fig. 8.39. A design model of a beam block (a fragment)

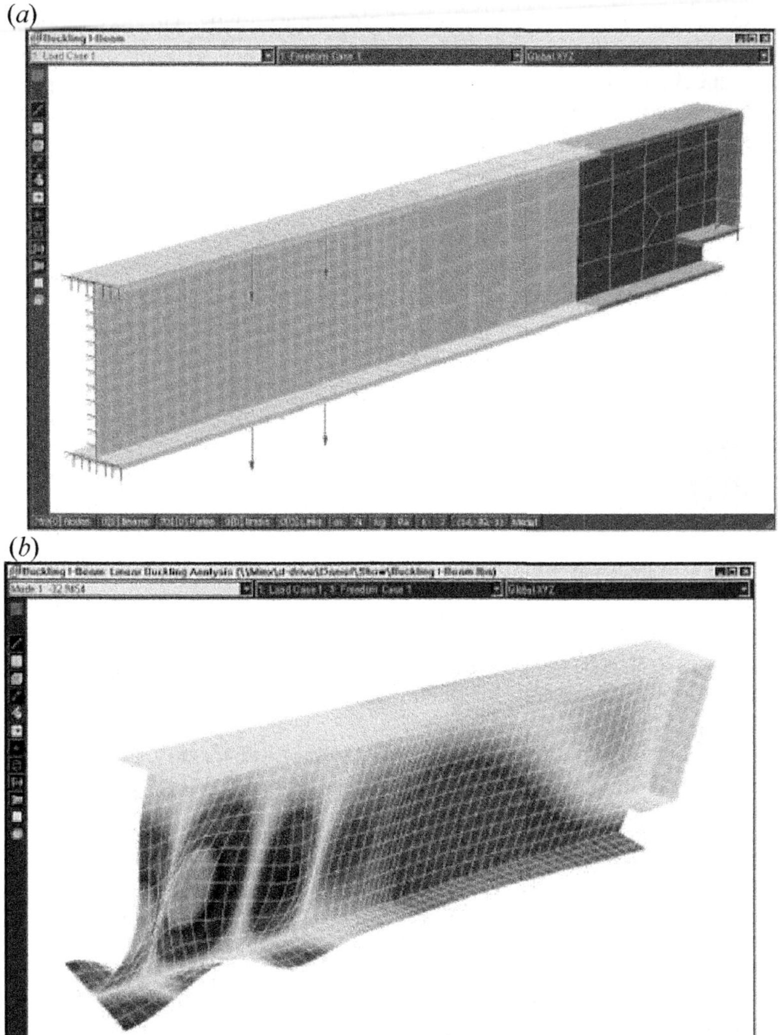

Fig.8.40. A finite element model of a welded double-tee beam: (*a*) its design model; (*b*) its mode of buckling.

8.9 Design Load Combinations

Any structure is subject to a lot of loads of various kinds that arise during its erection and use. At the same time, the analysis is most often performed for particular (elementary) loading options assuming that any real loading on the system can be represented as a linear combination of elementary loads. This assumption is quite acceptable for a linear analysis because the principle of superposition holds for linear mechanical systems. Making use of "nonlinear finite elements" provided by many structural analysis programs renders this approach senseless. Unfortunately, users do not always remember this fact. Therefore a "good" CAD software system should have adopted an appropriate check function to generate a warning, at the least.

Further we assume that the linearity takes place. Our intention is to search for linear combinations of elementary loads which are worst in some sense and enable us to verify that imposed limitations of the strength, stability, deformability etc. hold. Such are called *design load combinations*, and coefficients at the elementary loads in those are referred to as *intensities*.

Some load combinations can be recognized as unfeasible, for example, because they just fail to emerge or the probability of their emergence is negligibly small. The latter is often regulated by design codes where one can find statements such as: "*...Horizontal loads caused by masses on flexible suspensions, temperature climatic influences, dynamic effects from equipment and vehicles, braking and lateral forces caused by moving cranes must not be taken into account*".

There can be a variety of ways to choose admissible combinations. One of most general ways is suggested in [2] that uses a special oriented graph without loops having one start and one end vertex. The graph is built in such manner that each path from the start to the end vertex corresponds unambiguously to one admissible combination of elementary loads, and no inadmissible combination can be represented by such a path.

Let's consider how to build a graph of this kind for an example loading of a single-storied industrial building with cranes, subject to the following elementary loads:

- 1 — dead weight;
- 2 — snow;
- 3 — wind from the left;
- 4 — wind from the right;
- 5 — maximum crane pressure on the left column;
- 6 — maximum crane pressure on the right column;
- 7 — crane braking to the left transferred to the left column;
- 8 — crane braking to the right transferred to the left column;
- 9 — crane braking to the left transferred to the right column;
- 10 — crane braking to the right transferred to the right column.

When deciding whether a combination is admissible, we should taken into account the following limitations:

- the dead weight load cannot be absent;
- the wind, snow and crane loads are independent from one another;
- the wind of one direction excludes that of the other, one braking excludes all others, one vertical pressure excludes the other;

- the braking cannot occur without a vertical pressure.

Fig. 8.41 presents a graph that expresses these relationships. Together with arcs 1 to 10 corresponding to the respective elementary loads, there are four more arcs there (the figure shows them as dash lines) that describe the zero load intensities. These additional arcs enable one to come round the loads that do not have to be included in load combinations (as they cause an unloading effect).

There are some dependencies that can be of help to build a graph of admissible load combinations. These include the following rules:

- if a load has to be included in a combination, the respective arc in the graph must form a neck breaking which splits the graph into two parts, one including the start vertex and the other the end one (arc 1 in Fig. 8.41);
- if two loads are mutually exclusive, their respective arcs in the graph must connect the same vertices in a parallel manner (in Fig. 8.41 these are arcs 3 and 4; arcs 5 and 6; arcs 7, 8, 9 and 10);
- if a load can be absent in a combination, there must be an extra arc in the graph to enable coming round this load (in Fig. 8.41, arcs 2 and 12; arcs 3, 4 and 13 etc.);
- if loads can be applied at the same time, then the respective arcs must be connected successively (in Fig. 8.41 those are arcs 1, 2, 3 or 4 etc.);
- the condition that a load A cannot be taken without taking a load B at the same time, is provided by the absence of a zero arc passing the same vertices that the B arc (see Fig. 8.41 for 7–10 as A arcs and 5–6 as B arcs).

Every arc of the graph built in this way can be assigned two numbers: allowable maximum and minimum values of the respective load intensity. Then the set of all thinkable paths across the graph corresponds to a set of real loads upon the structure in question. And a design combination refers to a certain "ultimate" path which should be chosen using some criterion function. Such criteria can be:

- a stress in some point of a bar's cross-section;
- a displacement of a node in a given direction;
- a response of an elastic foundation;
- an ultimate value of a stress with respective values of other stresses.

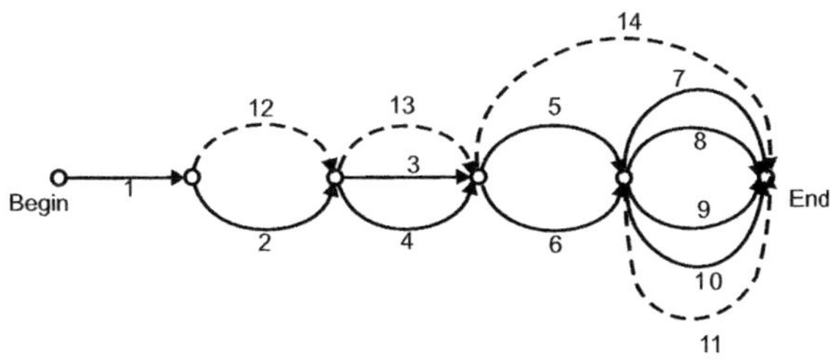

Fig. 8.41. A graph of a logical loading structure

In linearly elastic systems all these criteria can be expressed as linear combinations of load intensities, and the extreme values of those are achieved, generally, at different admissible combinations.

Factors with which the elementary load intensities participate in a criterion function often depend on geometrical properties of cross-sections. For example, if the criterion is a maximum stress at some point of a cross-section of a bar member, then one has to know beforehand the size of this cross-section or at least its core distances. But the search for an ultimate load combination is often intended to select the proper cross-section. Thus there is a contradiction to be overcome only by successive approximations in a loop like this:

specify section properties — calculate — find a design load combination — select a section — calculate — find a design load combination etc.

The search for a design combination being a phase of this iteration sequence (most often, the only one because it is a rare case that one revises one's source data after selecting the section to search for the load combination again) is implemented by most industrial structural analysis software applications. Also, one should remember that the very logic of defining admissible or inadmissible combinations, either as graphs or in another equivalent form, is based on design code recommendations. Thus the user should make sure his software implements recommendations of the code he needs to comply with.

Also note that a set of results the software shows to the user includes answers to questions of this kind: "What stresses are caused in the structure by such and such load?" or "What maximum stress may appear in this element at a worst possible load combination?". The important thing is that the developers of the software have formulated these questions beforehand, and the user has just the choice between asking them (demanding a result of a certain type) and not asking.

It seems better to enable the user to formulate some questions by himself. For example, an interesting question is "What elements' load combinations include such and such loads?". This question is sensible when the designer can influence the load and make a decision whether to do it. Unfortunately, no software known to the authors of this book implements this function.

As for the above stated contradiction between the source data for the design load combination search and the way the results of this search are used, it can be overcome to some extent by using an approach defined in [16]. There they suggest to extract a subset from the whole set of admissible combinations such that it contain design combinations for a class of criteria rather than a single one. This class is formed by all criterion functions possessing the convexity property. Due to the linearity of structures being discussed, we search for a design combination of stresses caused by loads applied to the structure rather than a combination of the loads themselves.

Let's denote as n the number of force factors acting simultaneously in a cross-section of an element. Every i^{th} combination of loads causes an n-dimensional stress vector that defines a point X_i in the force factor space R^n. For example, if we are searching for a disadvantageous stress caused by a longitudinal force and a moment, then $n = 2$ and the points X_i are determined by couples $\{N_i, M_i\}$. Such points (Fig. 8.42) can be very numerous in the space R^n. For example, if all elementary loads are independent, there will be 2^k of them where k is the number of the loads.

It is known that any downward convex criterion function $f(x_1, \ldots, x_n)$ defined on a convex polyhedron $M \in R^n$ achieves its maximum in one of the vertices of the polyhedron [5]. If the polyhedron M is a convex envelope of a finite set of points X_i (in Fig. 8.42 it is shown by dash lines), then the criterion achieves its maximum in one of its vertices independent of what this criterion is as long as it is convex. The number of the vertices is much less than that of all points X_i, and one can perform a simple exhaustive search.

An algorithm for finding vertices of a convex shell of a finite set of points is based on the fact that the coordinates of a point which is not a vertex of the polyhedron M can be represented as a convex linear combination of the coordinates of other points from this set. The algorithm is described in detail in [16]. As for the convexity of the criterion function, it is seldom questionable: most practical criteria are linear. Nevertheless, the convexity does not always take place. If the criterion function is a load-carrying capacity of an element complying with all regulatory requirements (strength, stability, rigidity), then its convexity should be thoroughly investigated. Particularly, one should make sure the load-carrying capacity area of the element in question is convex. The matter is that this area can be a kind of a level curve of the criterion function, up to a common material strength type scaling factor, in the course of searching for a worst load combination. Evidently, if the load-carrying capacity area is not assuredly convex, then the criterion function of the worst possible load combination analysis will be non-convex, too.

One can easily make sure this danger exists. Consider a compressed/bent steel member with its load-carrying capacity area bounded by a stress interaction curve, and you can see that this area can be non-convex.

The non-convexity of the area itself may result in a lot of trouble. The most evident pitfall is that, according to the tradition of estimating disadvantageous stress combinations, engineers either do not allow for some stress combinations or allow for those in their full scale. Though, in a non-convex area the worst value can be achieved in an intermediate point.

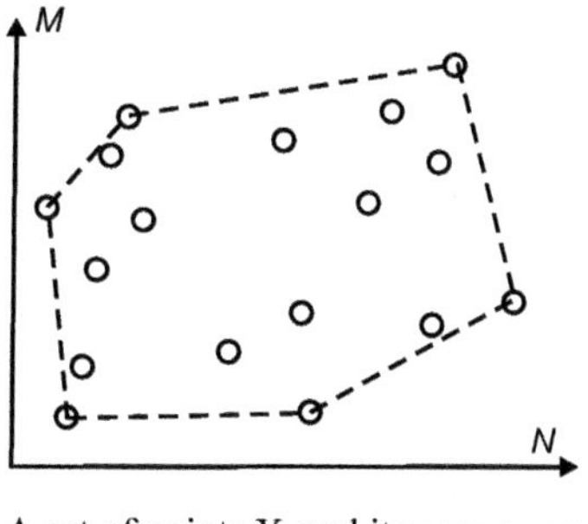

Fig. 8.42. A set of points X_i and its convex envelope

The following should be added to what was said above. Design codes of various countries recommend typical combinations of loads according to which the load-carrying capacity of structures should be validated. For example, a standard of USA [1] gives the following options:

$$\begin{aligned}
&1.4\ \mathrm{DL}\\
&1.2\ \mathrm{DL} + 1.6\ \mathrm{LL}\\
&0.9\ \mathrm{DL} \pm 1.3\ \mathrm{WL}\\
&1.2\ \mathrm{DL} \pm 1.3\ \mathrm{WL}\\
&1.2\ \mathrm{DL} + 0.5\ \mathrm{LL} \pm 1.3\ \mathrm{WL}\\
&0.9\ \mathrm{DL} \pm 1.0\ \mathrm{EL}\\
&1.2\ \mathrm{DL} \pm 1.0\ \mathrm{EL}\\
&1.2\ \mathrm{DL} + 0.5\ \mathrm{LL} \pm 1.0\ \mathrm{EL}\,,
\end{aligned}$$

where DL designates a *Dead Load*, LL a *Live Load*, WL a *Wind Load*, and EL an *Earthquake Load*. The similar Eurocode [11] recommendation is

$$\begin{aligned}
&1.35\ \mathrm{DL}\\
&1.35\ \mathrm{DL} + 1.50\ \mathrm{LL}\\
&1.35\ \mathrm{DL} \pm 1.50\ \mathrm{WL}\\
&1.00\ \mathrm{DL} \pm 1.50\ \mathrm{WL}\\
&1.35\ \mathrm{DL} + 1.35\ \mathrm{LL} \pm 1.35\ \mathrm{WL}\\
&1.00\ \mathrm{DL} \pm 1.00\ \mathrm{EL}\\
&1.00\ \mathrm{DL} + 0.30\ \mathrm{LL} \pm 1.00\ \mathrm{EL}\,.
\end{aligned}$$

Comparing to records of SNiP [32] where the requirements to design load combinations are represented in a more general (and complicated) form, the things stated above seem to be very simple and to define "the rules of thumb". Though, it is not hard to note that the simplicity is spurious. Just ask yourself a question: what direction of wind should be used to determine the value of WL, or which one of possible options of the LL application should be chosen. And it becomes clear immediately that the simple "rules of thumb" are not a plaster for all sores, and they do not resolve the issue how to reasonably choose design load combinations.

Note that some finite-element analysis software systems such as **ANSYS** or **NASTRAN** include few or no options for determining design load combinations. The explanation for this is that the said programs are oriented at mechanical engineering where objects are operated for a shorter time and their possible modes of loading are not so various and numerous as those of load-carrying structural constructions.

References

1. AISC (1994) Manual of Steel Construction, Load & Resistance Factor Design, 2nd edn., American Institute of Steel Construction, Chicago, Ill.
2. Artemenko VV, Gordeyev VN (1967) A program for calculating design stress combinations in the case of a complex logical relationship between loads (in Russian) In: Computers and mechanization facilities in structural design, 2:10–14
3. Barvashov VA (1977) On an analysis of a sag of soil beds represented by different models (in Russian) In: Foundations, beds and soil mechanics, 4:25–27
4. Barvashov VA, Fedorovsky VG (1978) A three-parametric model of a soil bed and a piling to allow for irreversible structural deformations of the soil (in Russian) In: Foundations, beds and soil mechanics, 4:17–20
5. Berger M (1987) Geometry, vol. 1. Springer-Verlag, Berlin
6. Bezukhov NI (1953) Theory of elasticity and plasticity (in Russian). Gostekhteorizdat, Moscow
7. Birbraer AN (1998) Analysis of structures for seismic resistance (in Russian). Nauka, St.-Petersburg
8. Borodachiov NM (1975) On an opportunity to replace complex elastic foundation models by simpler ones (in Russian). In: Structural mechanics and analysis of structures, 4:37–39
9. Dinkevich SZ (1969) Some issues of analysis of roof trusses (in Russian). In: Design of steel structures, Inf.-abs. bull., Series VII, 9(17):51–70
10. Elsukova KP, Slivker VI (1976) Some peculiar features of FEM in analysis of structures on elastic foundations (in Russian). In: Finite element method and structural mechanics, Proc. Len. Polytech. Inst., No 349, pp. 69–80
11. ENV 1991-2-4. (1995) Basis of Design of Structures. Part 2–4: Wind Actions. Brussels: European Committee for Standardization
12. Filonenko-Borodich MM (1945) A simplest model of an elastic foundation capable of distributing the load (in Russian). Proc. MEMIIT, Issue 53
13. Geldfandbein AN, Shelest LA (1974) Field tests of sags of cohesive soils of a foundation loaded by a stamp (in Russian). In: Foundations, beds and soil mechanics, 2:133-137
14. Ghilman GB, Borisenko VS (1973) Analysis of spatial systems with their structure changing in the course of loading (in russian). In: Computers for research and design of structural objects. Issue III. KievZNIIEP, Kiev, pp.27–37
15. Gorbunov-Posadov MI, Malikova TA, Solomin VI (1984) Analysis of structures on elastic foundations (in Russian). Stroyizdat, Moscow
16. Gordeyev VN, Artemenko VV, Minkovich EI (1989) Choosing a worst possible load combination as a solution of a multi-criteria optimization problem (in Russian). In: Numerical methods of analysis and optimization of structural constructions: Proc. of Kucherenko Central Research Institute of Structural Constructions, Moscow, pp. 26–32
17. Kolosova GS (1985) Finite element analysis of foundation slabs on an infinite two-parametric bed (in Russian). In: Finite element method and structural analysis, Proc. Len. Polytech. Inst., No 405, Leningrad, pp. 48–51
18. Lebedev NN (1963) Special functions and their applications (in Russian). Fizmatgiz, Moscow-Leningrad

19. Lysmer J, Kuhlemeyer RL (1969) Finite Dynamic Model for Infinite Medis. Journal Eng. Mechanics Division: Proc. ASCE. EM4:859–877
20. Lysmer J, Waas G (1972) Shear Wawes in Plane Infinite Structures. Journal Eng. Mechanics Division: Proc. ASCE, EM1:85–105
21. Manvelov LI, Bartoshevich ES (1961) Choosing a design model of an elastic foundation (in Russian). In: Structural mechanics and analysis of structures, 4:14-18
22. Mitsui K, Kanemitsu T (1999) Construction process analysis for truss structure by genetic algorithm In: Proceedings of the IASS 40th Anniversary Congress, Madrid, 20-24 September 1999 «Shell and Spatial Structures: From recent Past to the next Millennium», vol 1, Section B. CADEX, Madrid, pp. 31–36.
23. Novozhilov VV (1953) Foundations of nonlinear theory of elasticity. Graylock Press, Rochester, N.Y.
24. Pasternak PL (1954) Basics of a new method for analyzing foundations on elastic beds using two bedding factors (in Russian). Gosstroyizdat, Moscow – Leningrad
25. Paton EO, Gorbunov BN (1935) Steel bridges. Vol. 1. Trusses of the beam framing (in Russian). State Sci. and Tech. Publishing House of Ukraine, Kiev
26. Perelmuter AV (1998) Control of behavior of load-carrying structures (in Russian). UFIMB, Kiev
27. Perelmuter AV, Slivker VI (1979) Analysis of an elastic system supported by hydraulic jacks (in Russian). In: Structural mechanics and analysis of structures, 6:65–69
28. Rzhanitsin AR (1982) Structural mechanics (in Russian). Visshaya shkola, Moscow
29. Shashkin AG, Shashkin KG (1999) Interaction of a structure and its foundation. Technique of analysis and practical applications to design (in Russian) // www.georec.narod.ru
30. Shulkin YB (1978) Kinematical analysis of bar systems used in the transportation industry (in Russian). Len. Inst. Railway Eng., Leningrad
31. Slivker VI (1981) On an assignment of properties of a two-parametric elastic foundation (in Russian). In: Structural mechanics and analysis of structures, 1:36–39
32. SNiP II-7-81* (1996) Construction in seismic regions (in Russian). Ministry of Construction of Russia, Moscow
33. Tseitlin AI, Guseva NI (1979) Statistical methods of analysis of structures under group dynamical effects (in Russian). Stroyizdat, Moscow
34. Uflyand YS (1967) Integral transformations in problems of elasticity (in Russian). Nauka, Leningrad
35. Vlasov VZ (1963) Thin-walled elastic bars: Selected works.Vol. 2 (in Russian). Publish. House of USSR Academy of Science, Moscow
36. Vlasov VZ, Leontiev NN (1960) Beams, plates, shells on elastic foundations (in Russian). Fizmatgiz, Moscow
37. Wilson EI (1997) Three Dimensional Dynamic Analysis of Structures. A Publication of Computer and Structures, Inc., Berkeley

9 Buckling Problems and Related Issues

We can see how a designer's using a symmetry and optimization may often result in building a system that has a lot of pathology and structural instability. For example, one can find no perfectly symmetric thing in the nature, though such appear on drawing boards of designers. The humans, themselves being a product of the nature, require things to be such as the nature cannot make them: they try to build such things, and the nature resists to the attempt. This is how the symmetry is violated by imperfections.

J.M.T.Thomson. Theory of catastrophes and its role in applied mechanics // Theoretical and applied mechanics: Proc. XIV International IUTAM Congress.

9.1 Validating an Equilibrium Stability

As it has been stated for many times, an engineering analysis is performed using idealized design models that represent a real structure in a more or less simplified form. Some peculiarities, being of minor importance from the viewpoint of the model's creator, may well fall out of consideration. Such assumptions always need validating because there are numerous well-known examples how small deviations (perturbations) can affect a structure's behavior very much, both at the quantitative and qualitative level. A real structure is different from its design model, and [2]:

"*... this difference is provided by numerous, more or less negligible, deviations from a design, faults and imperfections. An engineer needs to be confident that a real structure is going to behave in about the same way as its respective idealized model, in spite of the presence of those deviations. Having no such confidence, one should not design anything at all for it would not be sensible*".

The validation of the equilibrium's stability, that is, making sure that a small disturbance of the initial equilibrium state does not cause the equilibrium to fail and the system does not acquire new qualitative properties in its stress and strain distribution, should be an integral part of any practical design procedure. The book by V.I. Feodosiev [7], quoted here for a lot of times, says truly that:

"*... when solving stability problems, one should specify the load in much greater detail than that for a conventional strength analysis. It does not suffice to* *show just the magnitude and direction of the applied forces. It is necessary to specify their behavior in the course of the system's deviating from its initial equilibrium position.*".

The example is a set of options how to transfer a load to a simple cantilever bar which make the critical force several times different from the classical value.

An elastic cantilever bar (Fig. 9.1, *a*) subjected to a compressive force N is a standard object of study in the "Buckling" section of the strength of materials. The classical Eulerian solution of the problem is a critical force $N_{cr} = \pi^2 EI / (4l^2)$. Here possible deviations from the original rectilinear shape are assumed to keep the force N directed vertically and moving passively together with the column's end.

The author of the book [15] gives various examples of a more complex behavior of an applied force in the course of the same cantilever column's deformation, in the section with a peculiar title "What does a tracking force track?". Some characteristic examples are given in Fig. 9.1.

In the scheme of Fig. 9.1, *b* the force P transfers itself to the end of the column by means of a rope passed through perfect pulley blocks. Then during the bending of the bar the force N looks towards a fixed point A located at the distance a vertically from the top end of the column where the rope is attached. The respective critical value of the force will be $P_{cr} = \alpha^2 EI$. The parameter α will be determined as a minimal root of the equation (see [6])

$$\operatorname{tg} \alpha l = \alpha l \,(1 - a / l) \,.$$

Particularly, at $a = l$ the fixed pulley (point A) stands level with the column's fixation point which makes the bending moment in the fixation always equal to zero. In this case the critical force will be $P_{cr} = \pi^2 EI / l^2$, i.e. the same as one for a bar simply supported at both ends where the moments are zero just as those in the example considered.

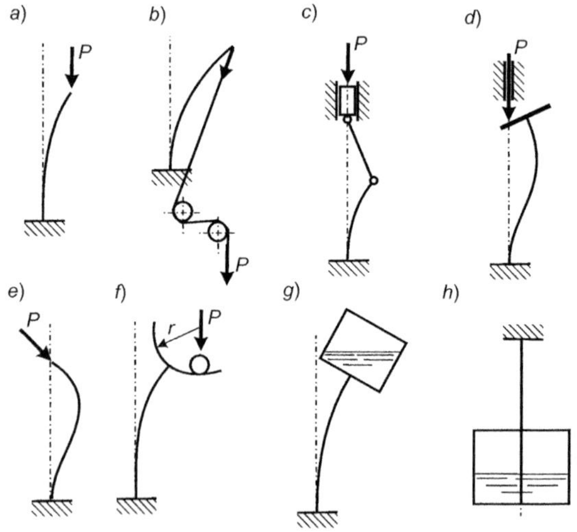

Fig. 9.1. Loaded cantilever options

In the scheme of Fig. 9.1, *c* the longitudinal force N that appears in a perfectly rigid rod AB tracks the position of the point A capable of moving vertically. Here the top end of the column is subjected to the compressive force P and a horizontal component of the force N. The critical value $P_{cr} = \alpha^2 EI$ is to be determined from the transcendental equation (see [7])

$$\operatorname{tg} \alpha l = \alpha L ,$$

the solution of which depends on the ratio between the column's length l and the total length of the column and the rod L.

A scheme shown in Fig. 9.1, *d* (a Reuth's problem) assumes that the force is transferred to the column via a rigid disk and is always directed along the original vertical axis. It has a fixed line of action, its critical value is $N_{cr} = 20.05EI / l^2$, just as in the scheme of Fig. 9.1, *e* where the force N tracks the deformation of the column's end (a Back's problem).

In the scheme of Fig. 9.1, *f* the force is applied to a lowermost point of a cup of the radius r rigidly attached to the end of a column with the length l. The critical force $N_{cr} = \alpha^2 EI$ is determined by the solution of the transcendental equation

$$\alpha l \operatorname{tg} \alpha l = l / r .$$

Finally, schemes of Fig. 9.1, *g* and 9.1, *h* present the same cantilever bar loaded by a weight of liquid. Here the behavior of the bar is noticeably different from Euler's problem. The respective solutions are given in [6].

In flexible bars where the displacements may be comparable with their length the load can also behave in an unusual way. A simplest example is shown in Fig. 9.2, *a* that presents a flexible bar loaded by a force P which remains permanently perpendicular to the deformed axis of the bar.

Figs. 9.2, *b* and 9.2, *c* show flexible systems with supporting rollers where the reactions in the supports change their direction. Solutions of these problems were obtained by V.M. Ovsianko using an electronic modeling technique [15].

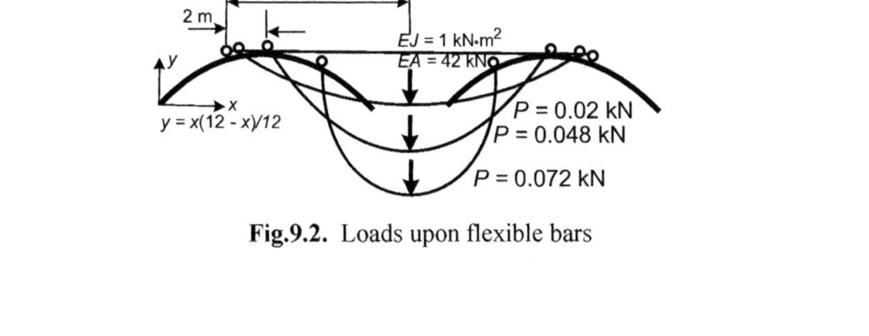

Fig.9.2. Loads upon flexible bars

It should be noticed that any force represents a measure of interaction between a system and its environment. It is the extraction of an object from its environment where errors may be introduced due to an incorrect treatment of the load's behavior when the structure deforms. If the rigid connecting rod in Fig. 9.1, *c* or the rope tie in Fig. 9.1, *b* is included in the design model, then the behavior of the force *P* (now simulated as an internal compression force of the column) will be simulated correctly.

It may seem the examples like those presented in Fig. 9.1 are just exercises of theorists and present no practical interest. This is untrue, at least regarding the cases *b* and *c* encountered in the design practice. The scheme 9.1, *b* is a typical one in the analysis of crane booms and guyed masts. The scheme 9.1, *c* appeared when a reinforcement was designed for a two-span cross-section of an existing industrial building (Fig. 9.3, *a*). Calculations showed that the A–A section of the middle column was strongly overstressed, and the clearance of bridge cranes made it impossible to reinforce the column in this section.

It was designed to solve the problem drastically: refuse to resist to the moment in the A–A section by means of embedding a hinge in it (Fig. 9.3, *b*). As all other elements had a sufficient reserve of strength, it seemed this design was to solve the problem. But the altered response of the lower part of the middle row's column presented by the example in Fig. 9.1, *c*, rendered this reconstruction option unfeasible because the critical force became 2.3 times less.

The classical buckling/stability theory taught in most courses of structural mechanics and commonly used to perform structural analyses assumes (unfortunately, most often without giving any reasonable explanation) that all internal stresses increase proportionally to a single parameter while the relation between those remains constant. Though, this behavior is not always shown by a structural system. A good deal of structures (especially nonlinear ones) make the researcher allow for the fact that the relation between the internal stresses may change as the intensity of the load grows. Therefore if a critical (in the sense of buckling) load intensity is to be found, and it is that which one uses to calculate the stability coefficient, then one has to perform the analysis by simulating a growing load rather than growing stresses.

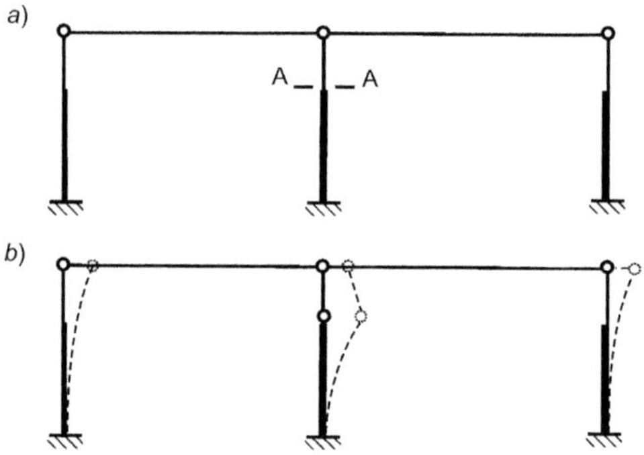

Fig. 9.3. A cross-section of a building:
(*a*) an existing design; (*b*) an option of its reconstruction

Even in linear systems the stability coefficient obtained in the classic manner may happen to lack a clear physical sense. To see this, let us imagine a structure with elements strongly compressed by a dead weight load G_0 and additionally loaded up by 20% by a live load P_0 (Fig. 9.4). The stability coefficient $k' = 1.25$ for the overall load assumes an utterly unreal growth of the dead weight by 25%. If we assume the possible growth of the weight by, say, 10% (that is, taking $k_g = 1.1$), then the live load must grow much higher to cause the critical state. It is natural that this consideration illustrated by plots in Fig. 9.4 represents the humble stability coefficients 1.25 in a totally different form ($k_p > 1.25$). Apparently, the result will depend strongly on the shape of the stability area: were its configuration different, all stability coefficients would turn to be such that the value k_pP_0 would be much less. It is important to notice the very fact of an insufficient accuracy achieved by the analysis with the conventional attitude towards the stability coefficient.

In responsible cases one should calculate values of the critical load for each pattern of loading (or, possibly, for a group of such) separately. Once such results have been obtained and the critical loads $P_{i,cr}$ ($i = 1,\dots,n$) have been determined for each of n loadings, we can use Papkovich's theorem [17] of a convex stability area for a linear system. This theorem implies particularly that the plane

$$\sum_{i=1}^{n} \frac{P_i}{P_{i,cr}} = 1 \tag{9.1}$$

is located not farther from the coordinate origin than the true boundary of the stability area. This is illustrated in Fig. 9.5 for a two-dimensional case. The line (9.1) is denoted as AB in the figure, and it replaces the true boundary of the stability area with a big reserve. The solution can be refined by finding the critical value of the system for a combination of loads (the point C on the ray OC) and approximating the AB curve by a polygonal line ACB.

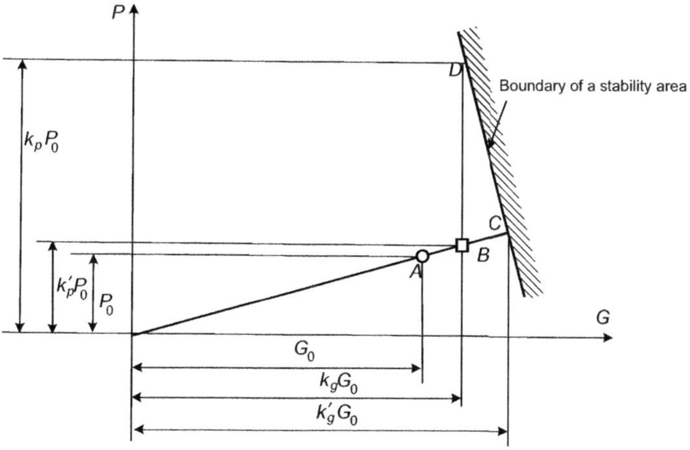

Fig. 9.4. A non-proportional growth of loads

Talking about the existing software, we find that the stability analysis implemented by those products does not include a simulation of an independent behavior of different loads. Unfortunately, standard options of CAD systems are oriented only at a one-parameter loading, such that all loads grow proportionally to a single parameter. The ideal choice would be to specify a multi-parametric loading with an automatic generation of the stability area's boundary. An intermediate alternative may include a classification of loads into *stable* and *growing* ones in the source data for a buckling analysis. Stable loads are to keep their fixed values, that is, do not depend on the loading parameter which is used to investigate the stability of a structure. As opposite to those, growing loads are proportional to the loading parameter. Varying the options of dividing the loads into stable and growing ones will enable the user to actually determine points of the stability area's boundary at any rays (see Fig. 9.5).

It should be said that the tradition of the engineering education, fostered also by requirements of existing design codes, makes an analyst engineer believe it is enough to check a structure for stability with respect to variations of loads and nothing else. Meanwhile in the real life of the structure it is not only loads that can vary. Indeed, together with loads, properties of materials, bar cross-section sizes, support conditions etc. may vary in pretty wide limits. How sensitive is a structure with respect to various disturbances comparing to its idealized model? Could small changes in other parameters of the system (besides loads) lead it to the state of an unstable equilibrium? It's a shame the standard software tools available to an analyst engineer do not answer these questions directly. We believe currently available CAD software products are substantially behind in the structural stability analysis comparing to their capabilities of finding stress/strain distributions in structures. Overcoming this discrepancy is a necessary and promising direction of the CAD systems development.

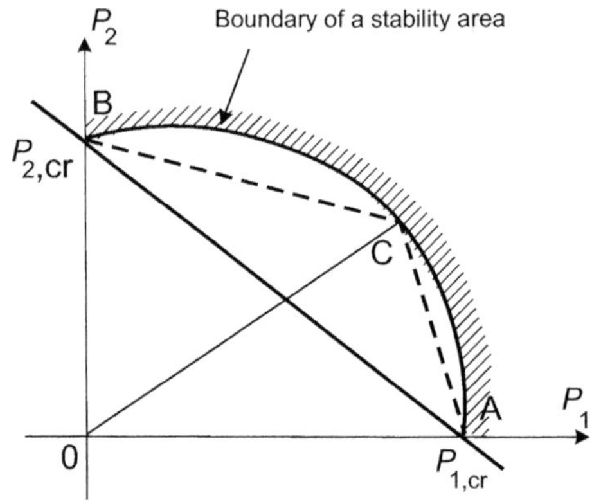

Fig. 9.5. An approximation of a stability area's boundary

9.2 Classic Problems of Equilibrium Stability

The classical statement of the stability problems implemented by nearly all software systems assumes commonly that all internal stresses grow proportionally to a single parameter caused by the same proportional growth of all loads applied to the system. Generally, this statement is not quite correct because the stiffness matrices of bars subjected to transverse-longitudinal bending are not constant. Instead, they depend on the longitudinal force intensity. Therefore the growth of the external load intensity may cause the longitudinal forces in the system to redistribute. But in a great deal of structures there is no change in relationships between their internal stresses in the course of loading (see the concept of a quasi-monocyclic system in Section 4.3 according to N.V. Kornoukhov [11]). A typical example includes conventional frame systems under a vertical load where the relation between stresses in columns remains almost the same. It is the case of a quasi-monocyclic system where the transition from the equations (3.132) to Eulerian ones like (3.134) is best founded. This transition was discussed before in Chapter 3.

Suppose the stability problem should be solved in the Eulerian formulation under the assumption that all applied external loads and internal stresses grow proportionally to a single parameter k. Then the minimal value (of positive ones) of the k parameter at which the stiffness matrix of the system $\mathbf{K}(k)$ ceases to be positive definite is a critical one, and its respective value $k = k_{cr}$ is the stability coefficient. Though, an analyst engineer should remember some systems are able to lose their stability at the opposite sign of the load. It corresponds to the case of negative values of the parameter k, and the negative critical value of k may be even different from its positive critical value in its magnitude (this situation is characteristic for asymmetrical systems, as a rule). A simplest example of this kind follow: a bar restrained at both ends from longitudinal displacements under a longitudinal force applied at the center of the span (Fig. 9.6).

The stiffness matrix $\mathbf{K}(k) = \mathbf{K}_0 + \mathbf{K}_G(k)$ consists of a "conventional" stiffness matrix $\mathbf{K}_0$ and a matrix of "pushing" reactions $\mathbf{K}_G(k)$ which are determined by compressive forces in bars, compressive stresses in shell-type finite elements etc.[1] Recall that the positive definitiveness of a stiffness matrix means any values of nodal displacements and rotations $\mathbf{u}$ leave the change of the system's potential energy positive (the energy grows, that is, certain work should be spent to deform the system, the latter resists to the deformation thus deserving the title of a *resistive* system).

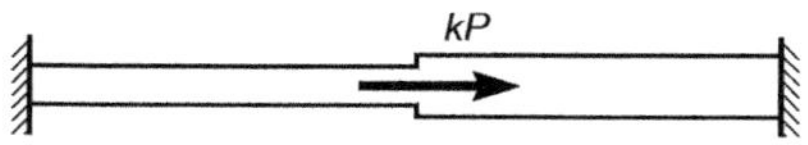

Fig. 9.6. An example of a system capable of buckling both at $k > 0$ and at $k < 0$

[1] The terminology commonly used in the finite element analysis theory refers to the matrix $\mathbf{K}_G(k)$ as a *geometrical stiffness matrix* which explains the use of the subscript $_G$ in its designation.

Once the system has buckled, its resistance vanishes and its stiffness matrix becomes degenerate (acquires the zero determinant). In its post-buckling state the system acquires the property of a negative resistance: when forcedly deformed, it gives out the energy of its "pushing" reactions previously accumulated, and its stiffness matrix becomes indefinite in its sign.

It should be noted that calculations based on the stiffness matrix make it possible to find only those critical states in which the buckling occurs in the mode where the nodal displacements and slopes are not simultaneously equal to zero (this is a so-called *explicit buckling mode*). One has to check also whether at the tried value of k (or earlier) a *hidden buckling mode* may occur. This one might be localized within a single finite element and cause neither nodal displacements nor slopes (the buckling mode causes only zero values of nodal displacements).

Generally, explicit and hidden buckling modes can be distinguished by comparing the given and the principal systems. If building the principal system includes the imposition of t_j constraints and removal of t_k ones (here we assume a mixed method of analysis, to be general), then the appropriate definitions [4] follow from Table 9.1.

Table 9.1.

Mode type	Given system	Principal system
	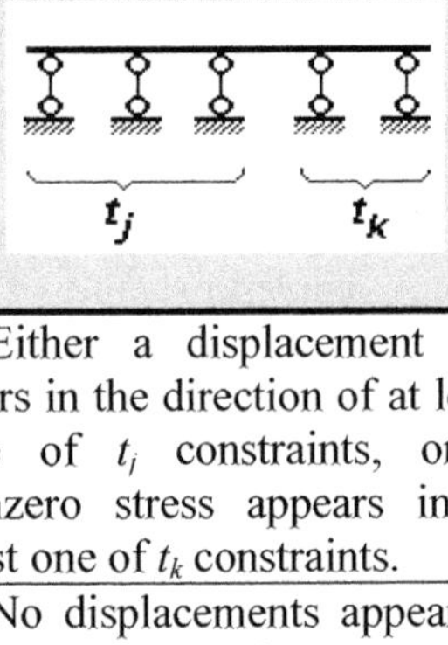	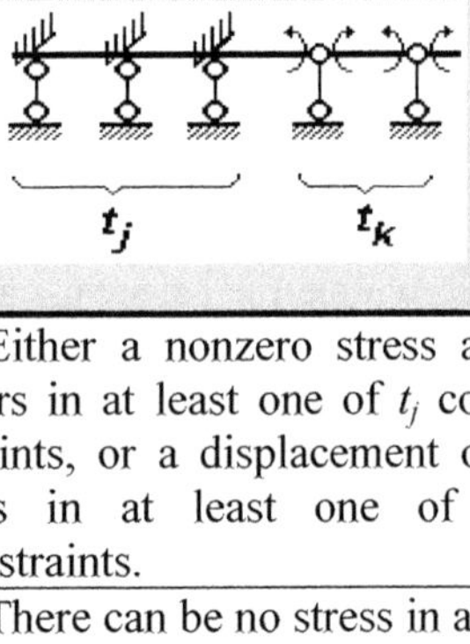
Explicit	Either a displacement appears in the direction of at least one of t_j constraints, or a nonzero stress appears in at least one of t_k constraints.	Either a nonzero stress appears in at least one of t_j constraints, or a displacement occurs in at least one of t_k constraints.
Hidden	No displacements appear in t_j constraints, there are no stresses in t_k constraints.	There can be no stress in any of t_j constraints, there can be no displacements in the directions of all t_k constraints.

Confining ourselves to the slope deflection method, as most contemporary CAD software systems do, we can notice that hidden buckling modes appear because additional constraints of the slope deflection method do not affect the critical load by being located at nodes of the buckling mode.

The check of the presence of hidden buckling modes is fairly simple and is performed by validating the following inequality for all finite elements:

$$k_{e\,cr} > k \tag{9.2}$$

where $k_{e\,cr}$ is the critical value of the loading parameter for a particular e^{th} finite element in the principal system of the slope deflection method, and k is the value of the system's loading parameter currently being tried[2].

If the equality (9.2) holds, it evidences there are no hidden modes of the system's buckling at the level of the load not exceeding the tried value of the parameter k. Otherwise there are hidden modes to be sought at a less load intensity. As the critical values $k_{e\,cr}$ are known for all types of finite elements and can be calculated by simple formulae, implementing the (9.2) condition's check in software should meet no difficulties.

Though, the developers of some software systems do not provide the determination of hidden buckling modes. Unfortunately, it is often the case that no warnings of this are included in the software manuals. From the viewpoint of a software user, the motive for this decision of the developers is not an issue of great interest, but the authors would like to present an example of an excusive consideration which they used to encounter:

"*As the principal system of the slope deflection method is formed from the given system by imposing some constraints, we can conclude that buckling modes hidden from the slope deflection technique correspond to a higher load intensity than the first (least) critical value of the loading parameter. It follows from the well-known Rayleigh theorem of the effect of additional constraints on the stability of a mechanical system. The practical need of a user engineer is such that he may be interested only with the first (lowest) buckling mode, so the detection of hidden (that is, higher) modes is of no practical importance*".

We cannot agree to this consideration due to the following reasons. First, software developments do not always follow the slope deflection method purely. Just recall the problem with poly-constraints discussed in Section 3 where the problem's solution includes an extension of the original space (in the variational sense it corresponds to the use of an auxiliary functional employing Lagrangian multipliers). The Lagrangian multipliers, being extra unknowns added to the desirable nodal displacements, have the sense of generalized forces. Thus we actually deal with a mixed principal system. Similarly it happens with null-compliant constraints. But this is not the only trouble. Is it really true that higher buckling modes are of no practical interest?

Let's present a simple example to demonstrate that the positive answer to this question is too much categorical, therefore incorrect.

One of practical design tasks was to determine the thickness of a ferroconcrete bottom of a deep underground structure. A simplified scheme of it is shown in Fig. 9.7. The side pressure on the walls of the basement was going to cause a strong compression of the bottom, so the latter's thickness was determined by stability/buckling conditions. Due to operational requirements, the internal spaces between the walls of the basement did not allow installing any additional partitions or other links to reinforce the bottom and increase the critical force.

---

[2] The inequality (9.2) implies that positive critical values of the parameter k are to be found.

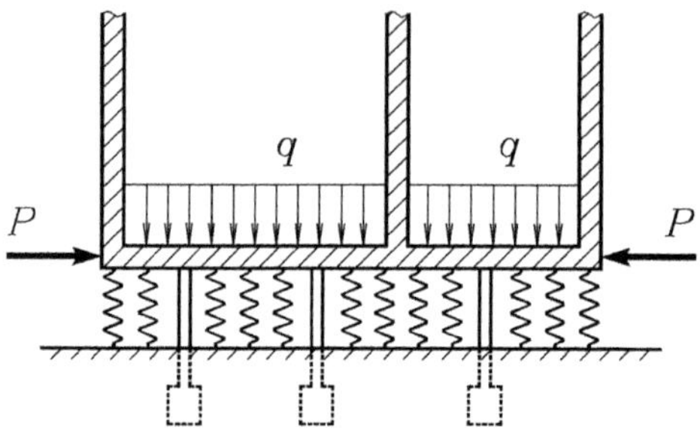

Fig. 9.7. Restraining a basement's bottom with anchors from buckling

A decision was made to set up special anchors and thus restrain the bottom from the buckling in order to increase the load-carrying capacity of it.

It was no easy technical task to provide the anchors, and this operation required a lot of effort. So it was necessary to solve the problem of minimizing the number of anchors and installing them reasonably along the bottom to maximize the critical compression force. It is easy to guess that having found the second, third and higher buckling modes for the anchor-free system would enable us to determine best locations for anchors if there must have been one, two etc. anchors. By installing a single anchor (that is, an external constraint) in the node of the second buckling mode we obtain the first critical force for the single-anchor system equal to the second critical force for the anchorless system. A higher level of the critical force just cannot be achieved with a single anchor! It follows from known estimates of the effect of additional constraints [14] on the critical force of a system. Similarly we can conclude that k anchors should be installed in nodal points of $(k+1)^{\text{th}}$ buckling mode of the original system.

To generalize this, let's make a conclusion that the investigation of higher buckling modes can be of a valuable practical help for an engineer to optimize structures with stability-related limitations. If a structure should be reinforced using additional constraints, these can be placed reasonably only if the first buckling mode is not the only one known to the engineer. This being the fact, the user is welcome to demand that a software developer provide the detection of both the explicit and hidden buckling modes.

The critical value of the stability coefficient k can be found by a dichotomy of the interval of its admissible values. The process is to continue until an interval (k_1, k_2) is found such that $k_2 - k_1 \le \varepsilon$ (ε is a small number that defines the accuracy of the critical load's calculation), and the matrix $\mathbf{K}(k_1)$ is positive definite while $\mathbf{K}(k_2)$ is not. The value of k_1 is declared critical. The algorithm to check the positive definitiveness of the stiffness matrix $\mathbf{K}(k)$ is very simple. It suffices to perform the direct pass of the Gaussian elimination, that is, decompose the stiffness matrix $\mathbf{K}$ into triangular matrix multipliers according to the formula (5.82). If this decomposition is possible and all diagonal elements of the matrix $\mathbf{U}$ are strictly positive ($U_{ii} > 0$, $i = 1,\dots, n$), then the matrix $\mathbf{K}$ is positive definite [23]. Otherwise the matrix $\mathbf{K}$ is not.

Assuming the stability coefficient $k \in (k_1, k_2)$ found to be accurate, one should solve the problem of finding the eigenvector **u*** for the known eigenvalue of the equation

$$[\mathbf{K}_O + \mathbf{K}_G(k)]\,\mathbf{u}^* = \mathbf{0}\,. \tag{9.3}$$

Note that the right parts of the system are zero, i.e. only such values of the nodal displacements and slopes **u*** are to be found that are caused solely by internal compressive stresses/forces. As it is known, lateral loads[3] do not affect the critical force values and buckling modes. The equation (9.3) is solved at the zero right part, so the buckling mode is determined up to a factor. Its increasing or decreasing any number of times does not violate the condition (9.3).

9.3 Free Lengths of Compressed Bars

Modern building standards, such as steel design codes, use widely the concept of a *free* (*effective, design, reduced* — various terms are in use) length of a bar. This means the length of an imaginable centrally compressed, simply supported bar of the same stiffness EI the critical compressive force of which is the same as that of the respective real bar. This concept has been introduced by F.S. Yasinsky [30] in order to generalize Euler's formula for the case of a centrally compressed, linearly elastic bar with any arbitrary fixation of its ends. It works all right to deal with simplest structures such as stand-alone columns. But its employment for analyzing bars included in a more complex system needs explaining.

The usual consideration is: if we know the critical value k^* of the load intensity parameter, then the longitudinal stress in j^{th} bar in the system's critical state will be $N_j^* = k^* N_j$ where the value of N_j is the stress in the same bar at the unit value of the load intensity parameter k. By equating N_j^* to the Eulerian critical force of a simply supported bar with the so far unknown length $\mu_j l_j$

$$N_j^* = \frac{\pi^2 EI_j}{(\mu_j l_j)^2}, \tag{9.4}$$

we obtain the *free length coefficient* μ_j. Further, all bars of the system are checked for stability with the working forces N_j assuming their free lengths are $L_j = \mu_j l_j$ $(j = 1, \ldots, n)$, n being the number of the compressed bars in the system.

---

[3] According to A.R. Rzanitsin [18], loads of this kind are called *active* unlike *parametric* loads directly related to longitudinal forces. Of course, these concepts are totally conditional though convenient, and they apply to loads only in simplest structural systems. One can easily imagine a load that causes longitudinal stresses in some bars and is a lateral load causing the flexure in other bars of the same system. So how should this load be classified now: an active or a parametric one?

It is not hard to prove that this check is insufficient. Let's take as an example a truss structure like that shown in Fig. 9.8 where an accurate value of the critical force is known [3]

$$P_{cr} = \frac{2(L/n)A_d\cos^2\alpha/\left[1+\cos(\pi/n)\right]}{1+\dfrac{A_d\sin^2\alpha}{2A\left[1-\cos(\pi/n)\right]}} \tag{9.5}$$

For the chord we have $N_j^* = P_{cr}/2$, and the free length coefficient according to (9.4) is

$$\mu = \frac{\pi}{l}\sqrt{\frac{2EI}{P_{cr}}}, \tag{9.6}$$

where EI is the flexural rigidity of the chord's elements.

Obviously, with the zero grid ($A_d = 0$) this coefficient is infinite. The critical force being zero with the zero grid is also unquestionable. To see this, consider that the system becomes geometrically unstable with the zero grid, thus it buckles at any small longitudinal force. This example is given here only to demonstrate that the coefficient μ for members of a truss can be unlimitedly big.

Note that if for truss bars we obtain the coefficient $\mu > 1$, it means the general buckling of the system occurs and increasing the members' cross-sections may prove ineffectual. For example, if the structure considered shows an insufficient stability of the chord that can happen at big free lengths, increasing the dimensions of its cross-section may affect the result far less than increasing the grid's stiffness, especially when the chord has a small area (Fig. 9.8, *b*). The conventional technique of increasing the inertia radius at a constant area of the cross-section which has been used successfully with simple bars, does not work here at all.

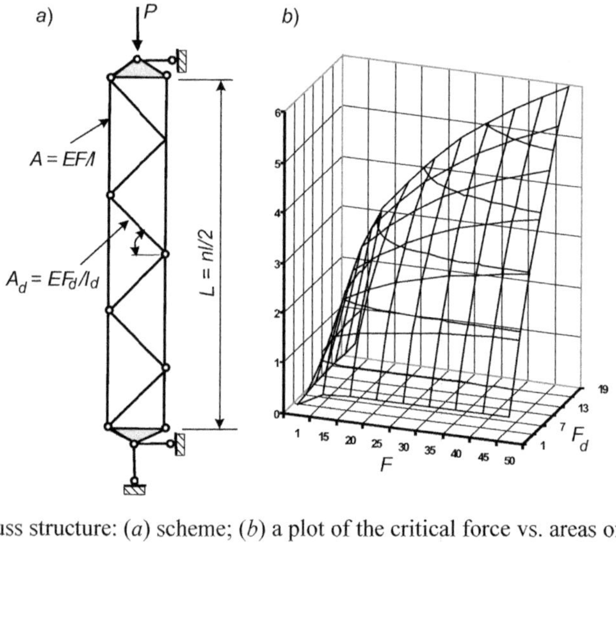

Fig. 9.8. A truss structure: (*a*) scheme; (*b*) a plot of the critical force vs. areas of elements

Thus the correct approach is such that the check of the system's stability in general, which is absolutely necessary, should be supplemented by a check of local stability of its particular members. This is different from the approach declared by design codes that require only the latter to be performed and do not mention the former. Note that the stability validation of members should allow for the possibility of initial imperfections. This is taken into account in steelwork design codes when building values of the longitudinal bending coefficient φ. Therefore the manual [27], recommends the use of a special reliability coefficient $\gamma_e = 1.3$ when determining the critical load for an idealized system by Euler's method.

The concept of the free length (or exactly the way it is used by design codes) has caused a whole bundle of inaccurate considerations, discussions and misapprehensions. Many people were confused by the fact that the calculated free lengths of the bars with a small compressive force were too big and thus conflicting with requirements of the design codes which regard the member slenderness limitations (see [8, 1] for examples). The attempt to increase the stiffness of such members in order to reduce the slenderness, first, contradicts to the design logic, second, does not yield a proper result because the free length of the under-loaded bars grows together with their stiffness.

The basis for the said contradictions is a confusion of concepts analyzed long ago [19] that has become a tradition still uncorrected in design codes. The matter is that the limitation of the slenderness introduced by the design codes is not related, generally speaking, to the concept of stability of a structure. To understand this, ask yourself a simple question: seeing that the free length from (9.4) changes from one loading pattern to another, what loading should be used to check it?

Here's a typical example of a description of the problem being discussed [22]:

"*Due to a high strength of steel, members of steel structures loaded by small forces have very small cross-section sizes which may result in a vibration of the constructions and their warping during the transportation and assemblage. In order to avoid this, the slenderness of bar members should be limited by ultimate values* $\lambda \le \lambda_{lim}$".

Hence it becomes obvious why the traditional approach to the slenderness limitation is so eclectic. For example, for a simply supported bar the relative deflection under a random lateral load P which should be minimized is equal to

$$f/L = \frac{PL^2}{48EI} = \frac{P\lambda^2}{48EA}, \tag{9.7-a}$$

and the period of natural oscillations T which should be minimized too is

$$T = \frac{2L^2}{\pi}\sqrt{\frac{m}{EI}} = \frac{2L\lambda}{\pi}\sqrt{\frac{\gamma}{E}}. \tag{9.7-b}$$

The formulae (9.7) use the following notation together with that introduced before: A is the area of a bar's cross-section; $\lambda = L/r$ is the bar's slenderness; $r^2 = I/A$ is the square of the radius of inertia of the bar's cross-section; m is the linear mass of the bar; $\gamma = m/A$ is the density of the bar's material.

As the limitations (9.7) are related to warping and vibration of bar members during their transportation and assemblage (see the quote from [22] above), the formulae (9.7) should treat L as the geometrical length of the bar l rather than its free length within a structure and under an external load imposed on this structure. It is assumed that during the transportation and assemblage the bar is fixed at its two ends as if being simply supported. Obviously, different relationships should be used instead of the formulae (9.7) to allow for other conditions of the bar's fixation during its transportation/installation. For example, if the temporary fixation of the bar is a cantilever clamping, then the appropriate formulae will be

$$f/L = \frac{PL^2}{3EI} = \frac{P\lambda^2}{3EA}, \tag{9.8-a}$$

$$T = \frac{8L^2}{\pi}\sqrt{\frac{m}{EI}} = \frac{8L\lambda}{\pi}\sqrt{\frac{\gamma}{E}}. \tag{9.8-b}$$

The slenderness λ is squared in the former case and linear in the latter, so its decreasing in different ways affects the compliance with requirements to the structure.

In any case, the temporary assemblage limitation of the bar's slenderness is in no way related to the stability of the structure during its operation. Therefore it would be more correct to formulate the requirements directly to f/L and T rather than in an indirect manner. But the design standards do not do it. We can only advise an analyst who does not know what to do in a particular situation that he should consider a similar structure and calculate the "rated value" of the quantities f/L and T for the simple support by the formulae (9.7). This is actually the way they reason in [20] where they suggest to establish the rated slenderness λ_p by comparing the sag f found by an analysis of the structure of interest under a special local self-balanced load (Fig. 9.9, *a*) with the sag f_0 of the same bar supported by hinges at its ends (Fig. 9.9, *b*).

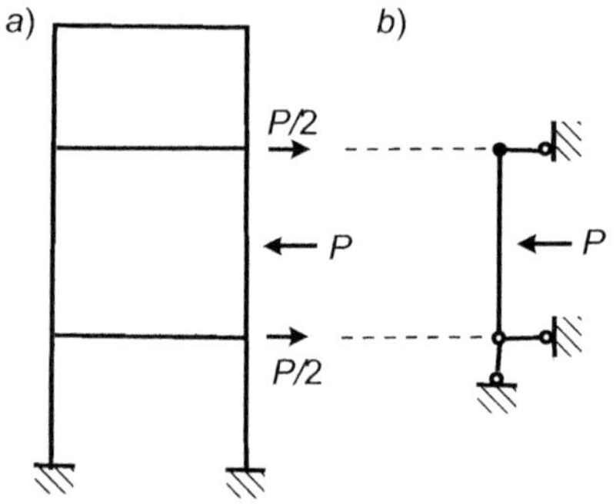

Fig. 9.9. To the analysis of allowable slenderness: (*a*) a system being checked; (*b*) an extracted element

This yields

$$\lambda_p = \lambda \sqrt{\frac{f_0}{f}\left(1 - \frac{1}{k^*}\right)}, \tag{9.9}$$

where $\lambda = l / r$ and k^* is the calculated stability coefficient.

9.4 Analysis of a Role Played by Particular Subsystems

When analyzing peculiarities of possible buckling phenomena, one often uses the concepts of a *constricted and forced* state of separate parts of a system when it buckles [21]. It is assumed that a part of a structure is in the constricted state when the structure itself buckles if this part, when separated from the rest of the system, buckles at a lower load transferred to this part than that under which it actually does. If this "isolated buckling" analysis of the selected part yields a higher load intensity that should be applied to the system so that the latter buckle, or if the system does not buckle at all, this situation is referred to as a forced buckling of the part of the system. Naturally these concepts affect the strategy of the system's reinforcement if an insufficiency of the total stability coefficient has been stated. One commonly begins with reinforcing that part of the structure which is in the constricted state. Though, it should be remembered this measure is effective only to a certain extent. The criterion is that the subsystem being reinforced acquires the same "power" as the rest of the structure. To perform further reinforcing efficiently, one should reinforce all compressed members together to keep the achieved state of a uniform stability of the system's members [21].

The concepts of the constricted and forced states with respect to selected fragments of a structure introduced into the engineering practice over half a century ago enable one to form a purposeful strategy of reinforcing the structure the load-carrying capacity of which is determined by its stability. Meanwhile, as far as the authors can judge, there has been developed no formal computational tool for this, one being especially needed in the computer-aided design. An apparent explanation for this is that the concepts themselves are conditional[4]. As A.F. Smirnov noticed [21]:

"*...the concepts of a constricted and forced buckling, and a phenomenon of uniform stability, cannot be unrelated to a constraint being removed*".

In this regard it seems reasonable to classify particular parts of a system (up to its separate elements) into one of two following classes: a class of *restraining* elements (or parts) and a class of *pushing* ones of the system. Restraining elements facilitate the stability of the equilibrium of the system while pushing elements play a negative role: it is they which force (push) the mechanical system to buckle. The

[4] A.F.Smirnov in his book [21] refers to N.V. Kornoukhov who was the first to use these terms — a constricted and a forced state of a structure's members when the structure buckles.

algorithmic technique of determining what class an element belongs to is surprisingly simple and self-evident.

Indeed, the role of particular subsystems can be checked by calculating the energy accumulated in different parts of the system as it deforms according to a buckling mode. For the whole system this energy is equal to zero [18]

$$\mathsf{E} = \tfrac{1}{2}\, \mathbf{u}^{*\mathrm{T}}[\mathbf{K}_0 + \mathbf{K}_G(k)]\mathbf{u}^* = 0\,. \tag{9.10}$$

If the system has been divided imaginarily into two parts, then a stiffness matrix can be composed for each of the parts, $\mathbf{K}_{0,i} + \mathbf{K}_{G,i}(k)$ $(i = 1,2)$, which we fill up with zeros to the dimension of the full stiffness matrix of the system. Calculating the energies accumulated by each of the two selected parts of the system

$$\mathsf{E}_1 = \tfrac{1}{2}\, \mathbf{u}^{*\mathrm{T}}[\mathbf{K}_{0,1} + \mathbf{K}_{G,1}(k)]\mathbf{u}^*\,, \tag{9.11}$$

$$\mathsf{E}_2 = \tfrac{1}{2}\, \mathbf{u}^{*\mathrm{T}}[\mathbf{K}_{0,2} + \mathbf{K}_{G,2}(k)]\mathbf{u}^* \tag{9.12}$$

will always give us values of opposite signs because their sum must be zero. This determines a subsystem (corresponding to $\mathsf{E}_i < 0$) which is reasonable to stiffen or brace additionally. Comparing these quantities with the concepts introduced above, it is not hard to guess that if $\mathsf{E}_i < 0$ then the respective i^{th} subsystem is a pushing one, otherwise at $\mathsf{E}_i > 0$ it is a restraining one. The case of $\mathsf{E}_i = 0$ is singular, so we will discuss it a bit later.

It is also a substantial fact that pushing elements of a system can be ranked by their "blame" for the critical state of the structure. Let the system be imaginarily divided into a few, say, m separate fragments, and let the value of the energy E_i $(i = 1,\ldots, m)$ be calculated for each of the parts. Then, the less the value of E_i (algebraically), the higher the responsibility of the respective i^{th} part of the system for its general buckling. In other words, the contribution of each element of the system to its common energy balance can be a convenient quantitative measure of this responsibility.

Let us illustrate things that have been said by an example from [21]. The schematic of a structure of interest is shown in Fig. 9.10 (up to its symmetry axis) where its linear rigidities $i = EI / l$ are indicated in frames.

The function $\varphi_2(\nu)$ depends on the argument ν determined by the compressive force N in the bar,

$$\varphi_2(\nu) = \frac{\nu(\mathrm{tg}\nu - \nu)}{8\mathrm{tg}\nu\left(\mathrm{tg}\dfrac{\nu}{2} - \dfrac{\nu}{2}\right)}, \tag{9.13}$$

and in our case

$$\nu_1 = \sqrt{\frac{N_1 l_1}{i_1}} = \sqrt{\frac{4k \times 6}{2}} = \sqrt{12k};\ \ \nu_2 = \sqrt{\frac{6k \times 8}{4}} = \sqrt{12k}\,. \tag{9.14}$$

The critical value of the loading intensity is determined by the value $k = 2.51$, and the buckling mode $\mathbf{u}^* = \|[Z_1, Z_2]\|^{\mathrm{T}}$ is determined by the slope values $Z_1 = 1$ and

$Z_2 = -1.366$. The values of the dimensionless parameters ν_1 and ν_2 that correspond to the critical state of the system happen to be equal:

$$\nu_1 = \nu_2 = \sqrt{12k} = 5.488 .$$

Let us divide the system ideally into three parts. The first one will include the left column, the second one will comprise the right column, and the crossbar of the frame will constitute the third part. The third stiffness matrix **K** of the system is

$$\begin{bmatrix} 22+8\varphi_2(\nu_1) & 8 \\ 8 & 28+16\varphi_2(\nu_2) \end{bmatrix},$$

and can be represented by a sum of three matrices

$$\begin{bmatrix} 8\varphi_2(\nu_1) & 0 \\ 0 & 0 \end{bmatrix} + \begin{bmatrix} 0 & 0 \\ 0 & 16\varphi_2(\nu_2) \end{bmatrix} + \begin{bmatrix} 22 & 8 \\ 8 & 28 \end{bmatrix},$$

which correspond to each of the three selected parts of the system.

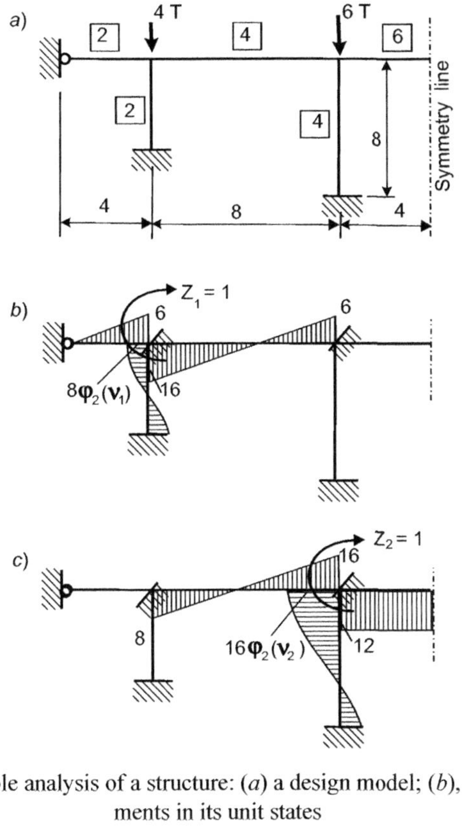

Fig. 9.10. An example analysis of a structure: (*a*) a design model; (*b*), (*c*) curves of moments in its unit states

Now it is easy to calculate that $\varphi_2(\nu_1) = \varphi_2(\nu_2) = -1.384$, thus

$$E_1 = ½\, Z_1^2 \times 8\varphi_2(\nu_1) = -5.536; \qquad E_2 = ½\, Z_2^2 \times 16\varphi_2(\nu_2) = -20.660;$$

$$E_3 = ½\,(22Z_1^2 + 28Z_2^2 + 2\times 8Z_1Z_2) = 26.196\,.$$

The balance of the energy values calculated using these matrices holds exactly as it should: $E_1 + E_2 + E_3 = 0$. The comparison of the energetic contribution of each of the three selected parts to its common energy accumulator evidences that the weak place of the system is its right column ($E_2 < E_1 < E_3$) which fact is in perfect accordance with results of an analysis performed in [21], though here we have done it much simpler from the algorithmic standpoint.

The calculation of expressions like (9.11) is a pretty simple operation once the critical value of the parameter k and its respective buckling mode have been found. It seems very useful and promising to supplement a software system with a function of determining E_i for any user-specified subset of elements of a system.

The software system should be capable of splitting a structure into any number of parts, down to separate elements, be able to estimate the contribution of each element to the positive and negative parts of the total energy, and generate recommendations on strengthening/weakening particular elements. This helps one find out the role played by particular elements in the general buckling behavior in more detail.

An additional useful information can be provided by an *energetic coefficient of stiffness fall* which can be calculated for each element of a system as

$$\eta_i = \frac{E_i(N)}{E_i(0)}\,, \tag{9.15}$$

where $E_i(N)$ denotes an energy accumulated by i^{th} element and calculated using the stiffness matrix $\mathbf{K}(k)$, and $E_i(0)$ is the energy of the same element at the zero longitudinal force when its full stiffness matrix is $\mathbf{K}_0$.

It is easy to see that

$$\eta_i = 1 + \frac{\mathbf{u}_i^{*\mathrm{T}} \mathbf{K}_{G,i} \mathbf{u}_i^*}{\mathbf{u}_i^{*\mathrm{T}} \mathbf{K}_{0,i} \mathbf{u}_i^*} \tag{9.16}$$

where $\mathbf{u}_i^*$ is a selection of components of the vector $\mathbf{u}^*$ referring to nodes of ith element of interest, $\mathbf{K}_{0,i}$ and $\mathbf{K}_{G,i}$ are respective matrices of usual and geometrical stiffness of the same element. It's clear that the coefficient η_i is an integral characteristic of the fall of stiffness of the element caused by the presence of compressive stresses in it. The less the coefficient η_i (algebraically), the more substantial is the effect of the longitudinal forces on the fall of stiffness of ith element. Also, the sign of the coefficient η_i allows us to classify the respective element of the system into one of the two classes: one of pushing elements (at $\eta_i \le 0$) or one of restraining elements (at $\eta_i > 0$) because the denominator of the formula (9.15) is always positive.

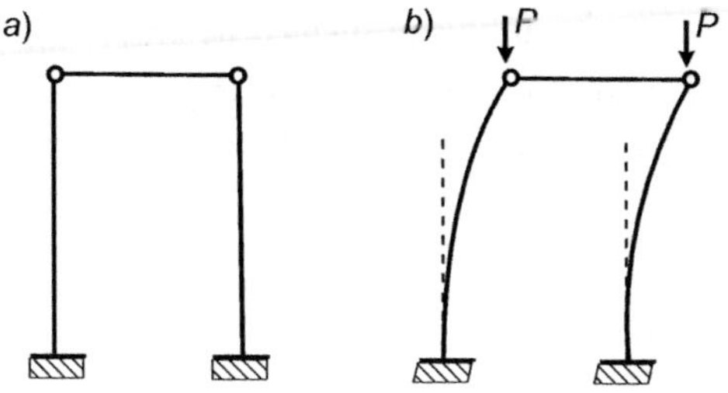

Fig. 9.11. A buckling of a Π-shaped frame

There's another important circumstance[5] due to which the employment of the energetic coefficient of stiffness fall is more informative than a mere comparison of the energetic contribution of each selected part of the system to its total energy in the course of its buckling. Let us illustrate this statement by a simple example.

Fig. 9.11 shows a simple Π-shaped frame with equal columns and a hinged attachment of the crossbar to the columns. It also shows the first buckling mode of the system.

Clearly the crossbar of the frame does not deform in the first buckling mode of the system when the latter buckles. Hence $\mathsf{E}_p = 0$ (the energy of deformation accumulated in the crossbar). But, as it is easy to see, in each of the two columns the respective part of the deformation energy is zero, too: $\mathsf{E}_c = 0$. This follows, for example, from such a consideration: the total energy of the system as it deforms in a buckling mode should be equal to zero, that is, $\mathsf{E} = \mathsf{E}_p + 2\mathsf{E}_c = 0$. It becomes clear that the mere comparison of the energies does not enable us to perform a formal division of elements into pushing and restraining ones. It is quite obvious that the columns are pushing elements while the crossbar of the frame needs to be classified into a separate category elements of which can be relevantly called *neutral*. Neutral members of a structure neither push it to the buckling, nor resist it. Those are just passive observers of the structure's behavior in the course of its buckling. Apparently, the analytical criterion $\mathsf{E}_i = 0$ states only that i^{th} member (or a selected part of the system) can be either pushing or neutral but by no means a restraining one.

To understand whether i^{th} member of a system is neutral or pushing, one has to find out the cause for the equality $\mathsf{E}_i = 0$. Here the role played by the above mentioned coefficients η for such elements is dramatic. Really, the equality

$$\mathsf{E}_i = \tfrac{1}{2}\, \mathbf{u}_i^{*\mathrm{T}} \left[\mathbf{K}_{0,i} + \mathbf{K}_{\mathrm{G},i}(k)\right] \mathbf{u}_i^* = 0$$

can appear in two possible situations: either $\mathbf{u}_i^* \neq \mathbf{0}$ or $\mathbf{u}_i^* = \mathbf{0}$. Clearly, in the first case the element in question is a pushing one and its coefficient η_i is equal to zero.

[5] Our attention was drawn to this circumstance (together with the example given) by Professor Zslot Gáspár from Budapest Technical University during a discussion on a report by the authors made at the second European Conference on Computational Mechanics (Poland, Krakow, June 26-29, 2001).

In the second case (at $\mathbf{u}_i^* = \mathbf{0}$) this element is neutral to the buckling mode of the system being considered: it just takes no part in the deformation of the system in its buckling mode. Its respective value of the coefficient η_i is undetermined because both the numerator and denominator of the formula (9.15) are equal.

This revised criterion restores the algorithmic capability of a full ranking of the system's elements. In the above example of the Π-shaped frame, the coefficients η for the crossbar and the columns will be equal, accordingly, to

$$\eta_p = \frac{0}{0}, \qquad \eta_c = 0,$$

thus the crossbar is neutral and the columns are pushing elements.

Of course, a more careful stability/buckling analysis of a rational reinforcement of a structure should be based on a calculation and comparison of sensitivity coefficients of the critical loading parameter value to variations of particular design parameters of the structure. This analysis can be performed in the same way it was done in Section 7.3 with the sensitivity coefficients of the natural oscillation frequencies of a system. Though, a bigger computational effort needed to determine the sensitivity coefficients in the buckling analysis makes it reasonable to employ the simpler algorithm based on an energetic ranking of the system's members.

We'd like to note one should be careful enough when using results of the above stated analysis to make a decision of modifying elements or subsystem. The matter is that the energy calculation is performed with the buckling mode $\mathbf{u}^*$ which has been determined for the system with an original stiffness and stress distribution. But this mode also changes when these parameters are modified. Therefore the result of such modification must necessarily be validated. Also, which is even more important, the effect of reinforcing particular elements is bounded by a certain limit above which this operation does not affect the result anymore. One can easily understand this statement by noticing that, for example, in the problem of Fig. 9.10 even an infinite growth of the stiffness of one of the columns does not result in the same growth of the stability coefficient because the system starts buckling in a "local" mode without deforming the over-stiffened column. Apparently, the first one who noticed this peculiarity of the buckling analysis of elastic systems was I.G. Bubnov who stated that there is a limit stiffness of supports in a problem of buckling of a beam on two such supports beyond which the further growth of it does not increase the critical force, and the supports with that limit stiffness behave like perfectly rigid ones [26].

9.5 On the Influence of Additional Constraints upon the Stability of a System

The point of view that additionally imposed constraints are unconditionally good for stabilizing a structure is widely popular and seldom subject to any discussion. Such a confidence is inspired by Rayleigh's theorem taught in most standard courses of theory of buckling or oscillations. But one should notice there is an important condition breaking which may render the whole theorem inapplicable: additional constraints may not change the state of equilibrium of a structure that is being analyzed, and the distribution of the stresses must be maintained together with the displacements of the structure. A.R. Rzhanitsin has called such constraints *immobile* and generalized Rayleigh's theorem to nonlinear systems with immobile constraints [19].

Examples can be easily built to illustrate a destabilizing effect of an additional constraint when this one alters the distribution of the longitudinal forces and transforms an absolutely stable structure (say, that in Fig. 9.12, *a*) into a system having a finite critical force (Fig. 9.12, *b*).

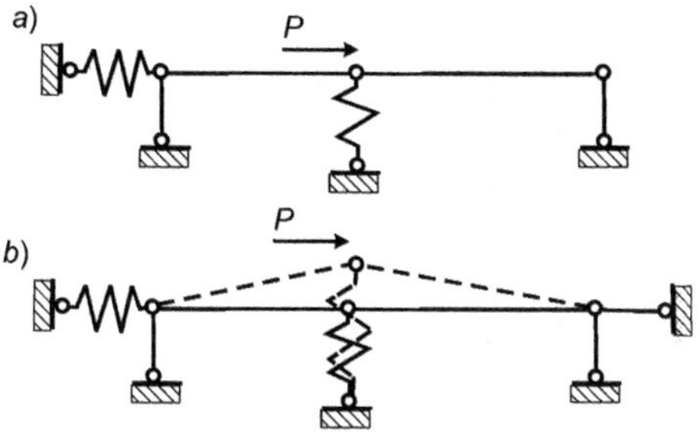

Fig. 9.12. A destabilizing additional constraint: (*a*) the original state; (*b*) the state with the additional constraint

When encountering such examples, mathematically educated engineers are inclined to find the situation paradoxical in a way. If a problem of a mechanical system's buckling is posed in a variational manner (as a minimum problem), then finding the first critical value of the loading parameter can be reduced to finding the minimum of a functional which defines the needed loading parameter's value. But in this case imposing any constraint on the mechanical system is equivalent mathematically to narrowing the set of functions to be compared: kinematically admissible displacements. In is clearly impossible that in the narrowed set the minimum of the functional can be less than that of the same functional in the original set!

Actually, there is no paradox here at all. The matter is that constraints which are not immobile in the sense of A.R. Rzhanitsin's definition can both narrow the space of admissible displacements of a system and alter the very functional. That's the principal difference between the immobile constraints and other ones.

9.5.1 On an Error of a Mysterious Nature that Happens in Software when Analyzing Mechanical Systems for Buckling

Let us begin with a simplest problem of buckling analysis of a mechanical system shown schematically in Fig. 9.13. The system includes a cantilever bar of length l with its flexural rigidity EI and a perfectly rigid insert of length h at its end.

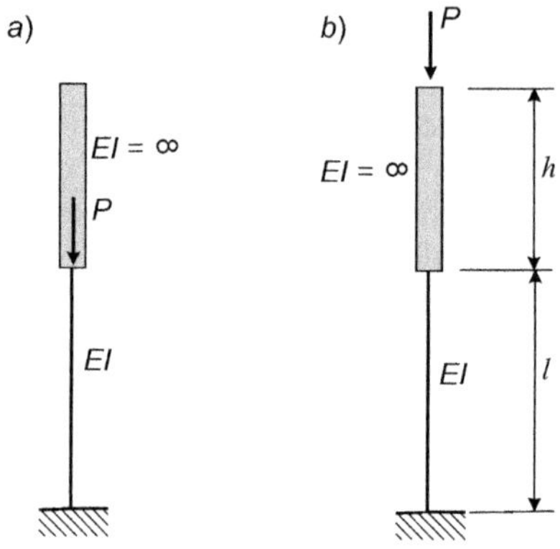

Fig. 9.13. A buckling analysis scheme of a cantilever bar:
(*a*) a force is applied directly to the deformable section of the bar;
(*b*) the force is applied to the vertex of the bar (at the end of the stiff insert)

In the problem in Fig. 9.13, *a* the critical force is a known value

$$P_{crA} = \frac{\pi^2 EI}{4l^2}, \tag{9.17}$$

because the stiff section of the bar is unstressed and has no effect on the critical load value. As for the problem in Fig. 9.13, *b*, its critical load is also not too hard to find.

Let's give an elementary derivation of appropriate formulae using the static method for finding the critical load: Euler's method. With this purpose in mind, let us consider the equilibrium of the system in its deviated position described by the displacement $w(x)$ as shown in Fig. 9.14.

In the state of buckling, the bending moment $M = EI\,w''$ that arises in any section of the bar is balanced by the respective moment M caused by the applied load P. Introducing the notation

$$w_l = w(l), \quad \theta_l = w'(l), \tag{9.18}$$

we have the following as seen in Fig. 9.14: $M = P(w_l + h\theta_l - w)$. Thus

$$EI\,w'' + Pw = P(w_l + h\theta_l)\,.$$

The general solution of this differential equation can be written as

$$w(x) = A\sin\alpha x + B\cos\alpha x + w_l + h\theta_l, \tag{9.19}$$

where

$$\alpha = \sqrt{\frac{P}{EI}}\,. \tag{9.20}$$

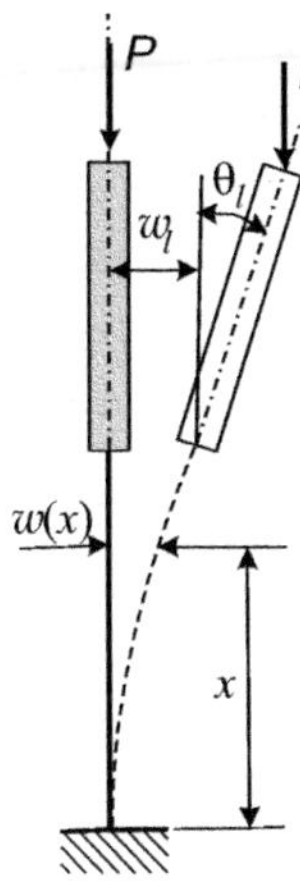

Fig. 9.14. To the determination of the critical load in the problem shown in Fig. 9.13, *b*

It immediately follows from the boundary conditions $w(0) = 0$ and $w'(0) = 0$ that $A = 0$ and $B = -(w_l + h\theta_l)$, so (9.19) is reduced to

$$w(x) = (w_l + h\theta_l)(1 - \cos\alpha x). \tag{9.21}$$

In this equation parameters w_l and θ_l are so far unknown. Using the definitions of these parameters according to (9.18) and introducing the designation of ν for a dimensionless parameter of the longitudinal force

$$\nu = \alpha l, \tag{9.22}$$

we find from (9.21) that

$$w_l = (w_l + h\theta_l)(1 - \cos\nu), \quad \theta_l = (w_l + h\theta_l)\frac{\nu}{l}\sin\nu.$$

After some elementary transformations we derive a system of two homogeneous linear algebraic equations with respect to the desirable parameters w_l and θ_l.

$$-w_l\cos\nu + h\theta_l(1 - \cos\nu) = 0,$$

$$w_l\nu\sin\nu + \theta_l(h\nu\sin\nu - l) = 0.$$

The existence of a nontrivial solution of this system requires that its determinant be equal to zero which gives us the equation (9.23) finally. The minimum root of the latter yields the critical value of the longitudinal force parameter ν_{cr}

$$\frac{l}{h} = \nu \operatorname{tg}\nu. \tag{9.23}$$

Fig. 9.15 below presents a plot that shows the dependence of ν_{cr} on the relative length of the stiff insert h/l. Particularly, this plot shows that at $h = 0$ the value of ν_{cr} is $\pi/2$ which corresponds to the solution of the problem in Fig. 9.13, *a*.

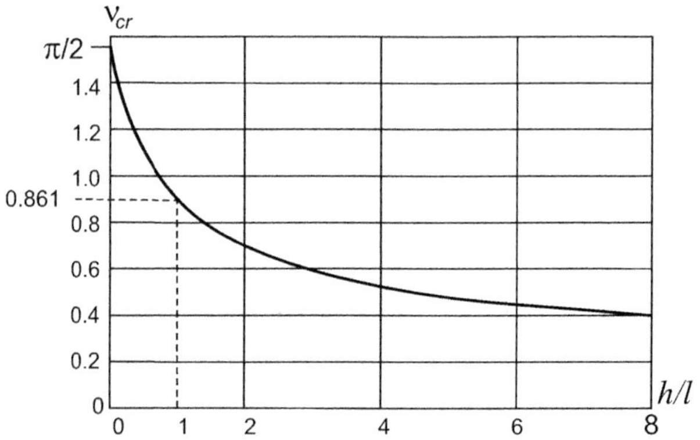

Fig. 9.15. A plot of function $\nu_{cr} = \nu_{cr}(h/l)$ built for the solution of equation (9.23)

Let us now try to solve the same two problems using a software tool. Let's take actual numerical values and assume $EI = 1$, $l = 1$, $h = 1$ for determinacy. Strange as it seems, but all computational programs checked by us have yielded the same result: a correct solution of the problem in Fig. 9.13, *a* and an incorrect one of the problem in Fig. 9.13, *b*: $P_{crA} = P_{crB} = 2.467$. The correct solution, as shows the graph in Fig. 9.15, must give $P_{crB} = 0.741$ ($\nu_{cr} = 0{,}861$). So, in the case considered the critical load for the problem in Fig. 9.13, *b* found by the software is greater than the proper value more than three times!

A dramatically more erroneous result is yielded by most commercial CAD programs for the problem shown in Fig. 9.16. Alas! but all programs listed in the footnote below, except for the **SCAD** software, stated *ex cathedra* that the system did not buckle at all[6].

Without considering the whole process of solution of this simple problem, let's point out the correct value of the critical load

$$P_{cr} = \frac{3EI}{lh}.$$

[6] Exactly speaking, we have checked the functionality of the following software products: **GTSTRUDL**, **Robot**, **Lira**, **Nastran**, **ANSYS**, **COSMOS**, **SCAD**. The only program that solved both problems correctly was the **SCAD** software, version 31. Sometimes the software being tested refuses to perform the analysis, in other cases (which is a greater danger) an incorrect result is given out without any warning. To be fair, we should note that when we specified stiff inserts in the **GTSTRUDL** environment for the problem (9.13-*b*) using its internal command END JOINT SIZE, the result was quite correct. Though, when using the perfectly rigid body handling tool (the RIGID BODY command) or the eccentricity tool (the ECCENTRICITIES command), **GTSTRUDL** failed to solve this problem just as well.

We would appreciate if our readers who have an access to other commercial software developments could inform us about same or similar tests of various programs with these examples.

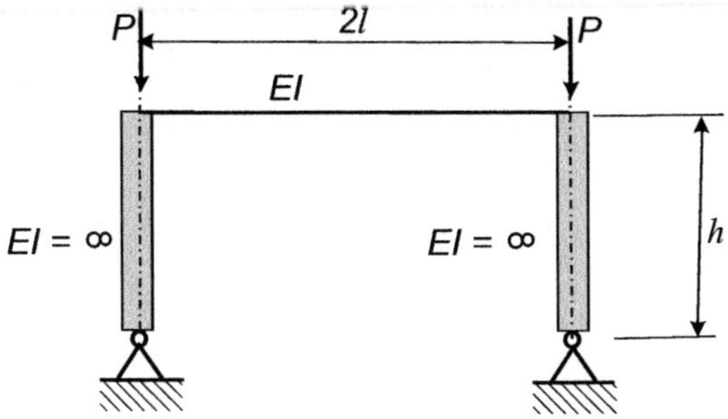

Fig. 9.16. A test problem for the buckling analysis

After some thinking one can easily conclude that the developers of most commercial programs have been fooled by the same trick, at the same time trying to have their users swallow that dangerous bait.

The matter is that a perfectly rigid body cannot be treated as a linear constraint in buckling problems. Indeed, all equations related to rigid bodies and given in Section 2.6 are linear equations, so what they describe is solely the behavior of mechanical systems in a geometrically linear formulation. At the same time, in the buckling analysis the equilibrium equation systems must be formulated as linearized (rather than linear!). In other words, the equilibrium equations should be derived for a deformed state of the system, thus corresponding to the first floor of geometrical nonlinearity.

So, let a mechanical system include a perfectly rigid body. Let's assume that a transition from the given mechanical system to its discrete design model has been done already. If now we extract the rigid body from the said model, then the action of the removed part of the model upon this body can be replaced by appropriate reactive forces and moments. In order to think generally, let's assume that besides the reactive forces the rigid body is subjected to some given (active) concentrated forces and moments. Let's denote the total number of points (or nodes, to one's liking) of the rigid body as m, vector forces $\boldsymbol{P}_i$ and vector moments $\boldsymbol{M}_i$ $(i = 1,\ldots,m)$ being applied to the points.

Now let's choose an arbitrary point O in this body and call it a *pole*. Particularly, though not obligatorily, the pole may coincide with one of m already selected nodes. Let $\boldsymbol{r}_i$ be a radius vector that goes from the pole to i-th node. Recalling that the stability of some **balanced** state of the system is under consideration, we can conclude that the extracted rigid body under all applied (both active and reactive) forces $\boldsymbol{P}_i$ and moments $\boldsymbol{M}_i$ $(i = 1,\ldots,m)$ must be in the state of equilibrium also. The latter fact means that the general force vector and the moment vector referred to the pole must be equal to zero:

$$\sum_{i=1}^{m} \boldsymbol{P}_i = \boldsymbol{0}, \quad \sum_{i=1}^{m} \left(\boldsymbol{M}_i + \boldsymbol{r}_i \times \boldsymbol{P}_i\right) = \boldsymbol{0}. \tag{9.24}$$

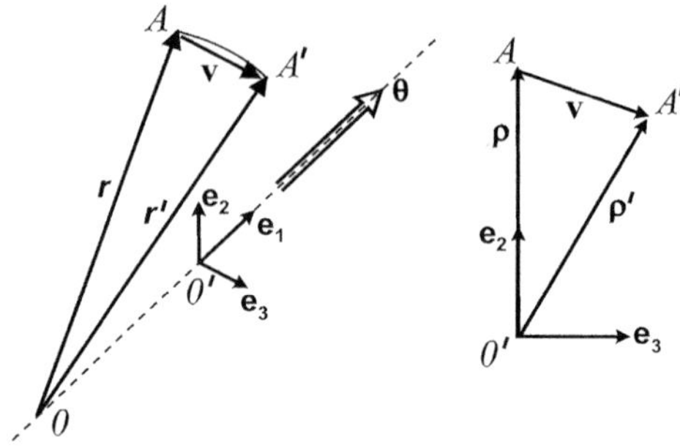

Fig. 9.17. To the derivation of a nonlinear constraint equation describing a perfectly rigid body

In the course of buckling, the displacements of a rigid body can be unambiguously determined by the pole displacement vector $\mathbf{u}_0$ and the body rotation vector $\boldsymbol{\theta}$, both beginning at the pole O.

Our primary task here is to derive an expression for the part of the rigid body's displacement caused by its rotation. With this purpose, let's take a particular point A of the rigid body that will move to the position A' after the body makes some revolution (Fig. 9.17).

Let $\boldsymbol{r}$ be the radius vector of point A, and $\boldsymbol{r}'$ be the radius vector of point A'. It means that displacement $\mathbf{v}$ caused by rotation $\boldsymbol{\theta}$ of the point in question will be expressed as

$$\mathbf{v} = \boldsymbol{r}' - \boldsymbol{r}\,. \tag{9.25}$$

Besides the initial Cartesian coordinate system (X,Y,Z), a global one for the system as a whole, with respective orts $\boldsymbol{i}_x$, $\boldsymbol{i}_y$, $\boldsymbol{i}_z$ of its axes, let's introduce another Cartesian coordinate system with orts $\mathbf{e}_1$, $\mathbf{e}_2$, $\mathbf{e}_3$. Ort $\mathbf{e}_1$ is directed along the axis of the body's revolution, and ort $\mathbf{e}_3$ is orthogonal to the plane formed by vectors $\boldsymbol{r}$ and $\boldsymbol{\theta}$. The direction of ort $\mathbf{e}_3$ is chosen so that the vector triple ($\mathbf{e}_1$, $\boldsymbol{r}$, $\mathbf{e}_3$) form a right-hand system of axes. Thus

$$\mathbf{e}_1 = \frac{1}{\theta}\boldsymbol{\theta}\,, \quad \mathbf{e}_3 = \frac{1}{r|\sin\alpha|}\,\mathbf{e}_1\times\boldsymbol{r}, \tag{9.26}$$

where θ and r are the lengths of the respective vectors expressed via components of these vectors with respect to axes (X,Y,Z) by formulae

$$\theta = \|\boldsymbol{\theta}\| = \sqrt{\theta_x^2+\theta_y^2+\theta_z^2}\,, \qquad r = \|\boldsymbol{r}\| = \sqrt{r_x^2+r_y^2+r_z^2}\,,$$

and α is an angle between vectors $\boldsymbol{\theta}$ and $\boldsymbol{r}$.

As for ort $\mathbf{e}_2$, it will be defined by the condition of the right-hand orientation of the Cartesian axis system (*1*, *2*, *3*), namely

$$\mathbf{e}_2 = \mathbf{e}_3\times\mathbf{e}_1 = \frac{1}{r|\sin\alpha|}\,(\mathbf{e}_1\times\boldsymbol{r})\times\mathbf{e}_1\,. \tag{9.27}$$

Vector calculus gives us the following well-known representation of a double vector product [9]

$$(\boldsymbol{a}\times\boldsymbol{b})\times\boldsymbol{c} = \boldsymbol{b}(\boldsymbol{a}\cdot\boldsymbol{c}) - \boldsymbol{a}(\boldsymbol{c}\cdot\boldsymbol{b})\,. \tag{9.28}$$

Here and further a dot denotes a scalar product of vectors. By applying equation (9.28) to the definition of ort $\mathbf{e}_2$ according to (9.27) we can transform this equality to the following form

$$\mathbf{e}_2 = \frac{1}{r|\sin\alpha|}[\,\boldsymbol{r} - \mathbf{e}_1(\mathbf{e}_1\cdot\boldsymbol{r})]\,. \tag{9.29}$$

Now let's build a plane through points A and A' that is perpendicular to the axis of rotation $\boldsymbol{\theta}$. Let O_1 be a point where the said plane crosses the axis of vector $\boldsymbol{\theta}$. Vector O_1A will be denoted as $\boldsymbol{\rho}$, and vector O_1A' as $\boldsymbol{\rho}'$. Apparently these two vectors have the same length ρ equal to

$$\rho = r\,|\sin\alpha|\,. \tag{9.30}$$

It's also apparent that the following equality can be used to determine vector $\mathbf{v}$ instead of (9.25):

$$\mathbf{v} = \boldsymbol{\rho}' - \boldsymbol{\rho}\,. \tag{9.31}$$

But vector $\boldsymbol{\rho}'$ has been formed from vector $\boldsymbol{\rho}$ by rotating the latter by angle θ around the axis $\boldsymbol{\theta}$ and, as one can see in Fig. 9.17,

$$\boldsymbol{\rho}' = \rho\cos\theta\,\mathbf{e}_2 + \rho\sin\theta\,\mathbf{e}_3, \qquad \boldsymbol{\rho} = \rho\,\mathbf{e}_2,$$

therefore

$$\mathbf{v} = \rho\,[(\cos\theta - 1)\,\mathbf{e}_2 + \sin\theta\,\mathbf{e}_3]\,. \tag{9.32}$$

After substituting the expressions for vectors $\mathbf{e}_2$ and $\mathbf{e}_3$ from (9.26) and (9.29) to (9.32), we will have the following, considering (9.30)

$$\mathbf{v} = (\cos\theta - 1)\,[\boldsymbol{r} - \mathbf{e}_1(\mathbf{e}_1\cdot\boldsymbol{r})] + \sin\theta\,\mathbf{e}_1\times\boldsymbol{r}\,. \tag{9.33}$$

The formula (9.33) is, in essence, a form of Rodrigues's formula known from the theory of finite revolutions of a solid body [13], convenient for our purpose.

But in our case of interest – the buckling analysis – equation (9.33) can be simplified. As it is known, the linearized formulation supposes that rotation angles are quantities the squares of which are negligible comparing to one. So, by expanding the trigonometric functions into Taylor series and keeping only second-order and lower terms we derive a simplified constraint equation in the following form:

$$\mathbf{v} = \theta\,\mathbf{e}_1\times\boldsymbol{r} - \tfrac{1}{2}\theta^2[\,\boldsymbol{r} - \mathbf{e}_1(\mathbf{e}_1\cdot\boldsymbol{r})]\,. \tag{9.34}$$

If now we assume point A of the rigid body to be its ith node, then the full displacement of this node $\mathbf{u}_i$, taking into account the $\mathbf{u}_0$ node displacement, will look as follows

$$\mathbf{u}_i = \mathbf{u}_0 + \theta\,\mathbf{e}_1\times\boldsymbol{r}_i - \tfrac{1}{2}\theta^2[\,\boldsymbol{r}_i - \mathbf{e}_1\,(\mathbf{e}_1\cdot\boldsymbol{r}_i)]\,. \tag{9.35}$$

Now let's calculate the work A of all forces applied to the perfectly rigid body. We have

$$\mathsf{A} = \sum_{i=1}^{m} (\boldsymbol{P}_i \cdot \mathbf{u}_i + \boldsymbol{M}_i \cdot \boldsymbol{\theta}) . \tag{9.36}$$

By substituting the expression for $\mathbf{u}_i$ from (9.35) to (9.36) we find that the work A consists of two quantities

$$\mathsf{A} = \mathsf{A}_1 + \mathsf{A}_2 ,$$

where A_1 depends linearly on the components of vectors $\mathbf{u}_0$ and $\boldsymbol{\theta}$, while A_2 is a homogeneous quadratic form with respect to the components of the rotation vector $\boldsymbol{\theta}$. Making use of a well-known vector calculus formula of a scalar triple product [9]

$$\boldsymbol{a} \cdot (\boldsymbol{b} \times \boldsymbol{c}) = \boldsymbol{b} \cdot (\boldsymbol{c} \times \boldsymbol{a}),$$

we can represent the linear component of the work, that is A_1, as

$$\mathsf{A}_1 = \sum_{i=1}^{m} [\boldsymbol{P}_i \cdot \mathbf{u}_0 + (\boldsymbol{M}_i + \boldsymbol{r}_i \times \boldsymbol{P}_i) \cdot \boldsymbol{\theta})] , \tag{9.37}$$

while its quadratic component, i.e. A_2, can be written as

$$\mathsf{A}_2 = \frac{1}{2} \theta^2 \sum_{i=1}^{m} [-(\boldsymbol{P}_i \cdot \boldsymbol{r}_i) + (\boldsymbol{P}_i \cdot \mathbf{e}_1)(\boldsymbol{r}_i \cdot \mathbf{e}_1)] . \tag{9.38}$$

First of all let us notice that the conditions of the A_1 stationarity lead expectedly to equilibrium equations (9.24) for perfectly rigid bodies.

Further, the components of the geometrical stiffness matrix $\mathbf{K}_G$ produced by the rigid body will be determined by double differentiation of the expression for work A. In other words,

$$\mathbf{K}_G = \begin{Vmatrix} g_{\theta_x\theta_x} & g_{\theta_x\theta_y} & g_{\theta_x\theta_z} \\ g_{\theta_y\theta_x} & g_{\theta_y\theta_y} & g_{\theta_y\theta_z} \\ g_{\theta_z\theta_x} & g_{\theta_z\theta_y} & g_{\theta_z\theta_z} \end{Vmatrix} = - \begin{Vmatrix} \dfrac{\partial^2 \mathsf{A}}{\partial\theta_x \partial\theta_x} & \dfrac{\partial^2 \mathsf{A}}{\partial\theta_x \partial\theta_y} & \dfrac{\partial^2 \mathsf{A}}{\partial\theta_x \partial\theta_z} \\ \dfrac{\partial^2 \mathsf{A}}{\partial\theta_y \partial\theta_x} & \dfrac{\partial^2 \mathsf{A}}{\partial\theta_y \partial\theta_y} & \dfrac{\partial^2 \mathsf{A}}{\partial\theta_y \partial\theta_z} \\ \dfrac{\partial^2 \mathsf{A}}{\partial\theta_z \partial\theta_x} & \dfrac{\partial^2 \mathsf{A}}{\partial\theta_z \partial\theta_y} & \dfrac{\partial^2 \mathsf{A}}{\partial\theta_z \partial\theta_z} \end{Vmatrix} . \tag{9.39}$$

It is this geometrical stiffness matrix which was not taken into consideration by programs that yielded the erroneous result in the examples given above.

After expanding the expression (9.38) of A_2 into its coordinate form we will have

$$A_2 = \frac{1}{2}\sum_{i=1}^{m}[-(\theta_x^2 + \theta_y^2 + \theta_z^2)(P_{ix}r_{ix} + P_{iy}r_{iy} + P_{iz}r_{iz}) +$$

$$+ (P_{ix}\theta_x + P_{iy}\theta_y + P_{iz}\theta_z)(r_{ix}\theta_x + r_{iy}\theta_y + r_{iz}\theta_z)]\,. \tag{9.40}$$

Now we can write expressions for the components of our matrix of interest $\mathbf{K}_G$ explicitly. So,

$$g_{\theta_x\theta_x} = \sum_{i=1}^{m}(P_{iy}r_{iy} + P_{iz}r_{iz}),$$

$$g_{\theta_y\theta_x} = -\sum_{i=1}^{m}(P_{ix}r_{iy} + P_{iy}r_{ix}),\quad .$$

$$g_{\theta_z\theta_x} = -\sum_{i=1}^{m}(P_{ix}r_{iz} + P_{iz}r_{ix}). \tag{9.41}$$

The components of the second and third columns of matrix $\mathbf{K}_G$ are not given here as they can be derived from (9.41) by a circular permutation of its indexes.

In a particular case of a plane mechanical system, the $\boldsymbol{\theta} = \theta\mathbf{e}_1$ rotation vector is orthogonal to the plane of the system of interest (say, plane X,Y) and thus has only one component θ_z. The order of the geometrical stiffness matrix of the rigid body becomes one, and its only component becomes especially simple:

$$g_{\theta_z\theta_z} = \sum_{i=1}^{m}(P_{ix}r_{ix} + P_{iy}r_{iy}) = \sum_{i=1}^{m}\boldsymbol{P}_i \cdot \boldsymbol{r}_i\,. \tag{9.42}$$

The same result can be derived directly from formula (9.38). As can be seen, in the plane case any vector $\boldsymbol{r}_i$ is orthogonal to the rotation vector $\boldsymbol{\theta}$, that is $(\boldsymbol{r}_i \cdot \boldsymbol{\theta}) = 0$. Thus formula (9.38) becomes as simple as

$$A_2 = -\frac{1}{2}\theta^2\sum_{i=1}^{m}(\boldsymbol{P}_i \cdot \boldsymbol{r}_i),$$

wherefrom (9.42) follows.

Let's demonstrate the technique of using the geometrical stiffness matrix of a perfectly rigid body by the example of an elementary problem shown in Fig. 9.18.

In this problem we assume that both springs satisfy a linear dependence between the stress in the spring and the respective displacements of the B point.

First of all, let us find the distribution of internal forces in the system assuming it to be linearly deformable. It is easy to see that this system has only one degree of freedom due to the presence of a perfectly rigid bar. The said degree of freedom will be the angle of rotation θ of the rigid bar around point A, the fixation point of this bar. Further, in the case of small rotations (the geometrically linear formulation of the problem is being considered) the horizontal u and vertical v displacements of point B will be equal, respectively, to

$$u = \frac{\sqrt{2}}{2} l\theta , \quad v = -\frac{\sqrt{2}}{2} l\theta , \tag{9.43}$$

where l is the length of bar AB.

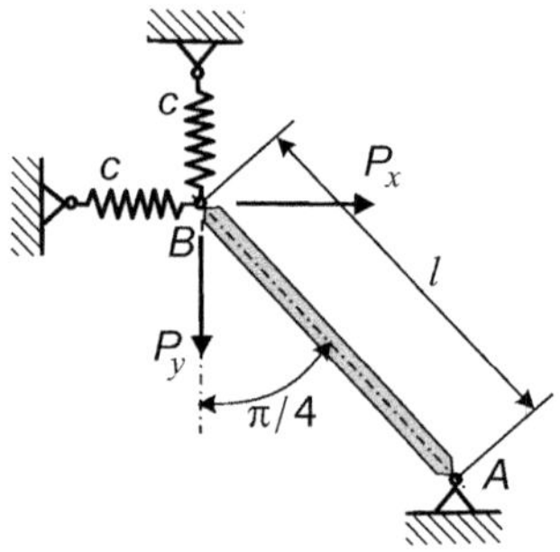

Fig. 9.18. An example of determining the critical load in a system including a perfectly rigid body

Assuming each spring's stiffness to be c, let's write out an expression for the full potential energy of the system

$$\mathsf{L} = \frac{cu^2}{2} + \frac{cv^2}{2} - P_x u - P_y v \tag{9.44}$$

that, after substituting the expressions of u and v via θ from (9.43), will become

$$\mathsf{L} = \mathsf{L}(\theta) = \frac{cl^2\theta^2}{2} - \frac{l\sqrt{2}}{2}(P_x - P_y)\theta . \tag{9.45}$$

The condition of stationarity of L as a function of θ will yield the following equilibrium equation

$$\frac{d\mathsf{L}}{d\theta} = 0 \text{ or } cl^2\theta - \frac{l\sqrt{2}}{2}(P_x - P_y) = 0 , \tag{9.46}$$

wherefrom

$$\theta = \frac{\sqrt{2}}{2cl}(P_x - P_y) . \tag{9.47}$$

The stresses in the horizontal spring N_x and vertical spring N_y will be respectively equal to

$$N_x = cu = \frac{P_x - P_y}{2} , \quad N_y = cv = -\frac{P_x - P_y}{2} . \tag{9.48}$$

If now we extract the rigid bar, then four forces will be applied to it in point B as shown in Fig. 9.19.

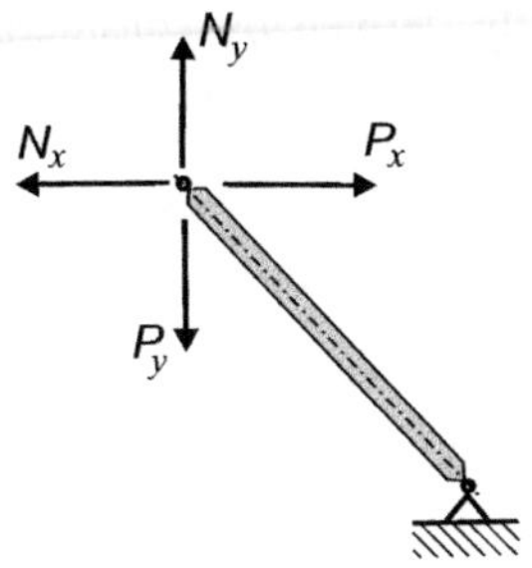

Fig. 9.19. Forces applied to the rigid bar

It is easy to understand that these four forces taken together can be reduced to one force N working on the extracted bar in the direction of the bar's axis

$$N = \frac{P_x + P_y}{\sqrt{2}}\,, \tag{9.49}$$

and the positive value of the sum of external forces ($P_x + P_y$) makes the longitudinal force N defined by formula (9.49) compressive.

Now, following formula (9.42), let us find the component of the geometrical stiffness matrix produced by the rigid bar. We have

$$g_{\theta\theta} = -\frac{P_x + P_y}{\sqrt{2}}\,l\,. \tag{9.50}$$

The conventional stiffness matrix **K** of the system consists here of only one element $\mathbf{K} = \|[k]\|$, and

$$k = \frac{\partial^2 \mathsf{L}}{\partial\theta\partial\theta} = cl^2\,. \tag{9.51}$$

The critical state of the system is determined by the condition

$$\det \mathbf{K} = \det(\mathbf{K}_0 + \mathbf{K}_G) = 0\,, \tag{9.52}$$

which gives us in this case

$$(P_x + P_y)_{cr} = cl\sqrt{2}\,. \tag{9.53}$$

This completes the analysis of stability of the discussed system in the linearized formulation. Though, to be more convincing, it makes sense to study peculiarities of the same system's behavior using a general geometrically nonlinear formulation.

If we begin originally with the geometrically nonlinear formulation, without confining ourselves to the condition of small rotations of the bar in the balanced state of the system, then the following equalities should be used instead of linear relations (9.43):

$$u = \frac{\sqrt{2}}{2}\,l(1 - \cos\theta + \sin\theta)\,, \quad v = \frac{\sqrt{2}}{2}\,l(1 - \cos\theta - \sin\theta)\,. \tag{9.54}$$

These relations can be easily derived by considering the full geometrical constraint conditions produced by the perfectly rigid bar AB. It is clear that small θ makes relations (9.54) as simple as (9.43).

To continue the consideration, it is convenient to change from external forces P_x and P_y to other force parameters N and T where N is defined by formula (9.49) and T is described by the expression

$$T = \frac{P_x - P_y}{\sqrt{2}} . \tag{9.55}$$

After substituting (9.54) to (9.44) with the use of designations of N and T we will have the following expression for the full potential energy of the system

$$\mathsf{L}(\theta) = cl^2(1 - \cos\theta) - lN(1 - \cos\theta) - lT\sin\theta . \tag{9.56}$$

In the state of equilibrium L acquires a stationary value, hence

$$\frac{d\mathsf{L}}{d\theta} = 0 \quad \text{или} \quad cl^2 \sin\theta - lN\sin\theta - lT\cos\theta = 0 . \tag{9.57}$$

Thus we obtain that in the equilibrium state

$$\mathrm{tg}\theta = \mathrm{s} , \tag{9.58}$$

where s is another (dimensionless) force parameter

$$\mathrm{s} = \frac{T}{cl - N} . \tag{9.59}$$

Fig. 9.20 shows curves of equilibrium states of the system in its configuration space. Notice that all branches of the equilibrium state curve shown in Fig. 9.20 are strictly ascending. It may seem that this evidences the stability of all equilibrium states of the system. But this conclusion would be erroneous because the force parameter s and kinematical parameter θ are not energy-coupled quantities, thus the curve in Fig. 9.20 is not a characteristic curve of the system (see Section 6.4 in this regard).

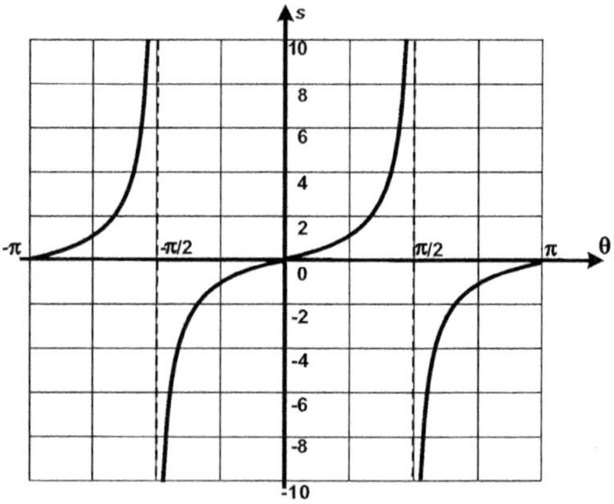

Fig. 9.20. Curves of equilibrium states in the plane (θ,s)

As one knows, the quality of equilibrium (whether it is stable or unstable) is defined by the sign of the second derivative of L with respect to θ calculated in the respective point of the equilibrium state curve. We have

$$\frac{d^2\mathsf{L}}{d\theta^2} = cl^2 \cos\theta - lN\cos\theta + lT\sin\theta = \frac{l\cos\theta}{cl-N}\left[(cl-N)^2 + T^2\right]. \quad (9.60)$$

It is clear that the sign of $d^2\mathsf{L}/d\theta^2$ is the same as that of the parameter χ

$$\chi = \frac{\cos\theta}{cl-N}, \quad (9.61)$$

wherefrom one can see that at small θ the sign of χ is defined by the denominator of formula (9.61). The change of the χ sign, thus the transition from stable modes of equilibrium at $\chi > 0$ to unstable ones at $\chi < 0$, occurs at $N = N_{cr}$ where $N_{cr} = cl$, which is expectedly the same as the conclusions that follow from formula (9.53).

To make the consideration complete, let us build a real characteristic curve of the system which, among other things, will enable us to easily separate areas of the system's stability from those of the instability, without confining ourselves to the search for critical load parameter values.

Let us take $q = \theta$ as a characteristic displacement. Then the respective generalized force λ will be defined as the coefficient at $d\theta$ in the expression for the increment of the work of external forces $d\Pi$. Here is a chain of transformations in the expression for the work $d\Pi$. We have

$$d\Pi = P_x du + P_y dv = l\,\frac{\sqrt{2}}{2}\,[P_x(\sin\theta + \cos\theta) + P_y(\sin\theta - \cos\theta)]\, d\theta =$$

$$= l\,(N\sin\theta + T\cos\theta)\, d\theta = l\cos\theta(Ns + T)\, d\theta = l\cos\theta\, \frac{clT}{cl-N}\, d\theta =$$

$$= cl^2 s\cos\theta\, d\theta = cl^2 \sin\theta\, d\theta.$$

Here we can see that the generalized force λ should be assumed equal to

$$\lambda = cl^2 \sin\theta\,. \quad (9.62)$$

Further, $d\lambda/dq = d\lambda/d\theta = cl^2\cos\theta$. As the condition $d\lambda/dq > 0$ holds in regions of the stable equilibrium, and the condition $d\lambda/dq < 0$ at unstable branches, we come to a conclusion that the displacement θ within the limits

$$-\pi/2 < \theta < \pi/2 \quad (9.63)$$

outlines the area of stability of the system. At $\theta > |\pi/2|$ the equilibrium is unstable. Boundary values $\theta_{cr} = \pm\pi/2$ are critical values for the displacement.

There is a delicate thing so far unconsidered in the characteristic curve analysis of the equilibrium quality that we've just made. The matter is that during this analysis we have used the condition $T \neq 0$ implicitly, or $P_x \neq P_y$ which is the same. But if the equality $P_x = P_y$ holds while the system is being loaded, then it means the external load is directed actually along the axis of the rigid bar and there are only two equilibrium states $\theta = 0$ и $\theta = \pi$, so the stability analysis based on curve

(9.62) does not work. The characteristic curve degenerates by turning into two isolated points. Though, as one can easily see, the critical value of the load from (9.53) remains true in this case as well. If the value of the load is less than the critical one, state $\theta = 0$ is stable while state $\theta = \pi$ is unstable. When the load passes the critical value, these states swap their equilibrium status: state $\theta = 0$ becomes unstable and state $\theta = \pi$ becomes stable.

By the way, if we assumed the linearized formulation rather than the geometrically nonlinear one (this is the first floor of the geometrical nonlinearity building) to look for equilibrium states of the system, then we'd have to expand trigonometric functions in the expression of L from (9.56) into Taylor series and keep second-order and lower θ terms. This operation would give us

$$\mathsf{L} = \frac{cl^2\theta^2}{2} - lN\frac{\theta^2}{2} - lT\theta\,. \tag{9.64}$$

The use of expression (9.64) for the full potential energy of the system corresponds to the so-called deformed state (second-order) analysis, and it yields the following linear dependence for equilibrium states of the system

$$\theta = \mathrm{s}\,. \tag{9.65}$$

It is clear that straight line (9.65) is a tangent to the trajectory of equilibrium states of the system according to (9.58) at the point $\theta = 0$. It is self-obvious that the use of (9.64) will again lead us to the same critical load of (9.53). To see this, substitute the expression of the parameter s from (9.59) to (9.65) and have

$$\theta = \frac{T}{cl - N}\,,$$

wherefrom $N_{cr} = cl$ follows.

Let us outline some algorithmic aspects of the use of the geometrical stiffness matrix for a perfectly rigid body. To make use of the terminology introduced in Section 2.6, it is reasonable to place the pole in the node assigned to be the master node of the body. As can be seen from Clause 4 of the master-slave relationships legislation given in Section 2.6, the master cannot be detached from the rigid body even partially. But this means that the master is **always** given rotational degrees of freedom. Consequently, the general stiffness matrix of the system will necessarily include all those components to which one has to sum elements of the geometrical stiffness matrix of a perfectly rigid body.

A singularity may arise if the original mechanical system includes finite elements (say, bars) that contain *stiff inserts* each of which is placed between a node of the system and the deformable part of the respective finite element (bar)[7]. From the mechanical standpoint a stiff insert is actually a partial (though very specific) case of a perfectly rigid body that includes just two nodes. One of the nodes of the

[7] There are no commonly accepted terms for a stiff insert. In literature one can encounter the term of a *stiff cantilever* as well. Some software systems entitle those as *eccentricities* (such as **GTSTRUDL**), other programs (say, **ANSYS**) use the concept of *offsets*.

stiff insert is a master node of the mechanical system being considered while the other one is attached to only one finite element (most often, to a bar). The stiff insert itself together with its node of attachment to the finite element is considered to be an integral part of this same finite element.

Its specifics is that the master node of a stiff insert is always a node of the system, that is, the node on which the deformable part of this particular finite element does not abut directly. In this situation it may occur that the master node lacks for its rotational degree of freedom in the general design model. An appropriate example is given in Fig. 9.21.

But the stiff insert has another specificity as compared to the general case of a perfectly rigid body. The matter is that in a general design model a stiff insert and its respective deformable part of the finite element are treated together as a complex finite element (or as a super-element if it's more to the reader's liking) rather than separately. Thus, the respective algorithm processes the stiff insert at the level of the stiffness matrix of the particular super-element rather than at the level of forming the whole stiffness matrix of the system. But during the formation of this super-element's stiffness matrix one can assume that the master node has been assigned all needed degrees of freedom (including rotational ones) which enables one to allow for the stiff insert's geometrical stiffness matrix algorithmically just as it would be done with a perfectly rigid body. If this element is attached to the system using hinges as shown in Fig. 9.21, then before summing into the system's general stiffness matrix one should perform an appropriate Jordan exclusion in the full stiffness matrix of this super-element ($\mathbf{K}_e + \mathbf{K}_{Ge}$) to get rid of the rotational degree of freedom of the node in this way (see Appendix A).

Discussing perfectly rigid bodies, we would like to draw the reader's attention to the following issue. From the course of theoretical mechanics an engineer should remember that "*without violating the equilibrium of a rigid body, one can transfer the point of a force's application along the line on which the force is located*" [12]. It is unquestionable and is not doubted here. Though, the knowledge of this simple principle could tempt an analyst to replace a system of forces applied to a perfectly rigid body by another system statically equivalent to the given one but possibly simpler from the standpoint of its description in the source data for a computational analysis. Certainly the equilibrium would not be violated. But the quality of the equilibrium state depends substantially on the way the load is applied. What we mean here by the quality of an equilibrium is a quantitative characteristic of a system's stability margin. Unfortunately, the students' attention is not always drawn to this important issue of the equilibrium of perfectly rigid bodies in common courses of theoretical mechanics, as it'd better be.

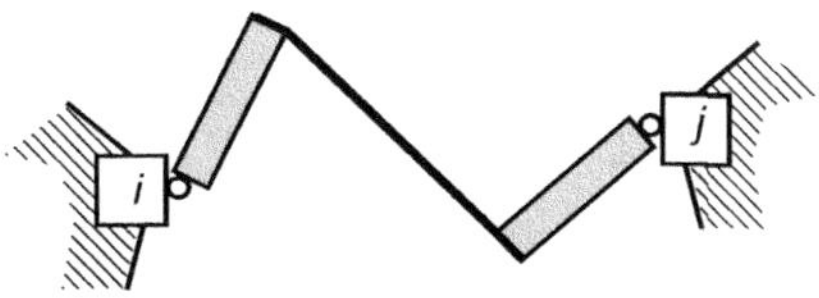

Fig. 9.21. A bar with stiff inserts connected to the system's nodes by hinges

To conclude this section, we'd like to notice that there is certainly no mystery at all in the way one poses and solves buckling problems for systems that contain perfectly rigid bodies. We have used this word in the title of the section only to emphasize a psychological aspect rather than physical one. What seems to us a mystery is the reason for which the developers of a number of software systems, who are surely beyond any suspicion of incompetence, have fell victims to a popular fallacy by forgetting the peculiarity of perfectly rigid bodies discussed here in the aspect of buckling. What a shame that the result of this fallacy should be a disoriented user of computational software. Hence our appeal to the community of developers that they should try to eliminate this fault in their commercial software as soon as possible, in order not to endanger the army of users by having them obtain incorrect results.

9.6 On a Paradox Encountered in a Bar Buckling Problem[8]

It is commonly known that analyst engineers should be especially careful when analyzing the stability of structures. Such problems, unlike conventional ones of structural stress/strain analysis, are particularly "insidious" because they may give birth to unexpected and seemingly preposterous results. One need not go far to get an example: a well-known book by Y.G. Panovko and I.I. Gubanova [16] confirms the fact well enough.

The objective of this section is to discuss results of a simple problem dedicated to a buckling analysis of a mechanical system. These results may well cause one to get bewildered or to distrust the software used.

So, let us consider the following problem, itself very simple in its formulation. Let a bar of a length l is subjected to a longitudinal force P as shown in Fig. 9.22.

The bar is assumed to be a perfectly elastic one of a constant cross-section fully clamped at its ends. As the right segment of the bar, with the length $l_2 = l - l_1$, is compressed, the system may buckle as soon as the load becomes large enough.

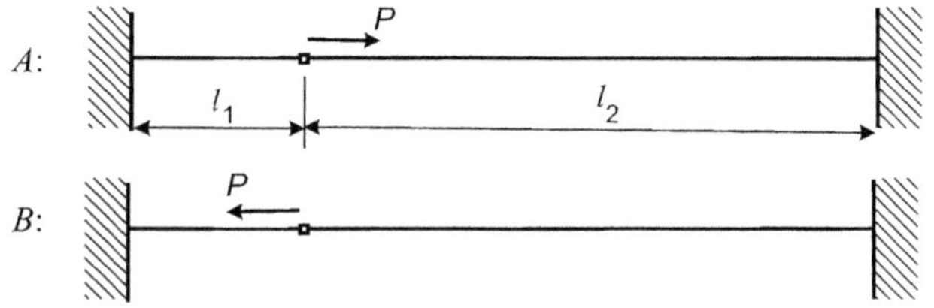

Fig. 9.22. A compressed/tensioned bar

The critical value of the working load that makes the buckling possible will be denoted as P_{Acr}. Along with this problem (entitled as Problem "*A*"), let us consider Problem "*B*" different from the original problem only in the direction of the

[8] Section 9.6 is written jointly with A.I. Kniga and M.V. Pashkovsky.

external force P (Fig. 9.22). Also, let us adopt particular numerical values for all parameters of the problems.

Let us assume, for example,

$$l_1 = 2, \qquad l_2 = 8, \qquad EI = 2.9\times10^5$$

in some consistent system of measurement units.

After making use of any of the available software systems, we have the following results[9]

$$P_{Acr} = 767000, \qquad P_{Bcr} = 580000.$$

This relation between the critical forces in the two problems should seem at least suspicious, intuitively. To see this, notice that the longitudinal forces N are distributed in the following way between the left and right segments of the bar:

$$N_1 = \frac{l_2}{l_1 + l_2} P, \qquad N_2 = \frac{l_1}{l_1 + l_2} P, \tag{9.65}$$

and the longitudinal force N_1 in Problem "*A*" tensions the left segment of the bar while the longitudinal force N_2 compresses the right one.

On the contrary, in Problem "*B*" N_1 is a compressive force while N_2 tensions. But, as the critical value of the longitudinal force in a compressed bar is inversely proportional to the square of its length, the ratio of the critical forces

$$\theta_P = \frac{P_{Acr}}{P_{Bcr}} = 1.32 > 1$$

seems weird and makes one reject the computed results because it is the shorter segment of the bar in Problem "*B*" that is compressed.

First of all, let's check the presence (or absence) of the detected effect at other relations between the parameters of Problems "*A*" and "*B*". With this purpose we introduce a dimensionless coordinate ξ that describes the relative position of the section where the external force P is applied on the bar. We assume

$$\xi = \frac{l_1}{l_1 + l_2}.$$

To find the critical value of the load P_{Acr}, we use a standard slope deflection technique. We can assume the design model to be a two-member system having a central node where the external force P is applied. We make the principal system of the slope deflection method by restraining the central node from the linear displacements Z_1 normal to the axis of the bar and from the slope Z_2 (Fig. 9.23).

[9] We have used two known programs: **SCAD** and **GTSTRUDL**. Both yield the same results presented in the text.

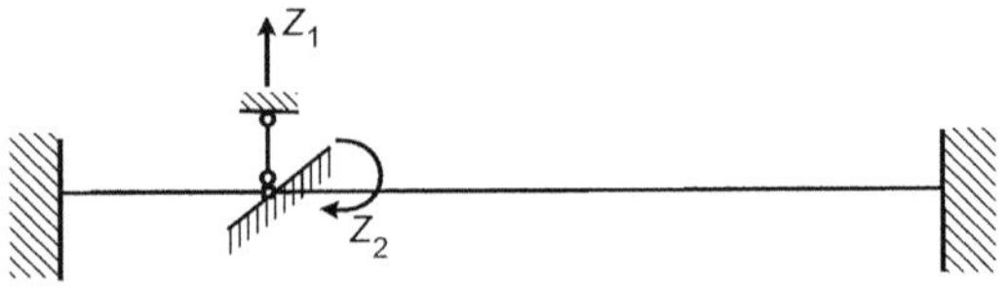

Fig.9.23. The principal system of the slope deflection method

By using ready-made formulae [11] for reactions of a single bar caused by unit displacements of its nodes and taking into account the longitudinal force, we obtain the following expressions of components of the system's stiffness matrix

$$r_{11}=\frac{EI}{l_1^3}\gamma_\mu+\frac{EI}{l_2^3}\gamma_\nu,\qquad r_{12}=r_{21}=-\frac{EI}{l_1^2}\delta_\mu+\frac{EI}{l_2^2}\delta_\nu,\qquad r_{22}=\frac{EI}{l_1}\alpha_\mu+\frac{EI}{l_2}\alpha_\nu,$$

where α_μ, α_ν, δ_μ, δ_ν, γ_μ, γ_ν are well-known N.V. Kornoukhov's functions [11]

$$\alpha_\nu=\frac{\nu}{\mathrm{tg}\nu}\frac{\mathrm{tg}\nu-\nu}{2\mathrm{tg}\frac{\nu}{2}-\nu},\qquad \delta_\nu=\frac{\nu^2\mathrm{tg}\frac{\nu}{2}}{2\mathrm{tg}\frac{\nu}{2}-\nu},\qquad \gamma_\nu=\frac{\nu^3}{2\mathrm{tg}\frac{\nu}{2}-\nu},$$

$$\alpha_\mu=\frac{\mu}{\mathrm{th}\mu}\frac{\mathrm{th}\mu-\mu}{2\mathrm{th}\frac{\mu}{2}-\mu},\qquad \delta_\mu=-\frac{\mu^2\mathrm{th}\frac{\mu}{2}}{2\mathrm{th}\frac{\mu}{2}-\mu},\qquad \gamma_\mu=-\frac{\mu^3}{2\mathrm{tg}\frac{\mu}{2}-\mu},$$

and functions with the subscript ν refer to the compressed bar while those with the subscript μ to the tensioned one. The parameters of the longitudinal force ν and μ are respectively equal to

$$\nu=(1\text{-}\xi)l\sqrt{\frac{N_2}{EJ}}=(1\text{-}\xi)l\sqrt{\frac{\xi P}{EJ}},\qquad \mu=\xi l\sqrt{\frac{-N_1}{EJ}}=\xi l\sqrt{\frac{(1\text{-}\xi)P}{EJ}},\tag{9.66}$$

where the longitudinal force N is assumed positive when the bar is compressed.

The critical state of the system is determined by the condition that the stability determinant should be equal to zero, or

$$\begin{vmatrix} r_{11} & r_{12}\\ r_{21} & r_{22}\end{vmatrix}=\left(\frac{EJ}{l_1^3}\gamma_\mu+\frac{EJ}{l_2^3}\gamma_\nu\right)\left(\frac{EJ}{l_1}\alpha_\mu+\frac{EJ}{l_2}\alpha_\nu\right)-\left(-\frac{EJ}{l_1^2}\delta_\mu+\frac{EJ}{l_2^2}\delta_\nu\right)^2=0,$$

which yields the following equation after introducing dimensionless quantities:

$$(\gamma_\mu+\eta^3\gamma_\nu)(\alpha_\mu+\eta\alpha_\nu)-(-\delta_\mu+\eta^2\delta_\nu)^2=0,\tag{9.67}$$

where

$$\eta=\frac{l_1}{l_2}=\frac{\xi}{1-\xi}.$$

Assuming a value of the parameter ξ and seeing that $\mu=\nu\sqrt{\eta}$, we use the equation (9.67) to find the critical value of the compressive longitudinal force (its minimal root) $\nu = \nu_{Acr}$. The critical load P_{Acr} corresponding to ν_{Acr} is equal to the following, according to (9.66)

$$P_{Acr} = \frac{1}{\xi(1-\xi)^2} \nu_{Acr}^2 \frac{EI}{l^2}. \tag{9.68}$$

To solve Problem "*B*", we can use the solution of the same equation (9.67) replacing the parameter ξ by (1-ξ). It becomes obvious when one notices that the translation of the *P* force application point to the symmetrical point with respect to the central section of the bar transforms Problem "*B*" to Problem "*A*" and vice versa. As it makes

$$P_{Bcr} = \frac{1}{\xi^2(1-\xi)} \nu_{Bcr}^2 \frac{EI}{l^2}, \tag{9.69}$$

we use (9.68) and (9.69) to find the ratio of the critical loads for Problems "*A*" and "*B*"

$$\theta_P = \frac{P_{Acr}}{P_{Bcr}} = \frac{\xi}{(1-\xi)} \left(\frac{\nu_{Acr}}{\nu_{Bcr}} \right)^2.$$

We can use any of the known available software products (such as ***Mathcad***) to find roots of the equation (9.67) and build a plot of the relation $\theta_P = \theta_P(\xi)$ – see Fig. 9.24.

This plot shows that, paradoxical as it is, there are two cross-sections of the bar corresponding to $\xi = \xi_1$ and $\xi = \xi_2$ such that at $\xi \in (\xi_1, \xi_2)$ the system buckles according to the "*A*" loading scheme occurs at a higher load intensity than that according to the scheme "*B*".

But actually there is no paradox here. The explanation for this unexpected effect is that the values of the critical forces are affected by not only the distribution of the longitudinal forces between the compressed and the tensioned bars but also by boundary conditions of the compressed bar at the point where it adjoins the tensioned bar.

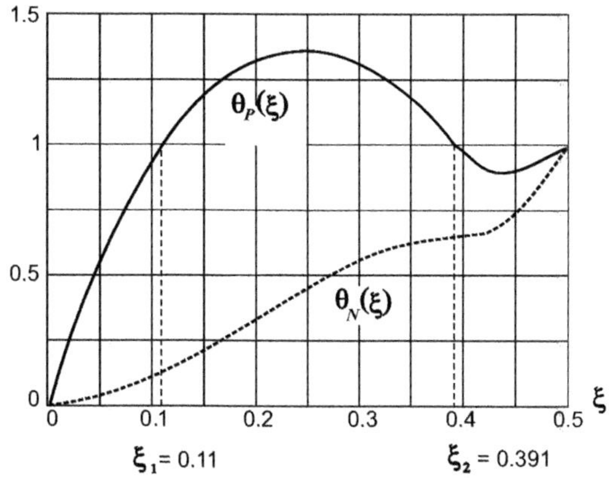

Fig. 9.24. A ratio of critical forces

Clearly the causer of the system's buckling is the compressed bar while the tensioned one is, roughly speaking, a kind of elastic fixation at one of the ends of the compressed bar. The more the length of the tensioned bar and the less the value of the longitudinal force in it, the weaker is the generalized stiffness of this fixation.

It's another matter that the ratio of the critical values of the longitudinal forces in the compressed elements of each of the discussed problems, which we denote as θ_N, is equal to

$$\theta_N = \frac{N_{Acr}}{N_{Bcr}} = \frac{\xi}{1-\xi}\theta_P .$$

A plot of the relation $\theta_N(\xi)$ is also presented in Fig. 9.24, and it is clear θ_N is less than one at $\xi < 0.5$. This deduction corresponds to intuitive expectations of engineers.

The question what position of the external force P provides the least stable state of the system is also relevant. This question can be answered by tracking simultaneously the change of the critical forces in Problems "A" and "B" as the point of application of the external force displaces, and by choosing then the least critical force. To accomplish this, let's introduce a conditional (Eulerian) critical force P_0

$$P_0 = \frac{\pi^2 EJ}{l^2},$$

which enables us to describe the critical load by dimensionless parameters k_A and k_B, assuming

$$k_A = P_{Acr}/P_0 = \frac{1}{\xi(1-\xi)^2}(\nu_{Acr}/\pi)^2, \quad k_B = P_{Bcr}/P_0 = \frac{1}{\xi^2(1-\xi)}(\nu_{Bcr}/\pi)^2 .$$

Plots of the functions $k_A(\xi)$ and $k_B(\xi)$ built after results of computation are shown in Fig. 9.25 where one can see the system is least stable when the force is located at about the quarter of the bar's length (exactly, at $\xi = 0.282$), and the external force compresses the shorter segment while tensing the longer one.

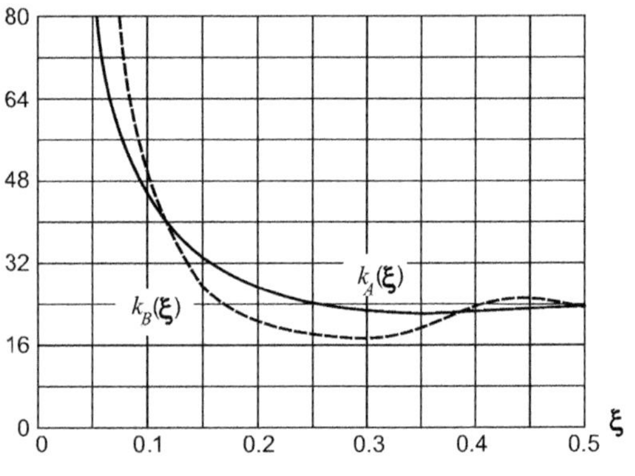

Fig.9.25. Dimensionless parameters of the critical load

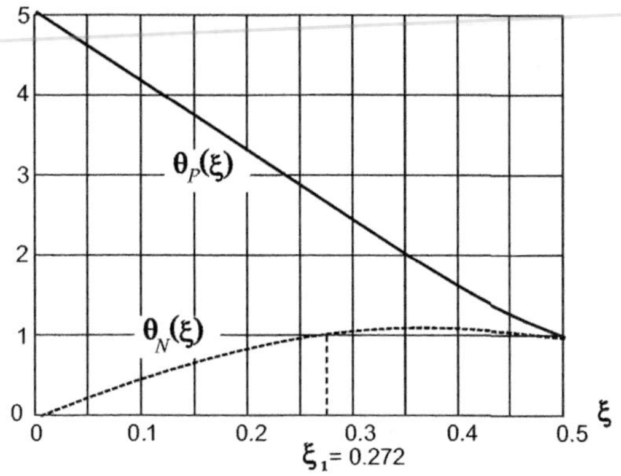

Fig. 9.26. A ratio of critical loads in a simply supported bar

It is interesting also to check the manifestation of this effect in other conditions of the bar's fixation, different from its being clamped. For example, in the case when the clamping shown in Fig. 9.22 is replaced by simple support conditions, the plots of the functions θ_P and θ_N look as shown in Fig. 9.26.

It turns out that the simple support of the bar's ends renders the system always more stable (with respect to the variation of the external force P) when the longer bar segment is compressed rather than the shorter one[10]. Moreover, it seems unexpected that the simply supported bar has an interval of ξ values (from $\xi_1 = 0.272$ to $\xi = 0.5$) where the function θ_N is also greater than one.

Of course, as Fig. 9.26 shows, the excess of θ_N over one is insignificant, but the very fact is important that reflects a qualitative relation between the critical longitudinal forces in Problems "*A*" and "*B*".

Note also that if the load is placed very closely to one of the ends of the simply supported bar (at a small value of ξ) and the force compresses the long segment of the bar, then the system is about 5 times more stable comparing to the opposite direction of the force P.

If we build plots of the functions $k_A(\xi)$ and $k_B(\xi)$ for the simply supported bar, these will look as shown in Fig. 9.27. It becomes obvious here that this system is least stable, too, when the force is applied at about a quarter of the bar's length (or exactly at $\xi = 0.22$) and the shorter segment is under compression.

To conclude the consideration, we'd like to note that the well-known book by V.I. Feodosiev [6] discusses a buckling of a simply supported bar with its end sections restrained from longitudinal displacements, as No. 101 in a section dedicated to non-traditional stability problems.

---

[10] Obviously, it is true everywhere except for the singular point $\xi = 0.5$ where the direction of the load does not affect the critical force $\theta_P = 1$.

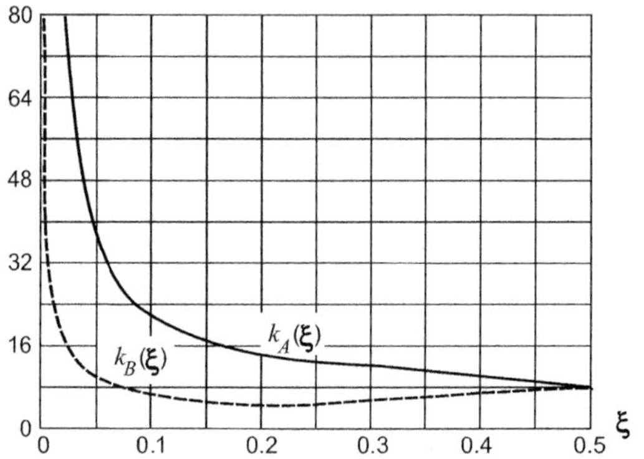

Fig.9.27. Dimensionless parameters of the critical load for a simply supported bar

Having considered the case of the longitudinal force P applied in the central section of the bar (our notation designates this case as the fixed value of $\xi = 0.5$), V.I. Feodosiev draws the reader's attention to the fact that this kind of fixation increases the critical force eightfold[11]. But this deduction, however peculiar, is an end of the discussion, and V.I. Feodosiev does not continue with an analysis of a more general situation where the coordinate ξ of the application point of the force P may vary.

9.7 Allowing for Imperfections in a Real Construction

V.V. Bolotin [2], when discussing methods alternative to a strict analysis of the equilibrium stability, indicates a *method of full integration* which means an investigation of a system's behavior under all possible classes of disturbances which could take place, or statistical methods of such investigation. Stating justly that a disturbed solution should be checked for stability too, V.V. Bolotin gives an elementary example of a simply supported elastic bar of a constant cross-section that has an anti-symmetric initial camber (Fig. 9.28). The deformed shape analysis of it gives an erroneous result (the second critical force is $4\pi^2 EI / L^2$) while the bar buckles in a symmetrical mode under a four times less critical force P.

Do we mean an analysis that allows for initial imperfections is senseless? Certainly we don't. Here we can give two considerations "pro" such analyses, without confuting the deduction that the disturbed solution should be checked for stability:

- these analyses allow one to more closely track the behavior of elements of the system not included in the response of an idealized model (constraints, zero bars etc.);

---

[11] The solution of the V.I. Feodosiev corresponds to the point $\xi = 0.5$ on the plots of Fig. 9.27 where one can see $P_{Acr} = 8P_0$.

- it becomes possible to estimate those allowances for the manufacturing and erection which should be formulated for a structure being investigated if it does not belong to any class of well-studied structures in this regard.

The imperfections themselves are random variables. A strict investigation of their effect require involving probabilistic methods and the availability of information regarding statistical properties of the imperfections. Though, practically engineers operate much poorer data on the basis of some assumptions regarding the expected or maximum possible imperfection magnitudes. Sometimes such data are rated by design codes such as those in Eurocode-3 [5]. It requires possible initial imperfections to be allowed for even in a general static analysis of a structure. Generally, the objects to be allowed for are imperfections of the whole system, those (non-idealities) of nodal joints, and those of structural members. In multistory frames such imperfection is a deviation from the vertical

$$\Psi = k_c k_s \Psi_0, \tag{9.70}$$

where

- Ψ_0 is an allowable skew of the column (for example, $\Psi_0 = 1/200$);
- $k_c = (0{,}5 + 1/n_c)^{1/2}$, but $k_c \le 1$;
- $k_s = (0{,}2 + 1/n_s)^{1/2}$, but $k_s \le 1$;
- n_s is the number of stories;
- n_c is the general number of columns passing through all stories where the load is not less than 50% of the average one per column.

The columns k_c and k_s take into account the small probability of the event that all columns are skewed in the same direction according to a simplest statistical assumption: all these skews are deemed statistically independent. This is the first idea implemented by Eurocode, and the second one is that initial geometrical deviations can be allowed for as equivalent horizontal forces such as shown in Fig. 9.29.

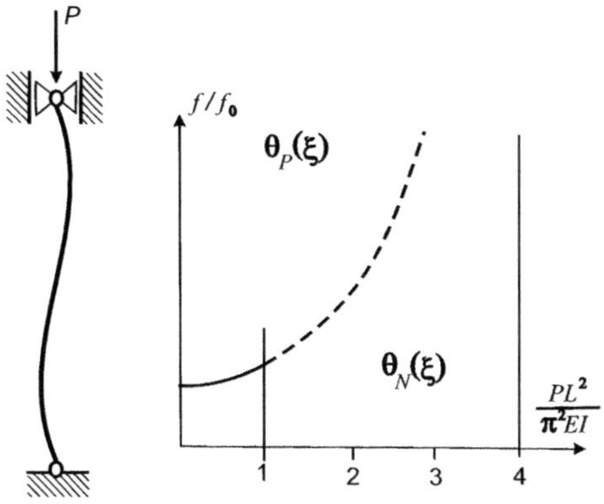

Fig. 9.28. A bar with an anti-symmetric camber

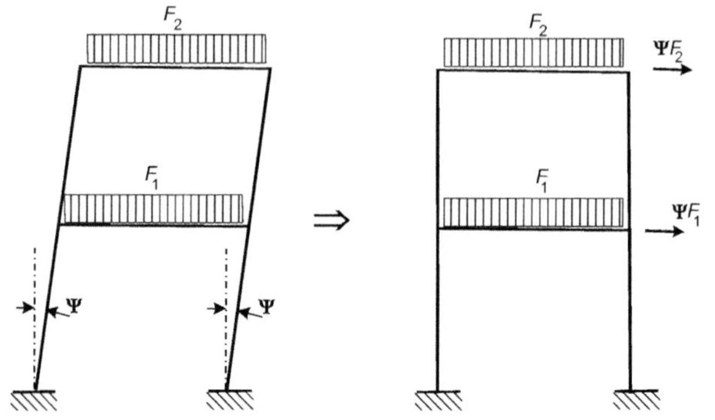

Fig. 9.29. Equivalent forces to take into account deviations from the vertical

Those force can be applied in all possible horizontal directions, and one should choose disadvantageous ones (for example, if the wind acts on the frame in the scheme of Fig. 9.29 from left to right, then the disadvantageous direction of the forces ψF_1 and ψF_2 will be the same).

A similar approach is used to analyze bracing systems that provide a side stiffness of a structure. It allows for initial imperfections of braced elements as cambers of the sag $e_0 = k_r L / 500$. The coefficient $k_r = (0.2 + 1/n_r)^{1/2}$, but not to exceed one; n_r is the number of reinforced elements. Here again the assumption of the cambers being statistically independent is involved.

The equivalent load upon the bracing system is (Fig. 9.30)

$$q = \frac{k_r + 500\delta_q / L}{62.5} \sum \frac{N}{L}, \qquad (9.71)$$

where δ_q is the deflection of the bracing system.

In cases when bracing a system stabilizes its flexural elements, the forces N can be determined as

$$N = M / h, \qquad (9.72)$$

where M is a maximal bending moment and h is the height of a flexural element.

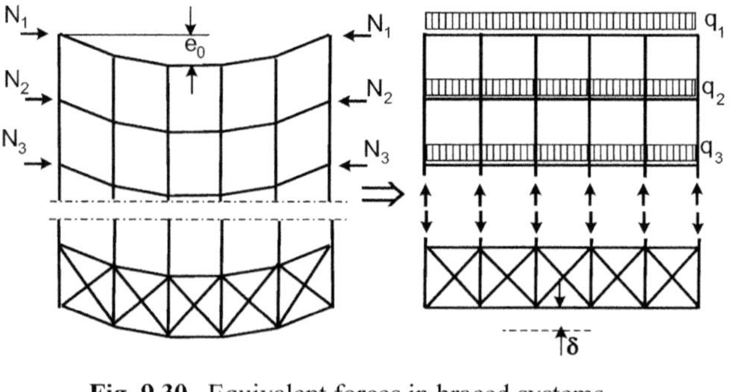

Fig. 9.30. Equivalent forces in braced systems

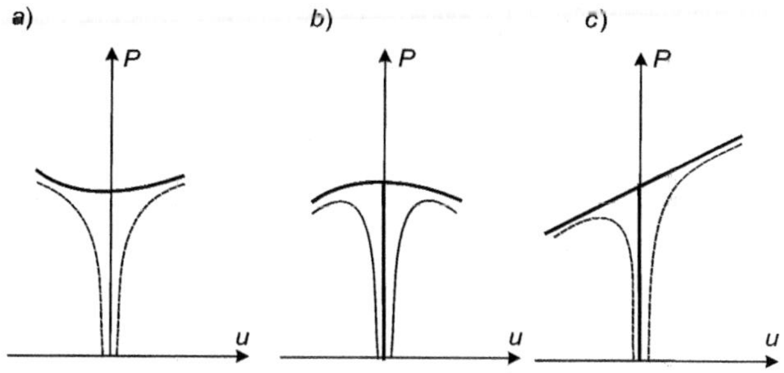

Fig. 9.31. Versions of the post-critical behavior: (*a*) soft; (*b*) hard; (*c*) intermediate

There are a lot of studies dedicated to a qualitative estimation of the effect of initial imperfections on the buckling. Such studies are being performed intensively within the so-called "theory of catastrophes". We refer the curious reader to literature on the subject (for example, see [24]) and outline here only some characteristic results.

Fig. 9.31 presents characteristic plots of dependence between the post-critical displacements and the load for some systems having a single degree of freedom.

If at the critical point of a perfect system its original stable state changes to another stable equilibrium mode (Fig. 9.31, *a*), then the behavior of the respective imperfect system is such that the critical load is passed in a fairly smooth manner. If the perfect system is such that at the critical point its stable state changes to an unstable one (Fig. 9.31, *b*), then the behavior of the imperfect system shows a substantial fall of the critical load. An intermediate behavior is also feasible (Fig. 9.31, *c*) when the system responses differently to disturbances of different signs. Note that the first behavior type is more characteristic for bar systems while the two others for shell structures.

Undoubtedly, the behavior of the system shown in Fig. 9.31, *b* is especially dangerous. Often the actual load limit of an imperfect structure is a few times different from that of its idealized model.

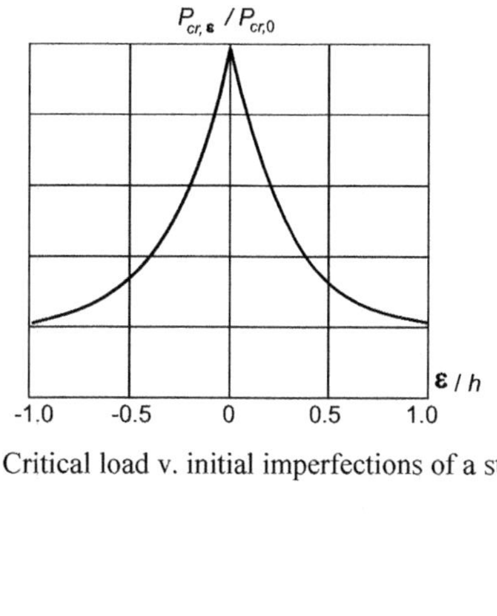

Fig. 9.32. Critical load v. initial imperfections of a structure

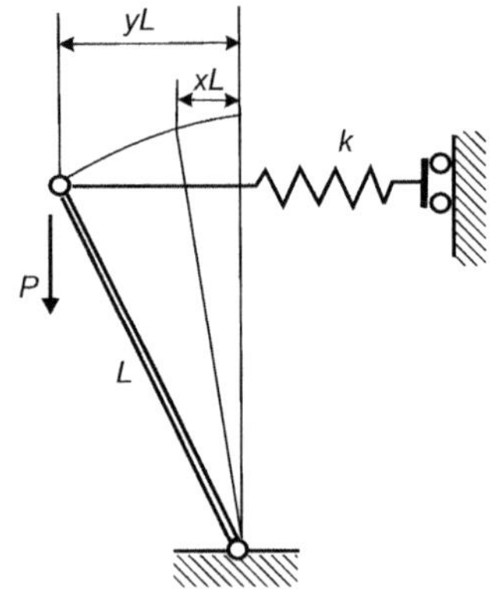

Fig. 9.33. A model to estimate initial imperfections

Thus, Fig. 9.32 presents a plot built in a classics by W. Koiter [10] to illustrate the ratio of the load limit of an imperfect system $P_{cr,\varepsilon}$ to the critical load of the respective perfect system $P_{cr,0}$ vs. the amplitude value ε of the imperfection divided by the thickness h of a closed cylindrical shell compressed along its generatrix. One should notice the values of ε / h greater than one are not rare exceptions.

The effect of initial imperfections on the magnitude of the critical load is easily illustrated by a simplest model (Fig. 9.33) for which an accurate solution is known [25].

This model system is a perfectly rigid bar of the length L attached to a foundation by a hinge and fixed by a linear spring of the stiffness k. The initial imperfection is simulated by a deviation xL of the bar from the vertical, the full displacement is yL. In the deviated state the tension in the spring is $R = k(y - x)L$, and the tilt angle α of the bar to the vertical is determined by the relationship

$$\cos\alpha = \frac{\sqrt{L^2 - (yL)^2}}{L} = \sqrt{1 - y^2}\ .$$

Then we derive this from the conditions of equilibrium:

$$\frac{Py}{\sqrt{1 - y^2}} = k(y - x)L\ .$$

The critical value of the load P_{cr} can be found from the condition $\partial P / \partial y = 0$ that occurs at $y = x^{1/3}$. The substitution of the value to the equilibrium condition yields a known expression

$$\lambda_{cr} = \frac{P_{cr}}{kL} = \left(1 - x^{2/3}\right)^{3/2}$$

for a dimensionless critical load [25]. The relation of λ_{cr} and the initial imperfection parameter x is shown in Fig. 9.34.

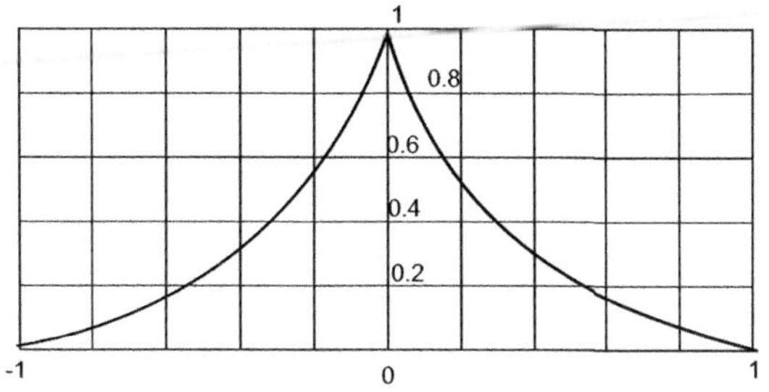

Fig.9.34. A dependence of the result on initial imperfections

As one can see in this figure, the qualitative character of the dependence of the critical load vs. the initial imperfections in this simple model problem resembles that obtained by Koiter for a circular cylindrical shell which is shown in Fig. 9.32.

9.8 Notes on Allowing for *P*-Δ Effects

The use of the geometrical stiffness matrix $\mathbf{K}_G$ according to (3.129) is intended, generally, to allow for secondary effects in static and dynamic structural analysis. In particular, there appear additional bending moments caused by the dead weight at noticeable horizontal displacements of a load-carrying framework. In English-language literature on structural design/analysis, such as Eurocode, this effect is usually referred to as a "P–Δ" analysis.

There have been proposed a good deal of methods to estimate these effects. The paper [29] analyzes the publications on the subject and proposes a simplified method to analyze these second-order effects. If one attempts to solve the problem as geometrically nonlinear, one will have to employ an iteration or stepwise process. Also, these effects will need to be allowed for by dynamical analysis.

When considering frameworks of buildings where the dead weight load is not changed by the deformation and full displacements can be assumed small comparing to the dimensions of the building, the P–Δ analysis problem can be linearized and its solution obtained without any iterations. The matter is that problems of this kind are quasi-monocyclic in terms introduced by N.V. Kornoukhov. It means that the distribution of the longitudinal forces between bars of a system (or that of membrane stresses in shell elements) can be found out at a practically sufficient accuracy "with a single blow": by solving a purely linear problem. But once this distribution is known, the full problem becomes linearized, actually. This approach has been implemented, for example, in the **SCAD** system. It is attractive because it does not require the user to make any further simplifications related to specifics of a particular problem. Therefore we do not go deeper into the analysis of various proposals how to allow for P–Δ effects in particular kinds of systems.

References

1. Belsky GE, Rivkin AM (1974) On a determination of free lengths in members of bar systems (in Russian). Structural mechanics and analysis of structures, 6:21–23
2. Bolotin VV (1965) A concept of stability in the structural mechanics (In Russian). In: Problems of stability in structural mechanics, Moscow, Stroyizdat, pp. 6–27.
3. Bovin VA (1963) Difference-variation methods of structural mechanics (in Russian). Stroyzdat USSR, Kiev
4. Dinkevich SZ (1977) Analysis of ciclic structures. Spectral method (in Russian). Stroyzdat Moskow
5. ENV 1993-1-1. Design of steel structures. General rules and rules for building (1995), Brussels
6. Feodosiev VI (1967) Selected problems and questions of strength of materials (In Russian), Nauka Publishing House, Moscow
7. Feodosiev VI (1969) Ten lectures-conversations on strength of materials (in Russian). Nauka Publishing House, Moskow
8. Gemmerling AV (1974) How to determine design length of columns in multistory buildings (in Russian). In: Structural mechanics and analysis of structures, 2:64–66
9. Kochin NE (1965) Vector calculus and basics of tensor calculus (in Russian). Moscow, Nauka
10. Koiter WT (1945) On the Stability of Elastic Equilibrium. Diss. Thesis. TH Delft, Amsterdam
11. Kornoukhov NV (1949) Strength and stability of bar systems (in Russian). Moscow, Gosstroyizdat
12. Loytsiansky LG, Lurie AI (1954) A course of theoretical mechanics (in Russian). Vol. I. Moscow, Gostekhteorizdat
13. Lurie AI (1961) Analytical mechanics (in Russian). Moscow, Fizmatgiz
14. Nudelman JaL (1949) Methods of determining natural frequencies and critical forces in bar systems (in Russian). Moskow, Gostekhteorizdat
15. Ovsianko VM (1999) A tracking force and thing around. A computer analysis of electronic models for deformable objects (in Russian). Polymia, Minsk
16. Panovko YaG, Gubanova II (1965) Stability and oscillations of elastic systems; paradoxes, fallacies, and new concepts. Consultants Bureau, New York
17. Papkovich PF (1963) Works on structural mechanics of ships.— Vol. 4. Stability of bars, floors and plates (in Russian). Sudpromgiz, Leningrad
18. Rzhanitsin AR (1955) Stability of equilibrium of elastic systems (in Russian). Gostekhteorizdat, Moscow
19. Rzhanitsin AR (1964) On some properties of a stability area of an equilibrium state (in Russian). In: Structural mechanics and analysis of structures, 1:1-8
20. Rzhanitsin AR (1975) On the question of equivalent lengths of bars (in Russian) In: Structural mechanics and analysis of structures, 5:74-76
21. Smirnov AF (1958) Stability and oscillations of structures (in Russian). Transzheldorizdat, Moscow
22. Steel constructions (1977) Vol.1: Members of steel constructions (in Russian) Ed.: Gorev VV, Uvarov BY, Filippov VV etc. Vyschaya shkola, Moscow
23. Strang G (1976) Linear algebra and its applications. Academic press, New York San Francisco London

24. Thompson JMT (1982) Instability and catastrophes in science and Engineering. John Wiley & Sons, Chichester New York Brisbane Toronto Singapore
25. Thompson JMT, Hunt GW (1973) A General Theory of Elastic Stability. John Wiley,London
26. Timoshenko SP (1936) Theory of elastic stability. McGraw-Hill, New York
27. USSR State Committee for Construction (1989) Manual on steelwork design to SNiP II-23-81*. Steel constructions (in Russian). Moskow, Stroyizdat
28. USSR State Committee for Construction (1990) SNiP II-23-81*. Steel constructions (in Russian). Stroyizdat, Moscow
29. Wilson EL Habibullah A (1987) Static and Dynamic Analysis of Multi-Story Buildings Including P-Delta Effects, Earthquake Spectra. In: Earthquake Engineering Research Institute, Vol. 3, No.3, May 1987.
30. Yasinsky FS (1952) Selected papers on stability of compressed bars (in Russian). Gostekhteorizdat, Moscow

10 Problems of Dynamics

The man asks the Nature a lot of questions senseless from the latter's "viewpoint", and wishes to get answers unambiguous and fitting to patterns he cherishes. In a word, it is a certain particular order rather than the universal Order that we seek to discover....

Stanislaw Lem. Summa technologiae

The system of education in high schools of civil engineering is organized in such manner that problems of dynamics occupy an undeservingly modest place in the curriculum. While there are numerous exercises intended to help students develop certain intuition of static analysis (apparently insufficient, but capable of being improved during their practical activities), their "dynamical sense" remains rudimentary. Hence a lot of mistakes and a totally formal attitude to the preparation of source data for a dynamical problem's solution by tools of computer-aided design.

The authors took this circumstance into account when preparing the text of this chapter. We have decided to redirect the presentment of the chapter so as to fill it up with reference and general scientific information. A prepared reader might think there is too much of it, so we apologize to professionals for our, possibly annoying, tutorial manner of presentment.

10.1 Design Models for Dynamical Analysis

A design model of a load-carrying framework that describes an elastic resistance of a structure in the course of analyzing its dynamical response is, most often, exactly the same as one used in the static analysis. Naturally this model is supplemented with inertial properties and data regarding forces of resistance to the movement. Also, external influences are described in more detail and represented as certain functions of time. Without discussing these supplements to the design model in too much detail, we'd like to notice certain danger of using the same descriptions of the structure for both static and dynamic analysis. This danger is not always realized even by experienced analysts who have solved static problems chiefly in their activities.

The matter is that there is a tradition of the static analysis oriented at limit states of systems that prescribes to remove elements from a design model that affect little the limit resistance of the structure. One removes partitions, elements of fencing, and other components of the structure which actually contribute little to the structure's response at a deformation corresponding to the level of the accepted static influence. Though, it may happen in the dynamical analysis, especially in the estimation of higher natural frequencies, that the respective oscillation amplitudes are much less than the displacements at which "non-load-carrying" structural elements cease to respond effectively.

Also, the rigidity properties of building materials (especially those of foundations) must be assumed to have other values in the dynamical analysis. The matter that these values could be adopted for the static analysis with taking into account rheological changes developed under long-term loads of, possibly, a nearly ultimate intensity. Hence a tradition to assume the compliance of a soil bed in accordance with its averaged "secant" modulus, and reference data regarding elasticity moduli of ferroconcrete allow for rheological processes of its hardening to some extent etc.

In the dynamical analysis at a medium level of operational loads and not very high frequencies, the dynamical stiffness is close to an "instantaneous" static stiffness (that is, determined by its tangential modulus rather than the secant one) and grows a bit as the frequency increases [35].

Thus, most often the use of the same elastic model for both static and dynamic analysis underestimates the value of the dynamical stiffness. As it is known, this results in underestimated natural frequencies.

Due to a practical difficulty to perform the dynamical analysis that involves the real value of the dynamical stiffness, one should regard the fact that the results are going to contain certain error of the natural frequency values. This error is even subject to regulation, depending on a type of a structure, in a number of structural design manuals that deal with operational dynamical influences [23, 24]. This approach is closely related to certain types of structures and requires a sufficiently detailed consideration before its applying to objects of a different build. Unfortunately, such consideration is missed in most cases.

Most software applications intended for dynamical structural analysis implement prescriptions of effectual design codes and operate measures of dissipation averaged over a structure (such as logarithmic oscillation decrements δ). This approach can be justified if the structure of interest is made of the same material with similar structural decisions and without clear locations where constructive friction develops. But its employment in analyzing systems made of materials and members of drastically different elastic, dissipative and inertial properties can result in noticeable inaccuracies.

In many real structures the values of δ for different subsystems differ by orders of magnitude: from thousandth fractions for steel constructions (rolled or welded members) to units for soil beds. Certainly these qualitative and quantitative differences in structural properties had better be allowed for by correct dynamical models.

A correct consideration of actual dissipative properties of a system is especially important when analyzing resonant modes: it is known that the height of a resonant peak is inversely proportional to the value of δ. It is important, too, to estimate the duration of exposing people to vibration on an oscillating structure [21].

The magnitude of the energy loss in a structure determined by internal friction and constructive damping can hardly be calculated theoretically, so it is usually found out by experiments and observations. Table 10.1 represents experimental data of this kind taken from recommendations on nuclear power plant design for the USA. The critical attenuation presented in the table $C_{cr} = 2(KM)^{1/2}$ where K is a stiffness and M is a mass corresponds to the case when a system deviated from its equilibrium returns to it without oscillating.

Table 10.1.

Type of structure	Attenuation in % of critical value of C_{cr} for stresses in fractions of rated resistance:	
	$0.67R_y$	**$0.90R_y$**
Steel constructions:		
welded	2	4
bolted	4	7
Concrete constructions:		
pre-stressed	2	5
conventionally reinforced	4	7

The above mentioned inaccuracy of calculating natural frequencies and errors in estimates of the logarithmic decrement have the strongest effect on the phase shift angle μ between a disturbing force and the structure's response. This circumstance may render questionable formally correct results of summation of the structure's response (displacements, stresses) calculated separately by natural oscillation modes because, for example, s^{th} term of such sum should participate there with the factor $\cos(\theta t - \mu_s)$, and

$$\operatorname{tg}\mu_s = \frac{\delta/\pi}{\sqrt{1+(\delta/2\pi)^2}}\,\frac{\theta/\omega_s}{1-(\theta/\omega_s)^2}, \tag{10.1}$$

where θ designates the frequency of the disturbing force and ω_s the natural frequency oscillation of sth mode of the non-damped system.

To allow for the said discrepancy, many authors have deemed it possible to assume the phase shift angle μ to be a random value distributed uniformly at the interval from zero to 180°. This has led to design formulae like a "root of sum of squares of responses that correspond to natural frequencies".

Pretty often dynamical analyses contain the following error: when dealing with structures on elastic beds, such beds are not assumed to have inertial properties. As special researches have shown [19], the error caused by this can be even of a qualitative character, not just a numerical discrepancy. For example, resonance curves of Fig. 10.2, *a* have been obtained for a structure shown in Fig. 10.1, for the case of an inertia-free semi-space. Three coefficients of inelastic resistance have been used: $\gamma = 0; 0.05; 0.1$.

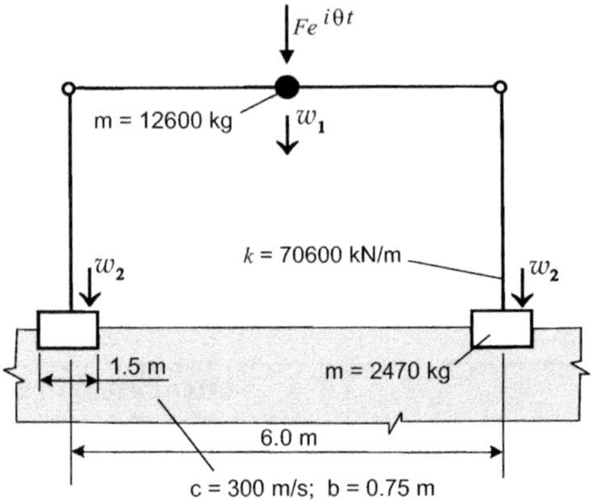

Fig. 10.1. A structure on an elastic semi-space

Similar curves obtained with taking the bed's inertia into account are presented in Fig. 10.2, *b*. The speed of propagation of S-waves (lateral) in the bed is c = 300 m/sec, and the sag is affected by the half-width of the foundation $b = 0.75$ m. The abscissa axis shows values of dimensionless frequencies $\chi = \omega b / c$.

It should be noted that there is no second resonant peak in plots that correspond to the inertial semi-space. Also, the amplitude remains finite even during the resonance and with no non-elastic resistance of the structure. The position of the first resonant depends little on the bed's inertial properties, but the amplitudes are much less when the inertia is allowed for. This effect is explained by the fact that the attenuation of the oscillations due to the energy radiation to the semi-space is comparable to (if not greater than) the loss due to an internal friction in the structure.

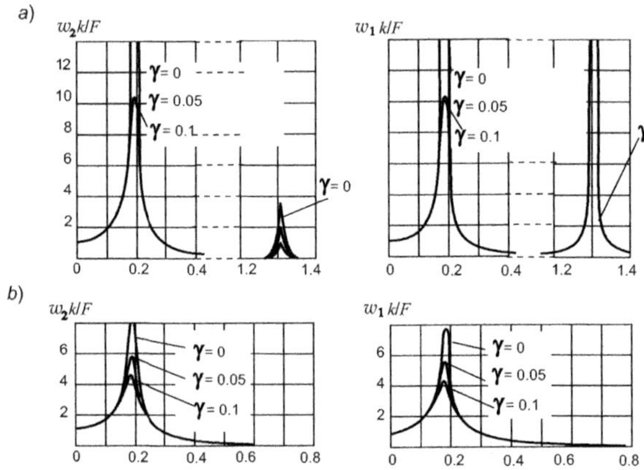

Fig. 10.2. Resonance curves: (*a*) the bed's inertia is not taken into account; (*b*) the bed's inertia is taken into account

The paper [13] presents comparative analyses of various types of elastic beds used to solve dynamical problems: a modified Winkler's model that allows for inertial properties of a bed; a two-parametric Pasternak's model, an elastic semi-space etc. It has been shown that a special selection of parameters yields close results but the inertia of the bed must necessarily be taken into account.

Practical analyses capable of allowing for effects of the oscillation energy radiation into a bed are based on the employment of equivalent rigidities and coefficients of energy loss in a simplified model where the foundation slab is attached to equivalent springs that simulate the resistance of the bed to linear and angular displacements and their respective dampers. Coefficients of specific damping are calculated by a formula like $b = k\sqrt{C_z}$ [kg×sec/cm^2]. Here k is a dimensionless coefficient that takes values from 0.07 to 0.14 depending on the oscillation direction, C_z [kg/cm^3] is a coefficient of elastic uniform resistance of the bed. As a real bed can have an immensely complicated build — for example, its jointing may hinder the energy removal — the magnitude of the energy dissipation in the bed should be limited by assuming the damping factor to be about 15...35% of the critical value.

10.1.1 Dynamical Degrees of Freedom

The computer solution of dynamical problems requires much more time than the static analysis of structures of the same complexity. Therefore, with the exception of simplest models, there is always a problem how to choose a proper number of dynamical degrees of freedom. There is a widely used empirical ruse that states: in a system having n dynamical degrees of freedom, about $n/2$ first natural frequencies and their respective natural oscillations can be calculated reliably. In this regard the book [5] says that the rule is related closely to Kotelnikov's theorem in the theory of information: to get a sufficient information about a continuous signal, the frequency of discrete measurements must be at least twice as high as the maximum frequency contained in the spectrum of the signal.

There is another consideration that fosters a suspicious attitude of an analyst to a structure's natural frequencies and modes belonging to a higher part of the spectrum. The matter is that the model of a dynamical system with a finite number of degrees of freedom does not appear in the world exactly as it is. Usually this one is a result of a discretization (by a finite element technique, for example) of an originally continuous model. For higher oscillation modes, an irreducible error which cannot be avoided but by changing the physical problem's statement might have appeared in the continuous model even before its transformation to a discrete model. For example, in a simple flexural Bernoulli – Euler beam of a rectangular section, when the ratio of the section's height to the half-wavelength of the respective oscillation mode begins exceeding a critical value (about 1/5), the very applicability of the Bernoulli – Euler theory becomes questionable. One has to switch to Timoshenko's beam bending theory and even allow for a rotational inertia. In shorter-wave oscillation modes this continuous model also becomes insufficient, so one needs to change the initial mechanical and mathematical statement of the problem employing, say, a plane dynamical elasticity theory. An analyst who studies modes of oscillations of a discrete model could estimate, at least qualitatively, whether it makes sense to take higher modes into account, and if it does then what highest mode should be included. Actually, this issue has been discussed by us before, at the beginning of Chapter 2 (see Fig. 2.1).

The requirement of using a double number of degrees of freedom comparing to the number of natural modes at the frequencies up to 33 Hz is included in regulations of the US Nuclear Regulatory Commission [36] as an alternative one. It requires that results of an analysis do not change more than by 10% as n increases. The same regulations contain another important requirement regarding the possibility not to allow for some subsystems of a structure in its dynamic analysis.

Let's denote the mass of a principal structure as M_1 and the mass of its secondary subsystem additionally associated with it (such as a suspended pipeline) as M_2. Also, we denote the respective natural frequencies of these two parts of the structure as ω_1 and ω_2. Then, the secondary subsystem can be excluded from the consideration when $M_2 / M_1 < 0.01$, or when $0.01 < M_2 / M_1 < 0.1$ and the ratio of the frequencies is $\omega_2 / \omega_1 < 0.8$ or $\omega_2 / \omega_1 > 1.25$.

Note that the minimal ω_1 and maximal ω_n natural frequency of a design model together define some "natural" scale of time $t_{\max} = 2\pi/\omega_1$ and $t_{\min} = 2\pi/\omega_n$ using which all processes can be classified into "slow", "normal" and "quick". This classification helps one avoid an analysis of inessential effects, for example, not to deal with the load's time history function if its total action time is substantially less than $t_{\min}$, or to ignore transient processes when considering intervals of time much longer than $t_{\max}$. If a space scale is introduced together with the time scale, one can consider the possibility to ignore wave processes which is common in the structural analysis practice. This approach is so deeply rooted among analysts that most often they even don't subject its applicability to any doubt. Meanwhile, contemporary structures may have dimensions so great that the time of a disturbance propagation in those (from an earthquake shock, for example) proves to be comparable with their natural oscillation periods. Apparently, ignoring the processes of the disturbance transfer is unacceptable in this case.

10.1.2 Dynamic Condensation — a Guyan's Procedure

The intention to reduce the number of dynamical degrees of freedom can be often implemented by grouping masses of a structure in a relatively small amount of nodes. The rest of the structure's nodes prove to be mass-free. Also, nearly always one can neglect some inertial properties of the system such as those related to rotational degrees of freedom. Thus there arises a problem of dynamic analysis of a *system with an incomplete set of masses* where the number of dynamical degrees of freedom taken into account can be, for example, 10% of that of the static ones [7]. Purely static degrees of freedom can be excluded using a static condensation procedure presented earlier by the equation (3.175). As for a partial exclusion of dynamical degrees of freedom, a condensation by R. Guyan is used for this purpose [17].

We deem it useful to turn again to a description of this procedure in order to analyze the error that arises in its application to a more realistic situation. The situation is that the inertial properties of degrees of freedom declared mass-free are not zero — they are just small. Then, instead of (3.174) we have

$$\left\|\begin{bmatrix} \mathbf{K}_{00} & \mathbf{K}_{01} \\ \mathbf{K}_{10} & \mathbf{K}_{11} \end{bmatrix}\right\| \bullet \left\|\begin{bmatrix} \mathbf{\Psi}_0 \\ \mathbf{\Psi}_1 \end{bmatrix}\right\| = \omega^2 \left\|\begin{bmatrix} \mathbf{m}_{00} & \mathbf{0} \\ \mathbf{0} & \mathbf{m}_{11} \end{bmatrix}\right\| \bullet \left\|\begin{bmatrix} \mathbf{\Psi}_0 \\ \mathbf{\Psi}_1 \end{bmatrix}\right\| \tag{10.2}$$

where the diagonal mass matrix $\mathbf{m}_{00}$ presents elements much less than those of the matrix $\mathbf{m}_{11}$.

Suppose that there can be established a linear dependence between the sub-vectors $\mathbf{\Psi}_0$ and $\mathbf{\Psi}_1$

$$\mathbf{\Psi}_0 = \mathbf{T}\,\mathbf{\Psi}_1 \,, \tag{10.3}$$

and the linear transformation matrix $\mathbf{T}$ is kept intact in the course of the system's oscillatory movement. Then the full vector $\mathbf{\Psi}$ can be represented as

$$\mathbf{\Psi} = \left\|\begin{bmatrix} \mathbf{\Psi}_0 \\ \mathbf{\Psi}_1 \end{bmatrix}\right\| = \left\|\begin{bmatrix} \mathbf{T} \\ \mathbf{I} \end{bmatrix}\right\| \mathbf{\Psi}_1 \tag{10.4}$$

where $\mathbf{I}$ is a unit matrix of the order equal to the dimension of the sub-vector $\mathbf{\Psi}_1$.

Substituting (10.4) to (10.2) and multiplying the obtained equality by the matrix $\|[\mathbf{T}^{\mathrm{T}}\,\mathbf{I}]\|$ from the left yields equations of free oscillations of a system having a shortened set of dynamical degrees of freedom (all dynamical degrees of freedom collected in the vector $\mathbf{\Psi}_0$ are excluded)

$$(\mathbf{K}_{11} + \mathbf{T}^{\mathrm{T}}\mathbf{K}_{00}\mathbf{T} + \mathbf{T}^{\mathrm{T}}\mathbf{K}_{01} + \mathbf{K}_{10}\mathbf{T})\,\mathbf{\Psi}_1 = \omega^2(\mathbf{m}_{11} + \mathbf{T}^{\mathrm{T}}\mathbf{m}_{00}\mathbf{T})\,\mathbf{\Psi}_1 \,. \tag{10.5}$$

To find the transformation matrix $\mathbf{T}$, we derive this from the first equation of the system (10.2):

$$\mathbf{\Psi}_0 = -\mathbf{K}_{00}^{-1}\mathbf{K}_{01}\mathbf{\Psi}_1 + \omega^2\mathbf{K}_{00}^{-1}\mathbf{m}_{00}\,\mathbf{\Psi}_0 \,. \tag{10.6}$$

If we neglect the second term in (10.6) due to small masses of $\mathbf{m}_{00}$, the desirable transformation matrix $\mathbf{T}$ will be written as

$$\mathbf{T} = -\mathbf{K}_{00}^{-1}\mathbf{K}_{01} \,, \tag{10.7}$$

and the equation (10.2) with this $\mathbf{T}$, after considering the symmetry conditions of the full stiffness matrix $\mathbf{K}_{01}^{\mathrm{T}} = \mathbf{K}_{10}$, will become

$$\mathbf{K}^*\mathbf{\Psi}_1 = \omega^2\mathbf{M}^*\mathbf{\Psi}_1 \,, \tag{10.8}$$

where $\mathbf{K}^*$ and $\mathbf{M}^*$ are *reduced* (sometimes referred to as *condensed*) stiffness and mass matrices, respectively,

$$\mathbf{K}^* = \mathbf{K}_{11} - \mathbf{K}_{10}\mathbf{K}_{00}^{-1}\mathbf{K}_{01} \,, \tag{10.9}$$

$$\mathbf{M}^* = \mathbf{m}_{11} + \mathbf{K}_{10}\,\mathbf{K}_{00}^{-1}\,\mathbf{m}_{00}\,\mathbf{K}_{00}^{-1}\,\mathbf{K}_{01}\ . \tag{10.10}$$

The equation (10.8) is a *condensed* equation of free oscillations of the system that features both the static condensation technique and the dynamic condensation according to Gayan.

If we keep only the first term in the expression of the reduced mass matrix (10.10) and omit the term of the small mass $\mathbf{m}_{00}$, we will have the static condensation procedure. Otherwise we'll have Gayan's dynamic condensation. The difference between these two procedures is that the static condensation simply drops out the masses $\mathbf{m}_{00}$ when analyzing natural oscillations of a system, while the dynamic condensation redistributes these masses between other inertial properties.

Let us compare solutions of the original problem (10.2) and the problem (10.8). Let ω_i $(i = 1,\ldots, n)$ be the spectrum of the original problem (10.2) in the ascending order, ω_{Si} and ω_{Di} $(i = 1,\ldots, n_1)$ be the respective frequency spectra of the statically and dynamically condensed problems. Here n is the number of dynamical degrees of freedom of the original problem, n_1 is the number of dynamical degrees of freedom left after the condensation. It is easy to notice that the dynamically condensed problem corresponds to oscillations of a mechanical system derived from the original one by imposing constraints (10.3) on its oscillation modes where the poly-constraint matrix $\mathbf{T}$ is determined by the relationship (10.7). Using Rayleigh's theorem, we immediately obtain the following estimate

$$\omega_i \le \omega_{Di} \quad (i = 1,\ldots, n_1)\ . \tag{10.11}$$

On the other hand, a mechanical system that corresponds to the statically condensed problem of natural oscillations has been derived from the dynamically condensed system by simply omitting some of the masses. Without involving formal considerations, we'd like to note that a reduction of any system's inertial properties can shift the spectrum only to the right. This statement can be proved generally by involving a variational formulation of the eigenvalue problem. The result is

$$\omega_{Di} \le \omega_{Si} \quad (i = 1,\ldots, n_1)\ . \tag{10.12}$$

By uniting the estimates (10.11) and (10.12) we obtain an important qualitative relationship which shows the dynamical condensation to be always more accurate than the static one with respect to the frequency spectrum.

Now let's use a relationship for determining corrections of first order of smallness to the natural frequencies and eigenvectors of perturbed matrices [2]. Dimensionless values of the corrections are proportional to elements of the matrix

$$\mathbf{C} = \mathbf{\Phi}_0^{\mathrm{T}}\Delta\mathbf{K}\,\mathbf{\Phi}_0 + \mathbf{\Phi}_0^{\mathrm{T}}\Delta\mathbf{M}\mathbf{\Lambda}_0\mathbf{\Phi}_0\ , \tag{10.13}$$

where $\Delta\mathbf{K}$ and $\Delta\mathbf{M}$ are respective perturbations of the stiffness matrix and the inertia matrix; $\mathbf{\Phi}_0$ is a matrix of eigenvectors of the non-perturbed system, and $\mathbf{\Lambda}_0$ is a diagonal matrix of eigenvalues of the non-perturbed system. The absolute value of the corrections is proportional to the norm of the matrix $\mathbf{C}$

$$\|\mathbf{C}\| \le -\|\mathbf{\Phi}_0\|^T\|\mathbf{\Phi}_0\| \,(-\|\Delta\mathbf{K}\| + \|\Delta\mathbf{M}\|\,\|\mathbf{\Lambda}_0\|) = P_\Phi(-\|\Delta\mathbf{K}\| + \|\Delta\mathbf{M}\|\,\|\mathbf{\Lambda}_0\|) \tag{10.14}$$

where P_Φ is a spectral conditionality number.

For the original problem it holds: $\|\mathbf{C}\| = \|\mathbf{m}_{00}\|$, and the problem (10.8) is different only by small-order terms ε in the inertia matrix, i.e. $\Delta\mathbf{K} = 0$. Then for this problem it holds that

$$\|\mathbf{C}^*\| = \|\mathbf{K}_{01}\mathbf{K}_{00}^{-1}\mathbf{m}_{11}\mathbf{K}_{00}^{-1}\mathbf{K}_{10}\| \le \frac{\|\mathbf{K}_{01}\|\;\|\mathbf{K}_{10}\|}{\|\mathbf{K}_{11}\|^2}\|\mathbf{m}_{00}\| < \|\mathbf{m}_{00}\| < \varepsilon\,. \tag{10.15}$$

Nearly the same error estimate can be obtained in the problem where the reduced mass matrix omits non-diagonal terms, so instead of the mass matrix from (10.10) the following is used

$$\mathbf{M}^{**} = \mathrm{diag}\left(\mathbf{M}_{11} - \mathbf{K}_{01}\mathbf{K}_{00}^{-1}\mathbf{m}_{11}\mathbf{K}_{00}^{-1}\mathbf{K}_{10}\right). \tag{10.16}$$

In this system the mass matrix is diagonal and thus convenient for the purpose of computation.

When solving dynamical problems, one does not have to consider all natural oscillation modes found because many of those are not actually excited by a particular external action. The issue of correct specification of a needed number of modes to be taken into account is solved generally by trial and error because an a priori estimate is too hard to get. Though, there are cases when design codes that regulate the analysis under a particular dynamic action recommend a concrete number of natural oscillation modes to be taken into account. For example, SNiP 2.01.07–85 limits the frequency range to be allowed for in the analysis of structures under wind gusts, and SNiP II–7–81requires the use of at least 10 oscillation modes in the seismic analysis of concrete dams while at least 15 modes for dams built of soil materials. The U.S. regulations are more consistent: those require that the sum of generalized masses by natural oscillation modes taken into account be not less than 90% of the total mass of the respective system [36].

Complying with this requirement reduces the risk of a situation when a very flexible part of a structure (a “whip”), that determines the lowest natural modes by its practically isolated oscillations, is taken into account while oscillations of the rest of the structure are ignored.

10.2 Integrating Motion Equations

In Section 3.7 we used the following expansion when considering motions of a mechanical system having a finite number of dynamical degrees of freedom:

$$\mathbf{u}(t) = \sum_{i=1}^{n} y_i(t)\,\mathbf{\Psi}_i = \mathbf{\Phi}\mathbf{y}(t)\,, \tag{10.17}$$

where $\mathbf{\Phi}$ is a matrix with its columns representing vectors of natural oscillations of the system $\mathbf{\Psi}_i$ ($i = 1, \ldots, n$).

The linear transformation (10.17) let us reduce the problem of analyzing a system with a finite number of dynamical degrees of freedom to an analysis of independent motions of a set of single-mass systems. The equation of motion of each single-mass system was supplemented by an additional term to describe a viscous resistance to the motion, one proportional to the velocity:

$$M_i \frac{d^2 y_i}{dt^2} + C_i \frac{dy_i}{dt} + \omega_i^2 M_i y_i = P_i(t) \qquad (i = 1, \dots, n). \tag{10.18}$$

An attentive reader might have noticed (good if he has) a kind of a trick here. Indeed, first we suggest to transform general equations of motion that do not allow for any resistance forces in such manner as to change from the original vector of generalized coordinates $\mathbf{u}(t)$ to so-called *principal coordinates* $\mathbf{y}(t)$ using the expansion of (10.17). Second, every equation of the already split system is supplemented by an independent resistance force[1]. In many cases this seemingly illegal step is quite feasible and can be justified by a lack of reliable source information regarding the forces of resistance to the motion of a system. In these circumstances almost any reasonable idea is equally applicable, including a simplifying hypothesis that viscous resistance forces can be introduced independently into each of the equations (10.18). Though, in cases when a structure is composed of inhomogeneous materials with drastically different dissipative properties, and also when there are constructive dampers in the structure, the introduction of the independent resistance forces into the equations (10.18) leads to noticeable errors in the analysis results. This effect is especially clear if the external actions are nearly resonant and the damping affects much the system's oscillation modes. Under these circumstances one has to involve a more careful formulation of the problem making it necessary to employ a numerical integration technique.

In a general case, the viscous resistance is to be introduced into the original motion equations (3.170) so the equations become

$$\mathbf{M}\ddot{\mathbf{u}} + \mathbf{C}\dot{\mathbf{u}} + \mathbf{K}\mathbf{u} = \mathbf{p}(t) \tag{10.19}$$

with a symmetric matrix of dissipative coefficients $\mathbf{C} = \|[c_{ij}]\|$ $(i, j = 1,\dots, n)$.

The theory of quadratic forms states [14] that there always exists a linear transformation of coordinates such that it reduces two symmetric matrices of the same order to a diagonal form simultaneously if at least one of the matrices is positive definite (in our case it is **M**). Though, it cannot be done in a general case with three matrices such as **K**, **M**, and **C** here. It is still possible in a particular case when the damping matrix **C** can be represented as a linear combination of the two other matrices. Let's assume, for example,

$$\mathbf{C} = a_M\mathbf{M} + a_K\mathbf{K}, \tag{10.20}$$

which describes the case of a so-called *proportional damping*. Then the equation

---

[1] Sometimes these are referred to as *physical coordinates* $\mathbf{u}(t)$ and *normal coordinates* $\mathbf{y}(t)$. The most frequently used designation for the principal coordinate vector is $\mathbf{q}(t)$, though this is not a canon. We break the tradition only because the symbol $\mathbf{q}$ was used by us earlier in Chapter 3 for a different purpose.

(10.19) will become

$$\mathbf{M}(\ddot{\mathbf{u}} + a_M \dot{\mathbf{u}}) + \mathbf{K}(\mathbf{u} + a_K \dot{\mathbf{u}}) = \mathbf{p}(t) \ , \tag{10.21}$$

and the transition to principal coordinates according to (10.17) yields a split system of equations again:

$$M_i \frac{d^2 y_i}{dt^2} + M_i\,(a_M + \omega_i^2\, a_K) \frac{dy_i}{dt} + \omega_i^2\, M_i\, y_i = P_i(t) \quad (i = 1, \dots, n)\,, \tag{10.22}$$

which is absolutely the same as (10.18) if we assume there

$$C_i = (a_M + \omega_i^2\, a_K)\, M_i\,. \tag{10.23}$$

Now let's turn to a more general case when the assumption of a proportional damping is too rough and cannot serve. Then the equations (10.19) cannot be solved by an expansion by natural oscillation modes, and a numerical integration should be involved.

There is another motivation for using the numerical integration procedures. When the external disturbance is a pretty complex function of time and is specified in a limited interval of time, the engineer can be interested with the behavior of the system in a transient process. Then, even with a proportional damping or without any, the numerical integration should be applied.

The main idea used by the numerical integration technique is surprisingly simple and easy to formalize algorithmically. An interval of time $[0,T]$ at which the engineer supposedly needs to know the behavior of a mechanical system is divided into relatively short sub-intervals Δt (usually assumed equal to one another to simplify the integration procedure). At each of the sub-intervals from an achieved moment t to $t+\Delta t$, derivatives included in the equation (10.19) are replaced by approximate difference relationships. Using the information of achieved displacements, velocities and accelerations of all masses of the system at the moment of time t and the difference analogue of the equations (10.19), one gets new values of the displacements, velocities and accelerations referring to the next moment of time $t + \Delta t$ which now become initial conditions for the next step of integration etc. This process can be interpreted mechanically as a stepwise satisfaction of the dynamic equilibrium equations expressed in increments with accumulating the resulted displacements and their first/second order derivatives.

There are a great deal of methods to build the numerical integration procedure for solving the equations (10.19) which are different, in essence, only in the way the difference relationships are introduced into the equations [3]. These methods are usually classified into *explicit* and *implicit* integration schemes. If a retarded (at a previous moment of time) dynamical equilibrium condition is used to determine the displacements $\mathbf{u}_{t+\Delta t} = \mathbf{u}(t + \Delta t)$, that is

$$\mathbf{M}\ddot{\mathbf{u}}_t + \mathbf{C}\dot{\mathbf{u}}_t + \mathbf{K}\mathbf{u}_t = \mathbf{p}_t\,, \tag{10.24}$$

then it is an explicit integration scheme. Otherwise, if the determination of the displacements $\mathbf{u}_{t+\Delta t}$ is based on an anticipating dynamical equilibrium condition at the interval $[t, t+\Delta t]$

$$\mathbf{M}\ddot{\mathbf{u}}_{t+\Delta t} + \mathbf{C}\dot{\mathbf{u}}_{t+\Delta t} + \mathbf{K}\mathbf{u}_{t+\Delta t} = \mathbf{p}_{t+\Delta t} , \tag{10.25}$$

then implicit schemes are involved. In (10.24) and (10.25) subscripts denote fixed moments of time to which the respective values refer.

The advantage of the explicit integration schemes is that they require less computing at each step of integration. Though, the practical application of the explicit schemes, as also proved by a theoretical analysis [3], shows that schemes of this kind suffer from an unpleasant disease: they accumulate the computational error too quickly[2]. Therefore contemporary industry-oriented computation software products use implicit schemes most widely. Of those we would like to distinguish two most popular ones: a β-method by Newmark and a ϑ-method by Wilson[3].

According to Newmark's method, the following approximations, depending on β parametrically, are to be used:

$$\ddot{\mathbf{u}}_{t+\Delta t} = \frac{1}{\beta \Delta t^2}\left[\mathbf{u}_{t+\Delta t} - \mathbf{u}_t - \Delta t\,\dot{\mathbf{u}}_t\right] - \frac{1/2-\beta}{\beta}\ddot{\mathbf{u}}_t ,$$

$$\dot{\mathbf{u}}_{t+\Delta t} = \dot{\mathbf{u}}_t + \frac{1}{2}\Delta t\left[\ddot{\mathbf{u}}_{t+\Delta t} + \ddot{\mathbf{u}}_t\right] ,$$

$$\mathbf{u}_{t+\Delta t} = \mathbf{u}_t + \Delta t\,\dot{\mathbf{u}}_t + (\Delta t)^2\left[(1/2-\beta)\ddot{\mathbf{u}}_t + \beta\ddot{\mathbf{u}}_{t+\Delta t}\right] . \tag{10.26}$$

Newmark's method includes the following schemes as particular cases depending on a value of the parameter β:

- $\beta = 0$ — a Euler's integration;
- $\beta = 1/6$ — a linearly accelerated integration;
- $\beta = 1/4$ — a constant average acceleration scheme;
- $\beta = 1/2$ — a scheme with central differences;
- $\beta = 1$ — a scheme with inverse differences.

Newmark proved the unconditional stability of his integration scheme at $\beta = 1/4$ which corresponds to the case of a constant average acceleration at each of the time intervals Δt.

Wilson's method assumes a linear dependence of the accelerations at the time interval $[t, t + \vartheta\Delta t]$ where $\vartheta \geq 1$ which gives the following approximations

---

[2] Experts in computing know well such concepts as conditionally stable integration schemes and unconditionally stable ones. These concepts relate to an accumulation of errors during the numerical integration. We are not going to discuss these issues here, referring the interested reader to other sources on the subject [3].

[3] We have decided not to use the original and already traditional designation of Wilson's θ-method in favor of another possible form ϑ of the same Greek letter "theta" in order to avoid a confusion of designations because the θ character in this chapter of the book denotes the frequency of an external harmonic action.

$$\ddot{\mathbf{u}}_{t+\vartheta\cdot\Delta t} = \frac{6}{\vartheta^2\cdot\Delta t^2}\left[\mathbf{u}_{t+\vartheta\cdot\Delta t} - \mathbf{u}_t\right] - \frac{6}{\vartheta\cdot\Delta t}\dot{\mathbf{u}}_t - 2\ddot{\mathbf{u}}_t \;,$$

$$\dot{\mathbf{u}}_{t+\vartheta\cdot\Delta t} = \dot{\mathbf{u}}_t + \frac{3}{\vartheta\cdot\Delta t}\left[\mathbf{u}_{t+\vartheta\cdot\Delta t} - \mathbf{u}_t\right] - 2\dot{\mathbf{u}}_t - \frac{\vartheta}{2}\Delta t\,\ddot{\mathbf{u}}_t \;,$$

$$\mathbf{u}_{t+\vartheta\cdot\Delta t} = \mathbf{u}_t + \vartheta\cdot\Delta t\,\dot{\mathbf{u}}_t + \frac{\vartheta^2\cdot\Delta t^2}{6}\left[\ddot{\mathbf{u}}_{t+\vartheta\cdot\Delta t} + 2\ddot{\mathbf{u}}_t\right] . \qquad (10.27)$$

An analysis of Wilson's scheme has shown its unconditional stability when the parameter $\vartheta \geq 1.37$. For practical purposes one chooses $\vartheta = 1.4$, as a rule.

From the viewpoint of a software user, the following critical decisions need to be made before performing the analysis itself. First of all, one needs to describe a history of a dynamical load, in other words, to specify the vector function of time $\mathbf{p}(t)$ in one way or the other. Next, one should get fully familiar with integration schemes available in the software environment, choose one and set a parameter of the scheme. Finally, one needs to assign the integration step Δt. For example, the **GTSTRUDL** software provides its users with both a flexible choice of an integration scheme for which its parameter can be assigned (by default if desired) and the possibility to divide the time interval into a sequence of sub-intervals with different integration steps at each of the sub-intervals.

Unconditionally stable integration schemes allow one to choose the step Δt by considerations of accuracy solely, without being afraid of a possible computing error accumulation. It is assumed commonly that satisfactory results can be obtained if this condition holds:

$$\Delta t \,/\, T_k \,\leq\, 0{,}01 \;, \qquad (10.28)$$

where T_k is the period of a characteristic k^{th} natural oscillation mode of a non-damped structure. The oscillation mode (k^{th} one) is considered to be a characteristic one if there is no need to take into account all following modes from $(k+1)^{\text{th}}$ on.

It is natural that the choice of the characteristic mode's ordinal number depends on a particular external action. Also, it should be clear that this is only a qualitative estimate that might need revising after results of calculations. If a loading history is relatively short, then a fairly rough time step can be enough. For example, U.S. regulations on nuclear plant design recommend choosing the step for a seismic shock analysis such that its halving should not make the result change by more than 10%, which is achieved usually by the step of about $(0.08 \ldots 0.10)T_k$.

To conclude this section where the problem of the numerical integration of motion equations has been just slightly outlined, we would like to emphasize that a great deal of computing is needed to implement these procedures. Therefore, each time one wants to perform a numerical integration to analyze the dynamical response of a structure one should make sure there is a sore need to do it.

10.3 Forced Oscillations under a Harmonic Action

Standards issued in the former USSR (such as [23, 24]) for the design of buildings, foundations under machines subjected to dynamic loads, other structural constructions, regulate the use of E.S. Sorokin's assumption how to allow for the internal resistance in a material when such structures are analyzed for a dynamic action in nearly resonant modes.

Without paying much attention to advantages and drawbacks of E.S. Sorokin's hypothesis (see [22] for examples), we concentrate here on computational aspects related to this issue. According to a methodology developed by Sorokin [31], an internal friction in a linearly oscillating system can be taken into account by assigning a complex elasticity modulus to the system's material. Equations of the oscillation are to be composed for the respective perfectly elastic system but in a complex form in which the elasticity modulus of the material should be formally multiplied by the coefficient $(1 + i\gamma)$, γ being a so-called coefficient of inelastic resistance of the material, $i = \sqrt{-1}$ an imaginary unit.

Suppose the displacements of the mechanical system included in the motion equations to be complex values, too. Then we have a complex equation of the mechanical system's motion. The real part of this equation that corresponds to the real part of the complex disturbing force, will be the solution of the problem posed.

Consider the case of an oscillating mechanical system with a finite number of dynamical degrees of freedom. First, we assume for simplicity that the structure is homogeneous. More exactly, we suppose that the non-elastic resistance coefficient γ is constant and the same in all members and parts of the structure.

After denoting the number of dynamical degrees of freedom as n, let's write the equations of motion of the linear system omitting the attenuation. We have

$$\mathbf{M}\ddot{\mathbf{u}} + \mathbf{K}\mathbf{u} = \mathbf{P}(t)\,. \tag{10.29}$$

Here $\mathbf{u}$ is a vector of displacements in the directions of dynamical degrees of freedom, $\mathbf{M}$ is a matrix of inertial properties of the system (a mass matrix), $\mathbf{K}$ is a stiffness matrix of the system that corresponds to the selected dynamical degrees of freedom. It's clear the left part of the equation (10.29) is a sum, by each dynamical degree of freedom, of an inertia force and an elastic restoring force which is to be equated to the external disturbing load $\mathbf{P}$ that changes with time, according to Dalembert's principle. This is how things go with a dynamical action of a load upon a perfectly elastic linear conservative system.

Components of the stiffness matrix $\mathbf{K}$ at a simultaneous and proportional change of the elasticity moduli of all structure's elements change proportionally to a common factor, a coefficient at the elasticity modulus. In the case being considered, this common factor is a complex value $(1 + i\gamma)$. Thus the equation (10.29) can be rewritten as follows after involving E.S. Sorokin's assumption of internal resistance simulation

$$\mathbf{M}\ddot{\mathbf{w}} + (1 + i\gamma)\mathbf{K}\mathbf{w} = \mathbf{F}(t)\,, \tag{10.30}$$

where $\mathbf{w}$ is a desirable vector of complex displacements, $\mathbf{F}(t)$ is a vector of complex external forces.

Both of these two complex vectors can be split into their real and complex parts

$$\mathbf{w} = \mathbf{u} + i\mathbf{v}\,, \qquad \mathbf{F} = \mathbf{P} + i\mathbf{Q}\,. \tag{10.31}$$

If the external action is harmonic, the force $\mathbf{P}$ can be written as follows, with its given frequency θ and its amplitude value $\mathbf{P}_0$

$$\mathbf{P}(t) = \mathbf{P}_0 \cos\theta t\,, \tag{10.32}$$

and, following E.S. Sorokin's method, the imaginary part of the external disturbance is assumed to change in time according to the same law but with its phase shifted by the angle $\pi/2$. In other words,

$$\mathbf{Q}(t) = \mathbf{P}_0 \sin\theta t\,. \tag{10.33}$$

The exponential form for the complex external load $\mathbf{F}$ is

$$\mathbf{F} = \mathbf{P}_0\, e^{i\theta t}\,. \tag{10.34}$$

In the case of forced steady-state oscillations, the solution of the vector differential equation (10.30) is assumed to be

$$\mathbf{w} = \mathbf{Z}\, e^{i\theta t}\,, \tag{10.35}$$

where $\mathbf{Z}$ is a vector which we will call the *vector of complex amplitudes* of the desirable displacements. This vector can also be represented by an expansion into its real and imaginary parts

$$\mathbf{Z} = \mathbf{X} + i\mathbf{Y}\,. \tag{10.36}$$

After substituting (10.34) and (10.35) to the differential equation (10.30) we will have a system of linear algebraic equations with respect to the unknown vector of complex amplitudes $\mathbf{Z}$

$$[(1 + i\gamma)\mathbf{K} - \theta^2\mathbf{M}]\mathbf{Z} = \mathbf{P}_0\,. \tag{10.37}$$

Now, separating the real and imaginary parts of the vector equality (10.37) gives us a system of equations with respect to the real $\mathbf{X}$ and imaginary $\mathbf{Y}$ parts of the complex vector $\mathbf{Z}$

$$\left\|\begin{bmatrix} (\mathbf{K} - \theta^2\mathbf{M}) & -\gamma\mathbf{K} \\ -\gamma\mathbf{K} & -(\mathbf{K} - \theta^2\mathbf{M}) \end{bmatrix}\right\| \cdot \left\|\begin{bmatrix} \mathbf{X} \\ \mathbf{Y} \end{bmatrix}\right\| = \left\|\begin{bmatrix} \mathbf{P}_0 \\ \mathbf{0} \end{bmatrix}\right\|\,. \tag{10.38}$$

As we are interested with resonant or nearly resonant modes of the structure's response, the matrix $(\mathbf{K} - \theta^2\mathbf{M})$ should be degenerate or nearly that. In a resonant mode, the disturbance frequency is equal to one of natural frequencies of the linear system ω_k $(k = 1,\ldots, n)$ determined by the secular equation

$$\det(\mathbf{K} - \omega^2\mathbf{M}) = 0\,. \tag{10.39}$$

We assume that the set of n dynamical degrees of freedom does not contain displacements of the structure as a whole rigid body. This implies that the spectrum of the eigenvalues of the matrix **K** does not include zero values (moreover, all eigenvalues of the matrix **K** will be strictly positive), meaning the matrix **K** itself will be non-degenerate and thus invertible

$$\mathbf{D} = \mathbf{K}^{-1} . \tag{10.40}$$

The **D** matrix is called a *compliance matrix* of the system.

Using the second equation of the matrix-vector system (10.38), we can express the vector **X** via the vector **Y**, namely

$$\mathbf{X} = -\gamma^{-1}\mathbf{D}(\mathbf{K} - \theta^2\mathbf{M})\mathbf{Y} . \tag{10.41}$$

Then, substituting (10.41) to the first equation of the system (10.38) will give us

$$\mathbf{Y} = -[\gamma^2\mathbf{K} + (\mathbf{K} - \theta^2\mathbf{M})\mathbf{D}(\mathbf{K} - \theta^2\mathbf{M})]^{-1}\gamma\mathbf{P}_0 , \tag{10.42}$$

and returning again to (10.41), we find finally

$$\mathbf{X} = \mathbf{D}(\mathbf{K} - \theta^2\mathbf{M})[\gamma^2\mathbf{K} + (\mathbf{K} - \theta^2\mathbf{M})\mathbf{D}(\mathbf{K} - \theta^2\mathbf{M})]^{-1}\mathbf{P}_0 . \tag{10.43}$$

To ensure the applicability of the formulae (10.42) and (10.43), one should make sure that the matrix in brackets in the right part of (10.43)

$$\gamma^2\mathbf{K} + (\mathbf{K} - \theta^2\mathbf{M})\mathbf{D}(\mathbf{K} - \theta^2\mathbf{M}) \tag{10.44}$$

does not degenerate at any θ if only $\gamma \neq 0$. To see this, first we note that under the assumption of the structure's being restrained from rigid-body displacement the stiffness matrix **K**, together with its inverse compliance matrix **D**, is not only non-degenerate but even positive definite. Then, due to the symmetry of the matrices **K** and **M**, the expression of the second term in (10.44) can be rewritten as

$$(\mathbf{K} - \theta^2\mathbf{M})^{\mathrm{T}}\mathbf{D}(\mathbf{K} - \theta^2\mathbf{M}) , \tag{10.45}$$

wherefrom the positive half-definiteness of the matrix (10.45) follows (all its eigenvalues are non-negative). As the sum of the positive definite matrix $\gamma^2\mathbf{K}$ and the positive half-definite matrix $(\mathbf{K} - \theta^2\mathbf{M})^{\mathrm{T}}\mathbf{D}(\mathbf{K} - \theta^2\mathbf{M})$ is assuredly a positive definite matrix, hence the admissibility of the inversion in the formulae (10.42) and (10.43).

One should not think the real part **X** of the complex amplitude vector **Z** and the vector of amplitudes of the real part **u** of the complex displacement vector **w** are the same (the same note holds also for the imaginary parts of the same vectors). To see this, we derive from (10.35) and (10.36) that

$$\mathbf{w} = (\mathbf{X} + i\mathbf{Y})e^{i\theta t} = (\mathbf{X}\cos\theta t - \mathbf{Y}\sin\theta t) + i(\mathbf{Y}\cos\theta t + \mathbf{X}\sin\theta t) , \tag{10.46}$$

consequently

$$\mathbf{u} = \mathbf{X}\cos\theta t - \mathbf{Y}\sin\theta t , \qquad \mathbf{v} = \mathbf{Y}\cos\theta t + \mathbf{X}\sin\theta t . \tag{10.47}$$

Let's introduce a vector of amplitudes **A** and a vector of phase shifts **μ** the components of which we determine via components of the vectors **X** and **Y** by the formulae

$$A_i = \sqrt{X_i^2 + Y_i^2}\,, \qquad \cos\mu_i = X_i / A_i\,, \qquad \sin\mu_i = Y_i / A_i\,. \tag{10.48}$$

Now, using well-known trigonometric formulae of sine and cosine of a sum of angles helps us derive this from (10.47) and (10.48):

$$\mathbf{u} = \left\| \begin{bmatrix} A_1\cos(\theta t + \mu_1) \\ A_2\cos(\theta t + \mu_2) \\ \vdots \\ A_n\cos(\theta t + \mu_n) \end{bmatrix} \right\|, \qquad \mathbf{v} = \left\| \begin{bmatrix} A_1\sin(\theta t + \mu_1) \\ A_2\sin(\theta t + \mu_2) \\ \vdots \\ A_n\sin(\theta t + \mu_n) \end{bmatrix} \right\|. \tag{10.49}$$

So we can see that the respective components of the real and imaginary parts of the displacement vector coincide by their amplitudes and are shifted by phase with respect to each other by the angle of $\pi/2$.

Particularly, in the case of a single-mass system (or exactly for a system having one dynamical degree of freedom, that is, at $n = 1$), we derive from (10.42) and (10.43) that

$$X = \frac{P_0}{k}\,\frac{1 - \theta^2/\omega^2}{(1 - \theta^2/\omega^2)^2 + \gamma^2}\,, \qquad Y = -\frac{P_0}{k}\,\frac{\gamma}{(1 - \theta^2/\omega^2)^2 + \gamma^2}\,, \tag{10.50}$$

where we take the fact into account that the natural frequency ω is calculated by the formula $\omega^2 = k/m$ in a single-mass system with its inertial characteristic m and the stiffness of the system in the direction of the mass's movement k. Now using the formulae (10.48), let us find the amplitude value A of the mass's displacements

$$A = \frac{P_0}{k}\,\frac{1}{\sqrt{(1 - \theta^2/\omega^2)^2 + \gamma^2}}\,, \tag{10.51}$$

that coincides with the known solution.

The formulae (10.42), (10.43) and (10.48) determine the solution of the problem of finding the dynamical displacements of masses in a structure, assuming that the system experiences steady-state oscillations, has a finite number of dynamical degrees of freedom, and internal friction forces in its material are taken into account.

Note that the book by R.A. Reznikov [29] presents an algorithm intended to solve the same problem but based on expanding the desirable displacements and the external action by natural oscillation modes of the respective perfectly elastic system with no internal friction. This requires a substantial amount of additional computing: R.A. Reznikov's algorithm requires the full eigenvalue/eigenvector problem to be solved beforehand for a matrix of the order n. At the same time, the algorithm presented above replaces this laborious (and insidious) operation by the operation of inverting the matrix (10.44) of the same order n.

Having determined the displacements of the structure's masses does not complete the solution because further one has to find all the rest of components of the stress/strain distribution in the structure that present an interest for an engineer. With this purpose we use the following chain of trigonometric transformations

$$a\cos(\theta t + \alpha) + b\cos(\theta t + \beta) =$$

$$= a(\cos\theta t\cos\alpha - \sin\theta t\sin\alpha) + b(\cos\theta t\cos\beta - \sin\theta t\sin\beta) =$$

$$= (a\cos\alpha + b\cos\beta)\cos\theta t - (a\sin\alpha + b\sin\beta)\sin\theta t = c\cos(\theta t + \varphi)\,, \qquad (10.52)$$

where

$$c = \sqrt{a^2 + b^2 + 2ab\cos(\alpha - \beta)}\,, \qquad (10.53)$$

$$\cos\varphi = (a\cos\alpha + b\cos\beta)\,/\,c, \qquad \sin\varphi = (a\sin\alpha + b\sin\beta)\,/\,c\,. \qquad (10.54)$$

This transformation chain corresponds geometrically to a summation of two vectors **a** and **b** specified in the complex plane by their magnitudes a and b (amplitude values) and phase shift angles α and β (Fig. 10.3).

In the case when we are interested with a sum of a few (say, m) quantities, and for each summand its amplitude-phase representation is known

$$c(t) = a_1\cos(\theta t + \alpha_1) + a_2\cos(\theta t + \alpha_2) + \ldots + a_m\cos(\theta t + \alpha_m)\,, \qquad (10.55)$$

we can use the geometrical interpretation of the trigonometric transformations (10.52) – (10.54) to immediately write the expressions of the amplitude c and the phase α_c

$$c(t) = c\cos(\theta t + \alpha_c)\,, \quad c = \sqrt{\left(\sum_{j=1}^{m} a_j\cos\alpha_j\right)^2 + \left(\sum_{j=1}^{m} a_j\sin\alpha_j\right)^2}\,,$$

$$\cos\alpha_c = \left(\sum_{j=1}^{m} a_j\cos\alpha_j\right)/\,c\,, \quad \sin\alpha_c = \left(\sum_{j=1}^{m} a_j\sin\alpha_j\right)/\,c\,. \qquad (10.56)$$

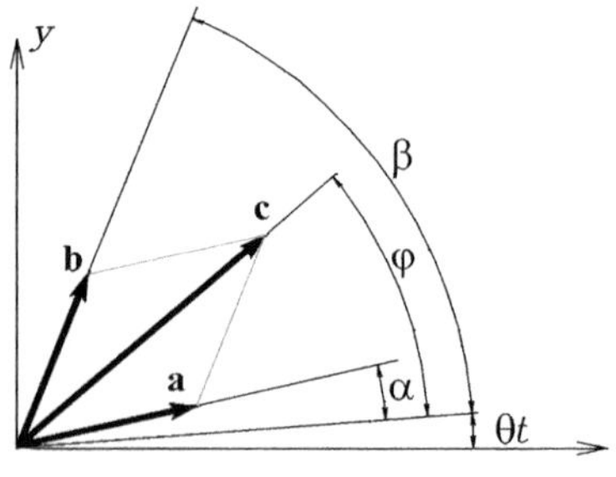

Fig. 10.3. A summation of complex vectors

Suppose we want to know the time history of a kinematical or force factor of the structure: it might be a displacement of a selected point of the structure in a specific direction, a slope angle of a bar's cross-section, a strain component, an internal stress etc. Denoting the factor of interest as H, we can represent it as follows

$$H(t) = \sum_{i=1}^{n} H_i R_i(t) = \mathbf{H}^{\mathrm{T}} \mathbf{R}(t) \,, \tag{10.57}$$

where H_i is the value of the desirable factor in the structure caused by a unit force $\bar{R}_i = 1$ applied at the location and in the direction of ith degree of freedom of the system. Components of the vector $\mathbf{R}$(t) are determined by the condition that the vector should cause the structure's mass displacements $\mathbf{u}$ when assumed to be as if an external load. In other words,

$$\mathbf{R} = \mathbf{K}\mathbf{u} \,. \tag{10.58}$$

Now we use an amplitude-phase representation of the displacement vector $\mathbf{u}$ in the form (10.49) and come to the following linear combination

$$H(t) = a_1\cos(\theta t + \mu_1) + a_2\cos(\theta t + \mu_2) + \ldots + a_n\cos(\theta t + \mu_n) \,, \tag{10.59}$$

where the vector $\mathbf{a}$ with the components $a_1, \ldots, a_n$ is found using the condition

$$\mathbf{a} = \mathbf{K}\mathbf{H} \,. \tag{10.60}$$

As we are interested with the amplitude-phase representation of the factor H, we apply the above stated trigonometric transformation (10.53)— (10.56) to (10.59). The result is

$$H(t) = H_0 \cos(\theta t + \mu_H) \ . \tag{10.61}$$

The ratio of the amplitude H_0 to the sum of amplitudes a_i of its constituent components in the formula (10.59) is called a *phase coincidence coefficient* η of the factor H, that is,

$$\eta = \sqrt{\left(\sum_{j=1}^{n} a_j \cos\mu_j\right)^2 + \left(\sum_{j=1}^{n} a_j \sin\mu_j\right)^2} \Bigg/ \sum_{j=1}^{n} a_j \ . \tag{10.62}$$

It's clear that the phase coincidence coefficient is always less than one. It is equal to one only in the case of in-phase oscillations of all structure's masses, in other words, when the phase shift angles of all masses are the same

$$\mu_1 = \mu_2 = \ldots = \mu_n \ . \tag{10.63}$$

Let's discuss the problem of stress calculation briefly. In this formulation of the dynamical problem it has certain peculiarities.

So, any kind of an internal force S in elements of a system (the bending moment in a bar, the longitudinal force, the chain stress in a plate etc.) is determined by two parameters: an amplitude A_s and a phase shift μ_s. Also,

$$S = A_s \cos(\theta t + \mu_s)\,. \tag{10.64}$$

When calculating the dynamical stresses being linear combinations of two or more stress factors, it is not enough to use only amplitude values of the stresses because those are not in phase, generally. For example, let a cross-section of a bar of a plane bar system experiences a dynamical bending moment M and a longitudinal force N so that

$$M = M_0 \cos(\theta t + \mu_M)\,, \qquad N = N_0 \cos(\theta t + \mu_N) \tag{10.65}$$

with respective amplitude (M_0, N_0) and phase (μ_M, μ_N) parameters. Then the normal stress σ in fibers of this cross-section will be determined by an elementary formula of the theory of strength of materials

$$\sigma = \frac{N}{F} \pm \frac{M}{W} = \frac{N_0}{F} \cos(\theta t + \mu_N) \pm \frac{M_0}{W} \cos(\theta t + \mu_M)\,, \tag{10.66}$$

where F and W is the section area and the section modulus of the fiber, respectively.

The use of the chain of the trigonometric transformations (10.52) — (10.54) with the formula (10.66) yields an expression for the stress σ via the amplitude σ_0 and the phase shift μ_σ

$$\sigma = \sigma_0 \cos(\theta t + \mu_\sigma), \tag{10.67}$$

Obviously, the amplitude value of the stress σ_0 and the phase μ_σ will be

$$\sigma_0 = \sqrt{\left(\frac{N_0}{F}\right)^2 + \left(\frac{M_0}{W}\right)^2 \pm 2\frac{N_0}{F}\frac{M_0}{W}\cos(\mu_N - \mu_M)}\,,$$

$$\cos\mu_\sigma = \left(\frac{N_0}{F}\cos\mu_N \pm \frac{M_0}{W}\cos\mu_M\right) / \sigma_0\,,$$

$$\sin\mu_\sigma = \left(\frac{N_0}{F}\sin\mu_N \pm \frac{M_0}{W}\sin\mu_M\right) / \sigma_0\,. \tag{10.68}$$

In a general case, when the stress (the normal one σ or the tangential stress τ) is a linear combination of a few kinds of stresses, that is,

$$\sigma = a_1 S_1 + a_2 S_2 + \ldots + a_m S_m\,, \qquad S_i = S_{0i} \cos(\theta t + \mu_{Si})\,, \tag{10.69}$$

where a_i are some constants ($i = 1,\ldots, m$), one can use the general case to determine the amplitude σ_0 and phase μ_σ, i.e. the chain of the transformations (10.55)–(10.56).

Now let's turn to the situation when the dynamical stress of our interest cannot be represented as a linear combination of the previously calculated dynamical stresses. The example of this is a problem of calculating the principal stresses σ_1 and σ_2 by known components of the stress tensor in the plane stress theory. In this and similar cases, the dynamical stress needed by an engineer is a nonlinear function $f(S_i)$ of the dynamical stresses S_i, that is,

$$\sigma = a_1 S_1 + a_2 S_2 + a_m S_m\,, \qquad S_i = S_{0i}\cos(\theta t + \mu_{Si})\,, \tag{10.70}$$

In a situation like this, an acceptable approach independent of a particular nonlinear function $f(S_i)$ can be a direct tracking of all values of σ as the time t changes discretely. As all arguments of the function $f(S_i)$ are periodical functions of time t with the same period $T = 2\pi/\theta$, $\sigma(t)$ is a periodical function of the same period, too. It lets us to perform the exhaustive search of values of the time parameter in the formula (10.70) at the interval from zero to T only. In this case it is reasonable to adopt the phase coincidence coefficient for the stress σ as a ratio of the calculated amplitude value of the $\sigma(t)$ function to the value of the same function at a point where its argument achieves its amplitudes.

Obviously, a graphical representation of a stress or kinematical quantity as plots, isolines etc., which engineers are accustomed to, is a mirror of a structure's static state. Meanwhile, under a dynamic action the same stress/displacement quantity (say, in different sections of a bar) possesses different phase properties. Due to this circumstance a proper graphical representation of this quantity in dynamics is better done by animating. Of course, amplitude values of the quantity in question can be still shown statically, that is, as plots or isolines etc. These plots are, actually, envelop curves of the respective animated pictures.

It should be said that there is a plethora of new capabilities here, yet to be implemented by software developers who have this chance to exercise their creativity. For example, animated pictures demonstrating a joint time history of both the direction and the magnitude of principal stresses would be most descriptive for a user to understand the behavior of a structure. There is no doubt highly qualified analyst engineers will appreciate innovations of this kind which improve the attraction and competitiveness of the software products.

Let's address a bit of attention to a practically important situation when the structure's material is inhomogeneous. All that is needed to allow for the heterogeneity is to replace the matrix $\gamma\mathbf{K}$ by the matrix $\mathbf{K}_\gamma$ in the equation (10.38). Evidently, $\mathbf{K}_\gamma$ is a stiffness matrix of the structure composed of elements with their rigidity coefficients multiplied by a γ factor, each element having its own particular factor. Further considerations are similar to the case of a homogeneous structure and should cause any difficulties.

10.3.1 A Model by Gordeyeva

All that was said in the section above is oriented principally at software developers. As for direct users of the software, over 30 years ago S.P. Gordeyeva [16] suggested a scheme of computation intended to obtain solutions of the problem in question while remaining within the scope of capabilities of a standard CAD software.

To validate Gordeyeva's model, we will use the system of linear algebraic equations (10.38) after transforming it to a form convenient for further considerations. We should stipulate here that the application of Gordeyeva's model is essentially limited by two requirements:

- the inertia matrix **M** must be diagonal;
- the non-elastic material resistance coefficient γ must be the same for all elements of the system.

The latter requirement is especially important because it does not allow the application of Gordeyeva's model to structures that include, for example, both ferroconcrete and steel members.

So, let's introduce the following block-diagonal matrix

$$\left\| \begin{bmatrix} -\mathbf{D} & \mathbf{0} \\ \mathbf{0} & -\mathbf{D} \end{bmatrix} \right\| . \tag{10.71}$$

By multiplying both parts of the equation (10.38) by this matrix from the left and considering the relationship (10.40) we obtain

$$\left\| \begin{bmatrix} (\theta^2 \mathbf{DM} - \mathbf{I}) & \gamma\mathbf{I} \\ \gamma\mathbf{I} & (\theta^2 \mathbf{DM} - \mathbf{I}) \end{bmatrix} \right\| \bullet \left\| \begin{bmatrix} \mathbf{X} \\ \mathbf{Y} \end{bmatrix} \right\| = \left\| \begin{bmatrix} -\mathbf{DP}_0 \\ \mathbf{0} \end{bmatrix} \right\| . \tag{10.72}$$

where **I** is a unit square matrix of the order n. Also, we introduce two new vectors of unknowns $\mathbf{J}_x$ and $\mathbf{J}_y$, assuming

$$\mathbf{J}_x = \theta^2 \mathbf{M}\,\mathbf{X}\,, \qquad \mathbf{J}_y = \theta^2 \mathbf{M}\,\mathbf{Y}\,. \tag{10.73}$$

Recall the physical sense of the vectors **X** and **Y**. The vectors $\mathbf{J}_x$ and $\mathbf{J}_y$ introduced in (10.73) can be treated as the respective real and imaginary parts of the complex amplitudes of the inertia force vector.

After substituting the expressions of **X** and **Y** from (10.73) to the equation (10.72) we have

$$\left\| \begin{bmatrix} (\mathbf{D} - \theta^{-2}\mathbf{M}^{-1}) & \gamma\theta^{-2}\mathbf{M}^{-1} \\ \gamma\theta^{-2}\mathbf{M}^{-1} & -(\mathbf{D} - \theta^{-2}\mathbf{M}^{-1}) \end{bmatrix} \right\| \bullet \left\| \begin{bmatrix} \mathbf{J}_x \\ \mathbf{J}_y \end{bmatrix} \right\| = \left\| \begin{bmatrix} -\mathbf{DP}_0 \\ \mathbf{0} \end{bmatrix} \right\| . \tag{10.74}$$

Now let us form a design model which implements what Gordeyeva has suggested. We assume this system to contain two parts called by us a *real* and *imaginary* parts of Gordeyeva's model. The real part will thoroughly duplicate the original mechanical system the oscillations of which are being considered. The imaginary part duplicates it, too, but with rigidities of all its constituent elements replaced by rigidities of opposite signs.

A package of three springs is introduced along the motion direction of each mass (in the direction of each dynamical degree of freedom), as shown in Fig. 10.4 where a circle indicates a node of the model's real part with its mass m_i, and a rectangle designates the respective node of the imaginary part of Gordeyeva's model[4].

---

[4] We hope a funny wordplay "Gordeyeva's node — Gordian knot" won't confuse the reader. It would be too rough to use Alexander's sword here.

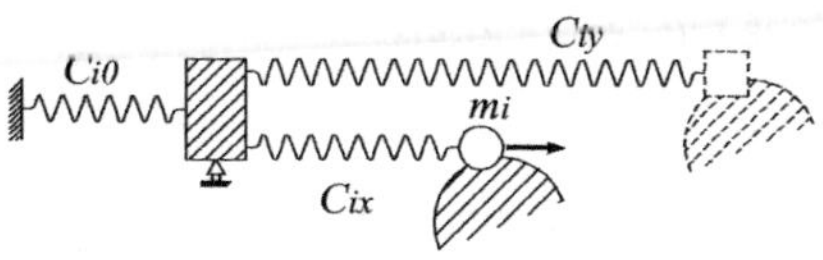

Fig. 10.4. A design model's principle of build

The real part of the system is indicated by a solid line in Fig. 10.4 while the imaginary one is shown by a dashed line. The rigidities of the springs that link the real and imaginary parts of Gordeyeva's model are selected as follows:

$$c_{ix} = -\frac{m_i\theta^2}{1+\gamma}, \qquad c_{iy} = \frac{m_i\theta^2}{1-\gamma}, \qquad c_{i0} = \frac{m_i\theta^2}{\gamma}. \qquad (10.75)$$

Now we are going to formally calculate the shown system that has been built under a static action of the external force $\mathbf{P}_0$. Then, the stress in the springs with the rigidities c_{ix} and c_{iy} will be the same as i^{th} components of the respective vectors $\mathbf{J}_x$ and $\mathbf{J}_y$ determined by the equation system (10.74). To see this, we cut these springs and treat the stresses in those as unknowns. In this way we build a system of canonic equations of the force method which is equivalent to the equation system (10.74).

To illustrate the application of Goreyeva's model, let us take an example of a cantilever column with a concentrated mass m placed at the free end of the cantilever, under a harmonic load $P(t) = P_0\cos\theta t$ (Fig. 10.5).

Let EI be the flexural rigidity of the column, l the height of the column. Following Gordeyeva's modeling scheme, we should choose the presented design model as shown in Fig. 10.5 in order to be able to employ an arbitrary static-analysis-oriented software. We use the force method and build its principal system by introducing cuts in the springs 1 and 2. The respective unknown stresses in the springs are denoted as J_x and J_y (Fig. 10.6)[5].

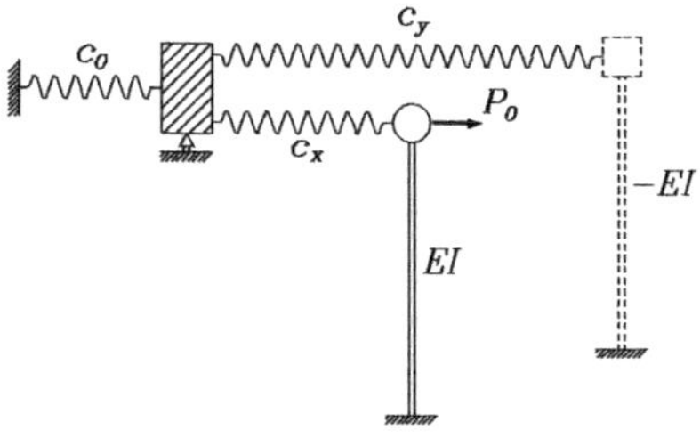

Fig. 10.5. Gordeyeva's model

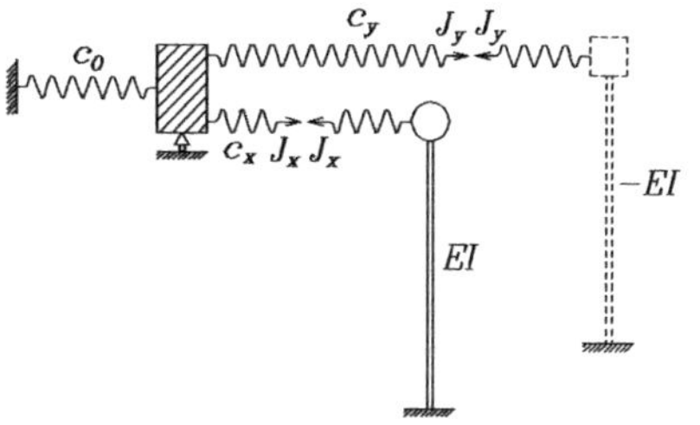

Fig. 10.6. Unknown stresses

[5] We deviate from commonly accepted designations of the force method's unknowns only to avoid a confusion: we have already used the X letter in this paragraph to denote the real part of the complex amplitudes Z. As for Gordeyeva's designations, she uses the respective symbols Z' and Z'' to denote the quantities J_x and J_y.

Now, let's introduce the designations

$$k = 3EI / l^3 , \qquad \omega^2 = k / m , \qquad p = \theta^2 / \omega^2 , \tag{10.76}$$

and write out expressions of the coefficients and the absolute terms in the force method's canonic system

$$\delta_{11} = \frac{1}{k} + \frac{1}{c_x} + \frac{1}{c_0} = \frac{1}{k} - \frac{1}{m\theta^2} = -\frac{1-p}{kp}, \quad \delta_{12} = \delta_{21} = \frac{1}{c_0} = \frac{\gamma}{kp},$$

$$\delta_{22} = -\frac{1}{k} + \frac{1}{c_y} + \frac{1}{c_0} = -\frac{1}{k} + \frac{1}{m\theta^2} = \frac{1-p}{kp}, \quad \Delta_{1P} = -\frac{P_0}{k}, \quad \Delta_{2P} = 0, \tag{10.77}$$

$$\delta_{11}J_x + \delta_{12}J_y + \Delta_{1P} = 0 \ ,$$

$$\delta_{21}J_x + \delta_{22}J_y + \Delta_{2P} = 0 \ . \tag{10.78}$$

Deriving the solution of the equation system (10.78) by Cramer's rule, we have

$$J_x = \frac{\Delta_{1P}\delta_{22} - \Delta_{2P}\delta_{12}}{\delta_{11}\delta_{22} - \delta_{12}\delta_{21}}, \qquad J_y = \frac{\Delta_{2P}\delta_{11} - \Delta_{1P}\delta_{21}}{\delta_{11}\delta_{22} - \delta_{12}\delta_{21}}, \tag{10.79}$$

which yields this after substitutions by the formulae (10.77) and elementary transformations:

$$J_x = \frac{P_0}{k}\frac{kp(1-p)}{(1-p)^2 + \gamma^2}, \qquad J_y = -\frac{P_0}{k}\frac{kp\gamma}{(1-p)^2 + \gamma^2}. \tag{10.80}$$

The displacements X and Y that correspond to the real and imaginary parts of the original problem's solution will be determined by the formulae (10.73) the form of which for this case is

$$X = \frac{J_x}{kp} = \frac{P_0}{k}\frac{1-p}{(1-p)^2 + \gamma^2}, \qquad Y = \frac{J_y}{kp} = -\frac{P_0}{k}\frac{\gamma}{(1-p)^2 + \gamma^2}, \tag{10.81}$$

being expectedly the same as the solution obtained before — the formulae (10.50).

Note the fact that the displacements of the real and imaginary parts of Gordeyeva's model are not equivalent to the real and imaginary parts of the displacement vectors of the original oscillating system, which is opposite to what one might expect at the first sight. For example, in the cantilever column problem the displacement X_Γ of the column's top for the real part and Y_Γ for the imaginary one of Gordeyeva's model are equal, respectively, to

$$X_\Gamma = (P_0 - J_x) / k = \frac{P_0}{k}\frac{(1-p)(1-2p) + \gamma^2}{(1-p)^2 + \gamma^2},$$

$$Y_\Gamma = J_y / k = -\frac{P_0}{k}\frac{p\gamma}{(1-p)^2 + \gamma^2}, \tag{10.82}$$

which is evidently different from the values of X and Y.

In this regard the recommendation by Gordeyeva that the dynamical stress amplitude should be determined by the formula

$$S = \sqrt{(S')^2 + (S'')^2}\ , \tag{10.83}$$

where S' and S'' are respective stresses in the real and imaginary parts of Gordeyeva's model, seems erroneous. Also, as it has been shown here, in a system with multiple dynamical degrees of freedom the amplitude values of stresses depend on both amplitude values of the masses' displacements and their mutual phase shifts.

It goes without saying that Gordeyeva's model is applicable only in cases when the user has no available software intended for a direct analysis of structures experiencing forced oscillations that allows for an energy loss in the material in accordance with Sorokin's assumption.

10.4 Decrement of Oscillations

The logarithmic decrement characterizes a process of attenuated oscillations in a system. It is equal to the natural logarithm of the ratio of an oscillating quantity's amplitudes measured with an interval of one period. For a simplest oscillator having a mass M, an elasticity coefficient K and a viscous resistance coefficient C the differential equation of its free motion is

$$M\ddot{y} + C\dot{y} + Ky = 0 \tag{10.84}$$

and it has a solution that describes an oscillatory process

$$y = A_0 e^{-\beta t} \cos\left(\bar{\omega} t + \mu\right) \tag{10.85}$$

with an amplitude (an arbitrary constant) A_0 and a phase shift (an arbitrary constant) μ, if the damping factor is lower than the critical value

$$C \le C_0 = 2\sqrt{KM}\ .$$

Here we denote

$$\omega^2 = K/M;\ \ 2\beta = C/M;\ \ \bar{\omega} = \sqrt{\omega^2 - \beta^2}\ .$$

A "period of oscillations" term will conventionally refer to the time $T = 2\pi/\bar{\omega}$, and the dimensionless product $\delta = \beta T$ will be a logarithmic decrement of the oscillations which corresponds to its definition given above.

Let's consider how the total energy of the system E changes in the course of damped oscillations. This energy is equal to a potential energy of the spring plus a kinetic energy of the moving mass

$$\mathsf{E} = \frac{1}{2}Ky^2 + \frac{1}{2}M\dot{y}^2\ . \tag{10.86}$$

Substituting data of the formula (10.84) to this, we have

$$\mathsf{E} = \frac{1}{2} K A_0^2 e^{-2\beta t} \left(1 + \frac{\beta \overline{\omega}}{\omega^2} \sin(2\overline{\omega} t + 2\mu) + \frac{\beta^2}{\omega^2} - 2\frac{\beta^2}{\omega^2} \sin^2(\overline{\omega} t + 2\mu) \right). \tag{10.87}$$

It is easy to check that the ratio of energies of the system E_0 at the moment of time $t = t_0$ and E_T at the moment of time $t = t_0 + T$ (after one period of oscillations ends) does not depend on the moment t_0 and is expressed by the formula

$$\frac{\mathsf{E}_T}{\mathsf{E}_0} = e^{-2\beta T} = e^{-2\delta}. \tag{10.88}$$

Hence, the logarithmic decrement can be also expressed easily via the energy

$$\delta = -\frac{1}{2} \ln \frac{\mathsf{E}_T}{\mathsf{E}_0}. \tag{10.89}$$

If the oscillator is subjected to an external periodic load

$$F = F_0 \cos \theta t = f_0 M \cos \theta t, \tag{10.90}$$

then the solution that corresponds to steady-state oscillations can be represented as

$$y = A \cos(\theta t - \alpha), \tag{10.91}$$

where

$$A = \frac{f_0}{\sqrt{\left(\omega^2 - \theta^2\right)^2 + 4\beta^2 \theta^2}}, \tag{10.92}$$

$$\alpha = \operatorname{arctg} \frac{2\theta\beta}{\omega^2 - \theta^2}. \tag{10.93}$$

The solution (10.91) can be also represented in a different form

$$y = A_s \sin(\theta t) + A_c \cos(\theta t), \tag{10.94}$$

where

$$A_s = \frac{2\beta\theta f_0}{\left(\omega^2 - \theta^2\right)^2 + 4\beta^2\theta^2}, \tag{10.95}$$

$$A_c = \frac{\left(\omega^2 - \theta^2\right) f_0}{\left(\omega^2 - \theta^2\right)^2 + 4\beta^2\theta^2}. \tag{10.96}$$

The constant A_s is called an amplitude of absorption and the constant A_c an amplitude of dispersion.

The energy absorbed by the system in one oscillation cycle can be calculated by the formula

$$\mathsf{E}_s = \pi F_0 A_s = \frac{2\pi M\beta\theta f_0^2}{\left(\omega^2-\theta^2\right)^2+4\beta^2\theta^2}. \tag{10.97}$$

The maximum value of the system's potential energy is

$$\mathsf{E}_y = \frac{1}{2}kA^2 = \frac{1}{2}\frac{M\theta^2 f_0^2}{\left(\omega^2-\theta^2\right)^2+4\beta^2\theta^2}. \tag{10.98}$$

It is easy to check that the logarithmic decrement in the case of steady-state oscillations can be expressed via the ratio of energies by the formula

$$\delta = \frac{\mathsf{E}_s}{2\mathsf{E}_y}. \tag{10.99}$$

The ratio $\psi = \mathsf{E}_s / \mathsf{E}_y$ is often called an *absorption coefficient*. Evidently, $\delta = \psi / 2$.

As one can see in the classical examples given here, the logarithmic decrement of oscillations can be easily expressed via the energy that the system absorbs in one oscillation cycle and the system's potential energy. The result is purer in the case of steady-state forced oscillations.

Using the energy expressions, we can try to construct an algorithm for finding a system with a conventional viscous friction equivalent to a real system by energy. The latter might have a damping mechanism quite different from the viscous friction. Moreover, it may change from one part (node, subsystem) of the system to another because in real systems such properties might vary by whole orders of magnitude.

The idea of this construction follows. Suppose a finite-element system in question, having n degrees of freedom, experiences steady-state oscillations with respect to a given, not necessarily zero, equilibrium position. The oscillation frequency θ, mode and amplitude of the oscillation are known (given). Actually, this case is a transition to a system with one degree of freedom. It is better if the oscillation frequency and mode are natural ones, but it is not a mandatory requirement.

Let an arbitrary deformed state of the system be described by a displacement vector $\|[u_1,..,u_n]\|^{\mathrm{T}}$, the equilibrium position around which the system oscillates be described by a vector $\|[u_1^0,..,u_n^0]\|^{\mathrm{T}}$, and the relative amplitude values of the displacements in this oscillation mode by a vector $\|[a_1^0,..,a_n^0]\|^{\mathrm{T}}$. Then the deformed state of the system at the moment of time t can be described by the formula

$$u_k = u_k^0 + a_k^0\cos(\theta t) \qquad (k=1,..,n). \tag{10.100}$$

If we manage to determine the loss of energy E_s in the system per one oscillation cycle and its elastic potential energy in its maximally deviated state E_y, then we will use the formula (10.99) to easily find the desirable logarithmic decrement of oscillations δ.

Obviously, the elastic energy is accumulated only in finite elements. Let us assume that the loss of energy occurs in finite elements only, too. If this is not the case, we can supplement the system with elements of a special type (dampers) not possessing any elasticity but performing dissipative functions, and thus make this assumption true. So, we should determine E_s and E_y for each particular finite element and sum the results.

It is pretty simple to do with the elastic energy. One should use the expression of a finite element's potential energy via its displacements (such expressions are derived during the development of finite elements) for each element to calculate the latter's energy after its nodes acquire the displacements $\|[a_1^0,..,a_n^0]\|^{\mathrm{T}}$. Having summed the results of the calculation over all elements, we have the elastic potential E_y for the whole system.

It should be noted that the elastic potential is calculated using the amplitude values of the oscillation mode $\|[a_1^0,..,a_n^0]\|^{\mathrm{T}}$ and by no means take into account the position described by the displacements $\|[u_1^0,..,u_n^0]\|^{\mathrm{T}}$ around which the oscillation itself takes place. We do it this way because the displacements $\|[u_1^0,..,u_n^0]\|^{\mathrm{T}}$ have been assumed to satisfy equilibrium conditions, that is, all partial derivatives of the potential energy with respect to the displacements $u_1,..,u_n$ are zero at this point. The potential energy's level at this point is assumed to be zero, too.

The energy loss in an element per one oscillation is a different thing. To obtain proper quantitative data, one has to study the dissipative properties of each particular element type. We suppose a finite element is referred to its own coordinate system, is characterized by m displacements $v_1,..,v_m$, and has a stiffness matrix with elements s_{ij} $(i,\ j = 1,..,m)$. The transition from the displacements of the system in question $u_1,..,u_n$ to the displacements of the finite element in its local coordinate system $v_1,..,v_m$ is performed using linear transformations by the formulae

$$v_i = \sum_{j=1}^{n} c_{ij} u_j \quad (i=1,..,m) \tag{10.101}$$

with known coefficients of the linear relationships c_{ij} $(i=1,..,m;\ j=1,..,n)$.

Now let us discuss simplest types of elements employed most often in design models.

10.4.1 Finite Elements of an Elastic Material

Having no well-founded dependencies, we will assume that each material is characterized by its particular absorption coefficient ψ that does not depend on the type, velocity, or intensity of deformation. This approach can be used with material finite elements widely used in finite-element analysis software. It is valid if the stresses do not exceed the rated strength.

In this case the energy dissipated in the material is described by the formula

$$\mathsf{E}_s = \frac{\psi}{2} \sum_{i=1}^{m} \sum_{j=1}^{m} s_{ij} b_i^0 b_j^0\,, \tag{10.102}$$

where the quantities

$$b_i^0 = \sum_{j=1}^{n} c_{ij} a_j^0 \quad (i=1,..,m) \tag{10.103}$$

are amplitude values of the displacements in a local coordinate system.

The absorption coefficients ψ are determined experimentally, and their values are given by reference manuals.

10.4.2 A Dry Friction Element

The element in question is a slider pressed against a rough surface so that the force that appears when it moves is equal to F (Fig. 10.7). The element has a single degree of freedom — a displacement v.

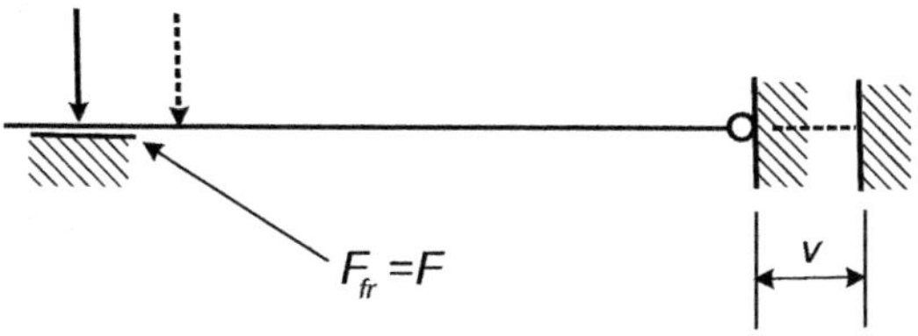

Fig. 10.7. A dry friction element

If the fixed end of the slider oscillates at the amplitude b^0, the loss of energy per one oscillation is described by the formula

$$\mathrm{E}_s = 4b^0 F \ . \tag{10.104}$$

10.4.3 A Viscous Friction Element

Similarly to a dry friction element, we can introduce a viscous friction element. The design of such element can include a cylinder filled by liquid in which a perforated or a non-tight piston moves (Fig. 10.8).

The element has a single degree of freedom, a displacement v. The viscous friction force in such device is proportional to the velocity v' of the piston's movement and can be expressed as $f = \alpha v'$ where α is a given factor.

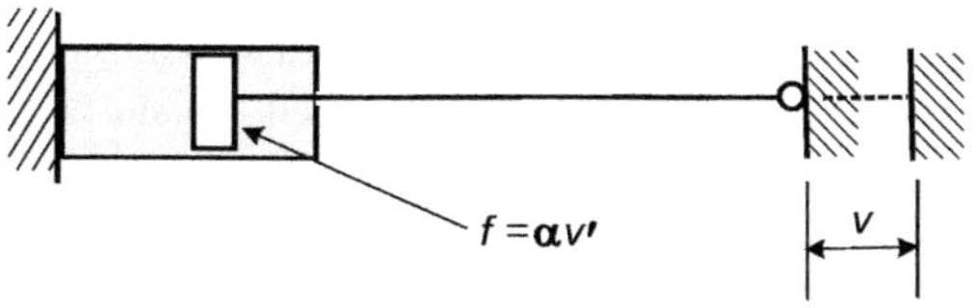

Fig. 10.8. A viscous friction element

If the piston moves according to the law

$$v = b^0 \cos(\theta t), \qquad (10.105)$$

then the friction force is

$$f = -\alpha\theta b^0 \sin(\theta t) \qquad (10.106)$$

and this energy is dissipated in one oscillation:

$$E_s = \pi\alpha\theta(b^0)^2. \qquad (10.107)$$

10.4.4 A Nonlinear Viscous Friction Element

The book by Y.G. Panovko [25] presents a solution of the problem how to determine an equivalent coefficient of the linear viscous resistance C_{eqv} for an element which has the following viscous friction:

$$F_d = C_n \text{sign}(\dot{y})\,\dot{y}^n. \qquad (10.108)$$

Results of this analysis can be represented as Table 10.2 [32] that presents also data regarding the energy dissipation per one oscillation according to the law $y = y_{\max}\cos(\theta t - \alpha)$.

Table 10.2.

Resistance type	n	Energy dissipation per oscillation E_s	Coefficient of equivalent viscous resistance C_{eqv}
Dry friction	0	$4C_0 y_{\max}$	$\frac{4}{\pi}\frac{C_0}{\theta y_{max}}$
Viscous friction	1	$\pi C_1 \theta y_{max}^2$	C_1
Quadratic	2	$\frac{8}{3} C_2 \theta^2 y_{max}^3$	$\frac{8}{3\pi}\theta y_{max}$
Cubic	3	$\frac{8}{3}\pi C_3 \theta^3 y_{max}^4$	$\frac{3}{4}(\theta y_{max})^2$

10.4.5 Damping by Radiation

Possibly one of principal mechanisms of structural damping is a radiation of the energy into the environment. For example, when a plate vibrates in a fluid it excites waves that cause a lot of the fluid to move. The energy transferred from the structure to the fluid (more generally, to the environment) is lost by the structure. The value of this loss per one cycle of oscillations is determined by the expression

$$E = \pi\theta\,\rho C r^2, \qquad (10.109)$$

where ρ is the density of a medium where waves arise, C is the sound speed in the medium, and r is an amplitude of the vibration. The energy dissipated in one cycle proves to be equal to that for a dry friction with the damping factor $B = \rho C$ in an area of a unit size. This is called a characteristic of a dynamical rigidity (impedance) of the medium.

Even in a low-density medium like air it contributes much to the damping within a structure. If the structure is surrounded by a denser medium such as water, then the radiation damping is more substantial. If the structure stands on a firm bed, the damping by radiation into this bed can be pretty high in their contact points. In the civil engineering it may be the main component of the damping. To illustrate the importance of the point, a non-damped model of a structure was analyzed [18] taking into account the damping by radiation into the bed. It yielded about 10% of the critical damping. The second example included an estimation of the damping magnitude for bridge piers standing in a river and a validation of various damping models. The models included the damping by radiation into the bed, the damping in the material of the piers, and the resistance of water in the river. The only mechanism that proved to be realistic comparing to the measured value of the damping, was the radiation damping. All other mechanisms yielded lower estimates by an order of magnitude. To determine damping factors of natural beds under an oscillating foundation, one can use approximate formulae that establish a relationship of these characteristics and the rigidity of the bed C_z

- $B_z = 0.014F\sqrt{C_z}$ — for vertical displacements (sags);
- $B_x = 0.007F\sqrt{C_z}$ — for horizontal displacements (shifts);
- $B_\varphi = 0.01I\sqrt{C_z}$ — for torsion.

Here F is an area of the foundation's foot, and I is a moment of inertia of this area.

Slightly different recommendations are given by Wilson [38]. He suggests the use of data presented in Table 10.3 and based on a solution for a round die oscillating on an elastic semi-space. This table also contains data regarding the associated masses and the rigidities of the bed.

Table 10.3.

Direction of oscillations	Rigidity	Damping	Associated mass
Vertical	$K = \dfrac{4Gr}{1-\nu}$	$1{,}79\sqrt{K\rho r^3}$	$1{,}50\rho r^3$
Horizontal	$18{,}2Gr\dfrac{1-\nu^2}{(2-\nu)^2}$	$1{,}08\sqrt{K\rho r^3}$	$0{,}28\rho r^3$
Torsion	$2{,}7Gr^3$	$0{,}47\sqrt{K\rho r^3}$	$0{,}28\rho r^3$

Notation: r is the radius of the die; G is the shear modulus; ν is the Poisson ratio; ρ is the density

10.5 Three Resonance Curves

Most engineers are aware of a resonant oscillation mode when an external action's frequency is coincident with a natural frequency and the amplitudes grow drastically. In an idealized system with no energy loss the resonance causes the amplitudes to grow infinitely, and in a real system this growth can be very high, too. Hence an understandable wish to avoid the resonance by setting up a system the natural frequency of which is different from that of the disturbing force.

But, how much different should it be? What is better: the natural frequency greater or less than the disturbance frequency? Answers to these questions can be obtained by analyzing so-called resonance curves.

Considering a simplest system with a single degree of freedom, we have the following equation similar to (3.177)

$$M\ddot{y} + C\dot{y} + Ky = F(t) \tag{10.110}$$

or, in a different form

$$\ddot{y} + 2\omega\zeta\dot{y} + \omega^2 y = F/M, \tag{10.111}$$

where $\omega = \sqrt{K/M}$ is a natural frequency of a non-damped system; $\zeta = C/C_{cr}$ is a relative damping; $C_{cr} = 2\sqrt{KM}$ is a critical damping, that is, such value of the C coefficient at which (and greater) the system returns to its equilibrium without an oscillation after being deviated; $\overline{\omega} = \omega\sqrt{1-\zeta^2}$ is a natural frequency of damped oscillations.

Under a harmonic disturbing force $F = F_0 \sin\theta t$ the system performs harmonic oscillations $y = y_0 \sin(\theta t - \varphi)$ the dimensionless amplitude of which depends on the ratio of the frequencies θ/ω and is determined by the expression

$$\frac{y_0 K}{F_0} = \left[\left(1 - \frac{\theta^2}{\omega^2}\right)^2 + \left(2\zeta\frac{\theta}{\omega}\right)^2\right]. \tag{10.112}$$

This amplitude has its maximum value at $\theta/\omega = \sqrt{1-2\zeta^2}$ when $y_0 K/F_0 = \left[2\zeta\sqrt{1-\zeta^2}\right]^{-1}$. A plot of the dimensionless amplitude vs. the ratio θ/ω is presented in Fig. 10.9.

It is useful to notice that from the standpoint of the oscillation amplitude reduction it is better to increase the natural frequency to make it greater than the disturbance frequency because the domain $\theta/\omega > 1$ presents smaller amplitudes.

A special case of excitation is shown in Fig. 10.10 where the oscillations are caused by an unbalanced mass m which has an eccentricity e and rotates at the angular speed θ. The disturbing force is $p = em\theta^2 \sin\theta t$, and the dimensionless amplitude

$$\frac{y_0 M}{em} = (\theta/\omega)^2\left[\left(1 - \frac{\theta^2}{\omega^2}\right)^2 + \left(2\zeta\frac{\theta}{\omega}\right)^2\right], \tag{10.113}$$

has the maximum value of $y_0 M/(em) = (1-2\zeta^2)/\left(2\zeta\sqrt{1-\zeta^2}\right)$. A plot of this function is presented in Fig. 10.10. It shows that here the recommendation how to avoid the resonance best is directly opposite to the previous case.

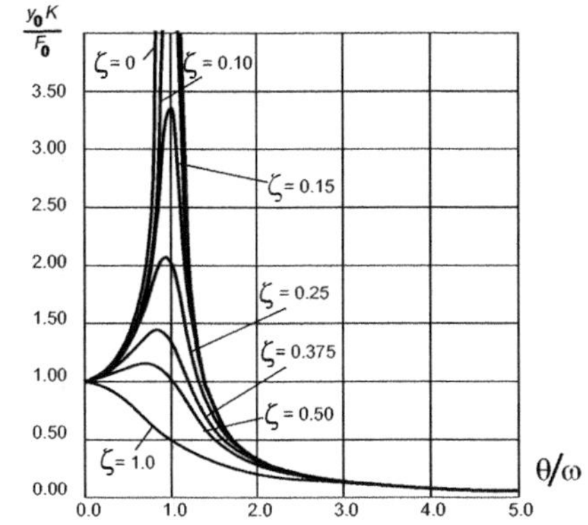

Fig. 10.9. A resonance curve for a harmonic excitation by a force

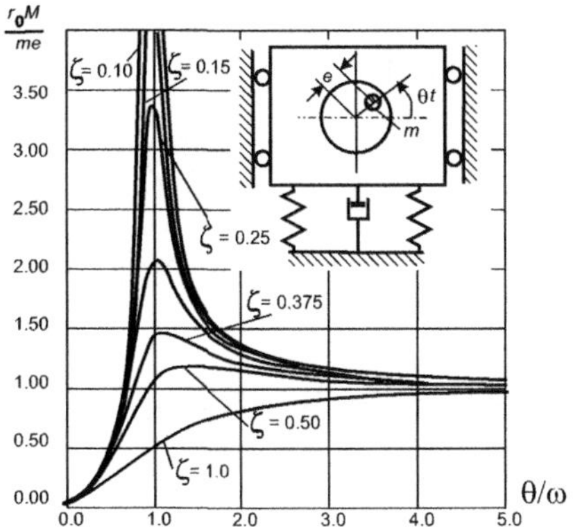

Fig. 10.10. A resonance curve for an excitation by an unbalanced mass

Finally, another case of a harmonic excitation can be considered: the foundation moves according to the law $z = z_0 \sin \theta t$. The dimensionless amplitude of the displacements is described by the expression

$$\frac{y_0}{z_0} = \sqrt{\frac{1+(2\zeta\theta/\omega)^2}{\left(1-\theta^2/\omega^2\right)^2 + (2\zeta\theta/\omega)^2}}\,. \tag{10.114}$$

A plot of this function is shown in Fig. 10.11. It resembles the plot in Fig. 10.9 but here all curves cross the point with the abscissa $\theta/\omega = 0.707$.

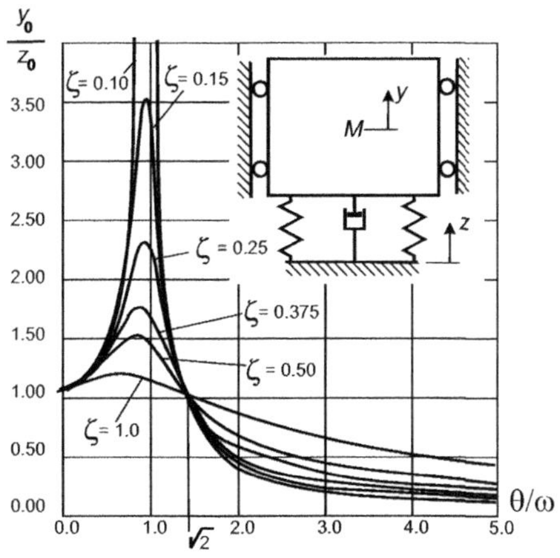

Fig. 10.11. A resonance curve for the case of a harmonically excited foundation

The scheme in Fig. 10.11 is used in vibration insulation systems. The ratio y_0 / z_0 characterizes the quality of the vibration control to some extent, so the plot shows it is efficient only at frequencies $\theta/\omega_0 > 0.707$, and the presence of attenuation in it is generally undesirable.

We should note that the stress in the elastic link is determined by the relative displacement $d = y - z$ for which the dimensionless amplitude is

$$\frac{d_0}{z_0} = (\theta/\omega)^2 \left[\left(1 - \frac{\theta^2}{\omega^2}\right)^2 + \left(2\zeta\frac{\theta}{\omega}\right)^2 \right]. \tag{10.115}$$

Its plot is the same as that in Fig. 10.10.

10.6 Analysis of Structures under Seismic Actions

10.6.1 Earthquake Spectra

Practical analyses for the action of seismic forces are regulated by standard codes based commonly on a so-called linear-spectral theory by M. Bio. Originally, the method was based on experimentally measured accelerations of pendulums with different periods of natural oscillations in conditions of an earthquake. Maximum values of such accelerations, obtained under the action of displacements of a foundation according to the time history of a real earthquake, are represented as functions of the natural oscillation period of the pendulum (trial oscillator). These form a *response spectrum*, a basis for the definition of a seismic load (Fig. 10.12).

The technique of the analysis is based on so-called *design spectra* determined as envelopes of a number of response spectra of various real earthquakes. Also, an assumption is usually made that the foundation of a structure behaves like a single rigid body (unfortunately, this hypothesis is not stated explicitly). It supposes that all points of the structure's support are translated according to the same law $\mathbf{X}_0 = \mathbf{X}_0(t)$.

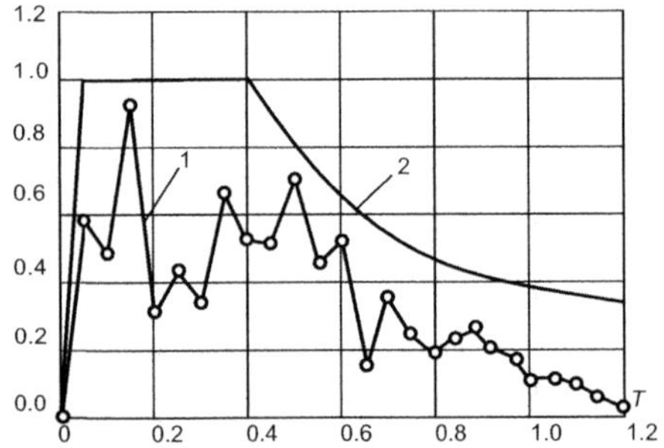

Fig. 10.12. Acceleration spectra:
(1) a measured one; (2) an approximated envelope from SNiP

Then, remembering that the displacements of the system are referred to a point of the foundation and the inertia forces are determined by the absolute acceleration rather than the relative one, we have this instead of (3.142):

$$\mathbf{M}\frac{d^2(\mathbf{u}+\mathbf{X}_0)}{dt^2}+\mathbf{K}\mathbf{u}(t)=\mathbf{0} \tag{10.116}$$

or this, after transferring the given acceleration of the foundation to the right part

$$\mathbf{M}\frac{d^2\mathbf{u}}{dt^2}+\mathbf{K}\mathbf{u}(t)=-\mathbf{M}\frac{d^2\mathbf{X}_0}{dt^2}\ . \tag{10.117}$$

10.6.2 Seismic Response

For a system with one degree of freedom, its mass being M and the rigidity K, we allow for the internal resistance force using Sorokin's assumption and have

$$M\ddot{w}+(1+i\gamma)Kw=-M\ddot{Z}_0\,, \tag{10.118}$$

where the displacement w measured with respect to a point of the foundation and the acceleration $\ddot{Z}_0$ of this foundation are complex quantities. To build $\ddot{Z}_0$, the given acceleration function $\ddot{X}_0$ is expanded into a Fourier series. This one and its conjugate Fourier series make the real and imaginary parts of the complex function $\ddot{Z}_0$.

At nonzero initial conditions, the real part $u(t)$ of the solution of the differential equation (10.118) is

$$u(t)=-\frac{1}{\bar{\omega}}\int_0^t\ddot{X}_0(\tau)\exp\left[-\frac{\gamma}{2}(t-\tau)\right]\sin(t-\tau)d\tau\,, \tag{10.119}$$

where $\bar{\omega}$ is a circular frequency of natural attenuated oscillations of the oscillator (allowing for the energy dissipation in the material)

$$\bar{\omega}=\frac{\omega}{\sqrt{1+\gamma^2/4}}\,,\qquad \omega=\sqrt{K/M}\ . \tag{10.120}$$

Understanding the seismic load as an inertia force acting on an elastic system

$$F(\mathrm{t})=-M(\ddot{Z}_0+\ddot{w})\,, \tag{10.121}$$

we differentiate (10.119) two times, perform easy transformations where the fact $\gamma \ll 1$ is essentially used, and get the real part of the complex inertia force as

$$P(t)=-M\,a(t,\gamma,T), \tag{10.122}$$

where

$$a(t, \gamma, T) = -\frac{2\pi}{T} \int_0 \ddot{X}_0(\tau) \exp\left[-\frac{\gamma\pi}{T}(t-\tau)\right] \sin\left(\frac{2\pi}{T}(t-\tau)\right) d\tau \qquad (10.123)$$

is the acceleration of a linear oscillator having the natural oscillation period T and the inelastic resistance coefficient γ. The maximum value of the seismic force will be

$$\max P(t) = M\, C_a(T)\,, \qquad (10.124)$$

where $C_a(T)$ is a so-called *acceleration spectrum*, i.e. a function that describes maximum values of absolute accelerations of the linear oscillator vs. its natural oscillation period at a fixed value of γ and a given kinematical action d^2X_0/dt^2.

The standards present the acceleration spectrum as

$$C_a(T) = K_c\, g\, \beta(T), \qquad (10.125)$$

where the seismicity K_c coefficient depends on the measure of the earthquake's points (it grows twice per each point) and the dynamical amplification factor $\beta(T)$ is specified by plots depending on a soil type (Fig. 10.13). The value γ is averaged and assumed equal to 0.1. The gravity acceleration g is introduced into (10.125) because the standards operate weights of masses Q rather than the masses M.

For systems with multiple degrees of freedom, the equations (10.117), after introducing an attenuation according to Sorokin, are solved by expanding into a series by natural oscillation modes. For each mode its modal seismic load is defined as one for a system with one degree of freedom.

The seismic load is distributed between the masses of the system (рис. 10.14) using the coefficients

$$\eta_{ik} = \frac{\Psi_{ik} \sum_{j=1}^{n} Q_j \Psi_{ij}}{\sum_{j=1}^{n} Q_j \Psi_{ij}^2}\,. \qquad (10.126)$$

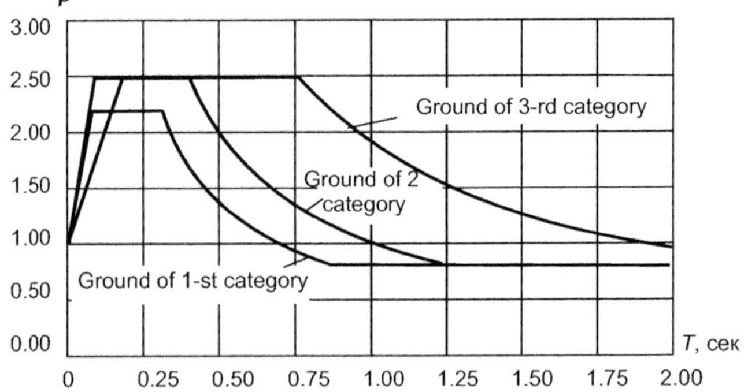

Fig. 10.13. A dynamical amplification factor according to SNiP

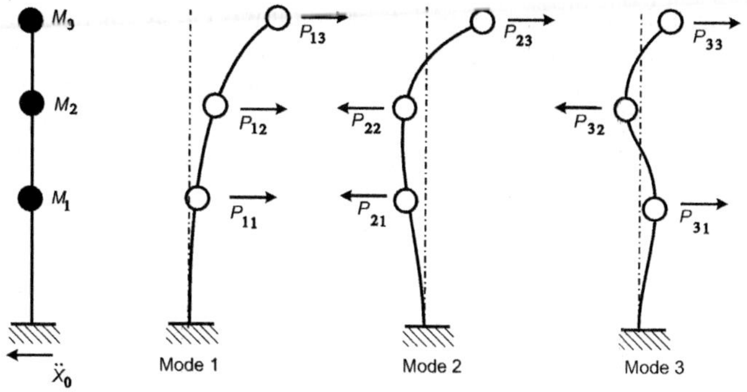

Fig. 10.14. Inertial seismic loads that correspond to various natural oscillation modes

The summation of contributions of particular oscillation modes is recommended by the Russian standard to be performed with the mean-square estimate by Rosenblum (3.160). This approach is based on the hypothesis that all modal responses are normally distributed stochastic variables with the same correlation coefficients, which is in a good accordance with many observations, though is not an established fact.

The internal stresses, determined by conventional rules for each of the natural oscillation modes taken into account, are summed by Rosenblum's formula, too. Though, there is a pitfall here: the use of absolute moments, longitudinal forces and lateral forces causes compressed-bent sections to disappear, and all of them become tensioned-bent. Similar effect of the sign loss may occur in shell-type elements, too.

To fight this effect, some of software systems give signs to Rosenblum-summed components of the internal stress vector same as the signs of their counterpart components in the first natural oscillation mode. This approach is implemented, for example, by the well-known **Lira** system, though it is pretty hard to validate even assuming that the first mode contributes most to the summed value of each component of the response vector.

Seeing that the estimated sum of the modal contributions contains a square root expression which should be preceded by the ± sign, we can assume that all components of the internal stresses may have arbitrary signs, and thus we have to consider all possible combinations of the signs. But it's better to remember that the found estimations of maximum stress values are not used roughly as they are to check the structure — they are substituted to a criterion formula that determines the validity of the structure (such as a stress calculation formula). So one needs to calculate the value of this criterion for each particular mode of natural oscillations taken into account, and only then perform the Rosenblum's summation.

It should be noted that the use of Rosenblum's formula suppresses the signs of modal components of a seismic response and may cause a confusion in some cases. A typical example of a situation like this is a case of multiple (duplicated) frequencies when natural modes corresponding to the frequency are determined up to an arbitrariness. For example, a vertically standing cantilever bar with the same principal rigidities of its cross-section has multiple modes of natural oscillations determined up to an arbitrary rotation around the axis Z as shown in plane in Fig. 10.15 where (*a*) natural oscillation modes are directed along the coordinate axes and (*b*) those are rotated by 45°. If the seismic pulse acts along the axis *X*, then in the case (*a*) there will be no response of the second mode and all displacements will belong to the plane (*X*,*Z*).

In the case (*b*) both oscillation modes will be excited, but the fact that the components along the axis *Y* have opposite signs and suppress each other will be lost once the rule of a "square root of sum of squares" is applied. The result will be that displacements are excited in the structure both in the direction of the seismic shock and laterally. As we can see, the result depends on the arbitrariness of the choice of a couple of natural modes that correspond to a multiple frequency.

A pretty serious issue in a seismic analysis is how to choose a proper design model. Simplest cantilever models widely popular in the design practice and recommended by many manuals are rather a tribute to the times of calculation by hands than a necessity. Moreover, those mask some real structural behavior effects and thus force one to introduce artificial eccentricities in order to allow for torsion oscillations. An illustration of this technique recommended by the codes is presented in Fig. 10.16. Naturally, a more realistic spatial model will allow for torsion modes of motion automatically if only the analyst does not choose to cut them off by specifying too few natural oscillation modes to be taken into account. This should be carefully tracked in the real design practice.

One should keep in mind that international codes on earthquake-resistant construction ISO [20] and Eurocode [10] use a coefficient of reduction of a design model K_d that reduces the rated seismic load by taking into account dissipative capabilities of structures, particularly, their ability to deform beyond the elastic limit under a cyclic action without serious deterioration of their strength. In this regard ISO gives warnings on the use of a conditional spectral seismic analysis method, and Eurocode represents these warnings as Table 10.4.

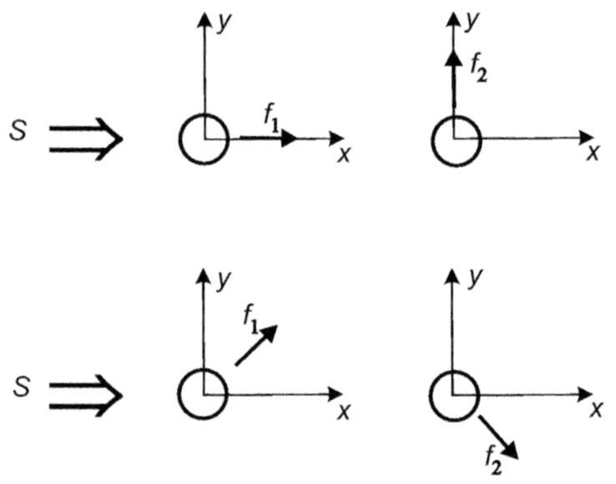

Fig. 10.15. The case of a multiple frequency

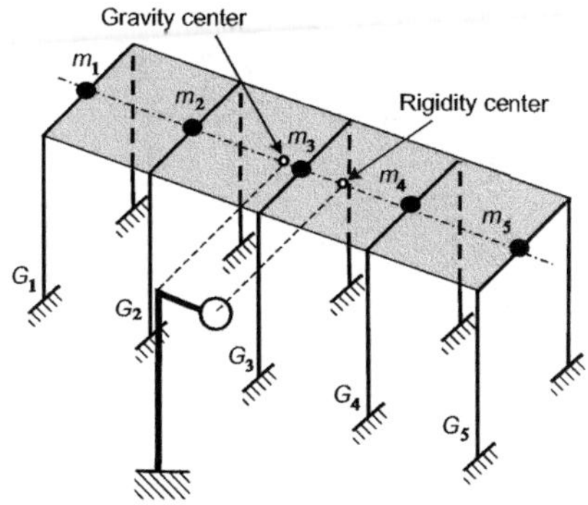

Fig. 10.16. Introducing a model of eccentricity into a cantilever design model between its stiffness center (SC) and mass center (MC)

Table 10.4.

Regularity of model		Design model	Value of reduction coefficient K_d
Plan	Elevation		
Yes	Yes	Plane	Standard
Yes	No	Plane	Reduced
No	Yes	Spatial	Standard
No	No	Spatial	Reduced

Two important points should be noticed in Table 10.4:

- Plane design models are allowed to use only with structures possessing the property of regularity in their horizontal plane.
- When the regularity of a structure in the vertical plane is violated, one has to reduce the reduction coefficients K_d recommended by international codes.

The paper [12] indicates that these warnings of the international regulatory codes present a serious reminder of a known circumstance: the conditional spectral method cannot yield reliable results for complex multi-modal oscillations of a structure as regards an ultimate state of destruction predicted to occur during a strong earthquake. Also, it should be noted that the technique from the standards presented in Fig. 10.16 compensates, to some extent, a pretty rough hypothesis that the structure's foundation moves as a single solid body. It is this hypothesis that eliminates the torsion components of the structure's movement when all frames have the same stiffness (Fig. 10.17).

Some techniques to allow for components of the seismic action in Fig. 10.17, *b* and 10.17, *c* have been suggested by V.K. Egupov and his followers in a series of papers [39].

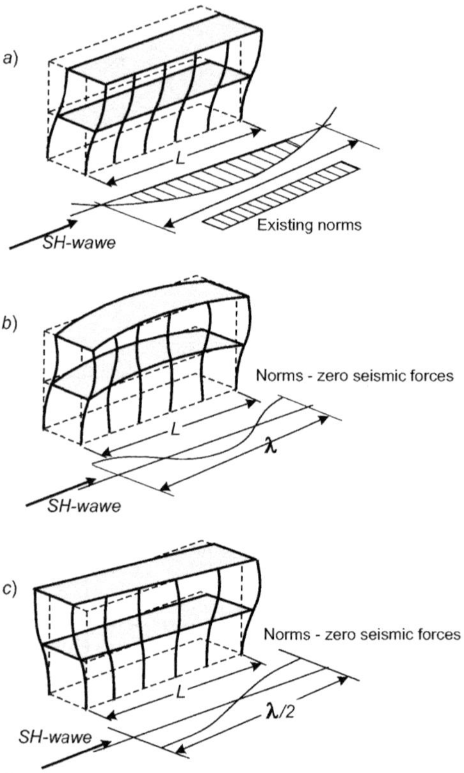

Fig. 10.17. Displacements in a structure when a seismic wave passes through: (*a*) translatory; (*b*) flexural; (*c*) torsional

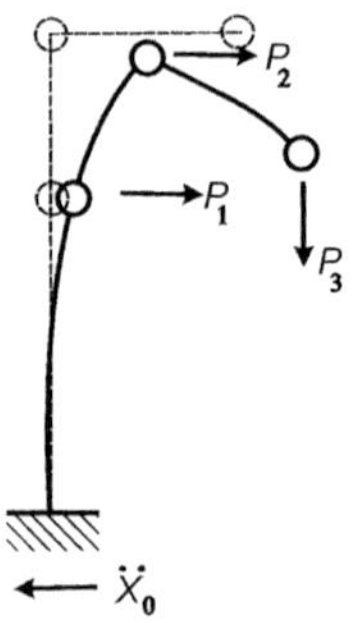

Fig. 10.18. A vertical seismic load component appearing during a horizontal seismic shock

The analysis in compliance with standards (using a software that implements these standards in an excessively straightforward manner) assumes that the foundation's acceleration occurs in the horizontal plane. Actually, during an earthquake the foundation oscillates in an oblique direction to both the horizontal and vertical axes causing the structure to move along all coordinate axes. The capability of specifying a seismic shock direction is a must for a CAD software. Also, one should remember that a horizontal seismic action applied to a spatial design model can cause vertical components of the seismic load (Fig. 10.18). This is a way to soften an inflexible approach of design standards. The latter indicate explicitly the necessity to allow for a vertical seismic component only in strictly defined cases.

10.6.3 Analysis with Accelerograms

Russian standards require that highly responsible objects and buildings over 16 stories high be analyzed using instrumental records of accelerations or synthesized accelerograms. Actually this is a requirement to determine the maximum value of an object's seismic response individually by (10.92). The construction site may happen not to have nearby seismic stations with records of known accelerations, therefore collections of accelerograms described in literature [15] are often used.

One should realize a risk of a serious error caused by choosing a record that does not correspond to one's construction site and its characteristic seismic disturbance. For example, the earthquake in Spitak [1] presented a peak of its spectrum in the vicinity of a natural oscillation period about 0.6 sec which was almost two times different from the spectrum adopted to analyze buildings (Fig. 10.19).

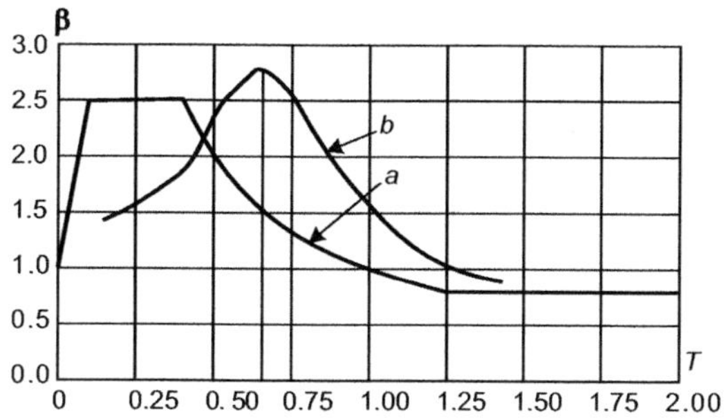

Fig. 10.19. Spectra of seismic response: (*a*) adopted for a design; (*b*) obtained after the earthquake by data of Institute of Earth Physics of the USSR Academy of Science

10.6.4 Story Response Spectra

One often needs to perform validating dynamical calculations of equipment or other additional components installed on a structure. One may think it suffices just to include those in a dynamical design model and perform the calculation, but this way is not really convenient. One thing is that the number of equipment pieces can be really great to complicate the model too much, and another thing is that such additional subsystems may be unknown beforehand or change in the course of operation.

The practical way is different: one uses the fact that the mass of the additional equipment is relatively small comparing to the building's mass to neglect the effect of the equipment on the dynamical properties of the building. Then one can obtain descriptions of motion of the equipment's installation locations — such descriptions are called story accelerograms if they refer to accelerations of the points. Then the equipment can be analyzed with the respective kinematical disturbance. But a detailed dynamical analysis of equipment units is performed only for most responsible machines and aggregates. In the majority of all other cases one uses story *response spectra* which represent values of a maximum possible amplitude of linear oscillators with different natural frequencies. In other words, a dynamical response is simulated for an imagined piece of equipment with its natural frequency f unknown beforehand, therefore the modeling is performed in a certain range of possible natural frequencies.

It should be noted that the procedure like this performs a double filtering of the external disturbance. First, the dynamical analysis of the building suppresses to a great extent a part of the input disturbance with its frequencies far from the natural frequencies of the structure. Then, this distorted input signal is again filtered by a linear oscillator, i.e. only its parts located near the natural frequency of the oscillator f are taken into account. It is natural that this procedure is very much sensitive to the accuracy of determining natural frequencies/modes. To allow for the error, one involves a procedure of smoothing and expanding the response spectra peaks [4]. An expanded peak is limited on its sides by a line parallel to a curve of the design spectrum (Fig. 10.20).

Generally, the magnitude of the expansion Δf should have been obtained by a sensitivity analysis. But in practice a relative fixed value of it is most often adopted, such as $\Delta f = \pm f_j$ where f_j denotes a frequency at which j^{th} calculated peak stands.

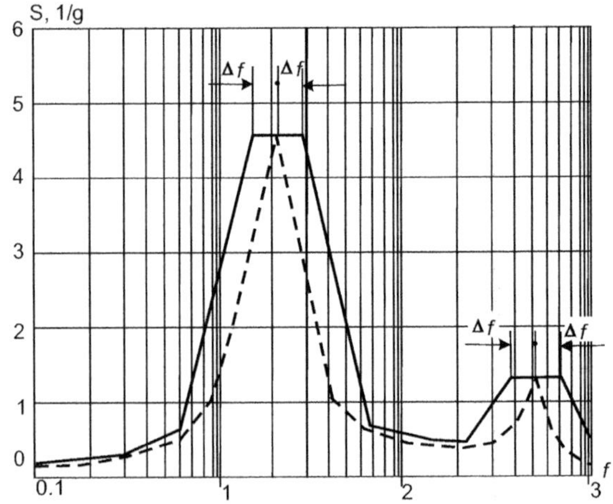

Fig. 10.20. Expanding calculated response spectra (the dash line shows calculated values)

As the value f_j is a kind of a random variable somehow, there is always a chance that it does not coincide with the oscillator's frequency. A detailed probability analysis has shown that the height of peaks of the probabilistic spectrum is much lower than the determined values [15]. This has been a background for including the following recommendation in U.S. nuclear power plant design codes: lower the maximums of the peaks by 15% together with expanding those.

It's useful to notice that the description of a structure's behavior as a response spectrum is reciprocal in some sense with the *amplitude-vs.-frequency characteristic* defined by an expression like (10.51). If we treat this expression as a function of the external disturbance's frequency, we will see that the amplitude-vs.-frequency characteristic defines a dependence of a quantity (a point's displacement, a stress etc.) taken in the given system on the varying disturbance frequency, while the response spectrum deals with a reaction of a frequency-variable oscillator to an unchangeable disturbance.

For a conclusion, we would like to remind once again that the use of the response spectra is possible only in the case the additional mass is relatively small and its effect on the general behavior of the whole "building – equipment" system can be neglected.

10.7 Action of Pulse and Impact Loads

Pulse loads discussed in this section are short-tern applied forces and/or impacts that do not cause macro-plastic deformations in elements of a structure. Micro-plastic deformations in places of contact with an impacting body or in other locations of the structure which cause the oscillation energy to dissipate, are allowed for in the analysis as an internal friction of the material.

Most often a pulse load is represented by an expression of this type:

$$P(\mathrm{t}) = \begin{cases} P_0 f(t) & at \ \ 0 \le t \le \tau \\ 0 & at \ \ t > \tau \end{cases}, \tag{10.127}$$

where $t = 0$ is the beginning of the load's action; P_0 is its maximal value; $f(t)$ is a time history of the load normalized so that the maximum $|f(t)| = 1$. The functions $f(t)$ most often used for an analysis are listed in Table 10.5.

Thus, a pulse load is described by its duration τ; its time history (shape of the pulse) $f(t)$ and its maximum value P_0. The shorter the time interval at which a constant-sign load acts, the less important is the time history of the load [28]. There remains only one quantity of key importance related to the load: its force momentum

$$S = P_0 \int_0^\tau f(t)dt \, . \tag{10.128}$$

For an instantaneous pulse when $\tau \to 0$, the maximum displacement of a simplest system with a single degree of freedom is determined by the formula

$$Z_{max} = \frac{S}{M\omega}, \tag{10.129}$$

where M is the mass; ω is the natural frequency. In other cases the value of Z_{max} should be refined: multiplied by a coefficient of pulsedness $\varepsilon(\tau^*)$ that depends on the relative duration of the pulse $\tau^* = \tau / T = \tau\omega / 2\pi$. The values of the coefficient decrease from 1.0 at $\tau^* = 0$ to zero (Fig. 10.21).

Table 10.5.

N	Plot	f(t) at $0 \le t \le \tau$	$\int_0^\tau f(t)dt$
1	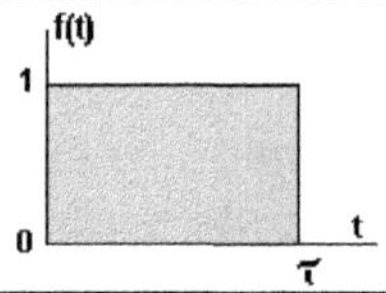	$f(t) = 1$	τ
2	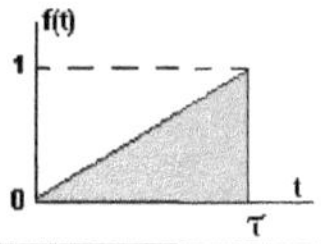	$f(t) = t/\tau$	$\tau/2$
3	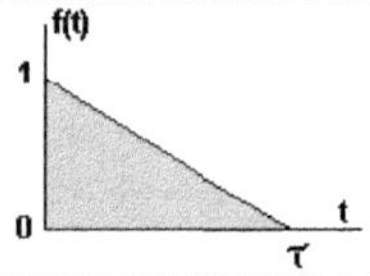	$f(t) = 1 - t/\tau$	$\tau/2$
4	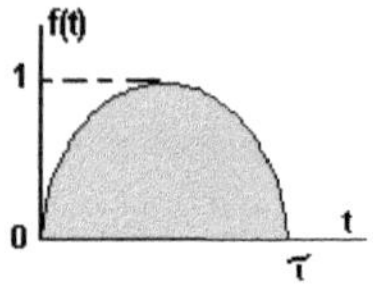	$f(t) = \sin(\pi t/\tau)$	$2\tau/\pi$
5	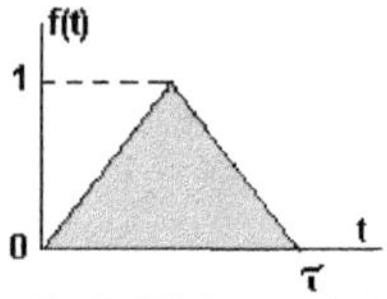	$f(t) = \begin{cases} 2t/\tau, & t \le \tau/2 \\ 2[1-(t/\tau)], & t > \tau/2 \end{cases}$	$\tau/2$
6	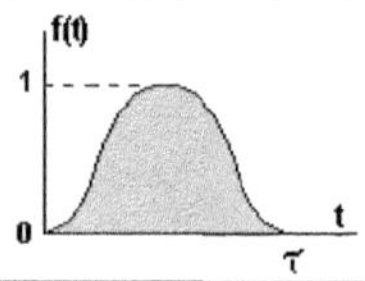	$f(t) = (1/2)[1 - \cos(2\pi t/\tau)]$	$\tau/2$

If a pulse is caused by an impact to the structure, its value can be estimated by the formula

$$S = mv_0(1 + k_0), \tag{10.130}$$

where m is the mass of the hitting body; v_0 is its velocity at the moment of contact with the structure; k_0 is a restoration (bounce) coefficient the value of which depends on properties of hitting bodies at the point of their contact and on their shapes (Table 10.6). This coefficient is usually determined experimentally as a ratio of a falling body's bounce height to that of its fall.

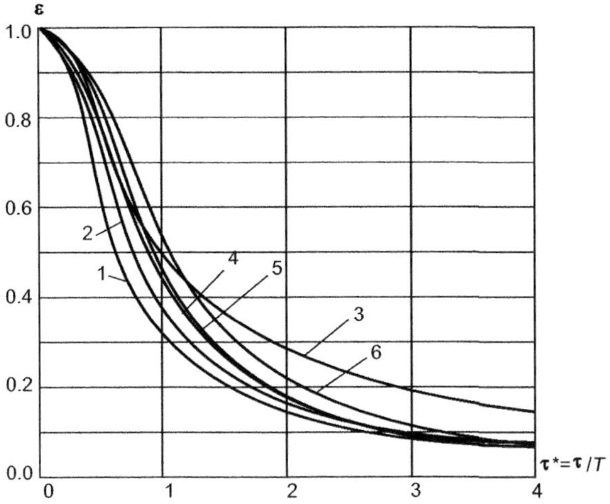

Fig. 10.21. A pulsedness coefficient for six shapes of a pulse from Table 10.5

Table 10.6.

Material of the structure's surface at the place of impact	Material and shape of hitting body				
	Steel and other hard materials		Copper, aluminum, hard plastics, stone, concrete etc.		Plastic and loose materials
	Round	Angular	Round	Angular	Any
Steel	0.60	0.35	0.40	0.25	0.00
Wood	0.55	0.30	0.40	0.20	0.00
Stone	0.40	0.20	0.30	0.15	0.00
Concrete	0.35	0.15	0.25	0.10	0.00
Xylolite	0.20	0.10	0.10	0.05	0.00
Asphalt	0.00	0.00	0.00	0.00	0.00

An important thing in the analysis of structures under pulse actions is a *loss coefficient* η that characterizes an attenuation of oscillations. It is related to the *energy absorption coefficient* ψ and the *internal friction coefficient* γ as

$$\eta = \frac{\psi}{2\pi} = \frac{\gamma}{(1-\gamma^2/4)} . \tag{10.131}$$

The latter coefficient also determines the logarithmic decrement of oscillations

$$\delta = \frac{\gamma\pi}{\sqrt{1-\gamma^2/4}} . \tag{10.132}$$

Reference data regarding the values of the coefficients η, ψ, γ or δ can be found in many sources including design codes [9] etc.

For systems with multiple degrees of freedom under an action of a set of instantaneous pulses S_k^0 applied in points $k = 1, \ldots , m$, the solution is sought by expanding into a series by natural oscillation modes (3.176) and is presented by the formula

$$u_j = \sum_{i=1}^{n} U_{ji} \exp(-\frac{\gamma}{2}\omega_i t)\sin(\omega_i t), \quad (j = 1, \ldots , n), \tag{10.133}$$

where U_{ji} is calculated using the natural modes $\mathbf{\Psi}_i$ and the formula

$$U_{ji} = \frac{\Psi_{ji}}{\omega_j} \sum_{r=1}^{m} \varepsilon\left(\frac{\tau_r \omega_j}{2\pi}\right) \Psi_{jr} S_r^0, \quad (j = 1, \ldots , n;\ i = 1, \ldots , n). \tag{10.134}$$

Of course, it is practically interesting to know ultimate values of displacements and stresses in the system in question. But, searching for a moment of time when the extremum is achieved by a sum like (10.133) is not just laborious from the standpoint of computing. There is a more serious issue: an unquestionably existing discrepancy between the calculated and actual natural oscillation frequencies of the system [30] leads to a noticeable deviation of the phase shift.

In the structural design and analysis practice there has been established a technique suggested by E.S. Sorokin [30] for estimating a sum of attenuated harmonics based on the following idea. For harmonics with maximum amplitudes, the sum includes their values corresponding to the moment of time $t = T_q / 4$ (q is the number of the harmonic at which a maximum amplitude U_{ji} is reached in the sum (10.133)), and for all the rest envelopes of the same harmonics at the same moment of time are taken. In other words, if the displacement at the point j satisfies the condition $|U_{jq}| > |U_{ji}|$ $(i \neq q)$, then this formula is to be used for calculation

$$|u_j| = \sum_{i=1}^{q-1} |U_{ji}| \sin\frac{\pi\omega_i}{2\omega_q} + \sum_{i=q}^{n} |U_{ji}| \exp\left(-\frac{\gamma\pi\omega_i}{4\omega_q}\right). \tag{10.135}$$

A graphical illustration for the case $q = 1$ is presented in Fig. 10.22.

The same should be said on the summation of internal stresses performed by a formula similar to (10.135) as we have said before about Rosenblum's summation for a seismic action: unpleasant sign loss effects arise.

Here, besides the approach stated earlier, we have an opportunity to directly estimate the maximum of the displacements (stresses etc.) determined by the formulae like (10.135). To do it, we have to perform an exhaustive search with the sum at the interval of time 0 to $T_q / 4$ with a fine step, for example, not greater than one eighth of the least natural oscillation periods taken into account.

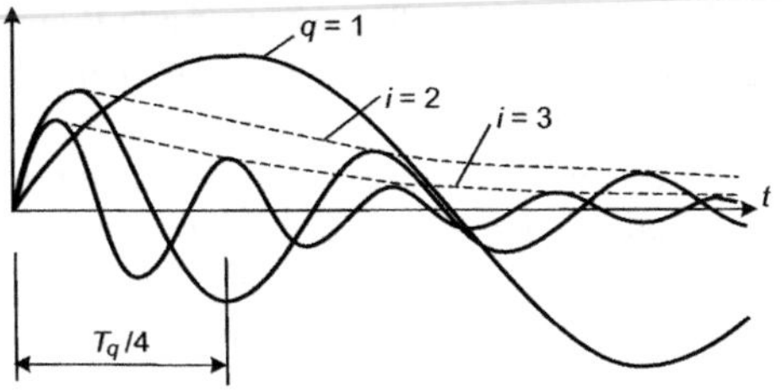

Fig. 10.22. A graphical interpretation of the extremum estimation technique

Also, we should consider the possibility that these periods can vary by ±5% or so, in order to allow for a potential inaccuracy of the determination of natural oscillation frequencies.

10.8 Oscillations under an Action of Wind Flow Pulsations

10.8.1 Modeling of Wind Actions

In compliance with standard requirements, a wind load can be treated as:

- a normal pressure w_e applied to an exterior surface of a structure or an element;
- a normal pressure w_i applied to interior surfaces of buildings that have penetrable fencings or openings.

In the first case the normal pressure w_e should be defined as a sum of an average and a pulsation components, while the second case w_i takes into account only the average component.

A wind load analysis is to determine displacements of nodes of a design model and stresses (internal forces) in elements of the system separately from the action of the average and the pulsation components of the wind. The full response of the structure (displacements, forces, stresses) is determined by the formula

$$X = X^s \pm X^d\,, \qquad (10.136)$$

where X is a quantity in question; X^s is a value of the same quantity defined by the static wind load component; X^d is that caused by the pulsation wind load component.

The sign ± before the dynamical component of the quantity X^d reminds that its rated amplitude is to be evaluated by the analysis, and the total wind action effect should include the most disadvantageous sign. This yields the formula

$$X = X^s + \mathrm{sign}\left(X^s\right)\cdot X^d\,. \qquad (10.137)$$

Unfortunately, in the popular reference manuals there is no ± sign preceding the dynamical component which fact sometimes causes a confusion. Also, the Russian code involves a concept of a *wind load pulsation component* without indicating explicitly that this load cannot be used by conventional linear analysis techniques to determine all other quantities (displacements, stresses etc.) if a system with more than one degrees of freedom is under consideration. It has been already stated in the previous edition of the Russian code but, unfortunately, does not appear explicitly in the newest code.

10.8.2 A Dynamic Action of the Wind Load's Pulsation Component

The purpose of a dynamical wind load analysis is to investigate steady-state oscillations of a structure subject to a wind flow. Probabilistic properties of nodal displacements, internal forces or stresses in finite elements are to be determined. The final goal is to find the dynamical component X^d of the design factor X.

Just as in the regulatory documents, here we use a quasi-steady-state flow model that supposes the characteristic dimensions of surfaces subjected to the wind to be small compared to lateral sizes of vortexes. This makes us confine the use of this approach to structures where the characteristic size of a solid surface under the wind is not greater than 100 m or so. If the plane sizes are greater than that, the quasi-steady-state model is hardly applicable because the principal assumption ceases to hold: the dimensions of the structure are no longer small compared to vortexes that contain the main energy of a turbulent flow. Apparently, this limitation is relevant to structures with solid exterior surfaces (buildings, cooling towers etc.) rather than open ones (masts, elevated ways etc.).

In Section 10.2 above we indicated how the general motion equations

$$\mathbf{M}\ddot{\mathbf{U}} + \mathbf{C}\dot{\mathbf{U}} + \mathbf{K}\mathbf{U} = \mathbf{q}(t) \tag{10.138}$$

can be transformed to independent equations by being expanded over natural oscillation modes

$$\mathbf{U} = \mathbf{\Phi}\mathbf{y} \tag{10.139}$$

The equations refer to generalized (normal) coordinates $\mathbf{y}(t)$

$$\ddot{y}_r(t) + \gamma\omega_r\dot{y}_r(t) + \omega_r^2 y_r(t) = Q_r(t) \quad (r = 1, \ldots, n), \tag{10.140}$$

where

$\gamma = \delta/\pi$ is an inelastic resistance coefficient;
δ is a logarithmic decrement of oscillations;
$Q_r(t)$ is a generalized force being r^{th} component of the vector

$$\mathbf{Q}(t) = \mathbf{\Phi}^{\mathrm{T}}\mathbf{q}(t) \tag{10.141}$$

or, when expanded,

$$Q_r = \sum_{j=1}^{n} \psi_{jr} q_j \,. \tag{10.142}$$

Presently the common approach is to assume the wind pulsation load to be a stationary and steadily correlated stochastic process. When a system with n degrees of freedom is subjected to such load $q_j(t)$ ($j = 1,...,n$), their description by spectral and reciprocal spectral densities $S_{ij}(\omega)$ is used[6].

The reciprocal spectral densities of generalized forces $Q_r(t)$ ($r = 1,... , n$), according to (10.142), are related to $S_{ij}(\omega)$ as

$$S_{Q_r Q_s}(\omega) = \sum_{i=1}^{n} \sum_{j=1}^{n} \psi_{ir} S_{ij} \psi_{js} \tag{10.143}$$

or in a matrix form

$$\mathbf{S}_Q = \mathbf{\Psi}^{\mathrm{T}} \mathbf{S}_q \mathbf{\Psi}, \tag{10.144}$$

where

$\mathbf{S}_q$ is a matrix of reciprocal spectral densities of loads $q_j(t)$ (j=1,...,n);
$\mathbf{S}_Q$ is the same for generalized loads $Q_r(t)$ ($r = 1,... , n$).

A response of the mechanical system $\mathbf{X}(t)$ to the action of the dynamical load $\mathbf{q}(t)$ will be a vector of nodal displacements, internal forces (stresses) in finite elements or other design factors. The only important thing is that the response $\mathbf{X}(t)$ is related to the generalized coordinates $\mathbf{p}(t)$ linearly:

$$\mathbf{X} = \mathbf{A}\mathbf{p} \,. \tag{10.145}$$

If $\mathbf{q}(t)$ is the above stated steady-state stochastic action, then components of the vector $\mathbf{X}(t)$ are stationary and steadily correlated stochastic processes described by two matrices:

- a matrix of spectral densities $\mathbf{S}_X$;
- a covariance matrix $\langle \mathbf{X}^2 \rangle$ elements of which are covariance values $\langle X_i X_j \rangle$ of the $\mathbf{X}$ vector's components (here and below we use angle brackets to denote an operation of averaging).

Due to the linear dependence (10.145), the following holds for the covariance matrix

$$\langle \mathbf{X}^2 \rangle = \mathbf{A} \langle \mathbf{p}^2 \rangle \mathbf{A}^T \,. \tag{10.146}$$

The covariance of the generalized coordinates is calculated by the formula

[6] The spectral density $S_i(\omega)$ value is proportional to a fraction of the total cardinality of the stochastic process $Q_i(t)$ lying in the frequency range ω to $\omega + d\omega$. The reciprocal spectral densities $S_{ij}(\omega)$ characterize the degree of a correlation between the processes $Q_i(t)$ and $Q_j(t)$ in the same frequency range.

$$\langle y_r y_s \rangle = \frac{1}{2\pi} \int_{-\infty}^{\infty} \frac{S_{Q_r Q_s}(\omega) d\omega}{(-\omega^2 + i\gamma\omega\omega_r + \omega_r^2)(-\omega^2 - i\gamma\omega\omega_s + \omega_s^2)} . \tag{10.147}$$

When analyzing the pulsation component of the wind load, it suffices to keep only first m columns in the matrix $\mathbf{\Psi}$ (most often, m is much less than n) for which the natural frequencies do not exceed a limiting value ω_{lim}. The latter is recommended by the Russian code depending on the wind region and the oscillation decrement. Accordingly, the number of normal coordinates y_r and the number of equations in (10.140) are reduced to m.

10.8.3 Representation of a Pulsation Component of a Wind Load

Let us suppose a structure standing in a wind flow is referred to a system of coordinates (X,Y,Z) with its origin at the point O such that:

- the point O lies on the Earth's surface;
- the Z axis is vertical and looks upwards;
- the direction of X is the same as the wind's.

Let us denote the vector radius of a point at the surface of the structure, with respect to the pole O, as $\mathbf{r}$. The pulsation vector component can be determined by the formula

$$\mathbf{W}(\mathbf{r},t) = \mathbf{W}_0(\mathbf{r})\zeta(z)\nu^P(\mathbf{r},t) , \tag{10.148}$$

where

$\mathbf{W}_0(\mathbf{r})$ is the vector of the average (static) component of the wind load;

$\zeta(z)$ is a pulsation coefficient at the height of z; the value of it depends on a terrain type;

$\nu^P(\mathbf{r},t)$ is the pulsation component of the wind load.

When the wind load is specified as a discrete one applied to nodes of the model, the formula (10.116) becomes

$$\mathbf{P}_k^W(t) = \mathbf{P}_k^s\zeta(z_k)\nu_k^p(t) \qquad (k = 1, \dots, n_y) \tag{10.149}$$

where

n_y is the number of nodes of the design model in which the static component of the wind load is applied;

$\mathbf{P}_k^s$, $\mathbf{P}_k^W$ are vectors of the static and pulsation components of the wind load in a node k.

Elements of the matrix of reciprocal spectral densities of the dynamical action $\mathbf{q}(t)$ are calculated by the formula

$$S_{ij} = P_{kq}^s P_{lt}^s \zeta(z_k)\zeta(z_l) S_{\nu_k^p \nu_l^p} , \tag{10.150}$$

where

$i, j = 1,\dots, n$ are numbers of generalized coordinates;
$k, l = 1,\dots, n_y$ are numbers of nodes;
$q = 1,\dots, n_k$; $l = 1, \dots, n_l$ are numbers of degrees of freedom in respective nodes k and l.

For example, in the **SCAD** software environment the number of dynamical degrees of freedom can change from one node to another. It is defined by both inertial properties of a node and its imposed constraints. Up to six degrees of freedom in one node are allowed, i.e. in (10.150) it's going to be $0 \le n_k \le n_l \le 6$. Thus the double subscripts at static components of the load in (10.150) determine No. of a node and No. of a degree of freedom in that node.

If we use the same indexation with the fundamental matrix $\mathbf{\Psi}$ the matrix form (10.144) will acquire the following expanded form

$$S_{Q_r Q_s} = \sum_{k=1}^{n_y}\sum_{l=1}^{n_y}\sum_{q=1}^{n_k}\sum_{t=1}^{n_l} \psi_{kqr}\psi_{lts} P_{kq}^s P_{lt}^s \ \zeta(z_k)\ \zeta(z_l) S_{v_k^p v_l^p} \quad (r, s = 1,\dots, m). \qquad (10.151)$$

Re-grouping the terms yields

$$S_{Q_r Q_s} = \sum_{k=1}^{n_y}\sum_{l=1}^{n_y} A_{kr} A_{ls} S_{v_k^p v_l^p}\,, \qquad (10.152)$$

where designations have been introduced

$$A_{kr} = \zeta(z_k)\sum_{q=1}^{n_k}\psi_{kqr}P_{kq}^s;\ \ A_{ls} = \zeta(z_l)\sum_{t=1}^{n_k}\psi_{lts}P_{lt}^s\,. \qquad (10.153)$$

After substituting (10.152) to (10.147) we have the following expression for the covariance of the generalized coordinates

$$\langle y_r y_s\rangle = \sum_{k=1}^{n_y}\sum_{l=1}^{n_y} A_{kr} A_{ls}\xi_{klrs} \quad (r, s = 1,\dots, m)\,, \qquad (10.154)$$

where

$$\xi_{klrs} = \frac{1}{2\pi}\int_{-\infty}^{\infty}\frac{S_{v_k^p v_l^p}\,d\omega}{(-\omega^2 + i\gamma\omega\omega_r + \omega_r^2)(-\omega^2 - i\gamma\omega\omega_s + \omega_s^2)}. \qquad (10.155)$$

Also,

$$\frac{1}{(-\omega^2 + i\gamma\omega\omega_r + \omega_r^2)(-\omega^2 - i\gamma\omega\omega_s + \omega_s^2)} = \mathrm{Re}_{rs}(\omega) + i\mathrm{Im}_{rs}(\omega) \qquad (10.156)$$

where

$$\mathrm{Re}_{rs} = \frac{(-\omega^2 + \omega_r^2)(-\omega^2 + \omega_s^2) + \gamma^2\omega^2\omega_r\omega_s}{\left[(-\omega^2 + \omega_r^2) + \gamma^2\omega^2\omega_r^2\right]\left[(-\omega^2 + \omega_s^2) + \gamma^2\omega^2\omega_s^2\right]}\,, \qquad (10.157)$$

$$\mathrm{Im}_{rs} = \frac{\gamma\omega\left[\omega_s(-\omega^2+\omega_r^2)-\omega_r(-\omega^2+\omega_s^2)\right]}{\left[(-\omega^2+\omega_r^2)+\gamma^2\omega^2\omega_r^2\right]\left[(-\omega^2+\omega_s^2)+\gamma^2\omega^2\omega_s^2\right]}\,. \tag{10.158}$$

Considering (10.156), the integral (10.155) can be rewritten as a sum

$$\xi_{klrs} = \frac{1}{2\pi}\int_{-\infty}^{\infty} S_{v_k^p v_l^p}\,\mathrm{Re}_{rs}(\omega)d\omega + \frac{i}{2\pi}\int_{-\infty}^{\infty} S_{v_k^p v_l^p}\,\mathrm{Im}_{rs}(\omega)d\omega. \tag{10.159}$$

The spectral and reciprocal spectral densities of the wind velocity pulsations must be even functions of ω, that is, $S_{v_k^p v_l^p}(\omega) = S_{v_k^p v_l^p}(-\omega)$ $(k,\ l = 1,\dots,\ n_y)$.

Also, from (10.157) and (10.158) it follows that $\mathrm{Re}_{rs}(\omega)$ is even, and $\mathrm{Im}_{rs}(\omega)$ is an odd function: $\mathrm{Re}_{rs}(\omega) = \mathrm{Re}_{rs}(-\omega)$, $\mathrm{Im}_{rs}(\omega) = -\,\mathrm{Im}_{rs}(-\omega)$.

The stated even/odd properties imply that the imaginary part of the integral (10.159) is equal to zero, and in the real part the integral from $-\infty$ to ∞ can be replaced by a doubled integral from 0 to ∞. Then (10.159) will change to the form

$$\xi_{klrs} = \frac{1}{\pi}\int_{-\infty}^{\infty} S_{v_k^p v_l^p}\,\mathrm{Re}_{rs}(\omega)d\omega\ (k,\ l = 1,\dots,\ n_y;\ r,\ s = 1,\dots,\ m). \tag{10.160}$$

An analysis of (10.154) gives that

$$\mathrm{Re}_{rs}(\omega) = \mathrm{Re}_{sr}(-\omega), \qquad \varphi_{klrs} = \varphi_{klsr}\,. \tag{10.161}$$

Considering (10.154), we can infer a symmetry of the covariance matrix of the generalized coordinates $\langle y_r y_s\rangle = \langle y_s y_r\rangle$ $(r,\ s = 1,\dots,\ m)$.

10.8.4 Spectrum of Wind Velocity Pulsations

A reciprocal spectrum of pulsations of the longitudinal wind velocity component in the free atmosphere is a scalar function of the frequency ω and a couple of space points associated with nodes of a structure to which the wind load is applied

$$S_{v_k^p v_l^p} = \frac{u^2}{3f\left(1+u^2\right)^{4/3}}\exp\left(-u\chi_{kl}\right),\quad (k,\ l = 1,\dots,\ n_y)\,, \tag{10.162}$$

where

$f = \omega/2\pi$ is the frequency in Hz;

$u = Lf/v_{10}$ is a dimensionless frequency;

$L = 1200$ m is an integral scale of a turbulent flow;

v_{10} is an average wind velocity at the height of 10 m;

χ_{kl} is an equivalent distance between the points k and l $(\chi_{kl} = \chi_{lk})$.

$$\chi_{kl} = (20|x_k - x_l| + 8|y_k - y_l| + 8|z_k - z_l|)\,/\,L\,. \tag{10.163}$$

Substituting (10.162) to (10.160) yields

$$\xi_{klrs} = \frac{1}{\pi}\int_0^{\infty}\frac{u^2 \exp(-u\chi_{kl})}{3f(1+u^2)^{4/3}}\mathrm{Re}_{rs}(\omega)d\omega, \tag{10.164}$$

and $\xi_{klrs} = \xi_{lkrs}$ (k, $l = 1,\dots, n_y$; r, $s = 1,\dots, m$).

10.8.5 A Dynamical Component of a Design Factor

Design dynamical values of a factor X related to the generalized coordinates linearly

$$X = \sum_{r=1}^{m} \alpha_r p_r \tag{10.165}$$

are determined by the formula similar to that by Rosenblum (3.160)

$$X^d = \left(\sum_{r=1}^{m}\sum_{s=1}^{m} \alpha_r \langle p_r p_s \rangle \alpha_s\right)^{1/2}. \tag{10.166}$$

Here, too, a loss of signs of particular components occurs which we discussed above in connection with the seismic response, and there is no correspondence between such responses as the displacements and internal forces (stresses) because the transformation (10.166) is nonlinear. The latter circumstance is not clearly stated in reference manuals, and it often leads to confusions. A typical example is shown in Fig. 10.23 where one can see that the summed modes of natural oscillation and their respective inertia forces do not yield a correct final moment curve calculated by moments of particular oscillation modes. For every separate mode this correspondence takes place, as expected.

It should be noted that the effectual regulations require the assurance factor $\lambda = 3.0$ to be used when assigning the pulsation coefficients. On the other hand, there is a known recommendation by A. Davenport [8] according to which this factor becomes related to the period of the natural oscillations and calculated as follows for i^{th} tone:

$$\lambda_i = \sqrt{2\ln(T/T_i)} + \frac{0.577}{\sqrt{2\ln(T/T_i)}} \tag{10.167}$$

at $T = 3600$ sec that corresponds to the averaging time used to get the spectra of pulsation of the longitudinal wind velocity.

Using this recommendation, we have

$$X^d = \left(\sum_{r=1}^{m}\frac{\lambda_r}{3}\sum_{s=1}^{m} \alpha_r \langle p_r p_s \rangle \alpha_s\right)^{1/2}, \tag{10.168}$$

which is the desirable result.

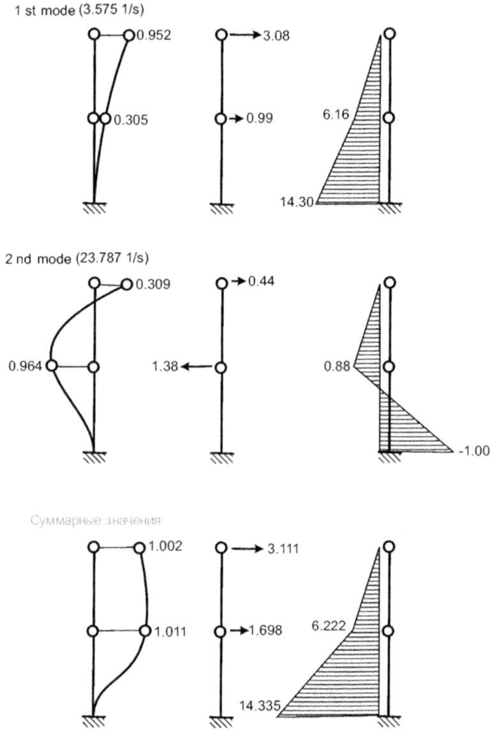

Fig. 10.23. Results of an analysis by natural oscillation modes

10.8.6 Issues of Numerical Implementation

All that was said shows the following key difficulty in the analysis of a dynamic action of a wind load's pulsed component (from the standpoint of computing): the necessity to determine an immense number of coefficients ξ_{klrs} , each of those requiring an improper integral to be calculated.

To reduce the computation effort, one can replace the discrete set of the quantities ξ_{klrs} by a continuous function $\xi_{rs}(\chi)$ for each couple of the natural frequencies ω_r , ω_s at the interval $[0, \chi_{\text{lim}}]$ where $\chi_{\text{lim}} \geq \chi_{kl}$ and is to be determined by the exterior dimensions of the building. The continuous function $\xi_{rs}(\chi)$ at the interval $[0, \chi_{\text{lim}}]$ can be well approximated by a cubic spline, and the desirable values of the integrals ξ_{klrs} are obtained by a substitution of values χ_{kl} to this spline.

The spline approximation itself is built in this way. A segment of the numerical axis $[0, \chi_{\text{lim}}]$ is divided into seven parts with their ratios 1:1:1:3:3:9:9 as shown in Fig. 10.24.

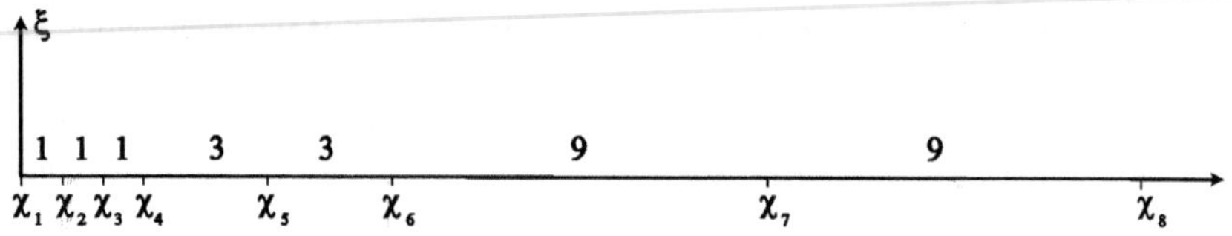

Fig. 10.24. Nodal points of a spline approximation

A cubic polynomial is then built by values specified in the nodal points. Some of its coefficients are calculated by conditions of matching based on properties of continuity of first and second derivatives of the approximant function $\xi_{rs}(\chi)$. An algorithm of interpolating with cubical splines is described in [11] in sufficient detail. The result of the approximation is that the respective cubic polynomial has been determined on each of the sub-intervals.

For each couple of natural frequencies $\omega_r \le \omega_s$ (r, s = 1,... , m) in each nodal point of a spline χ_i (i = 1,... , 8) one has to calculate the improper integral (10.164), which can be represented as a sum of integrals I_1 between 0 and ω_s and I_2 between ω_s and up to the infinity бесконечности. A typical plot of an integrand function is shown in Fig. 10.25.

The most complicated procedure is an integration along the segment [0, ω_s] because on this interval the function oscillates intensively and has sharp peaks. This behavior of the integrand impedes the numerical integration procedure. Most preferable integration procedures for such functions are adaptive quadrature techniques the purpose of which is to get a result with a prescribed accuracy and in a shortest time possible. A description of such procedure can be found in [11].

At the interval [ω_S, ∞] the integrand function decreases and is flat enough. The complexity of this integral is that it has an infinite top limit which must be replaced by a finite one in order to use the numerical technique. Also, one can use the property of an asymptotically exponential fall of the function to get a good estimation of the "real top limit" for the integration.

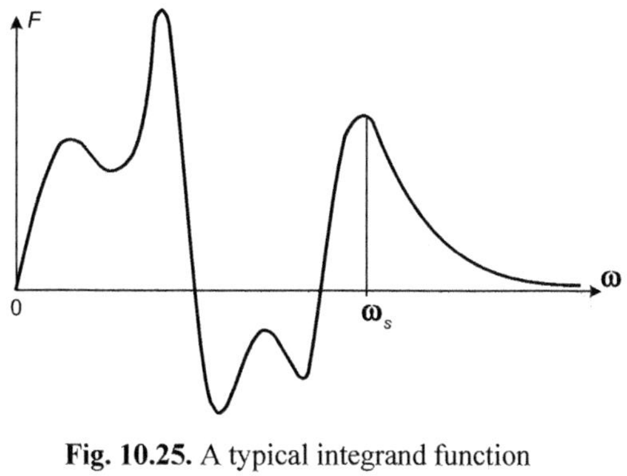

Fig. 10.25. A typical integrand function

Forming the generalized coordinate covariance matrix of the dimensions $m \times m$ (m is the number of generalized coordinates) needs a lot of computing. In order to reduce the amount of computing, it is useful to consider possible simplifications to the computational process. An analysis of this kind can be found in papers [26, 27], and it shows that the cross-correlation of rth and s^{th} modes of the natural oscillations $\langle y_r, y_s \rangle$ as compared to the principal terms $\langle y_r, y_r \rangle$ and $\langle y_s, y_s \rangle$ can be estimated using an approximate relationship

$$\langle y_r, y_s \rangle / \langle y_r, y_r \rangle \cong \frac{1}{1 + \frac{1}{\gamma^2}\left(1 - \frac{\omega_r}{\omega_s}\right)^2} = \Delta_{rs} . \tag{10.169}$$

The use of the formula (10.169) enables one to tentatively estimate the magnitude of side elements of the covariance matrix and not to calculate integrals for which the estimate (10.169) yields a small value negligible comparing to one. For example, at $\omega_r = 3\omega_s$ and $\gamma = 0.1$ the estimate (10.169) gives the value of 0.005, and the respective integral should hardly be taken into account.

References

1. Aisenberg JM (1999) Earthquake in Spitak on December 7, 1988. Some lessons and inferences (in Russian). In: Earthquake Engineering, Safety of Structures 1:6-9
2. Banah LJa, Perminov MD (1972) Investigation of complex dynamical systems using weak links between subsystems (in Russian). Mashinovedennie, 4:35–41
3. Bathe KJ, Wilson EL (1976) Numerical Methods in Finite Element Analysis. Prentice-Hall, Englewood Cliffs, NJ
4. Birbraer AN (1998) Analysis of structures for earthquake resistance (in Russian). Nauka Publishing House, Sankt-Petersburg
5. Blekhman II, Myshkis AD, Panovko YG (1983) Mechanics and applied mathematics: Logic and peculiarities of mathematical applications (in Russian). Nauka, Moscow
6. Bolotin VV (1984) Random vibrations of elastic systems. Martinus Nijhoffs, Hingham, MA
7. Clough RW, Penzien J (1975) Dynamics of structures. McGraw-Hill, New-York
8. Davenport AG (1962) The response of slender, line-like structures to a gusty wind. In:: Proc. Inst. Civil Eng, 23:389–408
9. ENV 1991-2-3. Eurocode 1 (1995) Basis of Design and Action on Structures. Part 2–3: Wind Actions. CEN/TC250/SC1, Brussels
10. Eurocode-8 (1994). Design provisions for earthquake of structures. European Prestandard. European Committee for Standardization, Brussels
11. Forsythe GE, Malcolm MA, Moler CB (1977) Computer methods for mathematical computations Prentice-Hall, Englewood Cliffs, N.J.
12. Gabrichidze GK (2001) How much the earthquake-resistant design regulations ensure the declared principles (in Russian).In: Earthquake Engineering, Safety of Structures, 1:40-42

13. Galchenko LA (1980) On a character of the deflection function of an infinite beam obtained by using various models of bedding (in Russian). In: Proc. Dnepropetrovsk Institute of Transport Engineers, Issue 210/27, pp.132–138
14. Gantmacher FR (2000) The theory of matrices, vol. 1-2. AMS Chelsea Pub. Co., New York
15. Goldenblatt II. et al. (1979) Models of earthquake resistance of constructions (in Russian). Nauka Publishing House, Moskow
16. Gordeeva SP (1966) . Analyzing bar systems for resonance using computers (in Russian). In: Studies on mechanics of bar systems and continua: Proc. Leningrad Civil Engineering Institute, vol.49. pp. 305–314
17. Guyan RJ (1965) Reduction of stiffness and mass matrices. AIAA Journal, vol.3, 2:380
18. Hitching D (Ed.) (1992) A Finite Element Dynamics Primer. NAFEMS, Glasgow
19. Ilyichyov VA (1970) Determining parameters of oscillation of a two-mass system supported by an elastic inertial semi-space (in Russian). In: Dynamics of structures: Proc. Kucherenko CRISC, pp.57–68.
20. ISO/DIS 3010 (2000). Bases for design of structures. Seismic actions on structures. International Organization for Standardization
21. Kazakevich MI, Kuliabko VV (1998) Relevant problems of structural dynamics (in Russian). Steel constructions, 1:65–74
22. Kolousek V (1965) Dynamics of structural constructions (in Russian). Stroyizdat, Moscow
23. Kucherenko CRISC (Central Research Institute for Structural Constructions) (1967) Instruction on analysis of floor panels of industrial buildings subject to dynamic loads Moscow: *Stroizdat*, 1967. (*In Russian*)
24. Kucherenko CRISC (Central Research Institute for Structural Constructions) (1970) Instruction on analysis of load-carrying constructions of industrial structures under dynamical loads (in Russian). Stroyizdat, Moscow
25. Panovko YaG (1980) Introduction to the theory of mechanical oscillations (in Russian). Nauka Publishing House, Moskow
26. Petrov AA, Basilevsky SV (1977) . On an allowance for a spatial correlation of speed pulsations in the determination of a dynamic component of a wind load (in Russian). In: Structural mechanics and analysis of structures, 5:67–71
27. Petrov AA, Bazilevsky SV (1979) An influence of a mutual correlation between generalized coordinates in random oscillations of linear systems (in Russian). In: Structural mechanics and analysis of structures, 4:52–56
28. Rabinovich IM, Sinitsyn AP, Luzhin OV, Terenin VM (1970) Analysis of structures subject to pulse actions (in Russian). Stroyizdat, Moscow
29. Reznikov RA (1971) Solution of structural analysis problems using digital computers (in Russian). Stroyizdat, Moskow
30. Sorokin ES (1968) . Key assumptions of an analysis of structures for pulse loads (in Russian). Research works of Moscow Institute of Transport Engineers. Problems of applied mechanics, Issue 260, Stroyizdat, Moscow
31. Sorokin ES (1972) Dynamical characteristics of building materials and constructions (in Russian). In: Korenev BD, Rabinovich IM (eds) Manual on dynamics of buildings. Stroyzdat, Moskow, pp. 38–61
32. Spence PW, Kenchington CJ (1993) The role of damping in finite element analysis. NAFEMS, Glasgow

33. Tseitlin AI (1975) To the theory of frequency-independent internal resistance during oscillations of elastic systems (in Russian). In: Dynamics of structures: Proc. Kucherenko CRISC, 56:5-23
34. Tseitlin AI, Bernstein AS, Guseva NI, Popov NA (1987) A new edition of the "Wind loads" section in the chapter of SniP "Loads and actions" (in Russian). In: Structural mechanics and analysis of structures, 6:28-33
35. Tseitlin AI, Guseva NI (1979) Statistical methods for analysis of structures under group dynamic influences (in Russian). Stroyizdat, Moscow
36. US Nuclear Regulatory Commission (1987) NUREG-0800. Standard Review Plan for the Review of Safety Analysis Reports for Nuclear Power Plants.
37. Vlasov VZ (1963) Thin-walled elastic beams. Izrael program of scentific translation, Jerusalem
38. Wilson EL (1997) Three dimensional dynamic analysis of structures. Computers and Structures, Inc., Berkeley CA
39. Yegupov VK et al. (1997) Simulation and automation of calculations of buildings (structures) on seismic effects. Computer & Structures, 6:1065–1083

11. A Word Instead of a Conclusion

A little boy
came to his dad
Asking him a question:
" What is good
and what is bad?"
That was his
Obsession.
There's no secret,
Not a thing, —
Hear me say, my dears,—
Father told him
Everything,
That's how it
Appears.

V.V. Mayakovsky

In the preface to this book we stated that it was our conscious intention not to give recommendations as to what particular software product to purchase. We have tried hard to keep this promise of ours. Though, we cannot leave this issue aside, therefore below we give some considerations to be taken into account when planning to buy a finite element computer program for structural analysis. Now that the reader has come to this place and thus has read the book as we do hope, he might already have the idea of the authors' opinion concerning "good and bad" CAD programs, so such considerations may prove relevant.

11.1 Our Recommendations to Potential Users

Historically, the finite element analysis once was a tool of an analyst. But a good deal of its successful applications have made it a principal tool of a designer, too. This transition has been enabled also by the possibility to simulate the "real world" using computers. Boundaries of the "real world" are now determined solely by the complexity of objects under consideration, and capabilities of a highly qualified analysis are growing all the time.

The key goal of the finite element analysis is to determine how an element or a whole system responds to a specific set of external actions. Results of the analysis can be used to verify a design solution. Also, those can be employed to improve and optimize the development. Of course, all this is possible assuming that a structure has been simulated correctly, its environment has been specified accurately, the software itself computes perfectly well, and results of the computation are interpreted and applied properly.

Most often, practical computations are made by software of an industry-oriented type, though we know examples of using single developments created specially to solve a particular problem. It is not a usual practice that such developments are built from the scratch. Commonly, these ones are based on an industry-oriented program modified by authors or third-party developers to fit special conditions of application. In the case of success (and if there's a demand) these modifications can be included afterwards in the next version of the software.

What do we mean by speaking of industry-oriented software? What is the difference of that from numerous, sometimes really sophisticated pieces of software often characterized in dissertations in a way like this: "*... scientific ideas of the candidate for the degree have been implemented in a software **AMATEUR-5** developed by him and recommended for a wide practical employment in the industry, and in other industries (?) too* ".

First of all, this must be a program that does not require the presence of its developer to be used successfully. That is, it must be at least properly documented. Just a manual printed on paper might prove not enough: there should be timely documentation upgrades, notifications of changes, online help etc.

Second, this software should not confine itself to the solution of a particular problem. Instead, it should solve a class of problems with clearly outlined limits. Within those limits, a system of input data verification, diagnostics and other "fool-proof"[1] measures should be designed.

Other important properties of an industry-oriented software product include its robustness with respect to hardware failures, the simplicity of installation and version upgrade, the capability of being customized for a particular user.

Finally, things that are a must for a contemporary industry-oriented software development are: upgrades on a regular basis, user support service, flexibility with respect to the ever-changing hardware and software operating platforms.

A user that chooses a software to buy should get familiar with it to such extent as to be able to say to himself:

- I understand this program, I realize what can be done with it and what environment it operates in;
- I understand what techniques can be employed to simulate an object of mine and its environment;
- I understand limitations of my software's capability to imitate the "real world".

What should be realized perfectly well is the need for validation of results. It can be provided by both a system of testing used during the software development and tools for analysis of results. In this regard we recall peculiar words by

[1] We do not intend to abuse a potential customer in any way by mentioning *fool proof systems*, but the facts of life are that software does need to be protected against fools.

R.A. Reznikov, our country's guru in development of software aids for structural design. At one of seminar schools on the finite element analysis, popular in days of yore, Dr. Reznikov prefaced his report with the following paradoxical statement:

"*The development of powerful computational methods such as FEM has given birth to various and numerous mistakes in numerical results, of both quantitative and qualitative nature. Contemporary computer-aided design tools help these mistakes to be propagated and widely implemented in the engineering practice by way of being transferred from journal reports to real structural projects*"[2].

Any modern industry-oriented software passes at least two phases of testing. A so-called *alpha testing* is performed by its developers to validate the serviceability and correctness of the program. During it specially selected problems are solved, not necessarily big ones but capable of revealing expected faults. Collections of test problems have been published on Internet and are available online. The most interesting thing is the second phase called *beta testing*. It is performed by a wide circle of qualified users related to the developers somehow. The objective of it is to verify both the serviceability and the convenience of the software for the solution of real world problems[3].

The beta testing is performed before the wide marketing of the product starts, and results of such testing are the basis for polishing up the software system. Many customers deem it important to know who (what experts and/or institutions) have done the beta testing, so respectable developers provide this information too.

Experience shows there is no testing system capable of ensuring the perfect fail-safety of a program. Faults and errors can be found after years of a software's operation. Though, if the software suggests a few ways to solve one problem, it prevents the danger of previously undetected errors affecting the solution. This multi-functionality is one of most essential features of a mature software product. Our gentle (but insistent) recommendation to developers, expressed many times in this book, that they should provide a variety of alternative ways to perform functions of their software, has been inspired exactly by this circumstance.

If we treat solutions of test problems obtained in the course of testing as an experimental validation of a software, we should recall at once that no sequence of experiments can be a proof of correctness of a theory being validated. There is a well-known statement by A. Einstein:

"*An experiment ... never says 'Yes' to a theory. In most advantageous cases it says 'Maybe', while most often 'No'... Every theory is going to hear its 'no', sooner or later*".

These words apply to the experimental validation of a software's functionality thoroughly.

---

[2] Quoted by us from memory.

[3] There is a jesting statement: alpha testing assures one that a program doesn't work in hands of its developer, and beta testing confirms that it does not work in hands of the user, too.

When choosing a software product, the users usually want to know whether it has been certified[4]. Often, by being "certified" one means that an authority should provide a guaranty that the software works correctly. This expectation is inspired also by some decisions of regulatory bodies which require a mandatory certification of software tools used for the purpose of design. Unfortunately, the hope of the user for this guaranty has no reasonable basis because the only thing really certified by a certificate of conformity issued by empowered authorities in our (former USSR) countries is that the software complies with codes (such as SNiP) as declared by its developers. Such certificate often omits the fact that the software may comply only with some provisions of a design code rather than with the whole code. Also, the expiration of the certificate after a certain time is a questionable issue. What can really change in the software during a year or two for which its certificate is issued, if the design codes remain the same? Very little[5]. Therefore, the certificate's validity limitation borrowed from industrial goods certification practice has little to do with information technologies.

In Western countries, another procedure of public recognition called "benchmarking" is widely popular. The sense of this procedure is that respectable and independent institutions (most often, associated with organizations of users of software products of particular purpose) perform a careful check of a software using rich collections of test problems. Such collections are widely known and are being expanded all the time [2, 6]. Procedure definitions, results of testing, test problems are published in reports and journals and are also available online on Internet.

The value of such collections is that they provide a thorough testing of all aspects of a software product. The completeness of the validation is very important, i.e. that the software should pass all problems of a collection. As a prominent physicist noticed nimbly [5]:

"*...if the clock is striking thirteenth hour, it is not just the thirteenth strike that one counts wrong. It reflects upon the preceding twelve strikes being right, too*".

A developer can be an initiator of testing, too. He can order a benchmark to be performed and demand that his product be tested in comparison with certain counterpart programs (by accuracy, required resources, convenience of interface, completeness and comprehensibility of reference manuals etc.). It may seem there is a trick here: one can point at an obviously weaker competitor and win the day easily. But the matter is that the price of the program will be assigned by the developer in comparison with that competitor program. So this is not the way to become a world champion by winning a village competition. Hence a natural aspiration for a worthy adversary. Also, one cannot contrive to deviate from the competitor's price too much downwards, which prevents a damping policy on the software market.

If the developer has ordered no special comparison, then the competitor programs are defined by an institution that performs the testing. Those institutions are

---

[4] We mean Russian users here.

[5] Or everything if the copy of the program used has proved corrupted.

usually too much concerned with their reputation to allow any conscious deviations from the objective estimation. They provide the customer with their report, and with the latter's permission the testing results can be published in a special journal such as The BENCHmark Magazine. The testing report can be presented to a buyer of the software at his request.

Having compared two testing systems described above, we state that the benchmarking is a more flexible and efficient procedure than our country's certification system performed by state bodies (but at the developer's expense). Also, it is much more formal in its idea.

In Russia systems of software testing and certification are often mandatory for some bodies and industries (see [7]), but results of such industry-oriented testing are, most often, unavailable to a wide circle of users, except for the very fact that a license has been issued by a particular body. It is very advantageous for users or potential users, as well as for developers, to have the complete information of the contemporary software market. Though, it is hardly possible to make a full list of finite element analysis programs currently in use: their number comes to hundreds.

There are a few most widely used families of such programs, well-known nowadays. The unquestionable leader among those is an **ANSYS** software complex which is used by over a million legal users. It was developed primarily as a mechanical strength analysis oriented software. Then, it expanded its scope by including thermal physics, electromagnetic fields and fluid/gas mechanics, thus having become a cross-disciplinary software (its developers declare it to be a multi-physics system). A group of finite element analysis programs called **NASTRAN** and owned by NASA is comparable with it by the number of legal users. A lot more systems are also known and used all over the world: **ABAQUS**, **COSMOS/M**, **GTSTRUDL**, **LUSAS**, **DIANA** etc. For these and other known software systems, an online journal Desktop Engineering Magazine available on the Internet occasionally publishes comparative tables presenting data of their functionalities and selling prices. A similar work is done by a known research company WyzeTek Inc. It has published a big comparative table on its site (***www.deskeng.com***) that contains data on the functionality of most popular software systems oriented at the finite element analysis. The objects of comparison include sets of functions of pre- and postprocessors, those of the solver kernel. The comparison involves a thorough list of parameters (over 200 items) developed by the authors of the book [1] for this purpose. Even more detailed analytical comparative reviews of the said developments have been published (see [3] for examples). Though, similar extended reviews of functionality and comparisons of quality are still to be made for most popular public software systems of our country's origin, including **Lira Windows**, **MicroFE** and **SCAD** for today. There's a natural wish to see a comparison of this kind performed by an equitable and really independent analyst. Though, the issue of organizing a panel of independent experts can hardly be solved because an absolute majority of experts are somehow related to certain teams of developers, either professionally or friendly. In this situation an algorithm proposed in [4] might prove useful. It involves an estimation by a panel of experts including representatives of interested parties.

Issues of testing and expert estimation of software (for reasons stated above we prefer the expertise rather than the certification) are related to another keen problem yet to be resolved: standardization in information technologies for structural design. In many other fields of information technologies the necessity to perform a standardization — even at the international level — has been recognized long ago[6]. A characteristic example is the standardization of programming languages. The recognition of the need for standardization and subsequent activities of appropriate standardization committees has been a real kick-off for creating truly industry-oriented software systems and eventually stimulated advancements in this area of information technologies.

Of course, the international standardization in the field of the structural design software is encumbered by substantial differences between structural design codes. Sometimes these differences between countries are even at the conceptual level. Nonetheless, there are many common things: laws of mechanics are the same in all parts of the world. Things like rules of signs, terms and definitions, requirements to preprocessor and postprocessor tools, interface with existing databases and graphical systems such as **AutoCAD**, the compliance with national regulations etc., can be and must be standardized in order to satisfy the users' interests as best one can. Once most authoritative experts in computer-aided structural design have developed and adopted standards of this kind, the certification will immediately have a clear sense as a confirmation that a software complies with the effectual standard.

Now it's time we should say goodbye to our patient reader who has guts enough to read up to these final lines. It cannot be emphasized enough that neither expertise nor certification can be a full indulgence for a software product. The biggest thing they can and must do is to help users orient themselves in the ocean of commercial proposals and make a responsible decision. We insist that it is you, our dear reader and user of CAD software, who is able to find out advantages/disadvantages of any advertised product regarding your particular needs better than anyone else, without heeding colorful labels like "*Certified by such and such big bug*" too much. So, now that we have to recommend an approach to the selection of software, it cannot be said better than by words of a song from a popular motion picture:

Think for yourself,
Decide for yourself,
What it's you wish to have!

* * *

And in our final note we would like to explain the title of this section, "A word instead of a conclusion". The matter is that a conclusion necessarily implies a completion, and the authors do not think their work has been completed together

[6] An example of this is a standard ISO/IEC 14882. Standard for the C++ Programming Language, adopted in August 1998.

with last words of this book. We hope for our new meetings with readers in the future.

References

1. Adams V, Askenazi A (1998) Building Better Products with Finite Element Analysis. OnWord Press, N.Y.
2. ASME Pressure Vessel and Piping Division, Committee on Computer Technology (1972) Program Verification and Qualification Library.
3. Avedian AB, Danilin AN (2000) Structural strength for non-experts (in Russian). In: CAD and graphics, 1:75-83, 2:38-40, 3:39-46, 4:62-69
4. Karimov IA (1987) On a group expert estimation of features measured by relationship scales (in Russian). In: Technical cybernetics, 6:13-21
5. Khurghin YI (1979) How to comprehend the incomprehensible (in Russian). Znanie, Moscow
6. MacNeal RH, Harder RC (1985) A Proposed Standard Set of Problems to Test Finite Element Accuracy. Finite elements in Analysis and Design. North-Holland
7. Russian State Supervisory Committee on Nuclear and Radiation Safety (1994) Provisions on certification of software tools used to validate or ensure the safety of nuclear- and/or radiation-dangerous objects and production processes (*In Russian*). *RD*-03-17-94.

Appendix

CAD software involves a great deal of mathematical procedures and often provide its user with multiple options of solution methods for a particular mathematical problem (such as a linear equation system solution method). In order to handle these capabilities properly, one should be familiar with those procedures at least generally. What is especially important, one should understand their mechanical interpretation. These appendices present some useful information of this kind.

A.1 Jordan Exclusions and their Role in Structural Mechanics

The technique of Jordan exclusions is a powerful tool widely applicable in structural mechanics. For example, it is easy to use the Jordane exclusions to build a universal procedure of forming a stiffness matrix of an arbitrary finite element with partially removed constraints in places where it's attached to nodes if the stiffness matrix is known for the same element with no constraints removed [3, 4]. The Jordan exclusions have a curious and unexpected application in the theory of systems with unilateral constraints [2].

A careful analysis helps one see that a static condensation procedure, one widely used both in the theory and applications of finite elements, can be interpreted mathematically as a sequential use of a few Jordan exclusion steps. Also, it can be proved that the Jordan exclusions are closely related to a known Legendre transformation employed to build and validate some types of variational problem statements. Therefore, the Jordan exclusions are worth studying by structural mechanics specialists.

At the same time, popular manuals on structural mechanics and modern courses of linear algebra don't even mention the Jordan exclusions which thus fall out of view of both professors and students. All this makes it reasonable to present both mathematical formulations and some applications of the Jordan exclusions here.

A.2 General Description

Consider a system of linear relationships

$$y_i = a_{i1}x_1 + a_{i2}x_2 + \cdots + a_{in}x_n \quad (i = 1, \cdots m) \tag{A.1}$$

with n independent variables $x_1, ..., x_n$. It can be represented as a table

	x_1	...	x_s	...	x_n
$y_1 =$	a_{11}	...	a_{1s}	...	a_{1n}
...	...	...	...	...	...
$y_r =$	a_{r1}	...	a_{rs}	...	a_{rn}
...	...	...	...	...	...
$y_m =$	a_{m1}	...	a_{ms}	...	a_{mn}

(A.2)

A step of Jordan exclusion [6] performed for Table (A.2) with a governing element $a_{rs} \neq 0$ which stands in r^{th} *governing row* and in s^{th} *governing column*, is an operation of solving r^{th} equation of the system (A.1)

$$y_r = a_{r1}x_1 + a_{r2}x_2 + \cdots + a_{rs}x_s + \cdots + a_{rn}x_n \tag{A.3}$$

with respect of the independent variable x_s and substituting the obtained expression of x_s to all other equations of (A.1)[1]. The mentioned condition $a_{rs} \neq 0$ ensures that such solution is possible. It follows from (A.3) that

$$x_s = -\frac{a_{r1}}{a_{rs}} x_1 - \frac{a_{r2}}{a_{rs}} x_2 - \cdots - \frac{a_{r,s-1}}{a_{rs}} x_{s-1} + \frac{1}{a_{rs}} y_r - \frac{a_{r,s+1}}{a_{rs}} x_{s+1} - \cdots - \frac{a_{rn}}{a_{rs}} x_n .$$

Substituting the expression to all equations of the system (A.1) yields new linear relationships in which the variables y_r and x_s have swapped over so that the Jordan exclusion step makes the table look as follows

	x_1	...	y_r	...	x_n
$y_1 =$	b_{11}	...	b_{1s}	...	b_{1n}
...	...	...	...	...	...
$x_s =$	b_{r1}	...	b_{rs}	...	b_{rn}
...	...	...	...	...	...
$y_m =$	b_{m1}	...	b_{ms}	...	b_{mn}

. (A.4)

In this table the elements of the governing row and governing column are shown darkened for better visual perception.

If we track changes made to coefficients of the table, we will find the process of building these new coefficients governed by the following simple rules that essentially describe a formal algorithm of a single Jordan transformation step:

- the governing element is to be replaced by the element $b_{rs} = 1/\,a_{rs}$;
- the other elements of the governing column are to be replaced by the elements $b_{is} = a_{is}/a_{rs}\ (i \neq r)$;
- the other elements of the governing row are to be replaced by the elements $b_{rj} = -a_{rj}\,/a_{rs}\ (j \neq r)$;
- "common" elements, i.e. elements belonging to neither the governing row nor the governing column, are to be calculated by the formula

$$b_{ij} = a_{ij} - \frac{a_{is} a_{rj}}{a_{rs}} \quad (i \neq r,\ j \neq r) .$$

Assume that $m = n$ in the system (A.1) and the matrix of coefficients $\mathbf{A} = \|[a_{ij}]\|$ written in Table (A.2) is not degenerate. Under these conditions one can prove that n Jordan exclusion steps can always be performed successively with (A.2) to obtain the table

	y_1	y_2	...	y_n
$x_1 =$	c_{11}	c_{12}	...	c_{1n}
$x_2 =$	c_{21}	c_{22}	...	c_{2n}
...	...	...	...	...
$x_n =$	c_{n1}	c_{n2}	...	c_{nn}

. (A.5)

in which all dependent variables $y_1,\ldots,y_n$ have been made independent (moved to

[1] In Table (A.2) the governing element a_{rs} is shown greyed intentionally.

the top of the table) while all independent variables $x_1,\ldots,x_n$ have been made dependent (moved to the left part of the table).

If governing elements at some Jordan exclusion steps were taken from other place than the main diagonal of the matrix, then the initial order of the variables will be broken and one will have to move some columns/rows of the table to make (A.5) of it. Let us assume that those have been already moved and the initial order of the variables $x_1,\ldots,x_n$ and $y_1,\ldots,y_n$ has been restored. Obviously, elements of the table transformed to (A.5) make an inverse matrix $\mathbf{A}^{-1} = \|[c_{ij}]\|$.

To solve the system of n linear equations with n unknowns

$$a_{i1}x_1 + a_{i2}x_2 + \cdots + a_{in}x_n = a_i \quad (i = 1, \cdots n) \tag{A.6}$$

provided the matrix $\mathbf{A} = \|[a_{ij}]\|$ is non-degenerate, we will rewrite this system as follows

	x_1	x_2	...	x_n	1
$0 =$	a_{11}	a_{12}	...	a_{1n}	$-a_1$
$0 =$	a_{21}	a_{22}	...	a_{2n}	$-a_2$
...	...	...	...	...	...
$0 =$	a_{n1}	a_{n2}	...	a_{nn}	$-a_n$

. (A.7)

Let us make n Jordan exclusion steps with governing columns other than the free term column. After each step, zero comes to the top of the table, and the respective column becomes unneeded (all its values are multiplied by zero) therefore it can be striken out. The result is that only one column will remain (where the free term column used to be) which will contain the solution of the system of equations.

The well-known *Gaussian method* is different from this one only in that each Jordan exclusion step removes a governing row together with a governing column, and an expression of the respective x_i is written explicitly. After one of the unknowns is found at the last step, values of the others will be calculated by a successive substitution of the obtained unknown values to the expressions of x_i (in the inverse order to the exclusion).

As for structural mechanics problems, the most interesting Jordan exclusions are those that involve governing elements from the main diagonal of the matrix, i.e. at $r = s$. This is the type of the Jordan exclusions we will deal with further.

A.3 Jordan Exclusions with the Stiffness Matrix of a Structure

Let's consider canonical equations of the slope deflection method represented as follows, according to the formula (3.8)

$$\mathbf{Ku} = \mathbf{P},$$

where $\mathbf{K}$ is a stiffness matrix of a structure, $\mathbf{u} = \|[u_1, u_2, \ldots, u_n]\|^{\mathrm{T}}$ is a vector of its nodal displacements, $\mathbf{P}$ is a vector of generalized nodal forces.

Let us write this system of equations as a table where each row, in its mechanical sense, defines a total reaction R_i in an additional i^{th} constraint of the slope deflection method's principal system.

	u_1	u_2	...	u_n	1
$R_1 =$	k_{11}	k_{12}	...	k_{1n}	$-P_1$
$R_2 =$	k_{21}	k_{22}	...	k_{2n}	$-P_2$
...	...	...	...	...	...
$R_n =$	k_{n1}	k_{n2}	...	k_{nn}	$-P_n$

. (A.8)

Here $k_{ij} = k_{ji}$ are elements of the symmetric stiffness matrix, u_j is an unknown generalized displacement (in a solid mechanics finite element model it is either a linear displacement or a slope of a node), P_i is an external load in the direction of ith additional constraint. All displacements u_j here are independent variables, and the goal of the solution is to find values of those that make the reactions R_i in the additional constraints zero. It corresponds to the fact that the constraints in question are present only in the imaginable principal system of the slope deflection method while absent in the real structure.

A Jordan exclusion step with Table (A.8), for example, with the governing element k_{11}, yields a table that corresponds to a certain form of a mixed method and looks like

	R_1	u_2	...	u_n	1
$u_1 =$	δ_{11}	δ'_{12}	...	δ'_{1n}	D_1
$R_2 =$	k'_{21}	k^*_{22}	...	k^*_{2n}	$-P^*_2$
...	...	...	...	...	...
$R_n =$	k'_{n1}	k^*_{n2}	...	k^*_{nn}	$-P^*_n$

, (A.8)

where we assume this, according to the general Jordan exclusion algorithm defined earlier

$$\delta_{11} = \frac{1}{k_{11}}, \qquad k'_{i1} = -\delta'_{1i} = \frac{k_{i1}}{k_{11}}, \qquad D_1 = \frac{P_1}{k_{11}},$$

$$k^*_{ij} = k_{ij} - \frac{k_{i1}k_{1j}}{k_{11}}, \qquad P^*_i = P_i - \frac{k_{i1}P_1}{k_{11}} \quad (i \neq 1,\ j \neq 1). \tag{A.9}$$

Recalling that the reaction R_1 must be equal to zero, we can strike out the first column. Also, for now we don't care about the displacement u_1, thus omit the first row. The transformed and truncated matrix of reactions $\mathbf{K}^* = \|[k^*_{ij}]\|$ (its order being lower by one after deleting the first row and column) will correspond to a structure with its first constraint removed, and the values $P^*_2 \ldots P^*_n$ will be its load vector.

Note that the formulae of the elements k^*_{ij} of the truncated stiffness matrix presented above are exactly the same as formulae of these components obtained by one step of the direct Gaussian factorization. It means that steps of the direct Gaussian exclusion performed with the stiffness matrix correspond to a removal of an extra constraint from the principal system of the slope deflection method.

By the way, elements of the governing row and governing column of the matrix transformed according to (A.8) also have a clear physical sense. The first equation of the system (A.8) expresses a geometrical condition that the summed displacement in the direction of the first constraint is equal to the value of u_1. Obviously, δ_{11} is the same displacement but caused by the unit force $\overline{R}_1 = 1$, while δ'_{1i} is the same displacement caused by the given unit displacement $\overline{u}_i = 1$. Similarly, the value k'_{i1} is a reaction in i^{th} constraint caused by the unit force $\overline{R}_1 = 1$. As one can see in (A.9), the Jordan exclusion gives an anti-symmetry relationship

$$k'_{i1} = -\delta'_{1i},$$

which is nothing but a known theorem of structural mechanics: Gvozdiov's theorem of reaction/displacement reciprocity [5].

Let's illustrate the presentment by a simple example. Fig. A.1 shows a system which we are going to analyze using four unknown displacement u_1, u_2, u_3 and u_4.

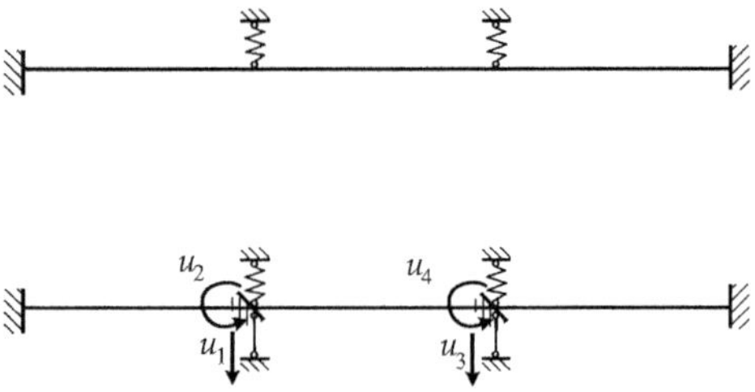

Fig. A.1. A structural analysis example

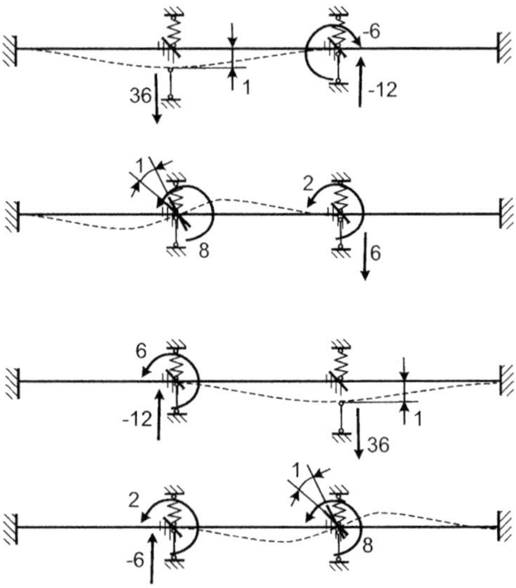

Fig.A.2 Reactions in constraints of the principal system of the slope deflection method caused by unit displacements of the constraints

Let us assume $L = 1$, $EI = 1$, $c = 12$ in a certain consistent system of units where: L is the length of one span of the beam, EI is its flexural rigidity, c is the stiffness of each spring. We can introduce additional constraints and find reactions in those by common methods of structural mechanics using unit displacements. The unit states of the system and their respective constraint reactions are shown in Fig. A.2.

Writing these reactions in the respective table will give us a stiffness matrix of this kind:

	u_1	u_2	u_3	u_4
$R_1 =$	36	0	-12	-6
$R_2 =$	0	8	6	2
$R_3 =$	-12	6	36	0
$R_4 =$	-6	2	0	8

. (A.10)

Now, performing four steps of Jordan exclusions with governing elements taken from the main diagonal will give:

after the first step

	R_1	u_2	u_3	u_4
$u_1 =$	1/36	0	1/3	1/6
$R_2 =$	0	8	6	2
$R_3 =$	-1/3	6	32	-2
$R_4 =$	-1/6	2	-2	7

, (A.11)

after the second step

	R_1	R_2	u_3	u_4
$u_1 =$	1/36	0	1/3	1/6
$u_2 =$	0	1/8	-3/4	-1/4
$R_3 =$	-1/3	3/4	55/2	-7/2
$R_4 =$	-1/6	1/4	-7/2	13/2

, (A.12)

after the third step

	R_1	R_2	R_3	u_4
$u_1 =$	7/220	-1/110	2/165	23/110
$u_2 =$	-1/110	8/55	-3/110	-19/55
$u_3 =$	2/165	-3/110	2/55	7/55
$R_4 =$	-23/110	19/55	-7/55	333/55

, (A.13)

finally, after the fourth step

	R_1	R_2	R_3	R_4
$u_1 =$	26/666	-14/666	11/666	23/666
$u_2 =$	-14/666	110/666	-23/666	-38/666
$u_3 =$	11/666	-23/666	26/666	14/666
$u_4 =$	23/666	-38/666	14/666	110/666

. (A.14)

It's easy to see how at each Jordan exclusion step there appears a mixed method matrix that corresponds to a mechanical system with the next constraint removed. The final matrix of (A.14) is a compliance matrix of a system where all four additional constraints have been removed. The check of the calculation is provided by the condition: the product of the initial stiffness matrix from (A.10) and the compliance matrix from (A.14) must be equal to the unit matrix.

Repeating all operations with a matrix of coefficients of canonical force method equations, we can see that each step in this case corresponds to restoring an extra constraint that has been removed to create a principal system of the force method. A direct mechanical interpretation of the Gaussian procedure (thus the Jordan exclusions, too) for the force method can be also found in [1].

A.4 Stiffness Matrix of a Finite Element Non-Rigidly Attached to its Nodes

A construction of the stiffness matrix of a particular finite elements is a necessary phase of analysis included in the process of forming the global stiffness matrix of a system **K** that describes the response of a structure with a discrete design model. The standard is that the stiffness matrix of a separate finite element is formed assuming nodal points of the element to be firmly fixed to nodes of the discrete model. It ensures the coincidence of displacements of the nodes and the finite element's nodal points. Though, this is not always the case. For example, a bar element can be attached by its nodes to those of a model via hinges or in such way that one of the bar's ends is capable of moving along some direction independently from the node's own displacement.

Generally, we will refer to a non-rigid connection between an element and a node in cases those are connected in such manner that some of the constraints between the node and the element have been removed. For the purpose of generality, we will also deal with a situation when a removed rigid constraint is replaced by a linearly responsive elastic link with its stiffness coefficient c.

At the fist view one may think this happens only with bar systems. But it is not true. One can recall a known problem of analysis of structures made of slabs connected to one another by hinges which arises in the aerodrome pavement design. Another example is a multi-wave shell with its separate waves connected non-rigidly.

A software for analysis of discrete systems might implement, potentially, a batch of different formulae describing components of the stiffness matrices and load vectors of finite elements for different methods of element-to-node connection. For example, bars supported in different ways within bar systems are often treated as different types of finite elements. Though, this approach violates the uniformity of the software development because it requires a lot of various formulae describing components of both the stiffness matrices and load vectors to be implemented by a program.

In this regard it seems interesting to build a general algorithm capable of yielding the stiffness matrix and load vectors of an arbitrary finite element non-rigidly connected to its nodes, provided the stiffness matrix and load vectors of the same element are known in the case it's connected rigidly. We present here this algorithm proposed in [4] and based on *modified Jordan transformations*.

So, let some finite element be connected to one of its nodes along its degree of freedom (let it be one No.l, to be particular) elastically rather than rigidly: we assume that there is an elastic link with its stiffness c between the node and the respective nodal point of the finite element. Next, let $q_1, \dots q_n$ be the set of displacements of all nodal points of the finite element (n is the full number of degrees of freedom of the finite element in question) and $u_1, \dots u_n$ be displacements of the system's nodes to which the element is attached. In the case of interest the components of the vectors $\mathbf{u} = \|[u_1, \dots u_n]\|^{\mathrm{T}}$ and $\mathbf{q} = \|[q_1, \dots q_n]\|^{\mathrm{T}}$ are related via the following kinematical relationships

$$u_l \neq q_l, \qquad u_i = q_i \quad (i = 1,\dots,n;\ i \neq l)\,.$$

The generalized stress T_l that conforms to the displacement q_l of the finite element's nodal point (a reactive force transferred from the node to the elastic spring) is defined as a product of the spring's stiffness by the difference of the displacements $u_l - q_l$, that is,

$$T_l = c(u_l - q_l). \tag{A.15}$$

Let $\mathbf{R} = \|[r_{ij}]\|$ and $\mathbf{R}_p = \|[r_{pi}]\|$ be the respective stiffness matrix and the load vector for the finite element in question when it is attached to its nodes rigidly. The vector of generalized inner stresses $\mathbf{T}$ that arise in nodal points of the finite element, which corresponds to the vector of the generalized displacements $\mathbf{q}$, will be defined by the formula

$$\mathbf{T} = \mathbf{R}\mathbf{q} + \mathbf{R}_p\,. \tag{A.16}$$

After finding the stress T_l from l^{th} equation of the system (A.16) and equating it to the expression of (A.15), we have

$$r_{ll}q_l + \sum_{i=1;\, i\neq l}^{n} r_{li}u_i + r_{pl} = c(u_l - q_l)\,,$$

wherefrom follows

$$q_l = \frac{cu_l - r_{pl} - \sum\limits_{i=1;\, i\neq l}^{n} r_{li}u_i}{r_{ll} + c}\,.$$

The potential energy E accumulated by the finite element and the elastic spring can be expressed as

$$\mathsf{E} = \frac{1}{2}\sum_{i=1}^{n}\sum_{j=1}^{n} r_{ij}q_iq_j + \sum_{i=1}^{n} r_{pi}q_i + \frac{1}{2}c(u_l - q_l)^2\,.$$

The potential E can be transformed once more by substituting the expressions of the displacements q_j via the displacements u_j and allowing for the symmetry of the matrix **R**. It will become this, up to an additive constant not depending on the displacement vector components

$$\mathsf{E} = \frac{1}{2}\sum_{i=1}^{n}\sum_{j=1}^{n} r_{ij}^{*} u_i u_j + \sum_{i=1}^{n} r_{pi}^{*} u_i ,$$

where designations are

$$r_{ij}^{*} = r_{ij} - \frac{r_{il} r_{lj}}{r_{ll} + c}, \qquad r_{pi}^{*} = r_{pi} - \frac{r_{il} r_{pl}}{r_{ll} + c}. \tag{A.17}$$

The matrix forms of these relationships follow:

$$\mathsf{E} = \frac{1}{2}\mathbf{u}^{\mathsf{T}}\mathbf{R}^{*}\mathbf{u} + \mathbf{u}^{\mathsf{T}}\mathbf{R}_{p}^{*}, \quad \mathbf{R}^{*} = \left\| \left[r_{ij}^{*} \right] \right\|, \quad \mathbf{R}_{p}^{*} = \left\| \left[r_{pi}^{*} \right] \right\|.$$

Obviously, the matrix $\mathbf{R}^*$ and the vector $\mathbf{R}_p^*$ are the desirable stiffness matrix and load vector for the element non-rigidly attached to the respective node of the structure.

Note that if the constraint is absent ($c = 0$) the formulae (A.17) become somewhat simpler and the same as those used for the common Jordan exclusion. It gives us the right to refer to the formulae (A.17) as those of *modified Jordan exclusions*.

Also, it is essential that the modified Jordan transformations introduced here are different from the usual ones in that the formulae (A.17) are applied to the whole matrix including its governing column and governing row. It is true at $c = 0$ too, which results in a simple zeroing of the governing column and governing row as one can notice easily.

Let us consider a set of all feasible finite elements derived from a certain standard finite element (rigidly connected to its nodes) by changing the method of its attachment to the nodes. Every element S of this set is determined by a couple: a stiffness matrix **R** and a load vector $\mathbf{R}_p$ so that

$$\mathrm{S} = \{\mathbf{R}, \mathbf{R}_p\}.$$

The modified Jordan transformation (A.17) maps the element S of this set to a new element $\mathrm{S}^* = \{\mathbf{R}^*, \mathbf{R}_p^*\}$ thus defining an operator which we will call a *J operator*. The symbolic notation of the *J* operator's action is

$$J[l,c]\mathrm{S} = \mathrm{S}^*.$$

Its parameters l and c that stand in the brackets are to emphasize the fact that the J operator depends on two arguments: No. l of the removed constraint and the stiffness coefficient c of the elastic constraint introduced.

Now let's turn to properties of the J operator.

1. The J operator is nonlinear because generally

$$J(S^{(1)} + S^{(2)}) \neq JS^{(1)} + JS^{(2)}.$$

2. It is necessary and sufficient for the J operator to exist that

$$r_{ll} + c \neq 0.$$

This property follows immediately from the formulae (A.17).

3. At $l_1 = l_2 = l$ a formula of product of J operators takes place

$$J[l,c_2]\,J[l,c_1] = J[l, \frac{c_1 c_2}{c_1 + c_2}]. \tag{A.18}$$

To see this, assume $J[l,c_1]S = S^{(1)}$, $J[l,c_2]S^{(1)} = S^{(2)}$. Then we have this, according to the formulae (A.17):

$$r_{ij}^{(1)} = r_{ij} - \frac{r_{il} r_{lj}}{r_{ll} + c_1}, \qquad r_{ij}^{(2)} = r_{ij}^{(1)} - \frac{r_{il}^{(1)} r_{lj}^{(1)}}{r_{ll}^{(1)} + c_2} =$$

$$= r_{ij} - \frac{r_{il} r_{li}}{r_{ll} + c_1} - \frac{\left(r_{il} - \dfrac{r_{il} r_{ll}}{r_{ll} + c_1}\right)\left(r_{lj} - \dfrac{r_{ll} r_{lj}}{r_{ll} + c_1}\right)}{r_{ll} - \dfrac{r_{ll} r_{ll}}{r_{ll} + c_1} + c_2} = r_{ij} - \frac{r_{il} r_{lj}}{r_{ll} + \dfrac{c_1 c_2}{c_1 + c_2}}, \tag{A.19}$$

and similarly

$$r_{pi}^{(2)} = r_{pi} - \frac{r_{il} r_{pl}}{r_{ll} + \dfrac{c_1 c_2}{c_1 + c_2}}. \tag{A.20}$$

Thus,

$$J[l, \frac{c_1 c_2}{c_1 + c_2}]S = S^{(2)},$$

and the formula (A.18) becomes evidently true.

4. The J operator is commutative, i.e.

$$J[l_1,c_1]\,J[l_2,c_2] = J[l_2,c_2]\,J[l_1,c_1].$$

At $l_1 \neq l_2$ its commutativity is obvious while at $l_1 = l_2 = l$ it follows from the formula of product of J operators (A.18) because the latter's right part includes the coefficients c_1 and c_2 symmetrically.

5. The J operator is invertible provided that $c \neq 0$, and the formula of its inversion is fairly simple

$$J^{1}[l,c] = J[l,-c]. \tag{A.21}$$

To see this, assume $c_1 = c$ and $c_2 = -c$ and use the formulae (A.19) and (A.20) to get

$$r_{ij}^{(2)} = r_{ij} - \frac{r_{il} r_{lj}(c-c)}{r_{ll}(c-c) - c^2} = r_{ij}, \quad r_{pi}^{(2)} = r_{pi} - \frac{r_{il} r_{pl}(c-c)}{r_{ll}(c-c) - c^2} = r_{pi},$$

thus

$$J[l,\text{-}c]\, J[l,c] = I\,,$$

where I is an identical operator.

At $c = 0$ the J operator prohibits the inversion because

$$J[l,0]\mathrm{S} = \mathrm{S}^0 = \{\, \mathbf{R}^0, \mathbf{R}^0_p \},\ \text{where } r^0_{ll} = 0\,,$$

therefore the repeated operation $J[l,0]\mathrm{S}^0$ is undefined according to the property 2.

The listed formal properties of J operators can be easily interpreted mechanically. Note that a successive action of two J operators (at $l_1 = l_2 = l$) is equivalent to a transfer to a finite element with two springs connected successively and replacing one l^{th} rigid constraint. Also note that the multiplication formula (A.18) is useful, for example, where one needs to repeat an analysis a lot of times varying stiffness properties of the springs (in nonlinear systems, optimizations etc.). If

$$\mathrm{S}^{(1)} = J[l,c_1]\mathrm{S}, \qquad \mathrm{S}^{(2)} = J[l,c_2]\mathrm{S},$$

then

$$\mathrm{S}^{(2)} = J\left[\, l, \frac{c_1 c_2}{c_1 - c_2} \right] \mathrm{S}^{(1)}\,. \tag{A.22}$$

It is easy to derive this from (A.18):

$$J\left[\, l, \frac{c_1 c_2}{c_1 - c_2} \right] J[l,c_1] = J[l,c_2]\,.$$

The formula (A.22) enables one to use just one direct modified Jordan transformation to get properties of the $\mathrm{S}^{(2)}$ finite element that has a spring with the stiffness c_2 in place of lth rigid constraint if one knows the properties of the $\mathrm{S}^{(1)}$ element with the spring of the stiffness c_1 in the same place.

The Jordan transformations can be also used efficiently to clear finite element stiffness matrices of components that correspond to *internal* degrees of freedom[2]. The exclusion of internal degree-of-freedom components from stiffness matrices is often referred to as a *static condensation procedure*. It is easy to notice that the formulae of recalculation of components of the condensed stiffness matrices and load vectors are identical to the formulae of the generalized Jordan transformations.

A.5 A Double Jordan Exclusion

Let's consider the case of a symmetric matrix $\mathbf{A} = \|[a_{ij}]\|$ of an order n and confine ourselves to Jordan exclusions with their governing elements at the main diagonal.

---

[2] See the footnote [1] in Chapter 4 for *internal and external* degrees of freedom of a finite element.

Choosing the governing element at the main diagonal helps keep the symmetry of the matrix (of course, up to signs of some elements) which lets us store only the upper triangle of the matrix. Assume the matrix **A** to be non-degenerate and not necessarily positive definite. Some diagonal element that we use for a Jordan exclusion, say a_{rr}, may well happen to be equal to zero. A situation like this is quite feasible in structural analysis because it conforms exactly to a system of equations of a mixed method applied to a discrete system — see the equations (3.75) and (3.77) to make sure of it. Without fearing to be misunderstood, we state that a direct execution of the Jordan exclusion with a zero governing element is "excluded".

In order not to violate a useful property of symmetry, we choose a certain index s such that $a_{rs} \neq 0$ and try to move independent elements y_r and y_s to the top of the table while x_r and x_s to the left column simultaneously. Let the original matrix in its tabular form be this:

	x_1	...	x_r	...	x_s	...	x_n
$y_1 =$	a_{11}	...	a_{1r}	...	a_{1s}	...	a_{1n}
...	...	...	...	...	...	...	...
$y_r =$	a_{r1}	...	$a_{rr} = 0$	...	a_{rs}	...	a_{rn}
...	...	...	...	...	...	...	...
$y_s =$	a_{s1}	...	a_{sr}	...	a_{ss}	...	a_{sn}
...	...	...	...	...	...	...	...
$y_n =$	a_{n1}	...	a_{nr}	...	a_{ns}	...	a_{nn}

. (A.23)

First we perform a Jordan exclusion with the governing element a_{rs}. Then the table (A.23) will become the table (A.24)

	x_1	...	x_r	...	y_r	...	x_n
$y_1 =$	b_{11}	...	b_{1r}	...	b_{1s}	...	b_{1n}
...	...	...	...	...	...	...	...
$x_s =$	b_{r1}	...	b_{rr}	...	b_{rs}	...	b_{rn}
...	...	...	...	...	...	...	...
$y_s =$	b_{s1}	...	b_{sr}	...	b_{ss}	...	b_{sn}
...	...	...	...	...	...	...	...
$y_n =$	b_{n1}	...	b_{nr}	...	b_{ns}	...	b_{nn}

, (A.24)

and the new element b_{sr} will keep its initial value according to general rules of the Jordan exclusion because

$$b_{sr} = a_{sr} - \frac{a_{ss} a_{rr}}{a_{rs}} = a_{sr} .$$

It is required that $a_{sr} = a_{rs} \neq 0$, which enables us to perform the next Jordan exclusion with the table (A.24) using the governing element a_{sr}. If now we swap the rows and columns with the numbers r and s, then we will have a final table that both keeps its symmetry and has the elements y_r and y_s moved to the top of the table.

References

1. Gantmacher FR (2000) The theory of matrices, vol. 1-2. AMS Chelsea Pub. Co., New York
2. Gordeyev VN (1971) On a finiteness of methods of solving a problem of quadratic programming (in Russian). In: Cybernetics, 1:85-88
3. Klempert YZ, Parikov VI, Slivker VI (1974) On a procedure of calculation of the stiffness matrix of a prismatic bar (in Russian). In: Analysis of spatial structures, issue XVI. Stroyizdat, Moscow, pp.179–189
4. Parikov VI, Slivker VI (1983) Stiffness matrix of a finite element non-rigidly connected to its nodes (in Russian). In: Bulletin of B.E. Vedeneyev *VNIIG*, collected scientific papers, vol. 164. Energoatomizdat, Leningrad, pp.20–28
5. Rabinovich IM (1954) A course in structural mechanics. Part 2. Gosstroyizdat, Moscow
6. Zukhovitsky SI, Avdeyeva LI (1967) Linear and convex programming (in Russian). Nauka, Moscow

Index

Foundations of Engineering Mechanics

Palmov	Vibrations of Elasto-Plastic Bodies (1998, ISBN 3-540-63724-9)
Babitsky	Theory of Vibro-Impact Systems and Applications (1998, ISBN 3-540-63723-0)
Skrzypek/ Ganczarski	Modeling of Material Damage and Failure of Structures Theory and Applications (1999, ISBN 3-540-63725-7)
Kovaleva	Optimal Control of Mechanical Oscillations (1999, ISBN 3-540-65442-9)
Kolovsky	Nonlinear Dynamics of Active and Passive Systems of Vibration Protection (1999, ISBN 3-540-65661-8)
Guz	Fundamentals of the Three-Dimensional Theory of Stability of Deformable Bodies (1999, ISBN 3-540-63721-4)
Alfutov	Stability of Elastic Structures (2000, ISBN 3-540-65700-2)
Morozov/ Petrov	Dynamics of Fracture (2000, ISBN 3-540-64274-9)
Astashev/ Babitsky/ Kolovsky	Dynamics and Control of Machines (2000, ISBN 3-540-63722-2)
Svetlitsky	Statics of Rods (2000, ISBN 3-540-67452-7)
Kolovsky/ Evgrafov/ Slousch/ Semenov	Advanced Theory of Mechanisms and Machines (2000, ISBN 3-540-67168-4)
Landa	Regular and Chaotic Oscillations (2001, ISBN 3-540-41001-5)

Foundations of Engineering Mechanics

Series Editors: Vladimir I. Babitsky, Loughborough University
Jens Wittenburg, Karlsruhe University

Muravskii	Mechanics of Non-Homogeneous and Anisotropic Foundations (2001, ISBN 3-540-41631-5)
Gorshkov/ Tarlakovsky	Transient Aerohydroelasticity of Spherical Bodies (2001, ISBN 3-540-42151-3)
Babitsky/ Krupenin	Vibration of Strongly Nonlinear Discontinuous Systems (2001, ISBN 3-540-41447-9
Manevitch/ Andrianov/ Oshmyan	Mechanics of Periodically Heterogeneous Structures (2002, ISBN 3-540-41630-7)
Lurie	Analytical Mechanics (2002, ISBN 3-540-42982-4)
Slepyan	Models and Phenomena in Fracture Mechanics (2002, ISBN 3-540-43767-3)
Nagaev	Dynamics of Synchronising Systems (2003, ISBN 3-540-44195-6)
Svetlitsky	Statistical Dynamics and Reliability Theory for Mechanical Structures (2003, ISBN 3-540-44297-9)
Neimark	Mathematical Models in Natural Science and Engineering (2003, ISBN 3-540-43680-4)
Babitsky/ Shipilov	Resonant Robotic Systems (2003, ISBN 3-540-00334-7)
Le xuan Anh	Dynamics of Mechanical Systems with Coulomb Friction (2003, ISBN 3-540-00654-0)
Perelmuter/ Slivker	Numerical Structural Analysis (2003, ISBN 3-540-00628-1)

GPSR Compliance

The European Union's (EU) General Product Safety Regulation (GPSR) is a set of rules that requires consumer products to be safe and our obligations to ensure this.

If you have any concerns about our products, you can contact us on ProductSafety@springernature.com

In case Publisher is established outside the EU, the EU authorized representative is:

Springer Nature Customer Service Center GmbH
Europaplatz 3
69115 Heidelberg, Germany

Batch number: 10755248

Printed by Printforce, the Netherlands